SYSTEM ENGINEERING MANAGEMENT

SYSTEM ENGINEERING MANAGEMENT

SECOND EDITION

Benjamin S. Blanchard
Virginia Polytechnic Institute
Blacksburg, Virginia
University of Exeter
United Kingdom

A WILEY-INTERSCIENCE PUBLICATION
JOHN WILEY & SONS, INC.
New York • Chichester • Weinheim • Brisbane • Singapore •Toronto

This text is printed on acid-free paper.

Library of Congress Cataloging in Publication Data:

Blanchard, Benjamin S.
System engineering management / Benjamin S. Blanchard.—2nd ed.
p. cm.
Includes index.
ISBN 0-471-19086-1 (alk. paper)
1. Systems engineering. I. Title.
TA168.B53 1997
620'.001.'1—dc21 97-3546

Printed in the United States of America

10 9 8 7 6 5 4 3 2

CONTENTS

FOREWORD

In the future, the competitiveness of modern technical systems will depend on characteristics like reliability, maintainability, operability, safety, and affordability. In order to achieve these characteristics, it is necessary to create mechanisms for dealing with mutually competing design requirements. It is fair to say that there is no optimal design; in essence, there is no design which can incorporate all optimal solutions proposed by specialist disciplines. Thus, the best design should represent the optimal compromise between competing requirements during the system integration process.

The traditional engineering and management education process, and consequently existing literature, does not encourage *systems* thinking, and even less prepare project managers and system integrators the for task of managing the system integration process.

Thus, there is a real need for better literature in the area of system engineering management. Decision makers need to be equipped with knowledge based on sound technical and scientific principles rather than personal experience and belief.

This book, written by an experienced system integrator and teacher, provides a good review of the system engineering management process, and thus goes for toward filling that need. Professor Blanchard covers a majority of the management tools and techniques with which modern system engineering managers should be familiar, and demonstrates them with clearly illustrative examples.

I believe that this book will assist practitioners in the day-to-day decision-making process, as well as giving them insight into the integration management function, with the aim of achieving overall systems effectiveness. And those studying system management within the academy will benefit as well, from the over forty years of practical and teaching experience summarized in this indispensable volume.

DR. J. KNEZEVIC
Reader in reliability and logistics engineering,
School of Engineering, and director,
Research Centre for Management of
Industrial Reliability, Cost and Effectiveness
University of Exeter, United Kingdom

FOREWORD

The multidisciplinary domain of systems engineering is finally being recognized for its potential to facilitate development of affordable systems fully responsive to customer requirements. This recognition is even more pronounced in the case of complex, distributed, and software-intensive systems. The development and implementation of system engineering processes and practices notwithstanding, their effective and efficient management is imperative if their full potential is to be realized. Management and organizational issues become even more pronounced for projects involving a geographically distributed design team from multiple organizations.

Only an individual such as Ben Blanchard, with a vast and balanced experience within industry (seventeen years—Boeing. General Dynamics, Sanders, Bendix) and academia (twenty-five years—Virginia Tech) can harness, demystify, and then explain the issues relating to system engineering management. This is reflected in *System Engineering Management.* While providing the necessary focus on the core subject, the reader is also provided an insight into the vast literature supporting this field through numerous well-placed footnotes and a detailed bibliography.

System Engineering Management provides a proper context by first focusing on the need for systems engineering along with a discussion of the current development environment. Thereafter, the focus shifts to the systems engineering process itself along with a discussion of selected and pertinent system engineering tools. With this introduction, the author then discusses the organizational and management issues and concerns at length. The simple and lucid discussion of the subject matter will make this textbook an indispensable addition to the library of any professional involved in the implementation and management of the system engineering process.

DINESH VERMA, PH.D.
Lockheed Martin Federal Systems, Inc.

PREFACE

Current trends indicate that, in general, the complexity of systems is increasing with the introduction of new technologies, the length of time that it takes to develop and acquire a new system needs to be reduced, and many of those systems (or products) in use today are not meeting the needs of the consumer in terms of performance, quality, and overall cost-effectiveness. At the same time, there is a greater degree of "outsourcing" and the utilization of suppliers throughout the world, there is a need for greater cooperation and exchange, and competition is increasing. This is happening at a time when available resources are dwindling worldwide.

Given today's environment, there is an ever-increasing need to develop and produce systems that are robust in nature, reliable and of high quality, supportable, cost-effective, and that will respond to the needs of the consumer in a satisfactory manner. From past experience, the majority of the problems noted have been the direct result of not applying a "systems" approach in meeting the desired objectives. The overall requirements for the system in question were not well defined from the beginning; a bottoms-up (versus top-down) approach was used in the system development process; the perspective relative to meeting a consumer need was relatively "short-term;" and, in many instances, the philosophy has been to "design it now and fix it later!" In essence, the system design and development process has suffered somewhat from the lack of good early planning and the subsequent definition of *requirements* in a complete and methodical manner. This approach has turned out to be quite costly from the long-term viewpoint, particularly when assessing the risks associated with the decision-making process during the early stages of system development.

The combination of these and related factors has created a "critical" need, that is, the requirement for developing and producing (or constructing) well-integrated, high-quality, cost-effective systems with complete consumer (user) satisfaction in mind. In this highly competitive resourced-constrained environment, it is now more important than ever to ensure that the concepts and principles of *system engineering* are properly implemented in both the design and development of new systems and/or the reengineering of existing systems. System requirements must be well defined from the beginning. The system must be viewed in terms of all of its components on a total integrated basis; that is, prime equipment, software, operating personnel, facilities, data, the production process, and the elements of maintenance and support. A top-down integrated approach must be assumed, with the appropriate allocation of requirements from the system level down to its various elements. Further, the system must be viewed in terms of its entire life cycle, that is, from conceptual design through preliminary and detail design, production and/or construction, operational use, maintenance and support, and system retirement.

These concepts are not necessarily new or novel. System engineering has been a topic of interest since the late 1950s and early 1960s (and perhaps even earlier). The principles have been successfully applied on a few programs. However, in most instances, although we may believe that we utilize these methods successfully, we really do not implement them very well (if at all). The successful implementation of system engineering requires not only a *technical* thrust, but a *management* thrust as well. It is essential that one select the appropriate technologies, utilize the proper analytical tools, and apply the necessary resources to enhance the system engineering process. Additionally, the proper organization environment must be established in order to allow for the successful implementation of this process. Thus, it is necessary to first understand and believe in the process and second, to establish the proper management and organizational structure that will allow it to happen! This, in turn, provides a *cultural* challenge for the future.

This text has been developed with the preceding objective in mind. The basic concepts, the need for system engineering and its applications, and an introduction to some key terms and definitions are covered in Chapter 1. This leads to a comprehensive presentation of the "system engineering process" in Chapter 2. This process commences with the identification of a consumer need and extends through the definition of system operational requirements and the maintenance concept, the identification and prioritization of technical performance measures (TPMs), the description of the system in *functional* terms and the allocation of requirements to the various elements of the system, synthesis and design optimization, concurrent engineering and design integration, test and evaluation, production and/or construction, distribution and system utilization, maintenance and support, and retirement and material disposal and/or recycling. Key areas of emphasis for system engineering are noted, and a thorough understanding of this process is fundamental in dealing with the overall subject ares. The material in Chapter 2 serves as a "baseline" for the discussion in subsequent chapters.

Given the overview, it is appropriate to delve further into some of the objectives of system engineering. One goal includes the integration of a wide variety of key design disciplines into the total mainstream design effort. Chapter 3 provides an introduction to some of these disciplines; that is, reliability, maintainability, human factors, safety, software, producibility, quality, logistics, value/cost, and environmental engineering. Chapter 4 follows with some discussion pertaining to the application of design methods and tools, utilized in such a manner as to enhance the fulfillment of system engineering objectives. The appropriate application of computer-aided methods allows for more front-end analysis, leading to a better design definition at an earlier stage in the life cycle. Chapter 5 provides the "checks and balances" in the design process through the accomplishment of design review, evaluation, feedback and control, and the initiation of corrective action as necessary. An objective of system engineering is to provide a strong engineering leadership role relative to the initial definition of system requirements, the necessary *integration* of design activities to ensure effective and efficient results, and the follow-on measurement and evaluation functions to ensure that the initially specified requirements have been met.

The next step is to address the *management* issues pertaining to the application of system engineering requirements to different projects. Chapter 6 leads off with an in-depth discussion of planning and the development of the System Engineering Management Plan (SEMP). System engineering tasks, the development of a work breakdown structure (WBS), program schedules, and the preparation of cost projections are included. Chapter 7 deals with system engineering in the typical project organizational structure, highlighting the differences between functional, product-line, and matrix structures. The interfaces between the consumer (customer), the producer (contractor), and the suppliers are addressed. Chapter 8 includes some considerations pertaining to supplier evaluation, selection, management, and control. Given the trends associated with the increasing use of suppliers, it is essential that system engineering requirements be traced from the consumer, through contractor activities, and to the various suppliers of system components.

Finally, the appendixes include some supporting "case studies" (Appendix A), a description of the life-cycle cost-analysis process (Appendix B), an in-depth design review checklist (Appendix C), a supplier evaluation checklist (Appendix D), a glossary of selected terms (Appendix E), and a selected bibliography (Appendix F).

In summary, the intent of this text is to describe system engineering in terms of its objectives and applications, the steps in the system engineering process, and to provide a management perspective for the implementation of programs with a system engineering thrust! It is believed that this text can be effectively utilized in the academic classroom, in support of a continuing education seminar or workshop, and as a reference guide on the job. Questions and problem exercises are included at the end of each chapter to provide the necessary emphasis where required.

BENJAMIN S. BLANCHARD

1 Introduction to System Engineering

This text deals with "system engineering," or the orderly process of bringing a system into being. A system constitutes a complex combination of resources (in the form of human beings, materials, equipment, software, facilities, data, etc.) integrated in such a manner as to fulfill a designated need. A system is developed to accomplish a specific function, and may be classified as a natural system, man-made system, physical system, conceptual system, closed system, open-loop system, static system, dynamic system, and so on.

A system may vary in form, fit, and/or function. One may be dealing with a group of aircraft accomplishing a mission at a specific geographical location, a communication network distributing information on a worldwide basis, a power distribution capability involving waterways and electrical power generating units, a manufacturing facility producing "x" products in a designated timeframe, or a small vehicle providing the transportation of certain cargo from one location to another. A system may be broken down into subsystems and various smaller components, the level of detail being dependent on the function to be performed.

The purpose of this chapter is to address the subject of "systems" in general, to define some terms and definitions and the characteristics of systems, to identify the need for and the basic requirements for bringing systems into being, and to provide an introduction to system engineering and the management of activities inherent within the system engineering process.

1.1 THE CURRENT ENVIRONMENT

Current trends indicate that, in general, the complexity of systems is increasing, and many of those systems in the inventory today are not meeting the needs of the consumer in terms of performance, effectiveness, and overall cost. New technologies are being introduced on a continuing basis, while the life cycles for many systems are being extended. The length of time that it takes to develop and acquire a new system is getting longer, the industrial base is rapidly changing, and available resources are dwindling. At the same time, there is a greater degree of international cooperation, and competition is increasing worldwide. These factors, conveyed in Figure 1.1, are all interactive, creating the need to address "systems" from a different perspective.

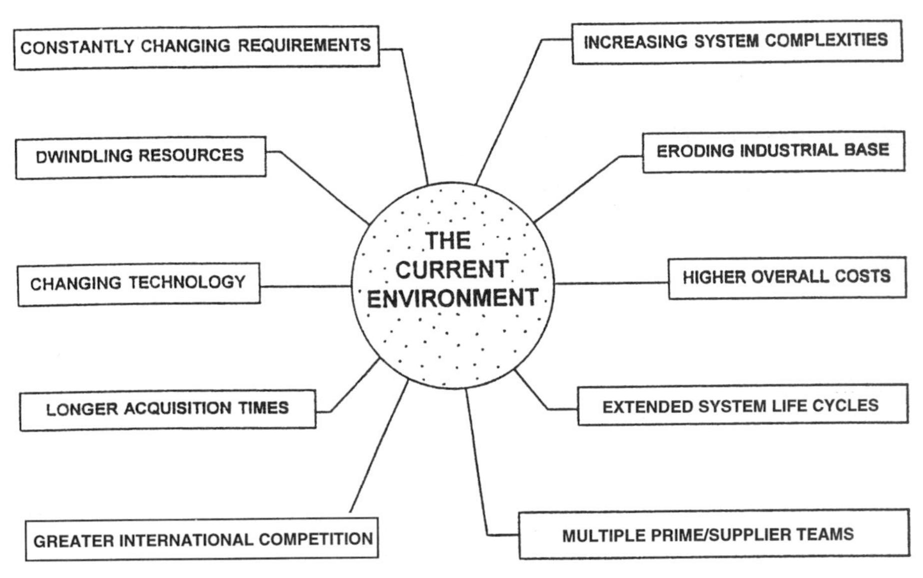

Figure 1.1 The current environment.

When past experiences with regard to the development of systems are evaluated, a majority of the problems noted have been the direct result of not applying a disciplined "systems approach" in meeting the desired objectives. The overall requirements for the system in question were not well defined from the beginning; the perspective in terms of meeting a consumer need has been relatively "short-term" in nature; and, in many instances, the approach followed has been to "deliver it now and fix it later!" In essence, the system design and development process has suffered somewhat from the lack of good early planning and the subsequent definition and allocation of requirements in a complete and methodical manner.

With regard to requirements, the trend has been to keep things "loose" in the beginning (i.e., develop a system-level specification that is very "general" in content), providing an opportunity to introduce the "latest and greatest" in technology developments just prior to going into the production/construction stage. Traditionally, engineers do not want to be forced into design-related commitments any earlier than necessary. Thus, in many instances, there are a lot of last-minute changes in the design. Sometime these changes are are actually incorporated at a later stage, which can be quite costly. Figure 1.2 conveys the impact of a *downstream* change and modification approach.

Past practices have had a great impact on the overall costs of systems. In fact, in recent years, there has been an *imbalance* between the "cost" side of the spectrum and the "effectiveness" side of the balance, as illustrated in Figure 1.3. The complexities of many systems have been greater and, with the ever-increasing emphasis on *performance* at the sacrifice of other key design parameters such as reliability and quality, the overall effectiveness of these systems has been decreasing and the

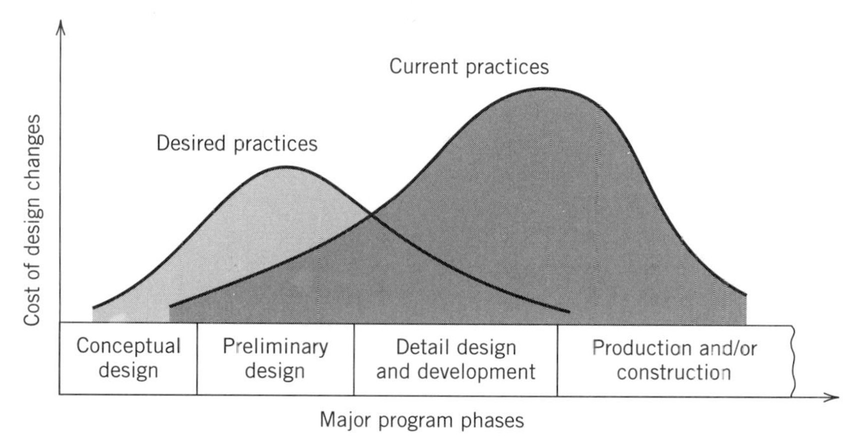

Figure 1.2 The cost impact due to changes.

costs have been going up. Thus, there is a need to provide the proper balance in the development of systems in the future.

When addressing the aspect of economics, one often finds that there is a lack of total *cost visibility,* as projected by "iceberg effect" illustrated in Figure 1.4. For many systems, design and development costs are relatively well known; however, the costs associated with system operation and maintenance support are somewhat hidden. In essence, the design community has been successful in dealing with the short-term aspects of cost, but has not been very responsive to the long-term effects.

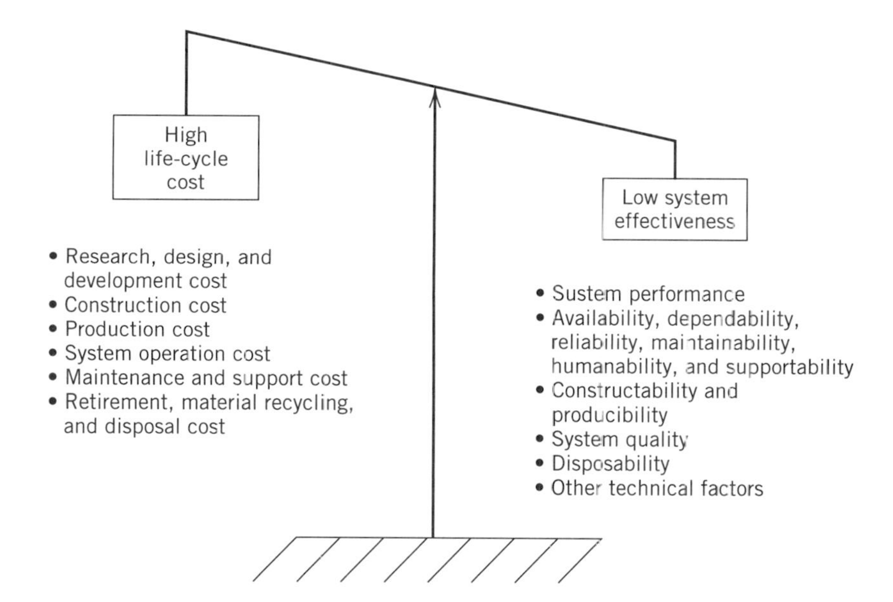

Figure 1.3 The imbalance between system cost and effectiveness factors.

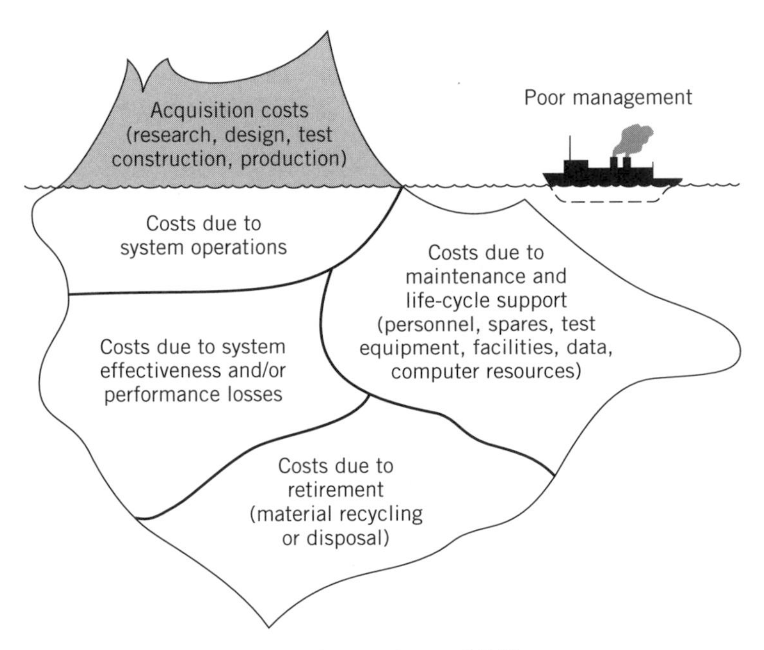

Figure 1.4 Total cost visibility.

At the same time, experience has indicated that a large segment of the life-cycle cost for a given system is attributed to the operational and maintenance support activities accomplished downstream in the life cycle (e.g., up to 75% of the total cost).

Additionally, when looking at "cause-and-effect" relationships, it has been determined that a major portion of the projected life-cycle cost for a given system stems from the consequences of decisions made during the early phases of advance planning and system conceptual design. These decisions relate to operational requirements (e.g., the number of consumer sites assumed, the selection of a given mission profile, the identified life cycle), maintenance and support policies (e.g., two versus three levels of maintenance, levels of repair), allocations associated with manual versus automation applications, equipment packaging schemes and diagnostic routines, hardware versus software applications, the selection of materials, the selection of a manufacturing process, and so on. Decisions made during the early phases of system development can have a great impact on total life-cycle cost, as illustrated in Figure 1.5.

Given the current environment as highlighted herein, there is an ever-increasing need to review our practices and methods for not only bringing new systems into being, but for operating and managing those systems already in use. A highly *disciplined* approach must be pursued in the design and development of new systems, in the evaluation of existing systems, and in the implementation of a *continuous process improvement* methodology. The objective is to provide the consumer (i.e., user)

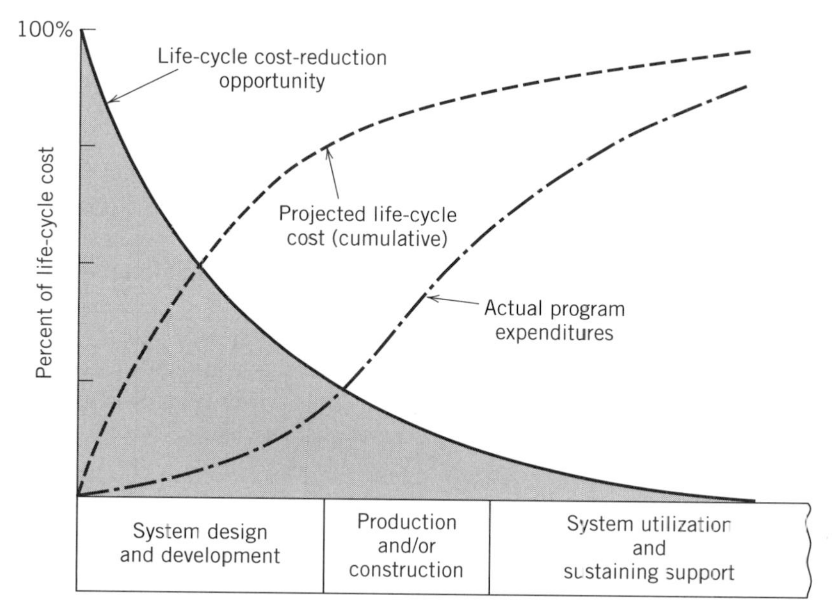

Figure 1.5 Commitment of life-cycle cost.

with a high-quality system that is cost-effective, considering a proper balance among the factors identified in Figure 1.3.

1.2 THE NEED FOR SYSTEM ENGINEERING

The trends and concerns conveyed in Section 1.1 represent only a sample of a few of the major issues that need to be addressed. The challenge is to be more effective and efficient in the development and acquisition of new systems, as well as in the operation and maintenance support of those systems already in being. This can be best accomplished through the implementation of system engineering concepts, principles, and methods.

In addressing topics such as "systems," "system engineering," "system analysis," and the like, one will find a variety of approaches in existence. These terms may be defined differently depending on individual backgrounds, experiences, and the organizational interests of practitioners in the field. When the literature is researched, the definitions used are quite different. Thus, with the objective of providing some clarification relative to the material provided throughout this text, it seems appropriate to direct some attention to a few key concepts and definitions. Additional definitions are introduced throughout subsequent chapters.[1]

[1] The bibliography presented in Appendix F includes a variety of publications covering system engineering and related areas.

1.2.1 Definition of a System

The term "system" stems from the Greek word "systēma," meaning an organized whole. Webster's Dictionary defines a system as "an aggregation or assemblage of objects united by some form of regular interaction or interdependence; a group of diverse units combined by nature or art as to form an integral whole, and to function, operate, or move in unison and, often, in obedience to some form of control; an organic or organized whole."[2] In essence, a system constitutes a set of interrelated components working together with the common objective of fulfilling some designated need.

Although this reflects a good initial overview, a greater degree of detail and precision is required to provide a good working definition acceptable for describing the concepts of system engineering. To facilitate this objective, a system might be defined further in terms of the following general characteristics:

1. A system constitutes a *complex combination of resources* in the form of human beings, materials, equipment, software, facilities, data, money, and so on. To accomplish many functions often requires large amounts of personnel, equipment, facilities, and data (e.g., an airline or a manufacturing capability). Such resources must be combined in an *effective* manner, as it is too risky to leave this to chance alone.

2. A system is contained within some form of *hierarchy.* An airplane may be included within an airline, which is part of an overall transportation capability, which is operated in a specific geographic environment, which is part of the world, and so on. As such, the system being addressed is highly influenced by the performance of the higher-level system, and these external factors must be evaluated.

3. A system may be broken down into *subsystems* and related components, the extent of which depends on complexity and the function(s) being performed. Dividing the system into smaller units allows for a simpler approach relative to the initial allocation of requirements and the subsequent analysis of the system and its functional interfaces. The system is made up of many different components, these components *interact* with each other, and these interactions must be thoroughly understood by the system designer and/or analyst. Because of these interactions among components, it is impossible to produce an effective design by considering each component separately. One must view the system as a whole, break down the system into components, study the components and their interrelationships, and then put the system back together.

4. The system must have a *purpose.* It must be functional, be able to respond to some identified need, and it should be able to achieve its overall objective in a cost-effective manner. There may be a conflict of objectives, influenced by the higher-level system in the hierarchy, and the system must be capable of meeting its stated purpose in the best way possible.

[2] Webster's *New International Dictionary of the English Language,* 3rd Ed., G. & C. Merriam Company, Springfield, MA, 1961.

1.2.2 Categories of Systems

When defining systems in terms of the general characteristics presented, it readily becomes apparent that some degree of further classification is desirable. There are many different types of systems, and there are some dichotomies in terms of similarities and dissimilarities. To provide some insight into the variety of systems in being, a partial listing of categories is noted.[3]

Natural and Man-Made Systems. Natural systems include those that came into being through natural processes. Examples might include a river system or an energy system. Man-made systems are those that have been developed by human beings, the results of which include a wide variety of capabilities. As all man-made systems are embedded in the natural world, there are numerous interfaces that must be addressed. For instance, the development and construction of a hydroelectric power system located on a river system creates impacts on both sides of the spectrum, and it is essential that the systems approach involving both the natural and man-made segments of this overall capability be implemented.

Physical and Conceptual Systems. Physical systems are those made up of real components occupying space. On the other hand, conceptual systems constitute an organization of ideas, a set of specifications and plans, a series of abstract concepts, and so on. Conceptual systems often lead directly into the development of physical systems, and there is a certain degree of commonality in terms of the type of processes employed. Again, the interfaces may be many, and there is a need to address these elements in the context of a higher-level system in the overall hierarchy.

Static and Dynamic Systems. Static systems include those having structure, but without activity (as viewed in a relatively short period of time). A highway bridge and a warehouse are examples. A dynamic system is one that combines structural components with activity. An example is a production capability combining a manufacturing facility, capital equipment, utilities, conveyors, workers, transportation vehicles, data, software, managers, and so on. Although there may be specific points in time when all system components are static in nature, the successful accomplishment of system objectives does require activity and the dynamic aspects of system operation do prevail throughout a given scenario.

Closed and Open-Loop Systems. A closed system is one that is relatively self-contained and does not significantly interact with its environment. The environment provides the medium in which the system operates; however, the impact is minimal. A chemical equilibrium process and an electrical circuit (with a built-in feedback and control loop) are examples. Conversely, open-loop systems interact with their

[3]This categorization follows the general format presented in B. Blanchard and W. Fabrycky, *Systems Engineering and Analysis,* 2nd Ed., Prentice Hall, Upper Saddle River, NJ, 1990. These categories represent only a few of those that could be described.

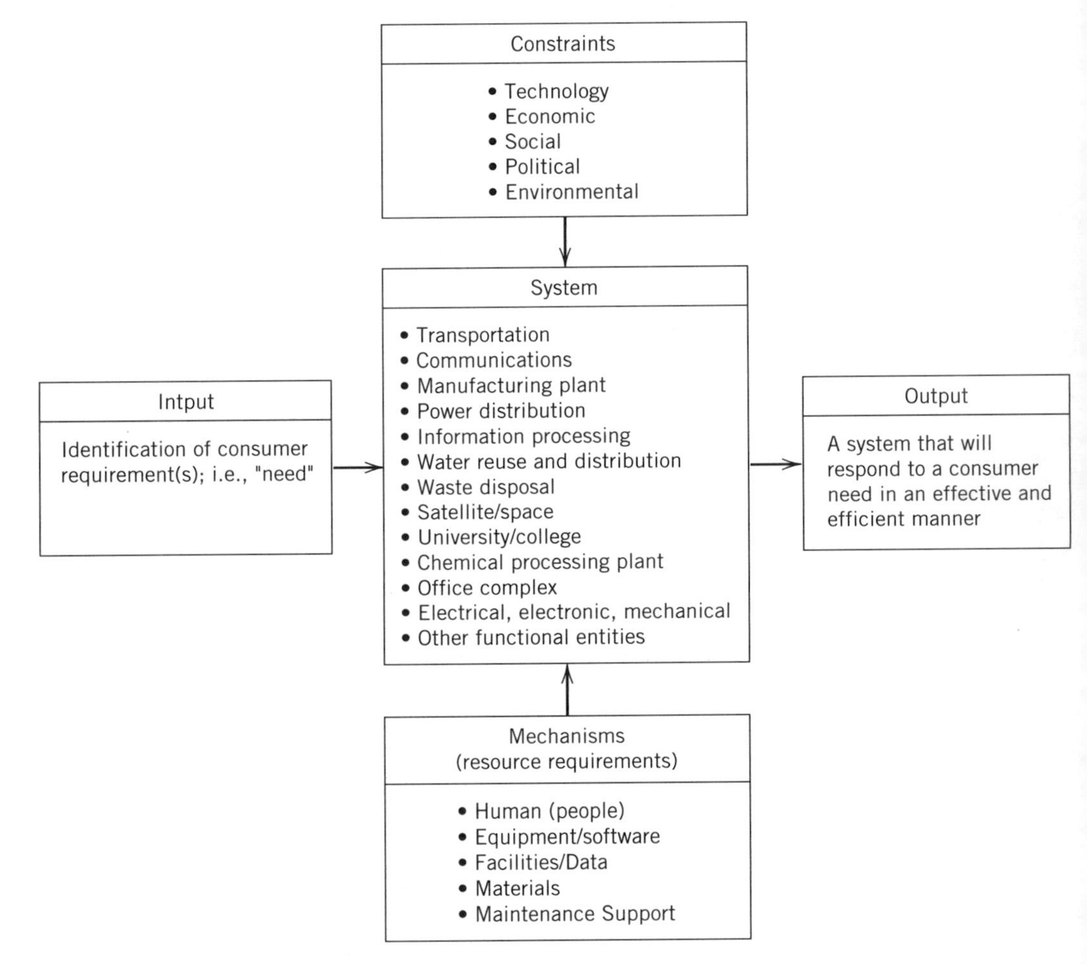

Figure 1.6 The system.

environment. Boundaries are crossed (through the flow of information, energy, and/or matter), and there are numerous interactions both among the various system components and up and down the overall system hierarchical structure. A system/product logistic support capability is an example.

The presentation of these categories is included to stimulate further thought relative to the definition of a system. It is not easy to classify a system as being either closed or open, and the precise relationships between natural and man-made systems may not be well defined. However, the objective here is to gain a greater appreciation for the many different considerations required in dealing with system engineering and its process. Emphasis throughout this text tends to favor *man-made* systems that are *physical* by nature, *dynamic* in operation, and of the *open-loop* variety.

Given the preceding, the systems addressed herein may include a wide variety of *functional* entities. There are transportation systems, communication systems, manufacturing systems, information processing systems, and so on, as indicated in Figure 1.6. In each instance, there are *inputs,* there are *outputs,* there are external *constraints* imposed on a system, and there are the required *mechanisms* necessary to realize the desired results. Within the framework of the "system," there are products and processes.

A system is composed of many different elements to include those that are directly utilized in the actual accomplishment of a mission (e.g., prime equipment, operating software, operating personnel) and the elements of maintenance support (e.g., maintenance personnel, test equipment, facilities, spares and repair parts). Although the support infrastructure is not often considered as an element of the system per se, the system may not be able to complete its designated function in the absence of such. Thus, the support infrastructure will be addressed as a major system element, presented in the context of the system life cycle. Figure 1.7 identifies the major elements of a system.

A system may be contained within some form of *hierarchy,* as shown in Figure 1.8. The question is are we addressing a *transportation* system, including many different types of vehicles (e.g., automotive, rail), a *vehicular* system, including many automobiles, or a *automobile,* with driver and associated support? It is not uncommon for a group of individuals to get together to discuss some issue, each having a different perception as to the "system" being addressed.

From Figure 1.8, there are "upward" and "downward" impacts that must be considered. Decisions pertaining to the *vehicular* system may have an upward impact

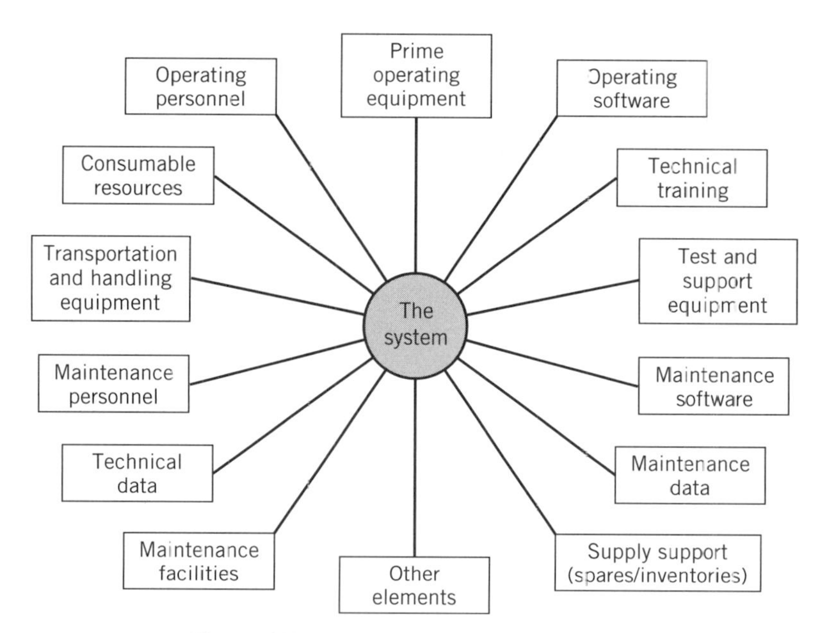

Figure 1.7 The major elements of a system.

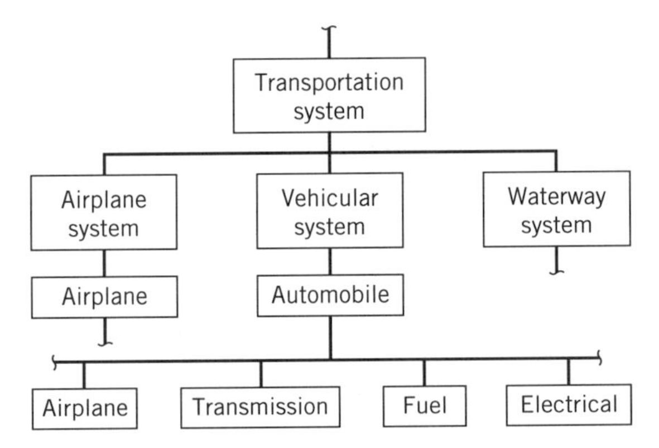

Figure 1.8 The hierarchy of systems.

on the transportation system, and certainly will have a downward impact on the automobile. For example, the maintenance support infrastructure for the vehicular system may have to be compatible with the maintenance concept specified for the transportation system. Additionally, this concept may also be imposed as a constraint in the design of the automobile. In any event, these interaction effects may be significant and must be addressed.

1.2.3 The System Life Cycle

From Figure 1.9, the *life cycle* includes the entire spectrum of activity for a given system, commencing with the identification of need and extending through system design and development, production and/or construction, operational use and sustaining maintenance and support, and system retirement and material disposal. As the activities in each phase interact with the activities in the other phases, it is essential that one consider the overall life cycle when addressing system-level issues, particularly if one is to properly assess the risks associated with the decision-making process throughout.

Although the life-cycle phases conveyed in Figure 1.9 reflect a more generic

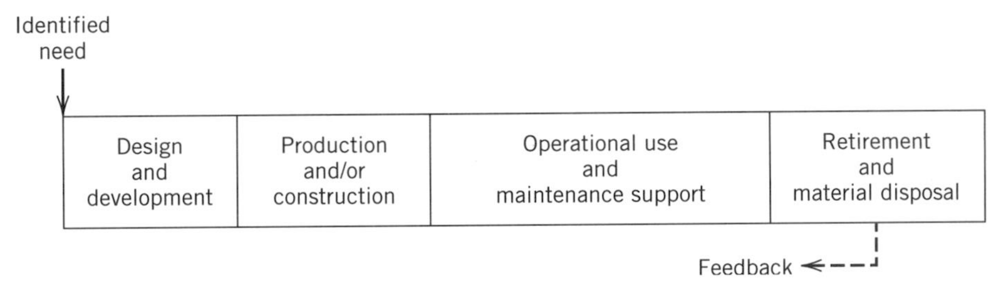

Figure 1.9 The system life cycle.

sequential approach, the specific activities (and the duration of each) may vary somewhat depending on the nature, complexity, and purpose of the system. Needs may change, obsolescence may occur, and the levels of activity may be different depending on the type of system and and where it fits in the overall hierarchical structure of activities and events. Additionally, the various phases of activity may overlap somewhat, as illustrated in the two examples presented in Figure 1.10.

The figure shows how an airplane, a ground transportation vehicle, or an electronic device may progress through conceptual design, preliminary design, detail design, production, and so on, as reflected through the series of activities for Example "A." When this example is evaluated further, the top row of activities is applicable to those elements of the system that relate directly to the accomplishment of the mission (e.g., an automobile). At the same time, there are two closely related life cycles of activity that must also be considered. The design, construction, and operation of the production capability, which can have a significant impact on the operations of the prime elements of the system, should be addressed concurrently along with the system maintenance and support activity. Further, these activities must be addressed early during the conceptual and preliminary design of those prime elements represented by the top row. Although all of these activities may be presented through an illustrated single flow, as conveyed in Figure 1.9, the breakout in Figure 1.10 is intended to emphasize the importance of addressing *all* aspects of the total system process and the various interactions that may occur.

Example "B" in Figure 1.10 is presented to cover the major phases associated with a manufacturing plant, a chemical processing plant, or a satellite ground

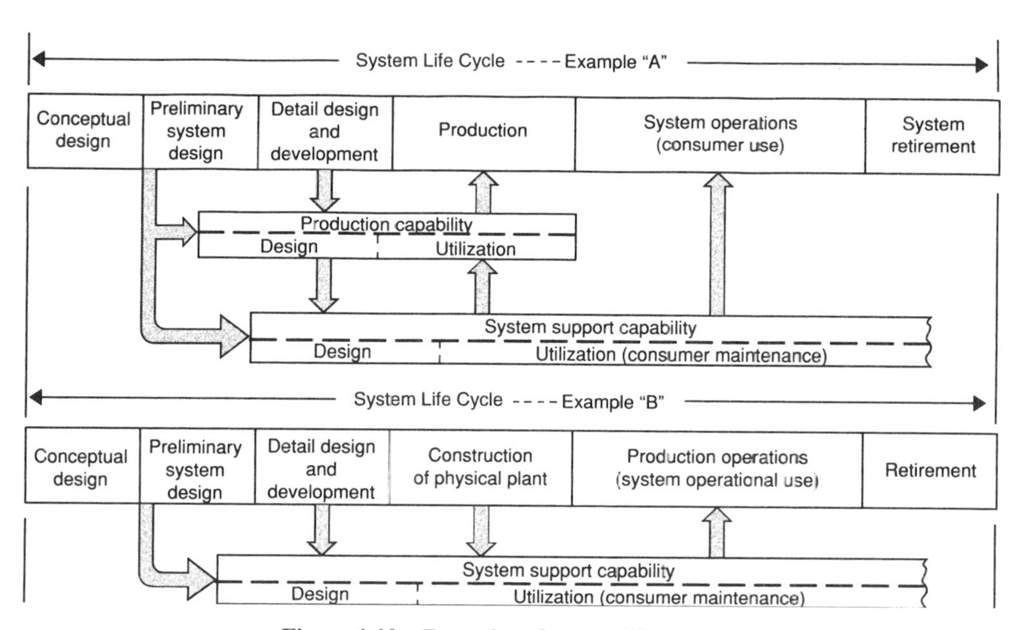

Figure 1.10 Examples of system life cycles.

tracking facility, where the construction of a "one-of-a-kind" system configuration is required. Again, the maintenance and support capability is identified separately in order to indicate degree of importance, and to suggest that there are many interaction effects that must be considered.

Although there may be variations in approaches, the nomenclature used, the duration of different phases, and so on, it is still appropriate that systems be viewed in terms of their respective life cycles. The past is replete with examples in which major decisions have been made in the early stages of system acquisition based on the "short-term" only. In other words, in the design and development of a new system, the consideration for *production/construction* and/or *maintenance and support* of that system was inadequate! These activities were considered later, and, in many instances, the consequences of this "after-the-fact" approach were costly as conveyed in Section 1.1.

1.2.4 System Engineering

System engineering may be defined differently depending on one's background and personal experience. The Inaugural Issue, "Systems Engineering," published by the International Council on Systems Engineering (INCOSE), describes a variety of approaches.[4] However, there is a basic theme throughout that deals with a top-down process, which is life-cycle-oriented, and involving the integration of functions, activities, and organizations.

Broadly defined, system engineering is "the effective application of scientific and engineering efforts to transform an operational need into a defined system configuration through the top-down iterative process of requirements analysis, functional analysis and allocation, synthesis, design optimization, test and evaluation and validation." The system engineering process, in its evolving of functional detail and design requirements, has as its goal the achievement of the proper balance among operational (i.e., performance), economic, and logistics factors.

The Department of Defense (DOD) defines systems engineering as the "process that shall:

1. Transform operational needs and requirements into an integrated system design solution through concurrent consideration of all life-cycle needs (i.e., development, manufacturing, test and evaluation, verification, deployment, operations, support, training and disposal);
2. Ensure the compatibility, interoperability, and integration of all functional and physical interfaces and ensure that system definition and design reflect the requirements for all system elements (i.e., hardware, software, facilities, people, data); and
3. Characterize and manage technical risks.

[4] Inaugural Issue, "Systems Engineering," *Journal of the International Council on Systems Engineering*, 1(1) (July-September 1994).

The key systems engineering activities that shall be performed are requirements analysis, functional analysis/allocation, design synthesis and verification, and system analysis and control."[5]

A slightly different definition (preferred by the author) is "the application of scientific and engineering efforts to (1) transform an operational need into a description of system performance parameters and a system configuration through the use of an iterative process of definition, synthesis, analysis, design, test and evaluation, and validation; (2) integrate related technical parameters and ensure the compatibility of all physical, functional, and program interfaces in a manner that optimizes the total definition and design; and (3) integrate reliability, maintainability, usability (human), safety, producibility, supportability (serviceability), disposability, and other such factors into the total engineering effort to meet cost, schedule, and technical performance objectives."[6]

Basically, systems engineering is *GOOD* engineering with certain designated areas of emphasis, a few of which are noted in what follows.

1. A top-down approach is required, viewing the system as a *whole.* Although engineering activities in the past have very adequately covered the design of various system components, the necessary *overview* and an understanding of how these components effectively fit together has not always been present.

2. A *life-cycle* orientation is required, addressing all phases to include system design and development, production and/or construction, distribution, operation, sustaining maintenance and support, and retirement and material phaseout. Emphasis in the past has been placed primarily on system design activities, with little (if any) consideration toward their impact on production, operations, and logistic support.

3. A better and more complete effort is required relative to the initial *identification of system requirements,* relating these requirements to specific design goals, the development of appropriate design criteria, and the follow-on analysis effort to ensure the effectiveness of early decision making in the design process. In the past, the early "front-end" analysis effort, as applied to many new systems, has been minimal. This, in turn, has required greater individual design efforts downstream in the life cycle, many of which are not well integrated with other design activities and require modification later on.

4. An *interdisciplinary* effort (or team approach) is required throughout the system design and development process to ensure that all design objectives are met in an effective manner. This necessitates a complete understanding of the many different design disciplines and their interrelationships, particularly for large projects.

[5] Department of Defense Regulation 5000.2-R, "Mandatory Procedures for Major Defense Acquisition Programs (MDAPs) and Major Automated Information System (MAIS) Acquisition Programs," Part 4, Paragraph 4.3, March 15, 1996.

[6] This is a modified version of the definition of systems engineering included in the original version of MIL-STD-499, "Systems Engineering," Department of Defense, Washington, DC, July 1969.

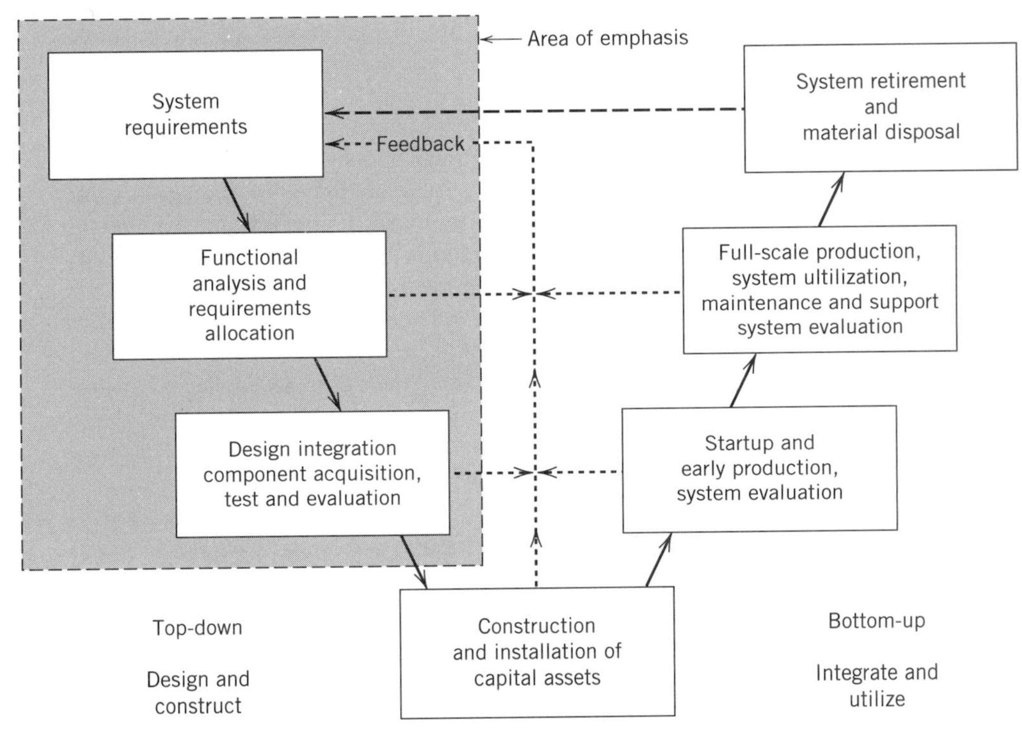

Figure 1.11 Top-down/bottom-up system development process.

Inherent within the systems engineering process is a "top-down/bottom-up" development approach, as illustrated in Figure 1.11. The emphasis throughout this text is the shaded area; that is, the front-end requirements analysis activity. Traditionally, the requirements have not been well defined from the beginning, resulting in some rather extensive and costly efforts during the final integration and test activity.

Figure 1.12 presents an extension of the basic life-cycle phases shown in Figure 1.9, describing typical activities that occur in each phase, identifying various configuration *baselines* that should be established as one progresses from the initial identification of need to the development of a fully operational system, and including the iterative steps inherent within the system engineering process. Although the presentation of information in the figure may lead the reader to believe that the system acquisition process is very complex, the objective is to show this as a *process* in itself. Every time that there is a newly identified *need,* there are certain steps through which one should evolve; that is, conceptual design, preliminary design, and so on. Even if the effort (in terms of the resources expended) is minimal, there is still the requirement for design activities at the *system level* and on down. The objective is to view these phase-related activities as a process within itself, and to identify the baselines where the design evolves from one level of definition to the next. "Tailoring" the activities in Figure 1.12 to the system in question is essential for the successful implementation of the system engineering process.

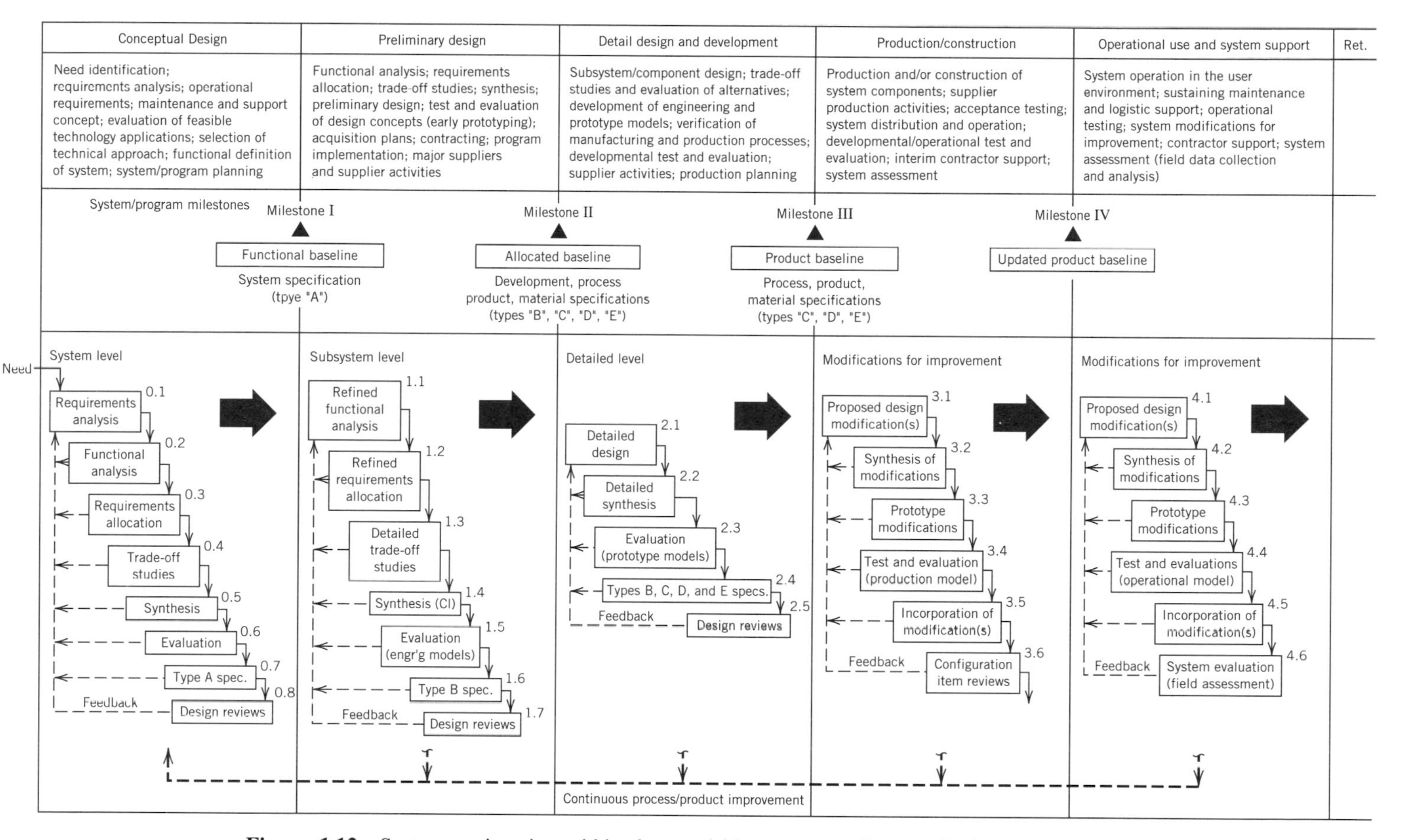

Figure 1.12 System engineering within the acquisition process. (*Source:* B. S. Blanchard, D. Verma, and E. Peterson, *Maintainability: A Key to Effective Serviceability and Maintenance Management,* John Wiley, New York, 1995.)

The system engineering process per se includes the basic steps of requirements analysis, functional analysis, requirements allocation, design optimization and trade-offs, synthesis, evaluation, and so on (refer to blocks 0.1, 0.2, 0.3, etc., in Figure 1.12). These steps are *iterative* by nature, evolving from the system-level definition to the subsystem level, detailed level, and on down to the component. Further, these steps are not necessarily accomplished in a serial sequence, but are interactive with the appropriate *feedback* provisions at each step in the process. Although the requirements may vary somewhat from program to program, the purpose of this figure is to provide a baseline for future reference as different topics are presented throughout this text.

In block 0.2 (Figure 1.12), the accomplishment of the *functional analysis* will lead to the identification of resources in terms of the need for hardware, software, people, facilities, data, and the like. The functional analysis identifies the "WHATs" from a requirements perspective, and this leads to the accomplishment of trade-offs and the description of the "HOWs" pertaining to the completion of functions. Figure 1.13 illustrates the identification of hardware, software, and human requirements (from the functional analysis), and the subsequent life cycles associated with the development of each of these resources. One of the goals of system engineering is to "justify" these resource requirements through a top-down approach, and to ensure the proper development of each through a fully integrated system as one evolves through the design of its various elements.

Figure 1.14 presents the systems engineering approach from a different perspective. As one evolves through the life cycle, there is a need to ensure the full "traceability" of requirements from the system level and on down to the component. As *technical performance measures* (TPMs), or the applicable *metrics,* are established for the system, these measures must be allocated or apportioned to the next level, appropriate *design criteria* are identified, and these criteria must be reflected and supportive from the top down. Further, the appropriate methods/tools must be applied in the design process to ensure that the overall objectives are met for the system. Inherent within the system engineering process is the need to ensure that this traceability is maintained, and to cause the integration of the appropriate techniques/methods/tools to facilitate the development process in an effective and efficient manner.

In summary, the system engineering process is continuous, iterative, and incorporates the necessary feedback provisions to ensure convergence. Figure 1.15 illustrates the *feedback* capability that must be built into the process, applied at the system level, to the subsystem level, and so on, as illustrated in Figure 1.12.

System engineering per se is not considered as an engineering discipline in the same context as civil engineering, mechanical engineering, reliability engineering, or any other design specialty area. Actually, system engineering involves the efforts pertaining to the overall design and development process employed in the evolution of a system from the point when a need is first identified, through production and/or construction and the ultimate installation of that system for consumer use. The objective is to meet the requirements of the consumer in an effective and efficient manner. The system engineering process is covered further in Chapter 2. Finally,

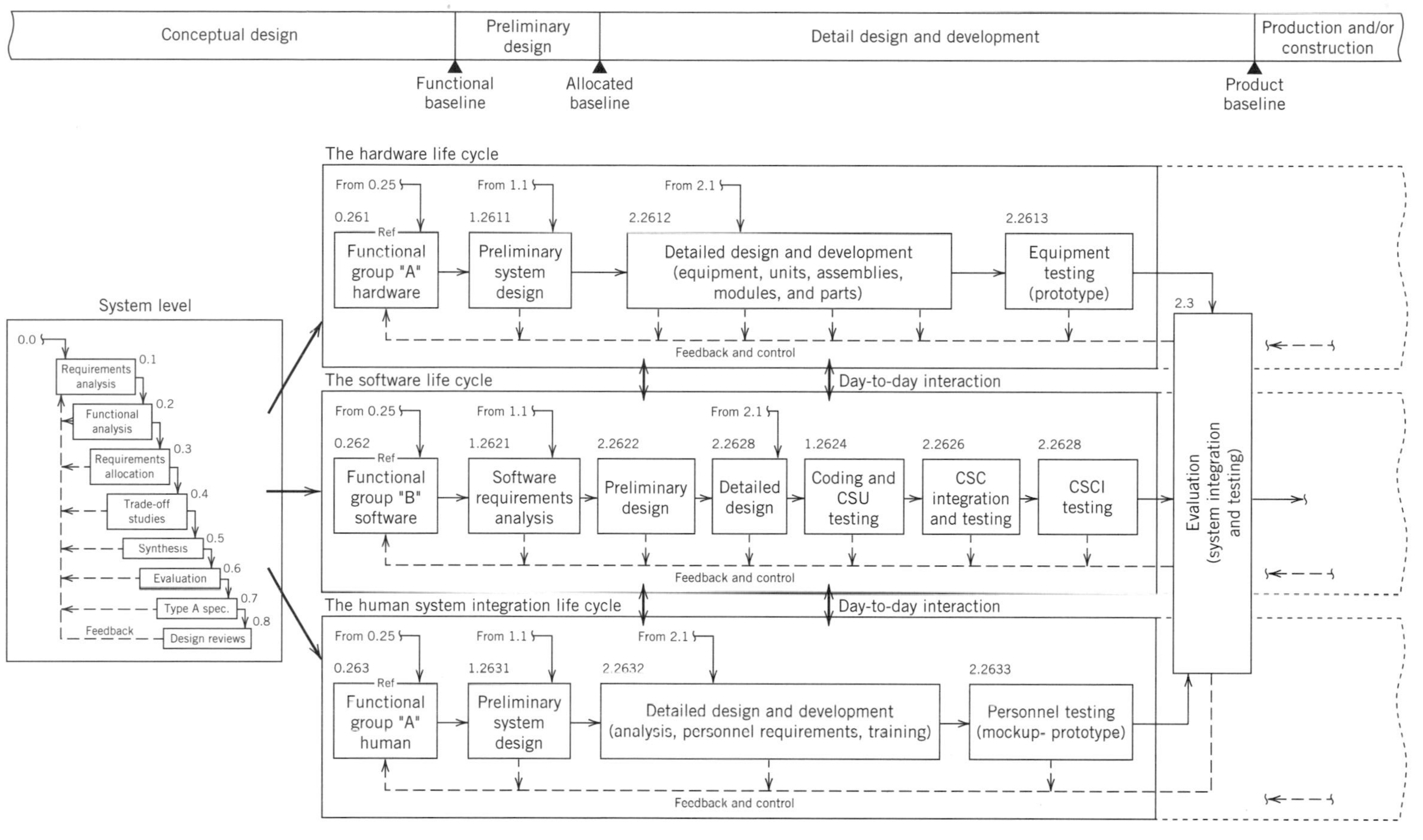

Figure 1.13 The integration of the hardware, software, and human life cycles.

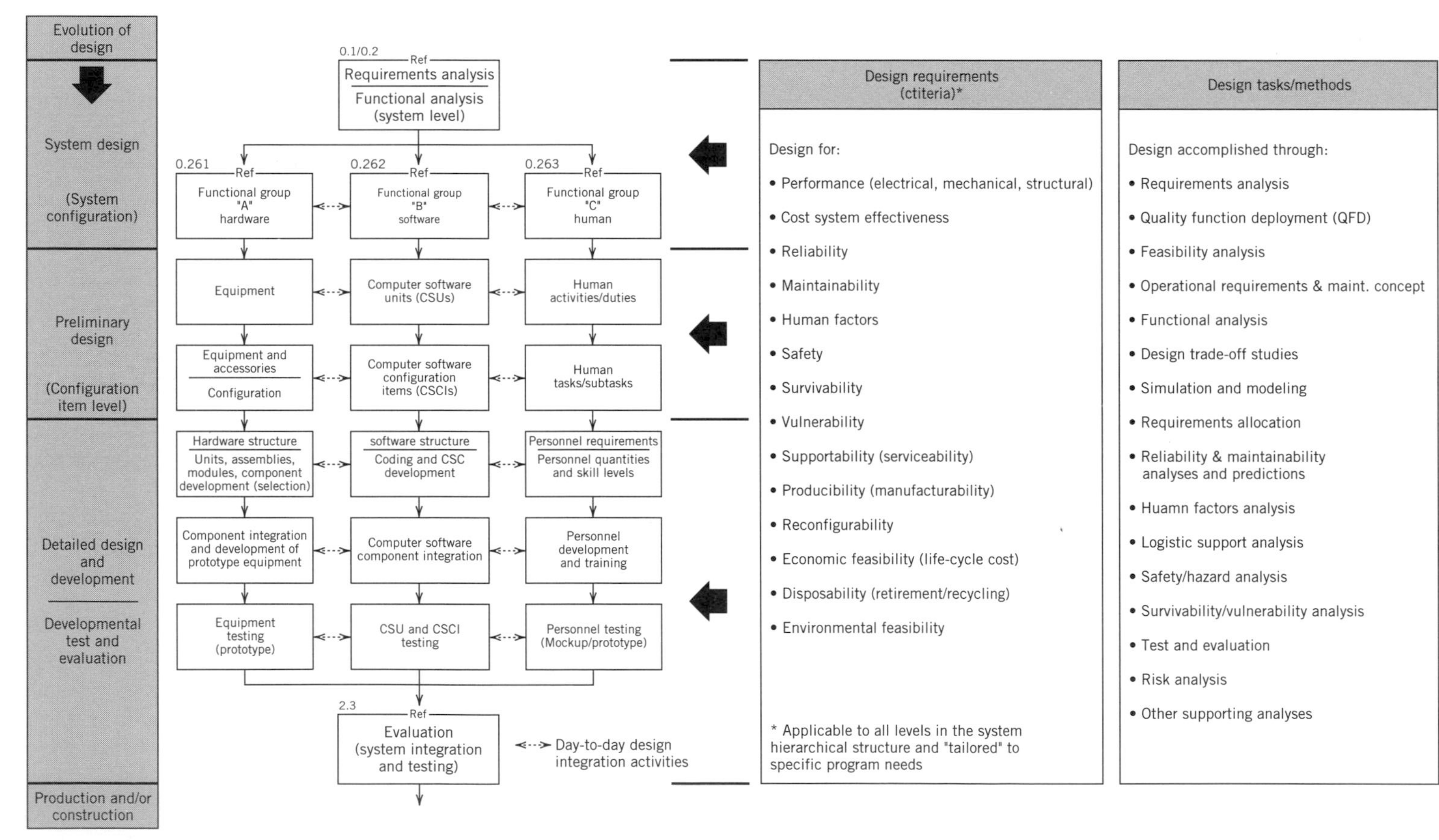

Figure 1.14 The top-down "traceability" of requirements.

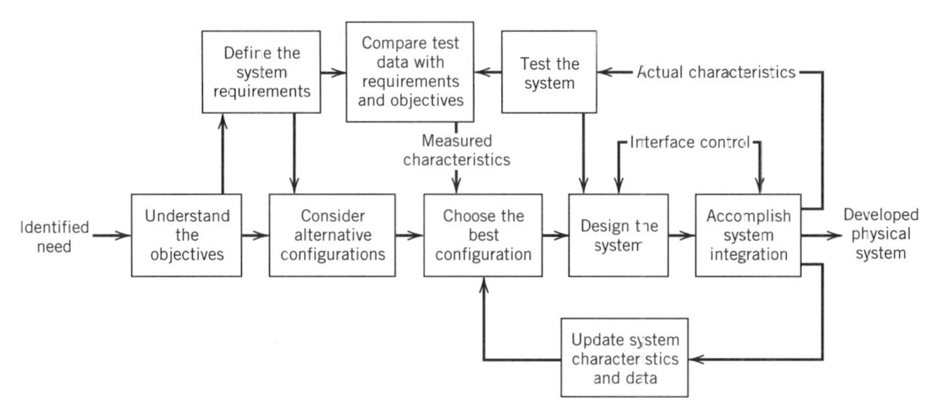

Figure 1.15 "Feedback" in the system engineering process.

the concepts and principles associated with system engineering are not necessarily new or novel. Review of some of the literature in Appendix F indicates that many of the principles identified herein were being promoted back in the 1950s and early 1960s. However, in many instances, the system engineering process has not been implemented very well (if at all). Yet, at this point in time, there is a need to emphasize these concepts more than ever! Although there are many sophisticated and high-technology systems being produced and/or are in use today, there are certain trends occurring that are of concern!

1.2.5 Some Additional Models

In the early 1980s when the "makeup" of systems became more *software-intensive,* there were a number of models developed with the objective of portraying the system life cycle. The "Waterfall Model" is probably the oldest and most widely used of the system development models in this category.[7] This model, shown in Figure 1.16, is based on a top-down approach for software development and includes the steps of initiation, requirements analysis, design, test, and so on. Often, in its implementation, the steps are viewed as being relatively independent from one another and must be executed in strict sequence. Additionally, the required interfaces with the other elements of the system (e.g., hardware, the human, facilities) are not usually considered.

In the mid-1980s, a generic "Spiral Model" was developed for software-intensive systems.[8] The model continually examines objectives, strategies, design alternatives, and validation methods. System development results through several iterations

[7] B. W. Boehm, *Software Engineering Economics,* Prentice Hall, Upper Saddle River, N.J., 1981, p. 36.
[8] The generic spiral model was presented by B. W. Boehm, "A Spiral Model of Software Development," in *Software Engineering Project Management,* ed. R. H. Thayer and M. Dorfman, IEEE Computer Society Press, Washington, DC, 1988. This was modified in Figure 1.17 and is included in A. P. Sage, *Systems Engineering,* John Wiley, New York, 1992, pp. 53–54.

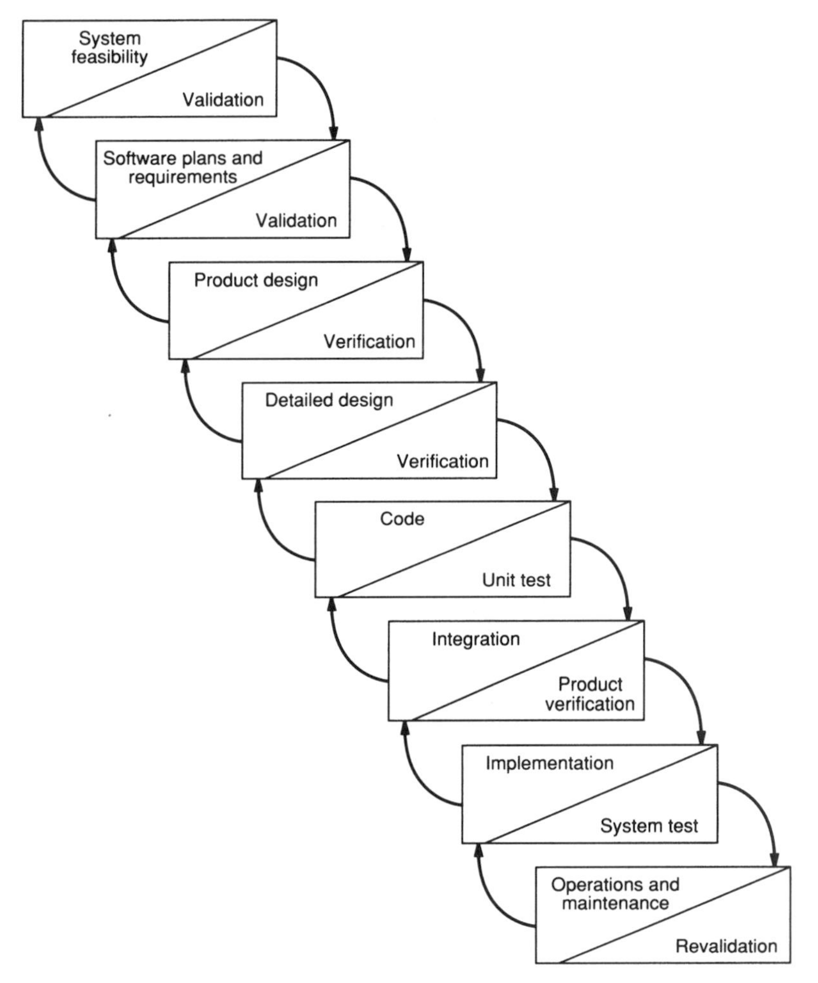

Figure 1.16 The "waterfall model" of the software life cycle. (*Source:* B. W. Boehm, *Software Engineering Economics,* Prentice Hall, Upper Saddle River, NJ, 1981.)

of this model. Figure 1.17 illustrates a modified version of the original generic approach, evolving from a prototyping model. Note that rapid prototyping is used in each cycle, and that the model emphasizes risk analysis. This approach is particularly useful in high-risk developments because design sometimes evolves as detailed requirements emerge.

The "Vee Model" was introduced in the early 1990s, and reflects a top-down and bottom-up approach to system development.[9] In Figure 1.18, the left side of the Vee represents the evolution of user requirements into preliminary and detail design, and

[9]K. Forsberg, and H. Mooz, "The Relationship of System Engineering to the Project Life Cycle," in *Proceedings of the Symposium of International Council on Systems Engineering (INCOSE),* INCOSE, Seattle, WA, 1991, p. 289.

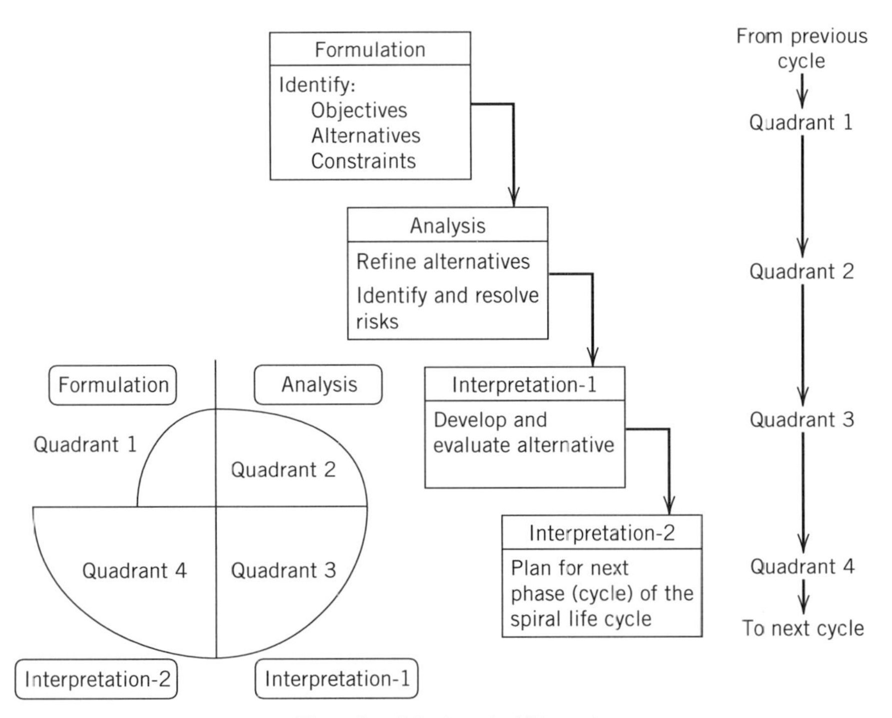

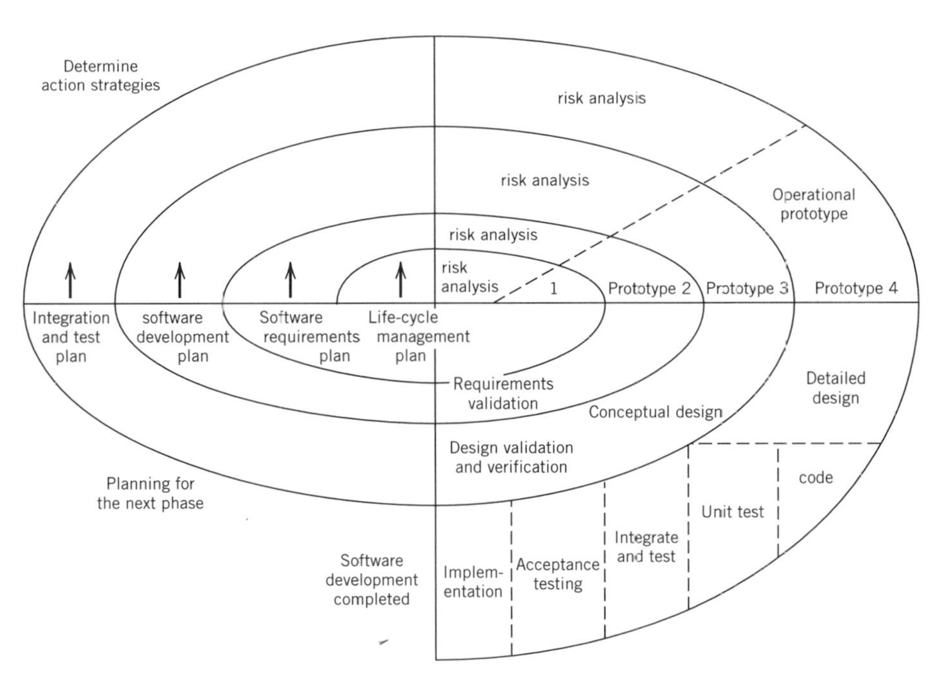

Figure 1.17 The spiral model for the software life cycle. (*Source:* A. P. Sage, *Systems Engineering,* John Wiley, New York, 1992.)

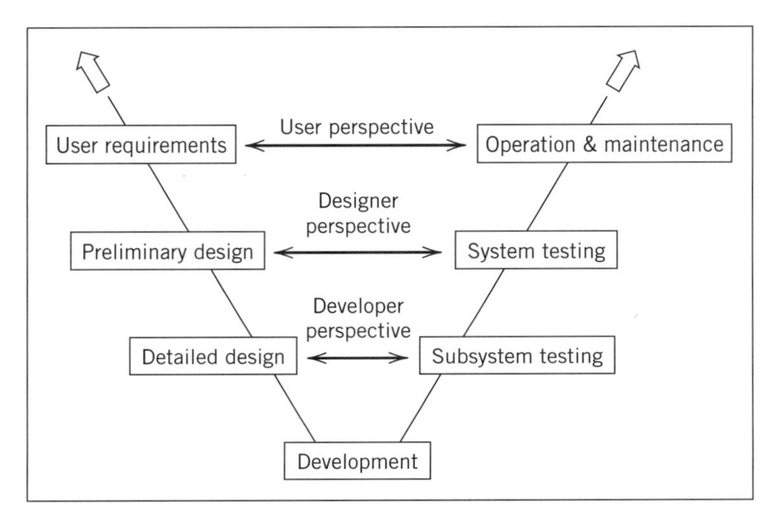

Figure 1.18 The generic "Vee" developmental model.

the right side represents the integration and verification of system components through subsystem and system testing. This model most nearly reflects the approach conveyed in Figure 1.11 (Section 1.2.4).

Figure 1.19 represents an extension of the Vee model concept.[10] Of particular note is the interface between the "system" and the "software subsystem." Quite often, individuals refer to "software systems." Although the software may be predominant within the structure of a system, it is not a "system" per se. It does not fulfill a functional requirement by itself. Software requirements are identified through the functional analysis and are subsequently developed through the steps illustrated in Figure 1.13. Figure 1.19 emphasizes the fact that there are system engineering activities that lead into the software development process.

There are numerous models that have been introduced through the past several decades, with the objective of providing a logical approach to the overall process of system design and development. The few identified here are only representative of the total population. Most of these models are directed primarily to the system acquisition process only and/or to some element of the system (such as software); hence, they lack a certain degree of "completeness!" If implemented properly, they are excellent in terms of accomplishing their intended objective. However, it should be recognized that their application may be limited unless utilized within the broader spectrum of system engineering described in Section 1.2.4.[11]

[10] B. G. Downward, "A Brave New World: Melding Systems and Software Engineering," in *Proceedings of the 4th Annual International Symposium, International Council on Systems Engineering (INCOSE)*, INCOSE, Seattle, WA, 1991, p. 157.

[11] There are numerous other models including prototyping models, the Sashimi Model, the Scrum Model, the Handcuff Model, the Hollywood Model, the Evolutionary Development Model, and so on. A good reference covering some of these models (in a summary manner) is R. S. Scotti and S. S. Gulu Gambhir,

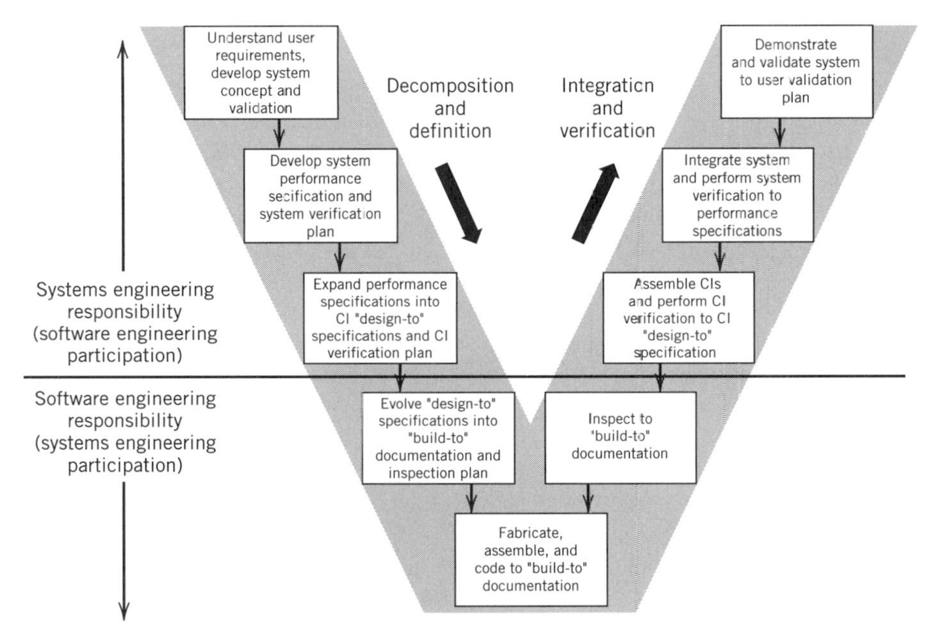

Figure 1.19 The systems versus software engineering boundary. (*Source:* B. G. Downward, "A Brave New World: Molding Systems and Software Engineering," in *Proceedings of the Symposium of the International Council on Systems Engineering,* INCOSE, Seattle, WA, 1991, p. 157.

1.2.6 System Science

Often, in addressing the subject of system engineering, one uses the terms "system science" and "system engineering" interchangeably. For the purposes of this text, *system science* deals primarily with the observation, identification, description, experimental investigation, and theoretical explanation of facts, physical laws, interrelationships, and so on, associated with natural phenomena. Science deals with basic concepts and principles that help to explain how the physical world behaves. In the more applied sense, the disciplines of biology, chemistry, and physics cover many of these relationships. In any event, system engineering includes the application of scientific principles throughout the system design and development process.[12]

"A Conceptual Framework for a Customer-Centered System Development Life-Cycle Model," *Proceedings of the 6th Annual International Symposium, International Council on Systems Engineering (INCOSE),* INCOSE, Seattle, WA, 1996, p. 547.

[12] Systems science is a major subject by itself, and the author appreciates the fact that adequate coverage is not included herein. Three excellent references are R. L. Ackoff, S. K. Gupta, and J. S. Minas, *Scientific Method: Optimizing Applied Research Decisions,* John Wiley, New York, 1962; G. M. Sandquist, *Introduction to System Science,* Prentice Hall, Upper Saddle River, NJ, 1985; and L. Von Bertalanffy, *General Systems Theory,* George Braziller, New York, 1968.

1.2.7 System Analysis

Inherent within the system engineering process is an ongoing analytical effort. In a somewhat puristic sense, analysis refers to a separation of the whole into its component parts, an examination of these parts and their interrelationships, and a follow-on decision relative to a future course of action.

More specifically, throughout system design and development there are many different alternatives (or trade-offs) requiring an evaluation effort in some form. For instance, there are alternative system operational scenarios, alternative maintenance and support concepts, alternative equipment packaging schemes, alternative diagnostic routines, alternative manual versus automation applications, and so on. The process of investigating these alternatives, and the evaluation of each in terms of some criteria, constitutes an ongoing analytical effort.

To accomplish this activity in an effective manner, the engineer (or analyst) relies on the use of available analytical techniques/tools to include operations research methods such as simulation, linear and dynamic programming, integer programming, optimization (constrained and unconstrained), and queuing theory to help solve problems. Further, mathematical models are used to help facilitate the quantitative analysis process.

In essence, *system analysis* includes that ongoing analytical process of evaluating various system design alternatives, employing the application of mathematical models and associated analytical tools as appropriate. Analytical methods and models are discussed further in Chapter 4.[13]

1.2.8 System Engineering in the Life Cycle

In Figure 1.12, the system engineering process is applicable in all phases of the life cycle. In the early stages of conceptual and preliminary design, the emphasis is on understanding the true needs of the consumer and in the determination of requirements. These requirements must be "traceable" from the top and on down to the component level as necessary. The top-down approach reflected in the left side of the Vee diagram in Figure 1.11 is critical for the successful implementation of a system engineering program.

Given the basic requirements, the emphasis then shifts to the iterative analysis, synthesis, design optimization, and validation process. Trade-offs studies are conducted with the objective of providing a well-balanced system design. System engineering activities continue through the construction and/or production processes to ensure that the designed system configuration is compliant with the initially specified requirements. Next, there is an ongoing iterative effort of assessement (or validation) throughout the operational use and maintenance support phases. Experience related to the evaluation and assessment of the system, operating and being main-

[13] System analysis is covered further in G. H. Fisher, *Cost Considerations in Systems Analysis,* RAND Corporation Report R-490ASD, American Elsevier, New York, 1971; F. S. Hillier, and G. J. Lieberman, *Introduction to Operations Research,* 6th Ed., McGraw-Hill, New York, 1995; and G. Majone, and E. S. Quade (Eds.), *Pitfalls of Analysis,* John Wiley, New York, 1980.

tained in the consumer's environment, must be captured. A baseline configuration (with the appropriate metrics) must be established for the purposes of benchmarking and the initiation of a continuous process improvement activity. This, of course, requires that a good comprehensive data collection, analysis, and evaluation capability be implemented to provide the necessary feedback.

Finally, when changes are being initiated (whether for corrective action or for product/process improvement), the consequences of such changes must be evaluated from a top-level system perspective; that is, assessing the impact of a change on the overall system. The principles of configuration management and change control must be implemented to ensure that the results are consistent with the basic requirements, in terms of both effectiveness and life-cycle cost. Such changes may be applicable to the prime mission elements of the system, the production/construction capability, and/or the maintenance and support infrastructure. The interaction effects, both upward and downward, must be properly addressed in a *systems* context.

1.3 SYSTEM ENGINEERING MANAGEMENT

The successful implementation of system engineering concepts and principles is dependent not only on *technological* issues but *management* issues as well. As illustrated in Figure 1.20, there are two sides of the spectrum, each dependent on the other. The best tools/models may be available to implement the process shown in Figure 1.12; however, there is no guarantee for success unless the proper organizational environment has been created and an effective management structure is in

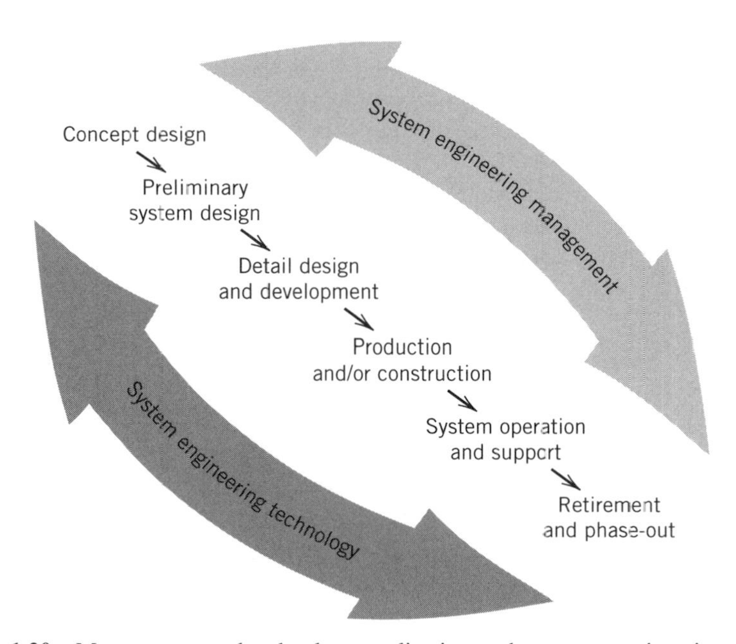

Figure 1.20 Management and technology application to the system engineering process.

place. Top management must first believe in and then provide the necessary support to enable the application of system engineering methods on in-house projects. Specific objectives must be defined, policies and procedures must be developed, and an effective reward structure must be supportive.

Although there are variations from one program to the next, Figure 1.21 presents a baseline for discussion. The major program phases and milestones are noted, along with a few selected activities and events that are significant from a system engineering perspective. These are summarized in what follows and are discussed in detail in Chapter 6.

1. During the early stages of conceptual design, it is essential that good communications between the producer and the consumer(s) be established from the beginning. Defining the true need, conducting feasibility analyses, developing operational requirements and the maintenance concept, and identifying specific quantitative and qualitative requirements at the system level are critical! These requirements must be properly conveyed through a well-prepared System Specification (Type "A"). This top-level system specification constitutes the most important *technical* document, from which all lower-level specifications evolve. Without a good "foundation" from the beginning, all subsequent lower-level requirements may be questionable (refer to Chapter 3).

2. During the latter stages of conceptual design, a comprehensive System Engineering Management Plan (SEMP) must be developed to ensure the implementation of a program that will lead to a well-coordinated and integrated product output. The SEMP, which evolves from the top-level Program Management Plan (PMP), provides the integration of all lower-level planning documents. It includes the design-related tasks necessary to enhance the day-to-day system development effort, the implementation of concurrent engineering methods, and the integration of the appropriate organizational entities into a "team" approach. The SEMP must directly support the requirements in the System Specification (Type "A") from a *management* perspective, and the two documents must "talk to each other!" The SEMP is addressed in detail in Chapter 6.

3. During the latter stages of conceptual design, a Test and Evaluation Master Plan (TEMP), or equivalent, must be developed for the purposes of assessment and ultimate validation. As requirements are initially specified in the System Specification (Type "A") and planned through the tasks described in the SEMP, the methods/techniques to be used for measuring and evaluating the system to ensure compliance with these requirements must be described. This plan must address test and evaluation activities on a fully *integrated* basis, employing the appropriate combination of simulation and other analytical tools, mockups, laboratory models, and prototype models. Test and evaluation is covered further in Chapter 2.

4. As system design and development progresses, there is a need to schedule a series of formal design reviews at discrete points where the design configuration evolves from one level of definition to another; that is, conceptual, system, equipment/software, and critical design reviews. The purpose of these reviews is to en-

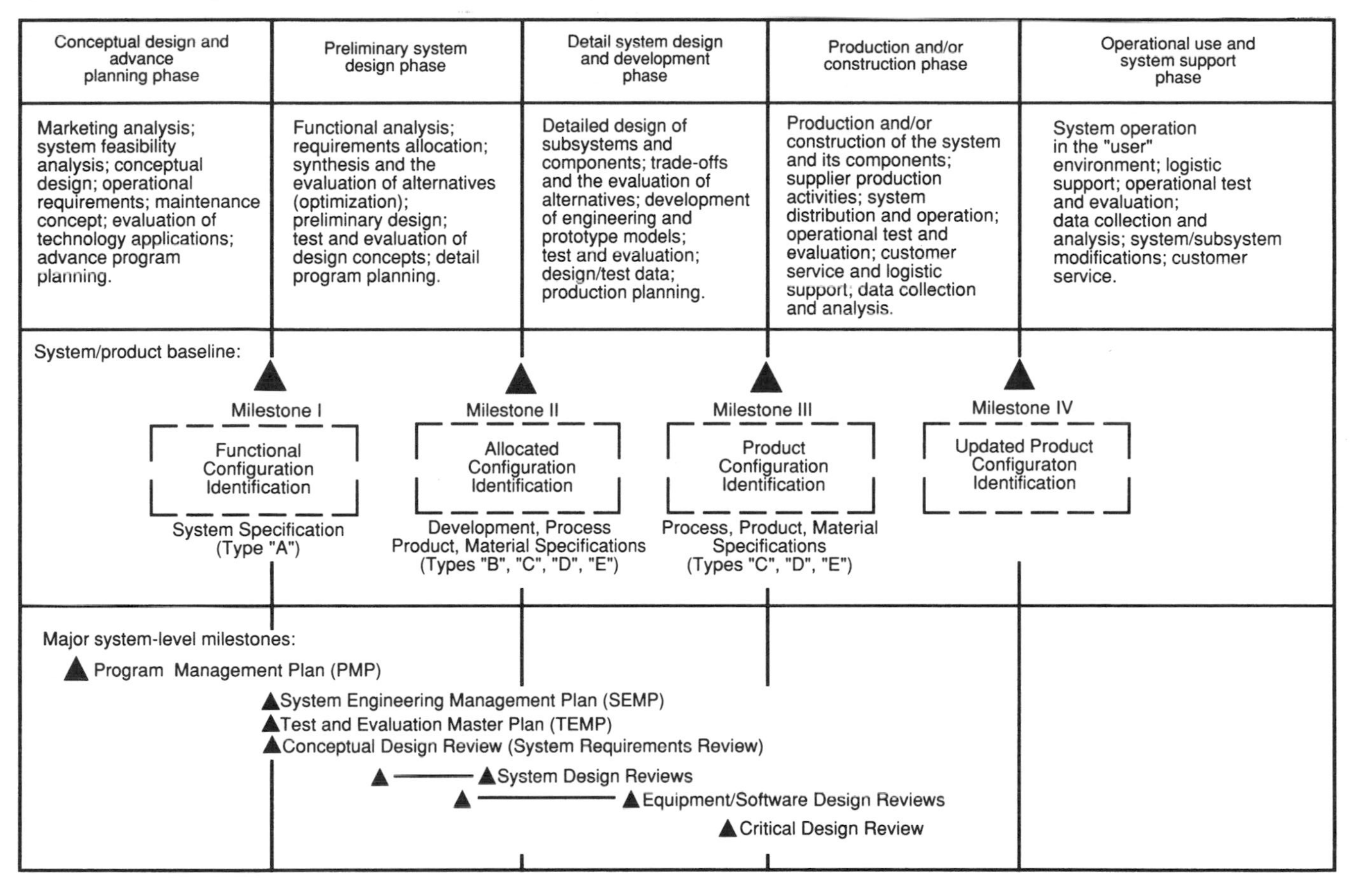

Figure 1.21 The system acquisition process and major milestones. (*Source:* B. S. Blanchard and W. J. Fabrycky, *Systems Engineering and Analysis,* 2nd Ed., Prentice Hall, Upper Saddle River, NJ, 1990.)

sure that the specified requirements are being met prior to entering into a subsequent phase of effort, and to ensure that the necessary communications exist across organizational lines. Refer to Chapter 5 for further discussion of design reviews and evaluation requirements.

5. Toward the latter stages of detail design, throughout the construction/production phase, and during the operational use and maintenance support phase, there is a need to provide an ongoing system assessment and validation effort. The objective is to ensure that the consumer requirements are being met, and to establish a "baseline" for the purposes of benchmarking and the initiation of a *continuous process improvement* activity. Design changes are initiated as required to correct any noted deficiencies.

The successful implementation of system engineering principles is highly dependent on properly managing the simplified process depicted in Figure 1.22. Inherent within this process is the application of the different technologies employed to facilitate the steps of requirements analysis, functional analysis and allocation, synthesis, design optimization, and validation.

1.4 RELATED TERMS AND DEFINITIONS

With the objective of enhancing the material presented thus far, a few related terms and definitions have been included. Although subsequent chapters introduce many additional concepts, the few covered in this section are considered to be closely related to system engineering.

1.4.1 Concurrent Engineering

One of the first definitions of *concurrent engineering* resulted from a Department of Defense study, and is "a systematic approach to the integrated, concurrent design of

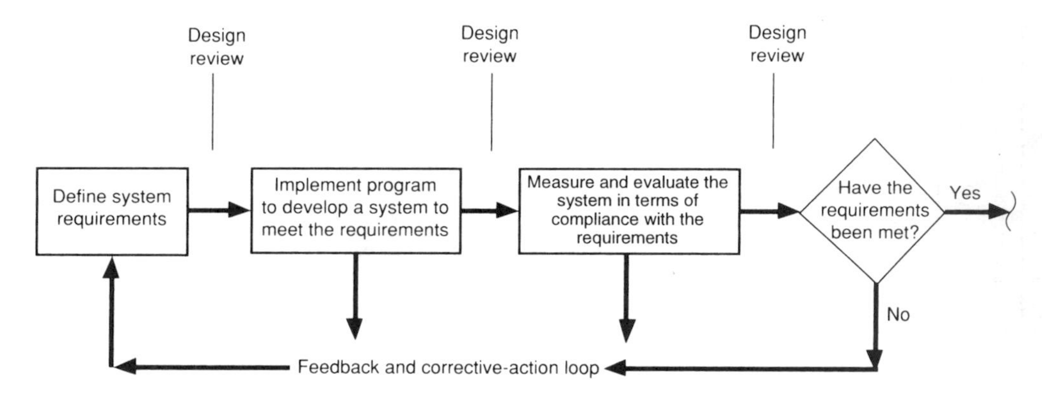

Figure 1.22 The basic system requirements, evaluation, and review process.

products and their related processes, including manufacture and support. This approach is intended to cause the developers, from the outset, to consider all elements of the product life cycle from conception through disposal, including quality, cost, schedule, and user requirements."[14] Like system engineering, the principles of concurrent engineering are inherent within the life cycles illustrated in Figure 1.10; that is, the design must consider the prime elements of the system, the manufacturing and/or construction process, and the maintenance and support process on an integrated and *concurrent* basis. As such, concurrent engineering should be included within the system engineering process.

1.4.2 Integrated Product and Process Development (IPPD)

With the objective of improving system design and acquisition, the Department of Defense recently initiated the IPPD concept, which can be defined as "a management technique that simultaneously integrates all essential acquisition activities through the use of multidisciplinary teams to optimize the design, manufacturing, and supportability processes. IPPD facilitates meeting cost and performance objectives from product concept through production, and including field support."[15] Inherent within the IPPD spectrum are a select number of *integrated product teams* (IPTs), which are composed of representatives from all appropriate functional disciplines working together to build a successful program and to design and develop an effective and efficient product for consumer use. Such teams may include representation from marketing, engineering, manufacturing, logistics, customer service, and the consumer. The principles of IPPD/IPT are directly supportive of the system engineering requirements for design integration and the required day-to-day communications.

1.4.3 Total Quality Management (TQM)

Total quality management (TQM) can be described as a totally integrated management approach that addresses system/product quality during all phases of the life cycle and at each level in the overall system hierarchy. It provides a before-the-fact orientation to quality, and it focuses on system design and development activities as well as manufacturing and production, maintenance and support, and related functions. TQM is a unification mechanism linking human capabilities to engineering, production, and support processes. Emphasis is placed on total customer satisfaction, the iterative practice of "continuous improvement," and a total integrated organizational approach. As part of the initial system design and development effort, consideration must be given to (1) the design of the processes that will be used to

[14] R. I. Winner, J. P. Pennell, H. E. Bertrand, and M. M. G. Slusarczuk, "The Role of Concurrent Engineering in Weapons Systems Acquisition," Institute of Defense Analysis, Alexandria, VA, Report R-338, 1988. Additional references are included in Appendix F.

[15] Department of Defense Regulation 5000.2-R, "Mandatory Procedures for Major Defense Acquisition Programs (MDAPs) and Major Automated Information System (MAIS) Acquisition Programs," Item C (Definitions) and Paragraph 1.6, March 15, 1996.

manufacture and produce the components of the system, and (2) the design of the support infrastructure that will provide the necessary ongoing maintenance of that system throughout its planned life cycle. In this regard, the principles of TQM must be inherent within the system engineering process.[16]

1.4.4 Configuration Management (CM)

CM is a management approach used to identify the functional and physical characteristics of an item in the early phases of its life cycle, control changes to those characteristics, and record and report change processing and implementation status. CM involves four functions that include (1) configuration identification, (2) configuration control, (3) configuration status accounting, and (4) configuration audits. CM is the concept of *baseline* management, which includes the *functional* baseline, the *allocated* baseline, and the *product* baselines identified in Figure 1.12. Successful implementation of system engineering requirements is heavily dependent on a good disciplined approach to baseline management.

1.4.5 Integrated System Maintenance and Support

As a result of the lack of consideration of system maintenance and support in the design process through the years (i.e., the third life cycle in Figure 1.10), the Department of Defense initiated the concept of *integrated logistic support* (ILS). ILS is a management function that provides the initial planning, funding, and controls that help to assure that the ultimate consumer (or user) receives a system that will not only meet performance requirements, but one that can be supported expeditiously and economically throughout its programmed life cycle. A major ILS objective is to assure the integration of the various elements of support; that is, manpower and personnel, training and training support, spares and repair parts and related inventories, test and support equipment, maintenance facilities, transportation and handling equipment, computer resources, and technical data.

Of a more specific nature, ILS can be defined as "a disciplined, unified, and iterative approach to the management and technical activities necessary to (1) integrate support considerations into system and equipment design; (2) develop support requirements that are related consistently to readiness objectives, to design, and to each other; (3) acquire the required support; and (4) provide the required support during the operational phase at minimum cost."[17]

Included within the concept of ILS is the element of "design for supportability," and the requirements in this area include the consideration for reliability, maintain-

[16] Two references are (1) DOD 5100.51G, "Total Quality Management: A Guide for Implementation," Department of Defense, Washington, DC, and (2) RAC SOAR-7, "A Guide for Implementing Total Quality Management," Rome Air Development Center, New York, 1990.

[17] This definition initially evolved from Department of Defense Instruction 5000.2, "Defense Acquisition Management Policies and Procedures," Part 7, 1991.

ability, human factors, safety, producibility, disposability, and related characteristics in design. Thus, from an organizational perspective, there is a very close relationship between integrated logistic support (ILS) and system engineering.

In the commercial sector, the maintenance and support issue has been addressed more recently through the concept of *total productive maintenance* (TPM). TPM, a concept originally developed by the Japanese, constitutes an integrated, top-down, system-oriented, life-cycle approach to maintenance, with the objective of maximizing productivity. TPM is directed primarily to the commercial manufacturing environment and

1. Promotes the overall effectiveness and efficiency of equipment in the factory. It includes *maintenance prevention* (MP) and *maintainability improvement* (MI), which consider the appropriate incorporation of reliability and maintainability characteristics in design.
2. Establishes a complete preventive maintenance program for factory equipment based on life-cycle criteria (similar to the reliability-centered maintenance approach used in establishing preventive maintenance requirements).
3. Is implemented on a "team" basis involving various departments to include engineering, production operations, and maintenance.
4. Involves every employee in the company, from the top management to the workers on the shop floor. Even equipment operators are responsible for the care and maintenance of the equipment they operate.
5. Is based on the promotion of preventive maintenance through "motivational management" (the establishment of autonomous small-group activities for the maintenance and support of equipment).

Total productive maintenance, often defined as "productive maintenance" implemented by all employees, is based on the principle that equipment improvement must involve everyone in the organization, from line operators to top management. The objective is to eliminate equipment breakdowns, speed losses, minor stoppages, and so on. It promotes defect-free production, just-in-time (JIT) production, and automation. The concept of TPM promotes *continuous improvement in maintenance.*[18]

Related to TPM is the concept of *total asset management* (TAM), which is being addressed by those industries and government agencies that are being faced with budgetary constraints, limited available resources, and greater international competition. TAM is directed toward "systems" where there are significant capital assets, and where there is a need to adopt a total life-cycle approach to resource management.

[18] The concept of TPM was initiated in Japan, through the Japan Institute for Plant Maintenance (JIPM), in 1971. A good initial reference is S. Nakajima (Ed.), *Total Productive Maintenance (TPM) Development Program,* Productivity Press, Portland, OR, translated into English in 1989. TPM has grown significantly in terms of implementation throughout the world. Refer to Appendix F for some additional references.

1.4.6 Total System Value

A system should be measured in terms of its total *value* to the consumer. For the purposes of discussion, one needs to consider both sides of the balance, as shown in Figure 1.23; that is, the *technical* factors and *economic* factors. In the early stages of conceptual design, the appropriate technical performance measures (TPMs) must be established and allocated to the various elements of the system, and there must be a top-down/bottom-up traceability of requirements. The development of TPMs and the allocation process is discussed further in Chapter 2.

Of particular interest within the domain of system engineering is the issue of life-cycle cost. Life-cycle cost involves *all* costs associated with the system life cycle:

1. *Research and development (R&D) cost:* the cost of feasibility studies; developing operational and maintenance requirements; system analyses; detail design and development; fabrication, assembly, and test of engineering models; initial system test and evaluation; and associated documentation.

2. *Production and construction cost:* the cost of fabrication, assembly, and test of operating systems (production models); operation and the sustaining maintenance and support of the manufacturing capability; facility construction; and the acquisition of an *initial* system support capability (e.g., test and support equipment, spare/repair parts, and technical documentation).

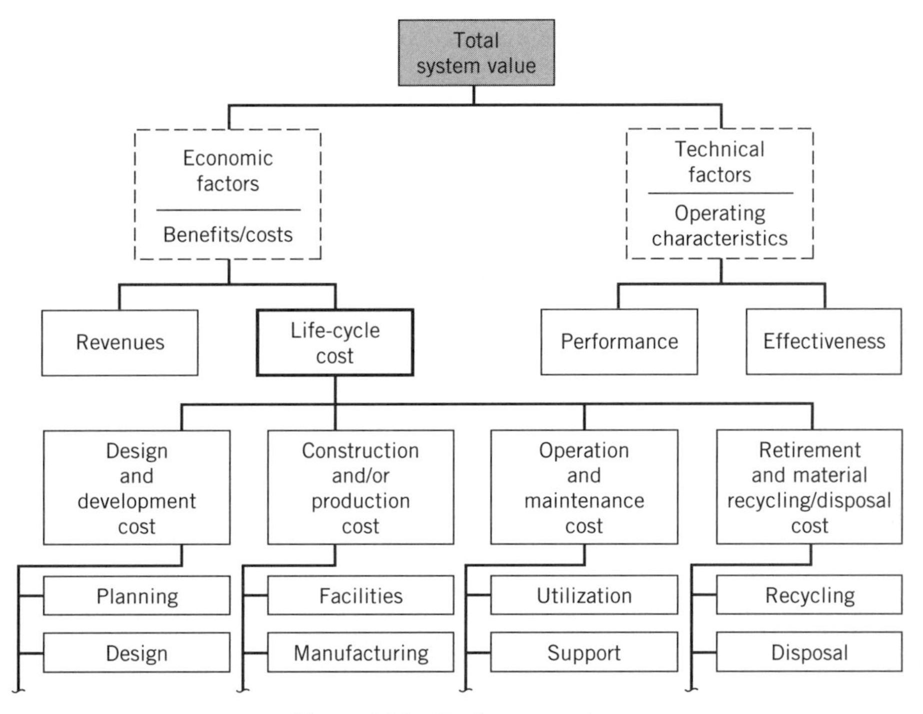

Figure 1.23 Total system value.

3. *Operation and maintenance cost:* the cost of system operation, and the sustaining maintenance and support of the system through its planned life cycle (e.g., manpower and personnel, spare/repair parts and related inventories, test and support equipment, transportation and handling, facilities, software, modifications, and technical data).

4. *System retirement and phaseout cost:* the cost of phasing the system and its components out of the inventory because of obsolescence or wearout; recycling of items for further use; condemnation and the disposal of materials.

Life-cycle costs may be categorized many different ways, depending on the type of system and the sensitivities desired in cost-effectiveness measurement. The objective is to provide *total cost visibility* (refer to Figure 1.4).[19]

1.5 SUMMARY

This chapter provides an abbreviated introduction to some of the key terms and definitions that are inherent in the discussions throughout subsequent chapters. Although there are many additional terms introduced as one progresses through this text, a brief discussion of "systems," "system analysis," "system science," "system engineering," and the "life cycle," will, hopefully, stimulate the "thought processes" for the material to come. The concepts presented herein, and particularly the information illustrated in Figures 1.12, 1.13, and 1.14, are a natural introduction to the system engineering process discussed in Chapter 2.

QUESTIONS AND PROBLEMS

1. Provide, in your own words, a definition of a "system." Include some examples.

2. Select a system of your choice and describe the system life cycle. Construct a detailed flow diagram "tailored" to your situation.

3. Define *system engineering.* What is included? Why is it important? How does system engineering differ from *system science* and *system analysis?*

4. What are the differences (or similarities) between "system engineering" and some of the more traditional disciplines such as civil engineering, electrical engineering, mechanical engineering, and so on?

5. Refer to Figure 1.10 (Example "A"). Describe the interrelationships between the three illustrated life cycles.

[19] Two references that address life-cycle cost in detail are W. J. Fabrycky and B. S. Blanchard, *Life-Cycle Cost and Economic Analysis,* Prentice Hall, Upper Saddle River, NJ, 1991; and J. V. Michaels, and W. P. Wood, *Design to Cost,* John Wiley, New York, 1989.

6. Refer to Figure 1.13. What are some of the key system engineering objectives that can be applied?
7. Refer to Figure 1.14. What are some of the key system engineering objectives that can be applied?
8. What is the significance of the *feedback* process illustrated in Figure 1.15?
9. What are the major system engineering functions in conceptual design? Preliminary design? Detail design and development? System operational use and life-cycle support?
10. Describe the basic differences among the Waterfall Model, the Spiral Model, and the Vee Model. How do they compare with the "model" proposed by the author?
11. The successful implementation of the system engineering process is dependent on both *technological* and *management* issues. Explain why? Provide an example of how one can impact the other.
12. Why is the System Specification (Type "A") important? Develop an outline for a system of your choice.
13. What is the purpose of design reviews?
14. What is *concurrent engineering?* How does it relate to system engineering?
15. What is *configuration management?* Why is it important in system engineering?
16. Why is ILS important? How does it relate to system engineering?
17. What is *life-cycle cost?* What is included? Why is it important to consider such in the decision-making process?

2 The System Engineering Process

The *system engineering process* is inherent within the overall system life cycle, as illustrated in Figure 1.12. The initial emphasis is on a top-down, integrated, life-cycle approach to system design and development, conveyed through the activities depicted in blocks 0.1 through 4.6. This includes the identification of consumer need, the conductance of feasibility analysis, the development of operational requirements and the maintenance and support concept, functional analysis, requirements allocation, and so on. Subsequently, there is the iterative process of assessment and validation, and the incorporation of changes for product/process improvement as required. Although the process is more directed to the early stages of system design and development, the activities in the latter phases of construction/production, operational use, and system maintenance and support are essential for understanding the consequences of earlier decisions and the establishment of benchmarks for the future. In other words, the *feedback loop* is critical and an integral part of the system engineering process.

This chapter addresses the system engineering process and the basic activities reflected in Figure 2.1. These activities represent a *process* that should be followed each time that there is a newly identified requirement for a system, and not to be interpreted as implying the expenditure of excess time and resources. The perception by many is that the implementation of system engineering requirements is costly and time-consuming. To the contrary, such implementation is likely to provide many benefits and savings in the long term; however, it does require a change in "thinking," a shift in emphasis and in the ways of doing business.

From the figure, the steps shown must, of course, be "tailored" to the system or program requirement. Also, there are many iterations that occur within! Analyses and trade-off studies are accomplished at each stage, functions are identified in several of the blocks, and so on, and it is impossible to show graphically everything that occurs throughout! However, for the purposes of discussion and for better understanding, the steps shown in Figure 2.1 are covered in this chapter, in the order indicated.

2.1 DEFINITION OF THE PROBLEM (CURRENT DEFICIENCY)

The system engineering process generally commences with the identification of a "want" or "desire" for something, and is based on a real (or perceived) deficiency. For instance, the current capability is not adequate in terms of meeting certain re-

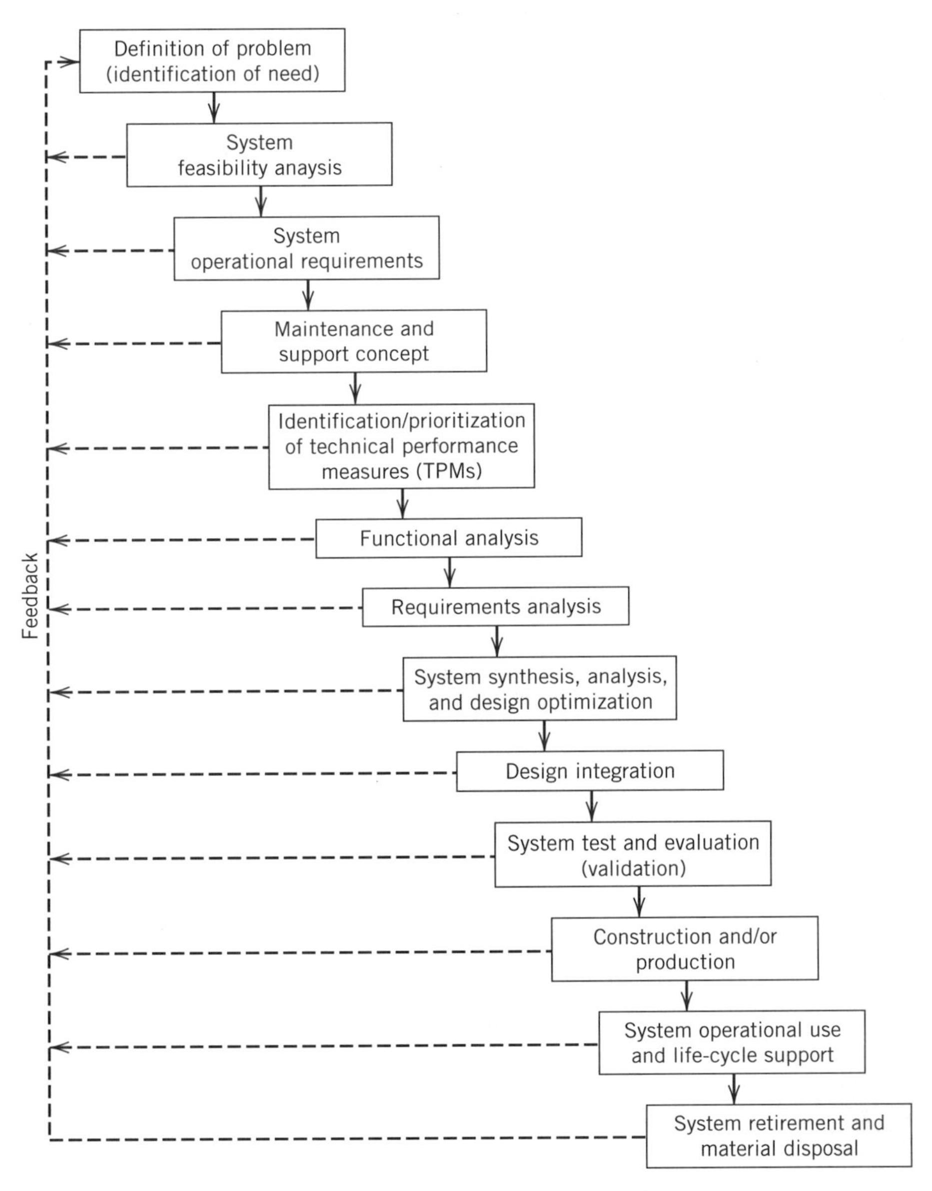

Figure 2.1 The system engineering process in the life cycle.

quired performance goals, is not available when needed, cannot be properly supported, is too costly in terms of operation, and so on. As a result, a new system requirement is defined along with the priority for introduction, the date when the new system capability is required for consumer use, and an estimate of the resources necessary for acquiring the new system capability. To ensure a good start, a "statement of the problem" should be presented in specific qualitative and quantitative terms, in enough detail to justify progressing to the next step.

The requirement for identifying the need (as a starting point) may seem to be rather "basic," or self-evident. However, one often finds that a design effort is initiated as a result of a personal interest or a political whim, without first having adequately defined the requirements for such! In the software area, there is the tendency to accomplish a lot of coding before identifying the need. Additionally, there are instances where the engineer sincerely believes to know what the customer needs, without having involved the customer in the process. The "design-it-now-fix-it-later" philosophy prevails which, in turn, can be rather costly.

Defining the problem is sometimes the most difficult part of the process, particularly if one is in a rush to "get going!" Yet, the number of "false starts" and the ultimate risks can be significant unless a good foundation is laid from the beginning. A complete description of the need, expressed in quantitatively-stated performance parameters where possible, is essential. It is important that the results reflect a true customer requirement, particularly in today's environment where resources are limited.

2.2 DEVELOPMENT OF CONSUMER NEED

Given the problem definition, a *needs analysis* must be accomplished with the objective of translating a broadly defined "want" into a more specific system-level requirement. The questions are: What is required of the system in *functional* terms? What functions must the system perform? What are the *primary* functions? What are the *secondary* functions? What must be accomplished to alleviate the stated deficiency? When must this be accomplished? Where is this to be accomplished? How many times must this be accomplished? There are many basic questions of this nature, and it is important to describe the customer (consumer) requirements in a *functional* manner in order to avoid a premature commitment to a specific design concept or configuration and, thus, the unnecessary expenditure of valuable resources. The ultimate objective is to define the "WHATs" and *not* the "HOWs!"

Accomplishing the needs analysis in a satisfactory manner can best be realized through a "team" approach involving the customer, the ultimate consumer (if different from the customer), the contractor or producer, and major suppliers as appropriate. The objective is to ensure that the proper communications exist between the parties involved. The "voice of the customer" must be heard, and the system developer must respond accordingly. Methods such as surveys, interviews, the use of checklists, the application of the "Quality Function Deployment (QFD)" method, and related techniques may be employed. As the definition of need is sometimes not apparent in the beginning, there may be several iterations of meetings, interviews, and so on, until full agreement occurs.[1]

[1] An excellent technique, often used as an aid in defining requirements and ensuring that the proper communications between the customer/consumer and producer exists, is the quality function deployment

2.3 SYSTEM FEASIBILITY ANALYSIS

Through the needs analysis, the functions that the system must perform are identified. There may be a single function such as "transport product *XYZ* from point *A* to point *B*," or "communicate among points *D, E,* and *F,*" or "produce *X* quantity of *Y* products by time *Z*." On the other hand, there may be a number of different functions to be performed, some primary and some secondary. To ensure a good design, all possible functions must be identified, the most rigorous functions being selected as the basis for defining system-level design requirements. It is important that *all* possibilities be addressed to ensure that the proper technologies and components are selected for design consideration.

The *feasibility analysis* is accomplished with the objective of evaluating the different technological approaches that may be considered in responding to the specified functional requirements. In considering different design approaches, alternative technology applications are investigated. For instance, in the design of a communications system, should one use fiber-optics technology, cellular, or the conventional hardwired approach? In designing an aircraft, to what extent should one incorporate composite materials? When designing an automobile, should one apply very high-speed integrated electronic circuitry in certain control applications, or should one select a more conventional electromechanical approach?

It is necessary to (1) identify the various possible design approaches that can be pursued to meet the requirements; (2) evaluate the most likely candidates in terms of performance, effectiveness, logistics requirements, and life-cycle economic criteria; and (3) recommend a preferred approach. The objective is to select an overall *technical* approach, and not to select specific components. There may be many different alternatives; however, the number of possibilities must be narrowed down to a few feasible options, consistent with the availability of resources (i.e., manpower, materials, and money).

It is at this early stage in the life cycle (i.e., the conceptual design phase) where major decisions are made relative to adopting a specific design approach. When there is not enough information available, a research activity may be initiated with the objective of developing new methods/techniques for specific applications. On some programs, the completion of applied research tasks and preliminary design activity is accomplished sequentially, whereas in other situations, there may be a number of different miniprojects underway at the same time.

The results of the feasibility analysis will have a significant impact not only on the operational characteristics of a system, but on the production and maintenance support requirements as well. The selection (and application) of a given technology has reliability and maintainability implications, may significantly impact the re-

(QFD) method. The QFD method was developed initially at the Kobe Shipyard of Mitsubishi Heavy Industries, Japan, and has evolved considerably since. It is used to facilitate the translation of a prioritized set of subjective customer requirements into a relevant set of system-level requirements during system conceptual design. The application of the QFD method is demonstrated further in Section 2.6.

quirements for spare parts and test equipment, may impact manufacturing methods, and will certainly impact life-cycle cost.

With the early feasibility analysis being so critical and having such a large impact on the follow-on system design and development activity, the role of the system engineer becomes important. In most situations, the detailed investigations and evaluation efforts leading to specific design approaches are highly technical, and are accomplished by specialists in a given engineering discipline. Often, these specialists are not oriented to the "system" as an overall entity, or its manufacturing process, or its maintenance and support capability, or the factors impacting life-cycle cost. Yet, major design decisions are made, the results of which end up in the system specification, and all subsequent design activity must comply. Thus, the need for a strong system engineering thrust at this early stage in the life cycle is critical.

2.4 SYSTEM OPERATIONAL REQUIREMENTS

With the identification of a need, combined with the selection of a feasible technical design approach, it is necessary to project this information in terms of anticipated operational requirements. Operational requirements reflect the needs of the consumer relative to system utilization and the accomplishment of a mission. The operational concept, as defined herein, includes the following information:

1. *Operational distribution or deployment:* the number of consumer sites where the system will be utilized, the geographical distribution and schedule, and the type

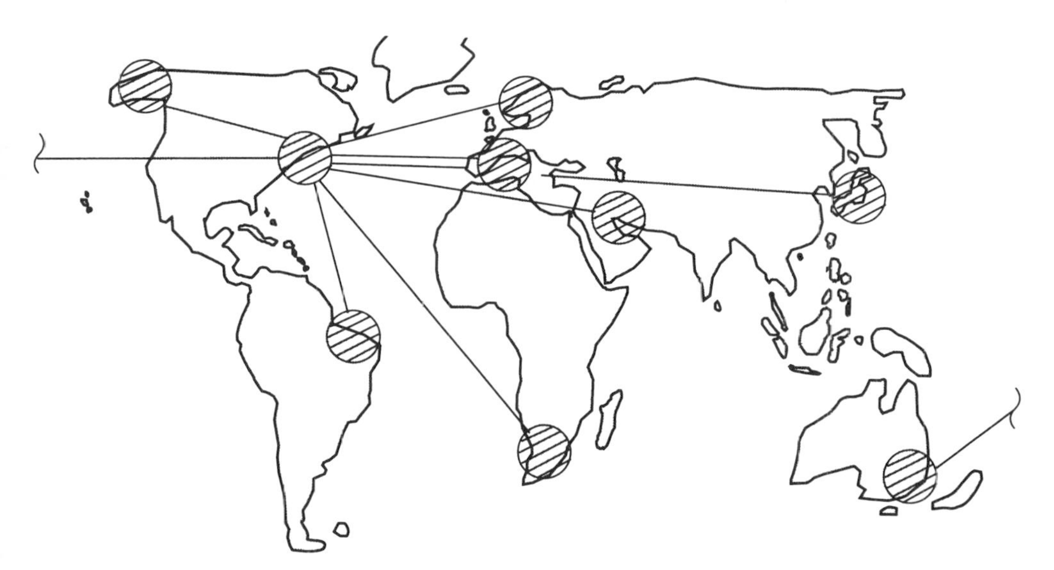

Figure 2.2 System operational requirements (geographical distribution).

and quantity of system components at each location. This responds to the question: Where is the system to be utilized? Figure 2.2 presents a sample worldwide distribution scheme.

2. *Mission profile or scenario:* description of the prime mission of the system, and its alternative or secondary missions. What is the system to accomplish in responding to the need? How will it accomplish its objective? This may be defined through a series of operational profiles, illustrating the dynamic aspects required in accomplishing a mission. An aircraft flight path between two cities, an automobile route, and a shipping route are examples. Figure 2.3 provides a simple illustration of possible profiles.

3. *Performance and related parameters:* definition of the basic operating characteristics, or functions, of the system. This refers to parameters such as range, accuracy, rate, capacity, throughput, power output, size, and weight. What are the critical system performance parameters necessary to accomplish the mission at the various consumer sites? These should be related to the profiles in Figure 2.3.

4. *Utilization requirements:* anticipated usage of the system, and its components, in accomplishing its mission. This refers to hours of system operation per day, duty cycle, on–off cycles per month, percentage of total capacity utilized, facility load-

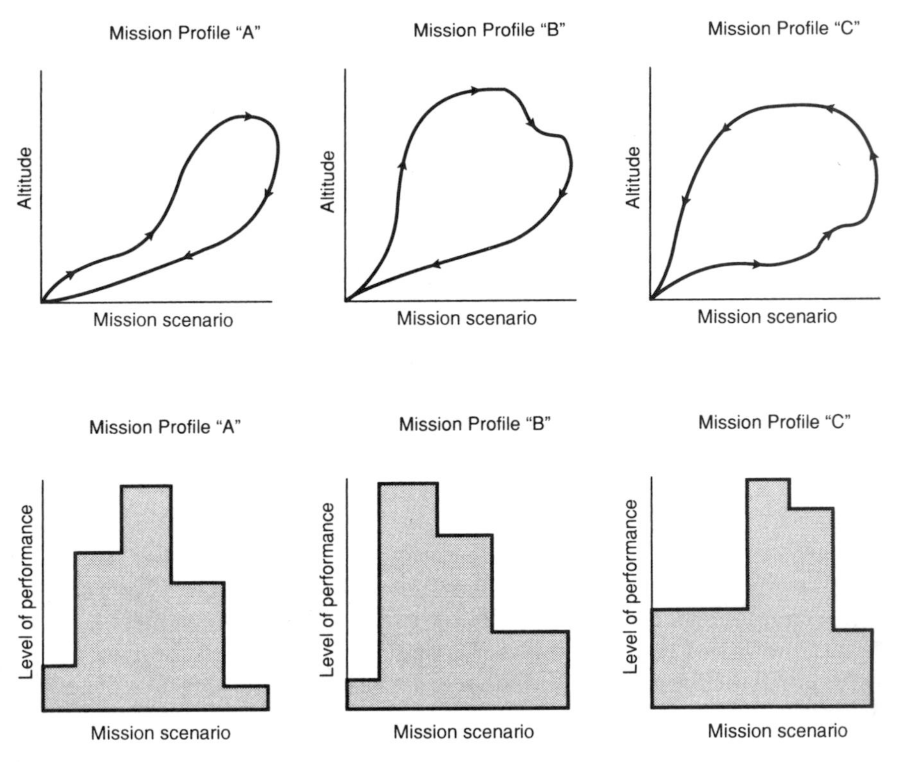

Figure 2.3 Sample system operational profiles. (*Source:* B. S. Blanchard, *Logistics Engineering and Management,* 4th Ed., Prentice Hall, Upper Saddle River, NJ, 1992.)

ing, and so on. How are the various system components to be utilized? This leads to determining some of the stresses imposed on the system by the operator and its environment.

5. *Effectiveness requirements:* system requirements, specified quantitatively as applicable, to include cost/system effectiveness, operational availability, dependability, reliability mean time between failure (MTBF), failure rate (λ), readiness rate, maintenance downtime (MDT), mean time between maintenance (MTBM), facility utilization (in percent), personnel skill levels, cost, and so on. Given that the system will perform, how *effective* or *efficient* will it be?

6. *Operational life cycle (horizon):* the anticipated time that the system will be in operational use. How long will the system be in use by the consumer? What is the total inventory profile for the system and its components, and where is this inventory to be located? One needs to define the anticipated system life cycle.

7. *Environment:* definition of the environment in which the system is expected to operate in an effective manner; for example, temperature, shock and vibration, noise, humidity, arctic or tropics, mountainous or flat terrain, airborne, ground, and shipboard. Following a set of mission profiles may result in specifying a range of values. To what will the system be subjected during its operational use, and for how long? In addition to system operations, environmental considerations should address transportation, handling, and storage modes. It is possible that the system (and/or some of its components) will be subjected to a more rigorous environment during transportation than during its operation.

The establishment of operational requirements forms the basis for system design. Obviously, one needs answers to the following questions before proceeding further:

1. *What* function(s) will the system perform?
2. *When* will the system be required to perform its intended function?
3. *Where* will the system be utilized and for how long?
4. *How* will the system accomplish its objective?

In responding to these questions, a baseline must be established. Although conditions may change, some initial assumptions are required. For example, system components will be utilized differently at different consumer locations, the distribution of system components may vary as the need changes, and/or the length of the life cycle may change as a result of obsolescence or the effects of competition. Nevertheless, the preceding information presented has to be developed in order to proceed with system design.

From a historical perspective, the operational requirements for many new systems in the past were developed (by a consultant, a marketing group, or some equivalent organizational entity), placed in a file while awaiting for a decision to proceed with preliminary design, and then forgotten when subsequent design activity finally did resume. At that point, with the need for this type of information readily apparent (but not available), individual design groups generated their own assumptions. Also, not all of the design functions were referencing the same baseline, and conflicting requirements evolved. This, in turn, led to systems being developed that did not meet consumer requirements, and the initiation of corrective action through costly modifications occurred. In other words, if the applicable operational requirements are not well defined and integrated into the design process, the results later on could turn out to be quite costly.

This is another critical area of activity where a strong system engineering thrust is necessary. The operational requirements for the system must be thoroughly defined and integrated, and the appropriate information must be disseminated in a timely manner throughout all applicable design organizations. Everyone involved in the design process must track the same baseline!

2.5 THE MAINTENANCE AND SUPPORT CONCEPT

In addressing system requirements, the normal tendency is to deal primarily with those elements of the system that relate directly to the "performance of the mission;" that is, prime equipment, operator personnel, operational software, and associated data. At the same time, there is very little attention given to system support. In general, the emphasis in the past has been directed toward only *part* of the system, and not the *entire* system. This, of course, has led to some of the problems discussed in Section 1.1.

To meet the overall objectives of system engineering, it is essential that *all* aspects of the system be considered on an integrated basis. This includes not only the prime mission-oriented segments of the system, but the support capability as well. System support must be considered from the beginning (e.g., during the feasibility analysis when new technologies are being evaluated for possible application), and a before-the-fact *maintenance concept* must be developed on how the proposed system is to be supported on a life-cycle basis.[2]

The maintenance concept, developed during conceptual design, evolves from the definition of system operational requirements, as illustrated by the flow chart in Figure 2.4. Initially, one must deal with the flow of activities and materials from design, through production, and to the consumer's operational site(s) where the system is being utilized. In addition, there is a flow involving the system support capability. Referring to the figure, a maintenance flow exists when items are returned from the operational site to the intermediate and depot levels of maintenance. A second flow involves the distribution of spare parts, personnel, test equipment, and data from the various suppliers to the intermediate and depot levels of maintenance,

[2]The author defines the "maintenance concept" as being a before-the-fact series of illustrations and statements on how the system is to be designed for supportability, and the "maintenance plan" defines the requirements for system support based on a known configuration and on the results of the logistic support analysis (or equivalent). The maintenance concept is an *input* to design, and the maintenance plan is the *result* of design. A complete coverage of the maintenance concept is presented in Chapter 4 of B. S. Blanchard, *Logistics Engineering and Management,* 4th Ed., Prentice Hall, Upper Saddle River, NJ, 1992.

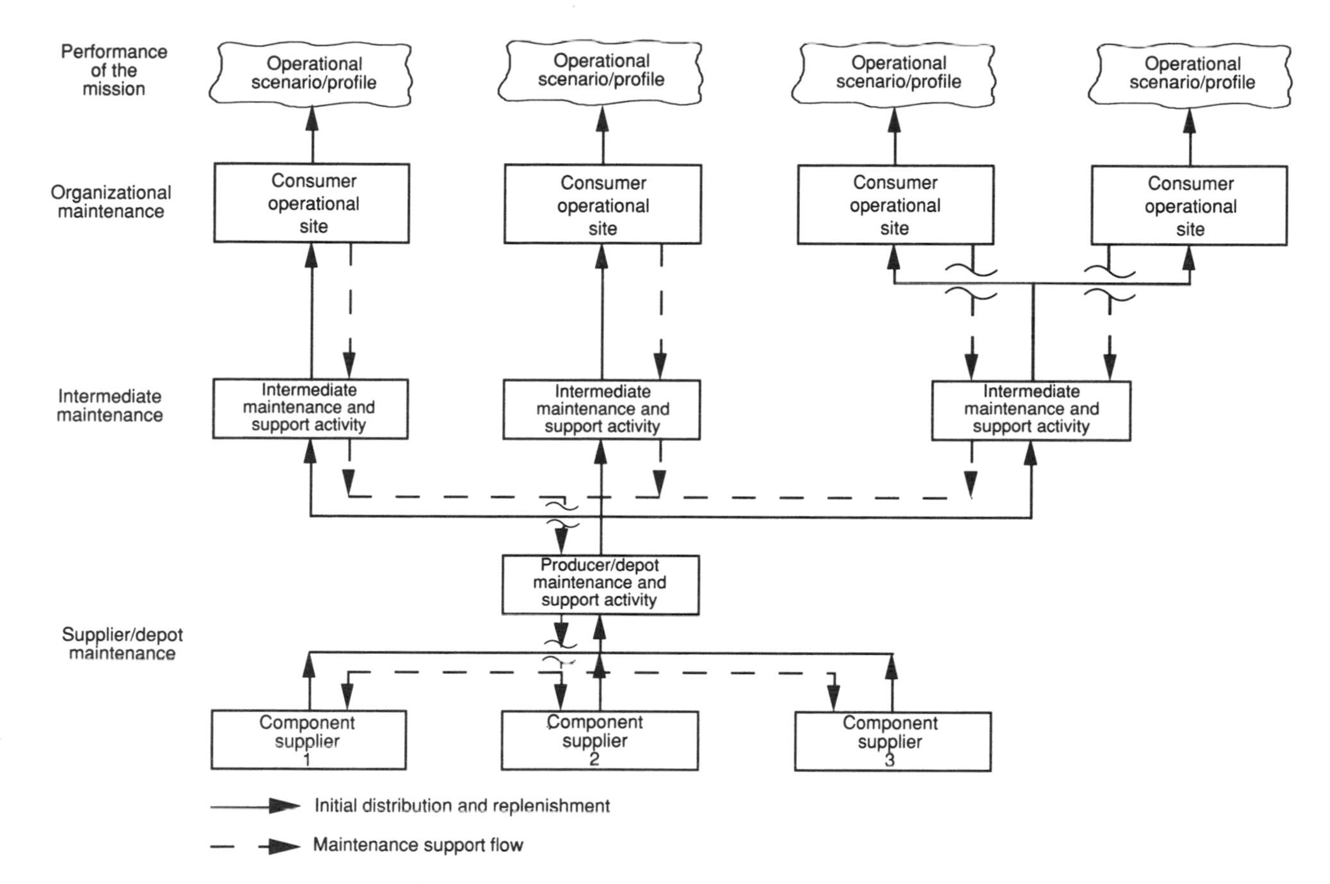

Figure 2.4 System operational and maintenance flow.

and to the operational sites as required. It is the flow chart in Figure 2.4 that reflects the activities that are related to the overall system support capability.

Although there are some variations as a function of the nature and type of system, the maintenance concept generally includes the following information:

1. *Levels of maintenance:* corrective and preventive maintenance may be accomplished on the system itself (or an element thereof) at the site where the system is used by the consumer, in an intermediate shop near the consumer, and/or at a depot or manufacturer's facility. Maintenance level pertains to the division of functions and tasks for each area where maintenance is performed. Anticipated frequency of maintenance, task complexity, personnel skill-level requirements, special facility needs, and so on, dictate to a great extent the specific functions to be accomplished at each level. Depending on the nature and mission of the system, there may be two levels, three levels, or four levels of maintenance. However, for the purposes of further discussion, maintenance may be classified as "organizational," "intermediate," and "supplier/depot."

a. *Organizational maintenance.* Organizational maintenance is performed at the operational site (e.g., airplane, vehicle, manufacturing production line, or communication facility). Generally, it includes tasks performed by the using organization on its own equipment. Organizational-level personnel are usually involved with the operation and use of equipment, and have minimum time available for detail system maintenance. Maintenance at this level normally is limited to periodic checks of equipment performance, visual inspections, cleaning of system elements, verification of software, some servicing, external adjustments, and the removal and replacement of some components. Personnel assigned to this level generally do not repair the removed components, but forward them to the intermediate level. From the maintenance standpoint, the least skilled personnel are assigned to this function. The design of equipment must take this fact into consideration (e.g., design for simplicity).

b. *Intermediate maintenance.* Intermediate maintenance tasks are performed by mobile, semimobile, and/or fixed specialized organizations and installations. At this level, end items may be repaired by the removal and replacement of major modules, assemblies, or piece parts. Scheduled maintenance requiring equipment disassembly may also be accomplished. Available maintenance personnel are usually more skilled and better equipped than those at the organizational level and are responsible for performing more detail maintenance.

Mobile or semimobile units are often assigned to provide close support to deployed operational systems. These units may constitute vans, trucks, or portable shelters containing some test and support equipment and spares. The mission is to provide on-site maintenance (beyond that accomplished by organizational-level personnel) to facilitate the return of the system to its full operational status on an expedited basis. A mobile unit may be used to support more than one operational site. A good example is the maintenance vehicle that is deployed from the airport hangar to an airplane parked at a commercial airline terminal gate and needing extended maintenance.

Fixed installations (permanent shops) are generally established to support both the organizational-level tasks and the mobile or semimobile units. Maintenance tasks that cannot be performed by the lower levels, due to limited personnel skills and test equipment, are performed here. High personnel skills, additional test and support equipment, more spares, and better facilities often enable equipment repair to the module and piece part level. Fixed shops are usually located within specified geographical areas.

Rapid maintenance turnaround times are not as imperative here as at the lower levels of maintenance.

c. *Depot or supplier maintenance.* The depot level constitutes the highest type of maintenance, and supports the accomplishment of tasks above and beyond the capabilities available at the intermediate level. Physically, the depot may be a specialized repair facility supporting a number of systems/equipments in the inventory or may be the equipment manufacturer's plant. Depot facilities are fixed and mobility is not a problem. Complex and bulky equipment, large quantities of spares, environmental control provisions, and so on, can be provided if required. The high-volume potential in depot facilities fosters the use of assembly-line techniques, which, in turn, permits the use of relatively unskilled labor for a large portion of the workload with a concentration of highly skilled specialists in such certain key areas as fault diagnosis and quality control.

The depot level of maintenance includes the complete overhauling, rebuilding, and calibration of equipment as well as the performance of highly complex maintenance actions. In addition, the depot provides an inventory supply capability. The depot facilities are generally remotely located to support specific geographical area needs or designated product lines.

The three levels of maintenance just discussed are covered in Figure 2.5.

2. *Repair policies:* within the constraints illustrated in Figures 2.4 and 2.5, there may be a number of possible policies specifying the extent to which repair of a system component will be accomplished (if at all). A repair policy may dictate that an item should be designed to be nonrepairable, partially repairable, or fully repairable. Repair policies are initially established, criteria are developed, and system design progresses within the bounds of the repair policy that is selected. An example of a repair policy for System XYZ, developed as part of the maintenance concept during conceptual design, is illustrated in Figure 2.6.[3]

3. *Organizational responsibilities:* the accomplishment of maintenance may be the responsibility of the consumer, the producer (or supplier), a third party, or a combination thereof! Additionally, the responsibilities may vary, not only with different components of the system, but as one progresses in time through the system operational use and sustaining support phase. Decisions pertaining to organizational responsibilities may impact system design from a diagnostic and packaging stand-

[3]Repair policies are ultimately verified through a level-of-repair analysis, the result of which leads into the maintenance plan. The level-of-repair analysis is usually accomplished as part of a maintainability analysis, a logistic support analysis, or both. Refer to Chapter 3 for additional coverage.

Criteria	*Organizational Maintenance*	*Intermediate Maintenance*		*Depot Maintenance*
Done where?	At the operational site or wherever the prime equipment is located	Mobile or semimobile units	Fixed units	Depot facility
		Truck, van, portable shelter, or equivalent	Fixed field shop	Specialized repair activity, or manufacturer's plant
Done by whom?	System/equipment operating personnel (low maint. skills)	Personnel assigned to mobile, semimobile, or fixed units (intermediate maintenance skills)		Depot facility personnel or manufacturer's production personnel (mix of intermediate production personel skills and high maintenance skills)
On whose equipment?	Using organization's equipment	Equipment owned by using organization		
Type of work accomplished?	Visual inspection Operational checkout Minor servicing External adjustments Removal and replacement of some components	Detailed inspection and system checkout Major servicing Major equipment repair and modifications Complicated adjustments Simple software maintenance Limited calibration Overload from organizational level of maintenance		Complicated factory adjustments Complex equipments repairs and modifications Overhaul and rebuild Detailed calibration Software maintenance (detailed modifications Supply support Overload from intermediate level of maintenance

Figure 2.5 Major levels of maintenance.

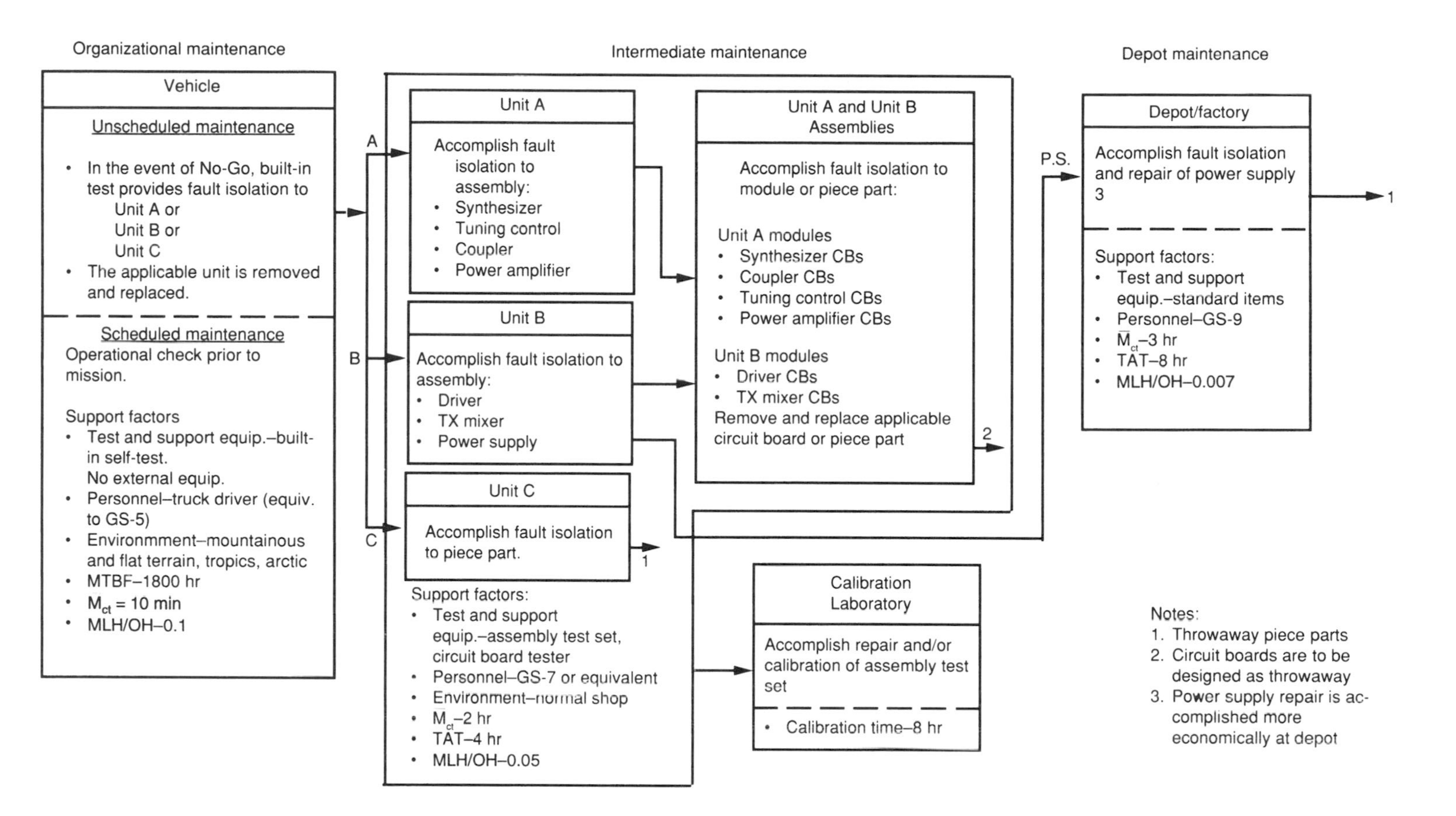

Figure 2.6 System maintenance concept flow (repair policy). (*Source:* B. S. Blanchard, *Logistics Engineering and Management,* 4th Ed., Prentice Hall, Upper Saddle River, NJ, 1992.)

point, as well as dictating repair policies, contract warranty provisions, and the like. Although conditions may change, some initial assumptions are required at this point in time.

4. *Maintenance support elements:* as part of the initial maintenance concept, criteria must be established relating to the various elements of maintenance support. These elements include supply support (spare and repair parts, associated inventories, provisioning data), test and support equipment, personnel and training, transportation and handling equipment, facilities, data, and computer resources. Such criteria, as an input to design, may cover self-test provisions, built-in versus external test requirements, packaging and standardization factors, personnel quantities and skill levels, transportation and handling factors and constraints, and so on. The maintenance concept provides some initial system design criteria pertaining to the activities illustrated in Figure 2.4, and the final determination of specific logistic and maintenance support requirements will occur through the completion of a maintenance engineering analysis as design progresses.

5. *Effectiveness requirements:* this constitutes the effectiveness factors associated with the support capability. In the supply support area, this may include a spare part demand rate, the probability of a spare part being available when required, the probability of mission success given a designated quantity of spares in the inventory, and the economic order quantity as related to inventory procurement. For test equipment, the length of the queue while waiting for test, the test station process time, and the test equipment reliability are key factors. In transportation, transportation rates, transportation times, the reliability of transportation, and transportation costs are of significance. For personnel and training, one should be interested in personnel quantities and skill levels, human error rates, training rates, training times, and training equipment reliability. In software, the number of errors per mission segment or per line of code may be important measures. These factors, as related to a specific system-level requirement, must be addressed. It is meaningless to specify a tight quantitative requirement applicable to the repair of a prime element of the system when it takes 6 months to acquire a needed spare part. The effectiveness requirements applicable to the support capability must complement the requirements for the system overall.

6. *Environment:* definition of the environment as it pertains to maintenance and support. This includes temperature, shock and vibration, humidity, noise, arctic versus tropical environment, mountainous versus flat terrain, shipboard versus ground conditions, and so on, as applicable to maintenance activities and related transportation, handling, and storage functions.

In summary, the maintenance concept provides the basis for the establishment of supportability requirements in system design. Not only do these requirements impact the prime mission-oriented segments of the system, but should provide guidance in the design and/or procurement of the necessary elements of logistic support. Additionally, the maintenance concept forms the baseline for the development of the detailed maintenance plan, prepared during the detail design and development phase.

2.6 IDENTIFICATION AND PRIORITIZATION OF TECHNICAL PERFORMANCE MEASURES (TPMs)

With the development of the operational requirements and the maintenance concept for the system, it is necessary for the designer to review these requirements in terms of relative degrees of importance, criticality from the standpoint of accomplishing the desired mission(s), and priorities in design in the event that trade-offs are necessary. In the design of a vehicle, is *speed* more important than *size?* For a manufacturing plant, is *production quantity* more important than *product quality?* In a communication system, is *range* more important than *reliability* or *clarity of message?* For a computer capability, is *capacity* more important than *speed?*

The number of objectives may be numerous, and the designer needs to understand which are more important than others and the relationships that exist between such! Additionally, it is desirable to express these objectives in quantitative terms where feasible. It is difficult (if not impossible) to proceed with the design in a satisfactory manner unless there are some "measurable" goals specified from the beginning. These goals, in turn, must reflect the customer's (consumer's) requirements.

In the development of system operational requirements and the maintenance concept, there are a number of measurable goals. The use of an "objectives tree," or something of an equivalent nature, may aid in facilitating the prioritization task. In Figure 2.7, requirements are often expressed in very general qualitative terms and

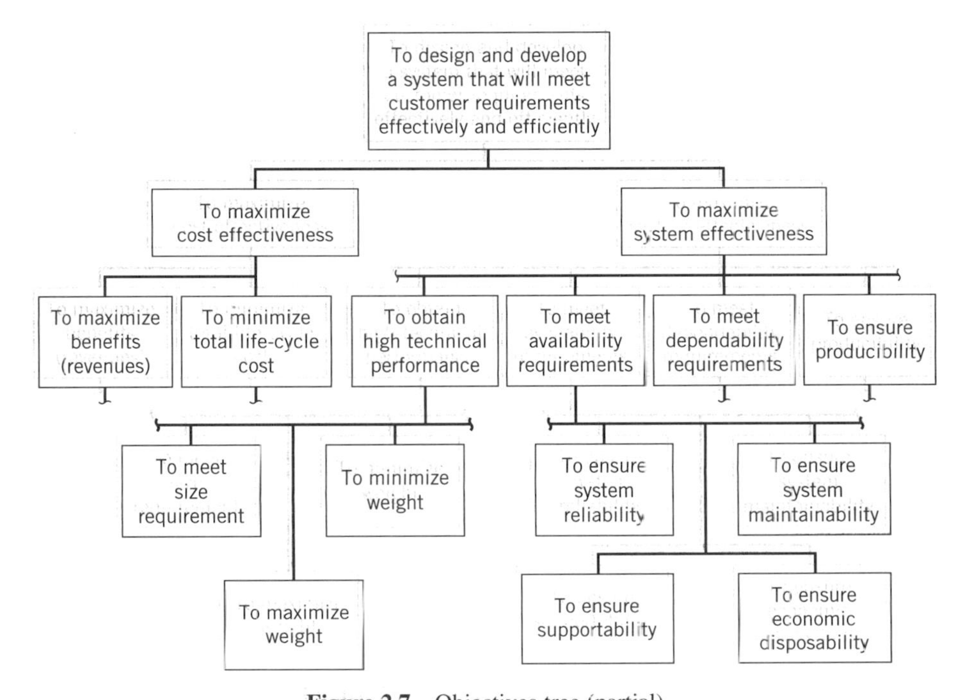

Figure 2.7 Objectives tree (partial).

included in a specification. The question is: How does one respond to a requirement such as "the system must be designed to meet customer requirements effectively and efficiently?" How does one measure the results for the purposes of validation?

In the absence of better guidance, the designer will need to interpret the specified requirements and make some assumptions relative to what is meant by "effectively" and "efficiently?" Although the objective is to design a system in response to consumer requirements, it may not always happen unless there is a good communications link between the designer and the customer. Through a "team" effort, the approach conveyed in Figure 2.7 can serve to help clarify the requirements. Initially, it may be necessary to express design objectives in qualitative terms, showing their relationships in a top-down hierarchical manner. Subsequently, an attempt should be made to establish *quantitative* measures for each block in the figure, and ensure that the appropriate "traceability" exists both downward and upward. Applying this to the system breakdown in Figure 1.14, what measures should be applied and to what level in the overall hierarchical structure for the system? Further, what design *criteria* should be established for each level? Is *reliability* more important than *maintainability?* Are *human factors* more important than *cost?* Establishing these relationships will, in turn, help the designer to identify areas where emphasis must be applied in the design process, and the areas that can be *traded-off* in the event that something has to "give!"

An excellent tool that can be applied to aid in establishing the necessary communications between designers and the consumer (i.e., the "customer") is the *quality function deployment (QFD)* method.[4] QFD constitutes a "team" approach to help ensure that the "voice of the customer" is refected in the ultimate design. The purpose is to establish the necessary *requirements* and to translate those requirements into technical solutions. Consumer requirements and preferences are defined and categorized as *attributes,* which are then weighted based on the degree of importance. The QFD method provides the design team an understanding of customer desires, forces the customer to prioritze those desires, and enables a comparison of one design approach against another. Each customer attribute is then satisfied by a technical solution.

The QFD process involves constructing one or more matrices, the first of which is often referred to as the "House of Quality (HOQ)."[5] A modified version of the HOQ is presented in Figure 2.8. Starting on the left side of the structure is the identification of customer needs and the ranking of those needs in terms of priority, the levels of importance being specified quantitatively. This reflects the "WHATs" that must be addressed. A team, with representation from both consumer and design organizations, determines the priorities through an iterative process of review, evalu-

[4] Two good references pertaining to the QFD process are (1) Yoji Akao (Ed.), *Quality Function Deployment: Integrating Customer Requirements Into Product Design,* Productivity Press, Portland, OR, translated into English, 1990; and (2) Lou Cohen, *Quality Function Deployment: How to Make QFD Work for You,* Addison-Wesley, Reading, MA, 1995. The QFD method was developed at the Kobe Shipyard of Mitsubishi Heavy Industries, Ltd., Japan, in the late 1960s and has evolved considerably since.

[5] J. R. Hauser and D. Clausing, "The House of Quality," *Harvard Business Review,* May–June 1988, pp. 63–73.

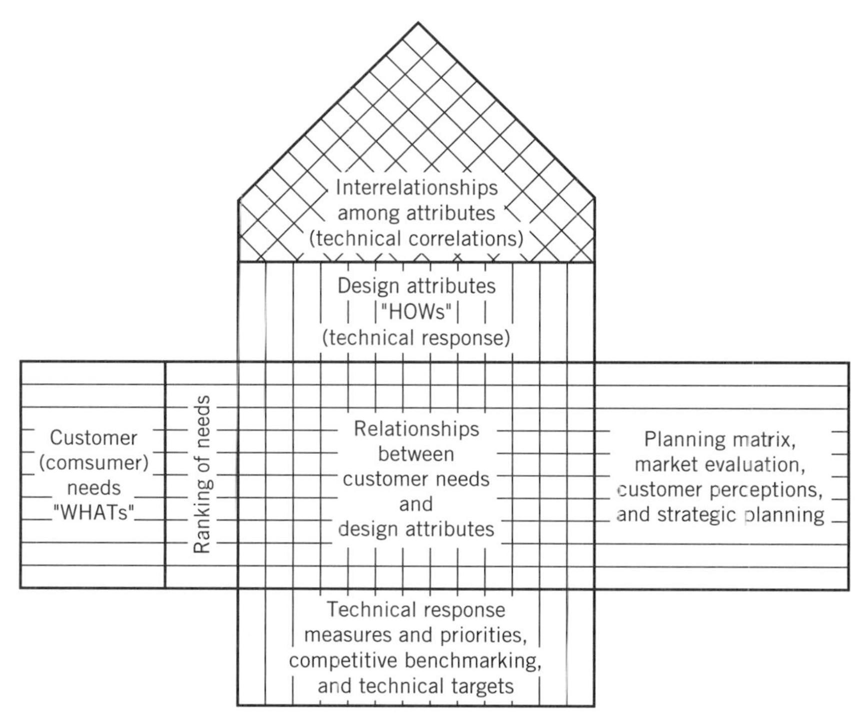

Figure 2.8 House of Quality (modified).

ation, revision, reevaluation, and so on. The top part of the HOQ identifies the designer's *technical* response relative to the attributes that must be incorporated into the design in order to respond to the needs (i.e., the "voice of the customer"). This constitutes the "HOWs," and there should be at least one technical solution for each identified customer need. The interrelationships among attributes (or technical correlations) are identified, as well as possible areas of conflict. The center part of the HOQ conveys the strength or impact of the proposed technical response on the identified requirement. The bottom part allows for a comparison between possible alternatives, and the right side of the HOQ is used for planning purposes.[6]

The QFD method is used to facilitate the translation of a prioritized set of subjective customer requirements into a set of *system-level* requirements during conceptual design. A similar approach may be used to subsequently translate system-level requirements into a more detailed set of requirements at each stage in the design and development process. In Figure 2.9, the "HOWs" from one house become the "WHATs" for a succeeding house. Requirements may be developed for the system, subsystem, component, the manufacturing process, the support infrastructure, and

[6] In order to gain a complete perspective of the QFD process and the advantages relative to its implementation, the reader is advised to review some of the literature in the field. The coverage presented herein is very cursory in nature and intended to provide only an "overview" of the process.

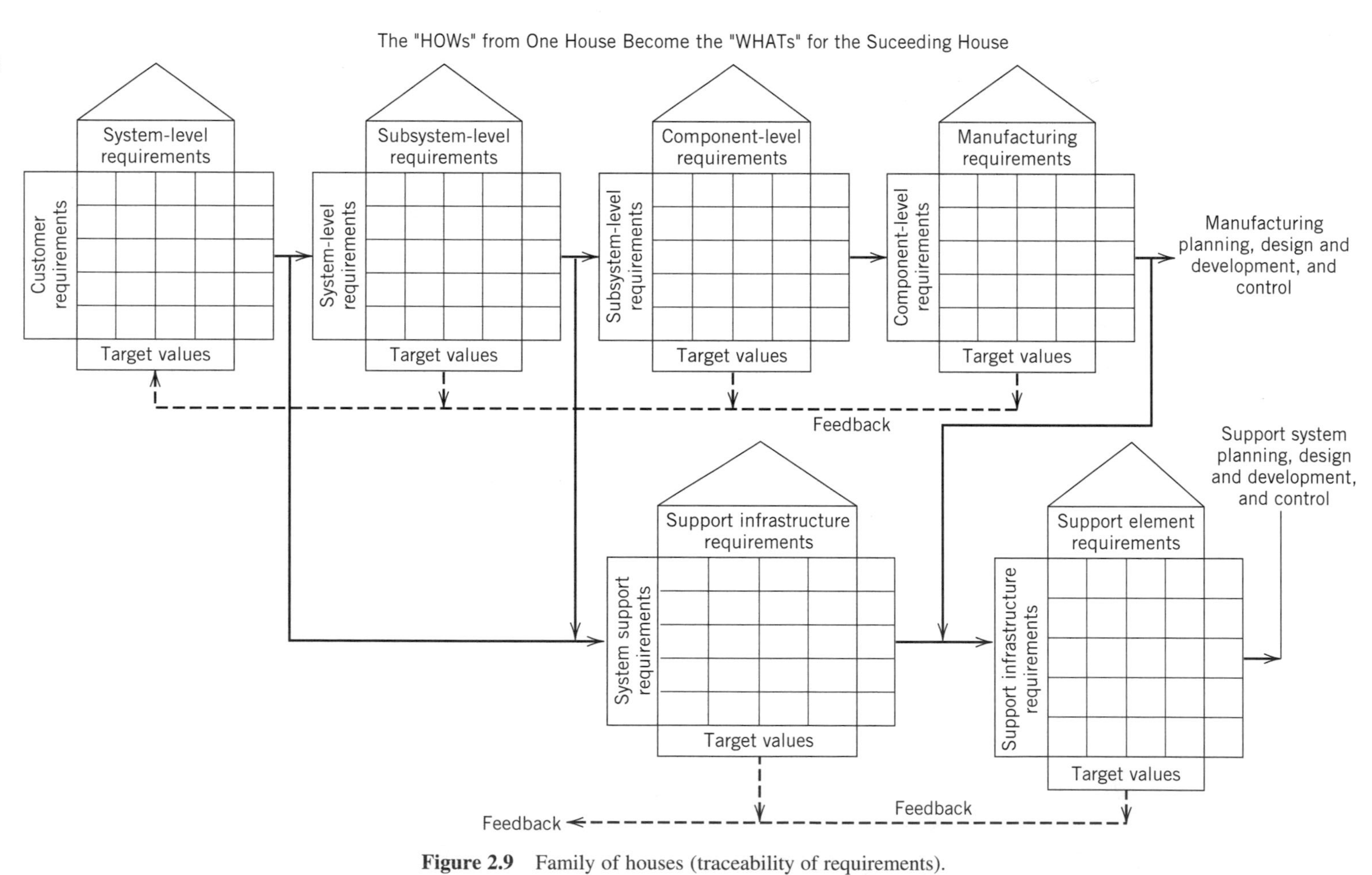
The "HOWs" from One House Become the "WHATs" for the Suceeding House
System-level requirements
Customer requirements
Target values
Subsystem-level requirements
System-level requirements
Target values
Component-level requirements
Subsystem-level requirements
Target values
Manufacturing requirements
Component-level requirements
Target values
Manufacturing planning, design and development, and control
Feedback
Support infrastructure requirements
System support requirements
Target values
Support element requirements
Support infrastructure requirements
Target values
Support system planning, design and development, and control
Feedback
Feedback

Figure 2.9 Family of houses (traceability of requirements).

Technical performance measure (TPM)	Quantative requirement ("metric")	Current "benchmark" (competing systems)	Relative importance (customer desires)
Process time (days)	30 days (maximum)	45 days (system "M")	10
Velocity (mph)	100 mph (minimum)	115 mph (system "B")	32
Availability (operational)	98.5% (minimum)	98.9% (system "H")	21
Size (feet)	10 feet long 6 feet wide 4 feet high (maximum)	9 feet long 8 feet wide 4 feet high (system "M")	17
Human factors	Less than 1% error rate per year	2% per year (system "B")	5
Weight (pounds)	600 pounds (maximum)	650 pounds (system "H")	6
Maintainability (MTBM)	300 miles (minimum)	275 miles (system "H")	9
			100%

Figure 2.10 Prioritization of technical performance measures (TPMs).

so on. The objective is to ensure the required justification and traceability of requirements from the top down. Further, requirements should be stated in *functional* terms.

Although the QFD method may not be the only approach used in helping to define the requirements for system design, it does constitute an excellent tool for creating the necessary visibility from the beginning. One of the largest contributors to "risk" is the lack of a good set of requirements and an adequate system specification. Inherent within the system specification should be the identification and prioritization of technical performance measures (TPMs), as illustrated in Figure 2.10. The TPM, its associated measure (i.e., "metric"), its relative importance, and "benchmark" objective in terms of what is currently available will provide designers with the necessary guidance for accomplishing their task. This is essential for establishing the appropriate levels of design emphasis, for defining the criteria as an input to the design, and for identifying the levels of possible risk should the requirements not be met.

2.7 FUNCTIONAL ANALYSIS

An essential element of early conceptual and preliminary design is the development of a *functional* description of the system to serve as a basis for the identification of the resources necessary for the system to accomplish its objective(s). A function refers to a specific or discrete action (or series of actions) that is necessary in order to achieve a given objective; that is, an operation that the system must perform to accomplish its mission, or a maintenance action that is necessary to restore the system to operational use. Such actions may ultimately be accomplished through the use of equipment, people, software, facilities, data, or combinations thereof. However, at this point, the objective is to specify the "WHATs" and *not* the "HOWs;" that is, *what* needs to be accomplished versus *how* it is to be done![5] The functional analysis is an iterative process of breaking requirements down from the system level, to the subsystem, and as far down the hierarchical structure as necessary to identify input design criteria and/or constraints for the various elements of the system.

In Figure 2.1, the functional analysis may be initiated in the early stages of conceptual design as part of the problem definition and needs analysis task, and functions that the system must perform in order to fulfill the needs of the consumer are identified. These *operating* functions are then expanded and formalized through the development of system operational requirements. Primary *maintenance and support* functions for the system, which evolve from the operational requirements, are identified as part of the maintenance concept development process. Subsequently, these functions must be expanded to include *all* of the activity from the initial identification of need to the retirement of the system.

The accomplishment of a functional analysis can be facilitated through the use of functional flow block diagrams, as illustrated in Figure 2.11. Block diagrams are developed primarily for the purpose of structuring system requirements into "functional terms." They are developed to illustrate basic system organization, and to identify functional interfaces. The functional analysis (and the generation of functional flow diagrams) is intended to enable the completion of the design, development, and system definition process in a comprehensive and logical manner. Top-level requirements are identified, partitioned to a second level, and on down to the depth required for the purposes of "definition." More specifically, the functional approach helps to ensure the following:

1. That all facets of system design and development, production, operation, support, and retirement are covered; that is, all significant activities within the system life cycle.
2. That all elements of the system are fully recognized and defined; that is, prime equipment, spare/repair parts, test and support equipment, facilities, personnel, data, and software.
3. That a means is provided for relating system packaging concepts and support requirements to specific system functions; that is, satisfying the requirements of good "functional" design.

[5] In applying the principles of systems engineering, not one piece of equipment, or software, or data item, or element of support should be identified and purchased, without first having justified the need for such through the functional analysis. On many projects, items are often purchased based on what is initially perceived as being a "requirement," but which later turns out not to be needed in the end. This practice, of course, can turn out to be quite costly.

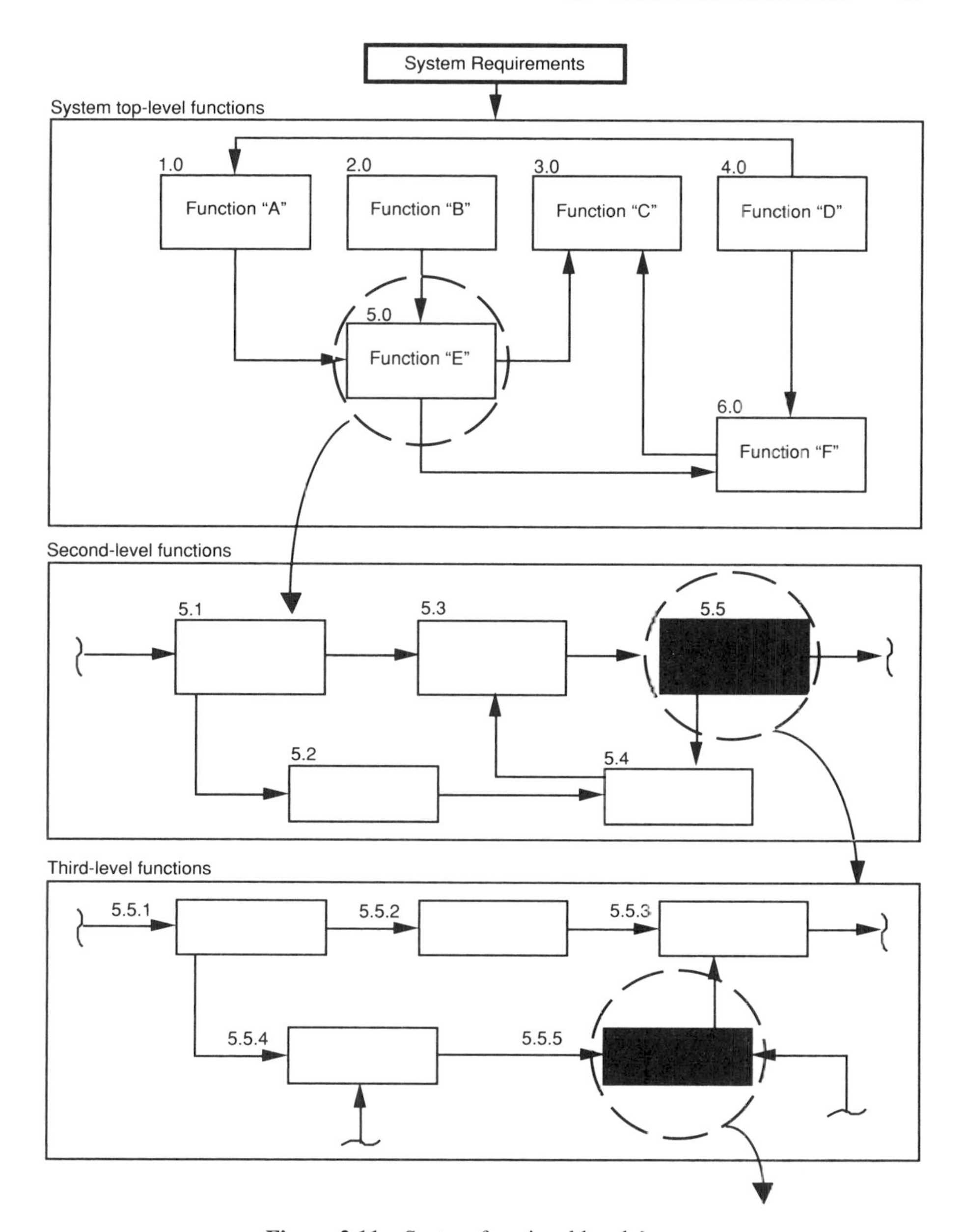

Figure 2.11 System functional breakdown.

4. That the proper sequences of activity and design relationships are established, along with critical design interfaces.

One of the objectives of functional analysis is to ensure traceability from the top system-level requirements down to the requirements for detail design. In Figure 2.12, it is assumed that there is a need for transportation between City "A" and

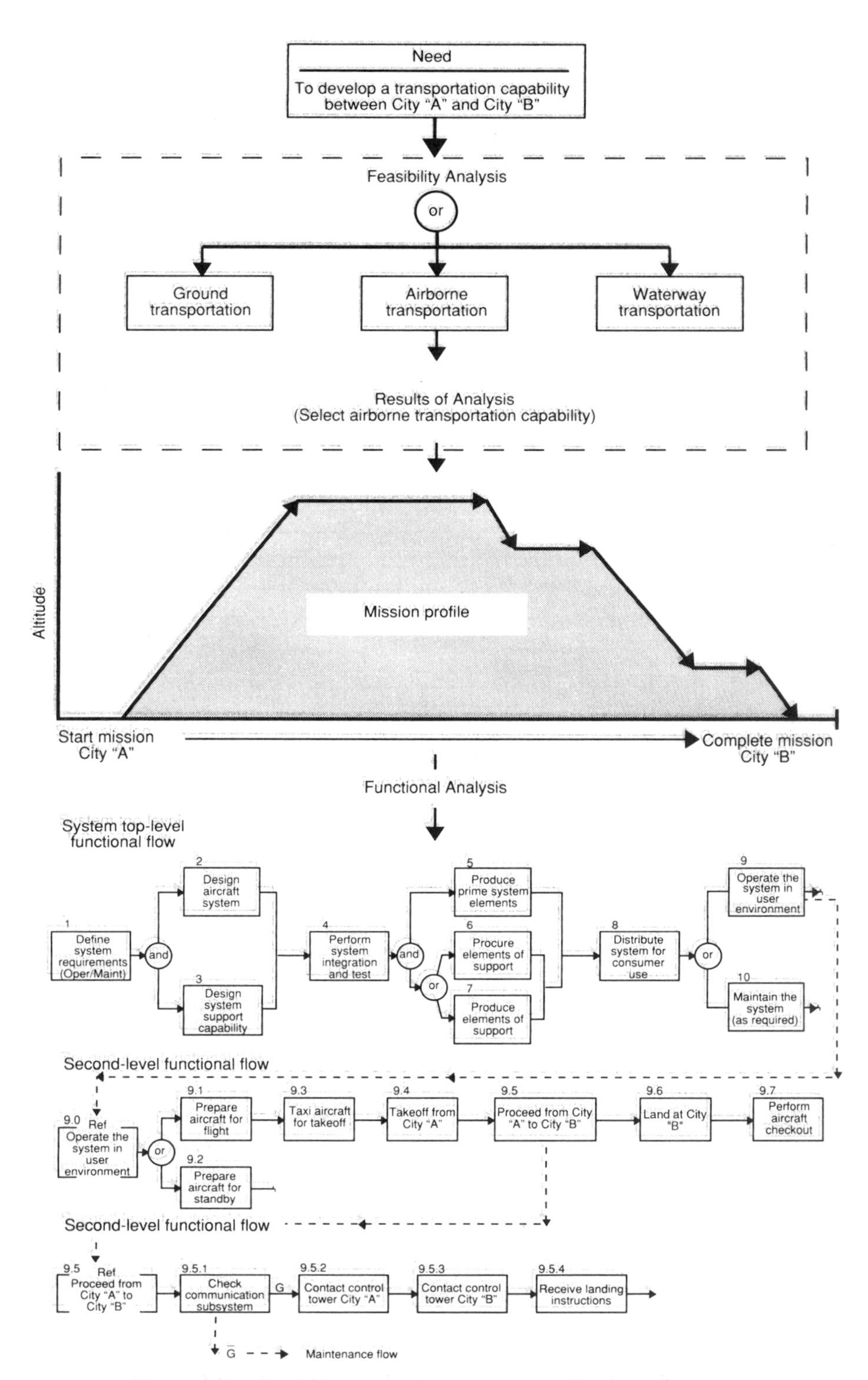

Figure 2.12 Evolutionary development of functional requirements.

City "B." Through the conductance of a feasibility analysis, trade-off studies are accomplished, and the results indicate that transportation by air is the preferred mode. Subsequently, through the definition of operational requirements, it was concluded that there is a requirement for a new aircraft system, demonstrating good performance and effectiveness characteristics with quantitative goals specified for size, weight, thrust, range, fuel capacity, reliability, maintainability, supportability, cost, and so on. An aircraft must be designed and produced that will accomplish its mission in a satisfactory manner, flying through a number of operational profiles such as the one illustrated in Figure 2.12. Further, the maintenance concept indicates that the aircraft will be designed for support at three levels of maintenance by the user, will incorporate built-in test provisions, and will be in operational use for a life cycle of 10 years.

With this basic information, following the general steps in Figure 2.1, one can commence with the structuring of the system in functional terms. A top-level functional flow diagram can be developed to cover the primary activities identified within the specified life cycle. Each of these designated activities can be expanded through a second-level functional flow diagram, a second-level activity into a third-level functional flow, and so on.

Through this progressive expansion of functional activities, directed to defining the "WHATs" (versus the "HOWs"), one can evolve from the mission profile in Figure 2.12 down to a specific aircraft capability such as "communications." A communications subsystem is identified, trade-offs are accomplished, and a detail design approach is selected. Specific resources that are necessary to respond to the stated functional requirement can be identified. In other words, one can drive downward from the system level to identify the resources needed to perform certain functions (e.g., equipment, people, facilities, and data). Also, given a specific equipment requirement, one can progress "upward" for *justification* of that requirement. The functional analysis provides the mechanism for "down-up" traceability.

2.7.1 Functional Flow Diagrams

In the development of functional flow diagrams, some degree of standardization is necessary, for the purpose of "communication," in defining the system. Thus, certain basic practices and symbols should be used, whenever possible, in the physical layout of functional diagrams. The following paragraphs provide some guidance in this direction:

1. *Function block:* each separate function in a functional diagram should be presented in a single box enclosed by a solid line. Blocks used for reference to other flows should be indicated as partially enclosed boxes labeled "REF." Each function may be as gross or detailed as required by the level of functional diagram on which it appears, but it should stand for a definite, finite, discrete action to be accomplished by equipment, personnel, facilities, software, or any combination thereof. Questionable or tentative functions should be enclosed in dotted blocks.

2. *Function numbering:* functions identified on the functional flow diagrams at

each level should be numbered in a manner that preserves the continuity of functions and provides information with respect to function origin throughout the system. Functions in the top-level functional diagram should be numbered 1.0, 2.0, 3.0, and so on. Functions that further indenture these top functions should contain the same parent identifier and should be coded at the next decimal level for each indenture. For example, the first indenture of function 3.0 would be 3.1, the second 3.1.1, the third 3.1.1.1, and so on. For expansion of a higher-level function within a particular level of indenture, a numerical sequence should be used to preserve the continuity of function. For example, if more than one function is required to amplify function 3.0 at the first level of indenture, the sequence should be 3.1, 3.2, 3.3, . . . , 3.*n*. For expansion of function 3.3 at the second level, the numbering shall be 3.3.1, 3.3.2, . . . , 3.3.*n*. Where several levels of indentures appear on a single functional diagram, the same pattern should be maintained. Whereas the basic ground rule should be to maintain a minimum level of indentures on any one particular flow, it may become necessary to include several levels to preserve the continuity of functions and to minimize the number of flows required to functionally depict the system.

3. *Functional reference:* each functional diagram should contain a reference to its next higher functional diagram through the use of a reference block. For example, function 4.3 should be shown as a reference block in the case where the functions 4.3.1, 4.3.2, . . . , 4.3.*n*, and so on, are being used to expand function 4.3. Reference blocks shall also be used to indicate interfacing functions as appropriate.

4. *Flow connection:* lines connecting functions should indicate only the functional flow and should not represent either a lapse in time or any intermediate activity. Vertical and horizontal lines between blocks should indicate that all functions so interrelated must be performed in either a parallel or series sequence. Diagonal lines may be used to indicate alternative sequences (cases where alternative paths lead to the next function in the sequence).

5. *Flow directions:* functional diagrams should be laid out so that the functional flow is generally from left to right and the reverse flow, in the case of a feedback functional loop, from right to left. Primary input lines should enter the function block from the left side; the primary output, or *GO* line, should exit from the right; and the *NO-GO* line should exit from the bottom of the box.

6. *Summing gates:* a circle should be used to depict a summing gate. As in the case of functional blocks, lines should enter and/or exit the summing gate as appropriate. The summing gate is used to indicate the convergence, or divergence, or parallel or alternative functional paths and is annotated with the term AND or OR. The term AND is used to indicate that parallel functions leading into the gate must be accomplished before proceeding to the next function, or that paths emerging from the AND gate must be accomplished after the preceding functions. The term OR is used to indicate that any of several alternative paths (alternative functions) converge to, or diverge from, the OR gate. The OR gate thus indicates that alternative paths may lead or follow a particular function.

7. *Go and no-go paths:* the symbols G and $\bar{G}$ are used to indicate go and no-go paths, respectively. The symbols are entered adjacent to the lines leaving a particular function to indicate alternative functional paths.

8. *Numbering procedure for changes to functional diagrams:* additions of functions to existing data should be accomplished by locating the new function in its correct position without regard to sequence of numbering. The new function should be numbered using the first unused number at the level of indenture appropriate for the new function.

The functions identified should not be limited strictly to those necessary for the operation of the system, but must consider the possible effects of maintenance on system design. In most instances, maintenance functional flows will evolve directly from operational flows.

2.7.2 Operational Functions

Operational functions, in this instance, constitute those that describe the activities that must be accomplished in order to fulfill the mission requirements. These may include both (1) those activities that involve the design, development, production, and distribution of a system for use; and (2) those activities that are related directly to the completion of a consumer mission scenario. In the second category, these may include a description of the various modes of system operation and utilization. For instance, typical gross operating functions may entail (1) "prepare aircraft for flight," (2) "transport material from the factory to the warehouse," (3) "initiate communications between the producer and the user," (4) "produce *'x'* quantity of units in a seven-day timeframe," and (5) "process *'a'* data to eight company distribution outlets, in *'b'* time, with *'c'* accuracy, and in *'d'* format." System functions necessary to successfully complete the identified modes of operation are then described.

Figure 2.13 illustrates a simplified operational flow diagram. Note that the words in each block are "action-oriented" and the block numbering allows for the downward-upward traceability of resource requirements. The functions are broken down to the depth necessary to describe the resources that will be required to accomplish the function; that is, equipment, software, people, facilities, and so on.

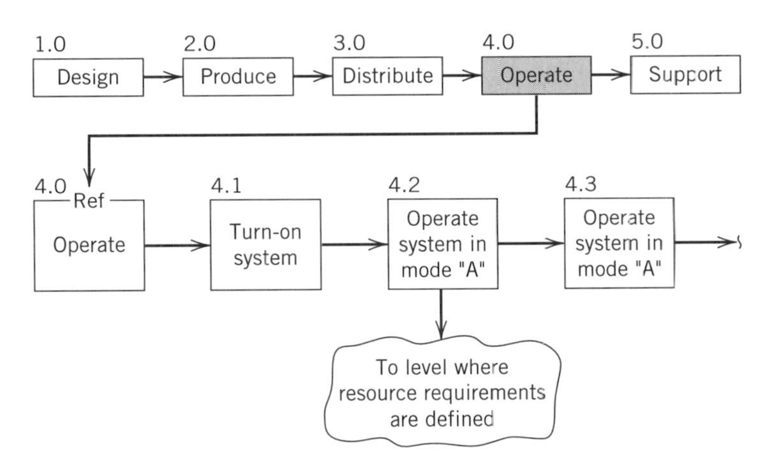

Figure 2.13 Functional block diagram (partial).

2.7.3 Maintenance and Support Functions

Once operational functions are described, the system development process leads to the identification of *maintenance and support* functions. For instance, there are specific performance expectations or measures associated with each block in an operational functional-flow diagram. A check of the applicable functional requirement will indicate either a "go" or a "no-go" decision. A "go" decision leads to a check of the next operational function. A "no-go" indication (constituting a symptom of failure) provides a starting point for the development of a detailed maintenance functional flow diagram. The transition from an operational function to a maintenance function is illustrated in Figure 2.14. Figure 2.15 presents a more in-depth functional flow diagram.

2.7.4 Application of Functional Flow Diagrams

The functional analysis provides an initial description of the system and, as such, its applications are extensive. Figure 2.16 illustrates a top-level operational functional flow diagram for a manufacturing system, commencing with the identification of need (block 1.0) and extending through system retirement (block 7.0). In areas where a greater degree of definition is desirable, the applicable block(s) may be broken down to a second level, third level, and so on, in order to gain the appropriate level of visibility necessary for the determination of resource requirements. In this instance, the ultimate manufacturing "operating" functions have been identified in the breakout of block 5.1.

For each of the blocks in Figure 2.16, the analyst should be able to specify *input* requirements, expected *outputs,* external *controls* and/or *constraints,* and the *mechanisms* (or resources) necessary to accomplish the specific function in question. In the process of identifying the appropriate resource requirements, there may be a number of different alternative approaches that should be considered. Trade-off

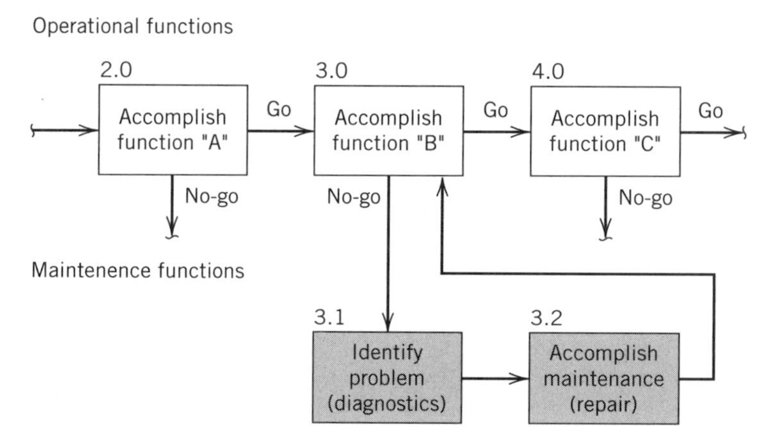

Figure 2.14 Transition from operational functions to maintenance functions.

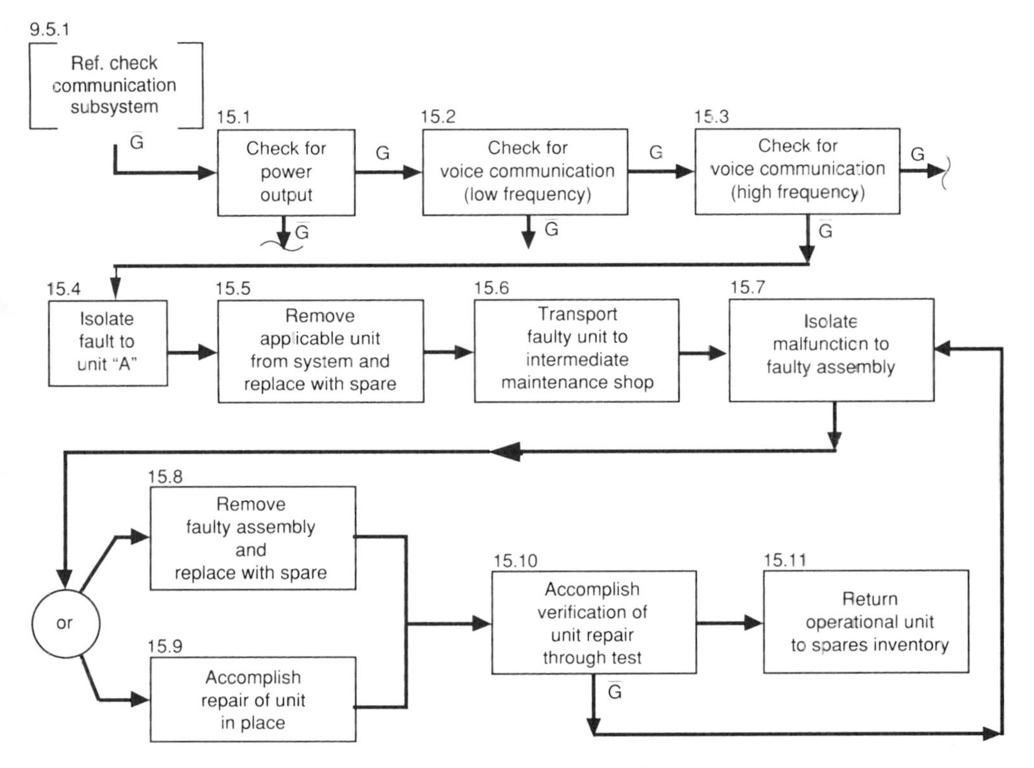

Figure 2.15 Maintenance functional flow diagram.

studies are conducted, alternatives are evaluated against criteria developed from the established technical performance measures (i.e., the TPMs derived in Section 2.6), and a preferred approach is recommended. It is at this point when one begins to identify the requirements for hardware, software, people, facilities, data, or combinations thereof. Figure 2.17 reflects the process that should be applied to each of the blocks in Figure 2.16.

In the evaluation of each functional requirement, the alternatives may include the selection of "commercial-off-the shelf (COTS)" items readily available from a number of different sources of supply, COTS items that may require some degree of modification, and/or "developmental" items that are unique to a particular application and where some new design is required. Past experience has indicated that extensive time and cost savings can be realized through the selection of readily available COTS equipment, reusable software, the utilization of existing facilities, and so on. Figure 2.18 illustrates the various options in this area.[6]

[6] In recent years, the Department of Defense (DOD) has placed considerable emphasis on the preferred use of COTS items versus the pursuit of new design and development efforts. The objectives are to reduce the time involved in the development and acquisition of new systems, improve system supportability/serviceability through the utilization of standard components that can be easily backed up with

Figure 2.18 shows that it is essential that a *good* definition of the inputs and outputs (and the applicable metrics) be established if one is to fully understand not only the *interfaces* between the different functions identified in Figure 2.16, but the precise requirements in the process of resource identification. If these input-output requirements are not well defined, the decision-making process as to a preferred approach becomes difficult, thus, leading to the possibility of initiating a new costly design and development effort when, in actuality, an existing off-the-shelf item could fulfill the need.

The functional analysis can facilitate an "open-architecture" approach to system design. A good comprehensive functional description of the system, with the interfaces well defined (both qualitatively and quantitatively), can lead to a structure that will not only allow for the rapid identification of resource requirements, but for the possible incorporation of new technologies later on. The objective is to design and develop a system that can be easily modified, through the insertion of new technologies, without causing a "costly" redesign of all of the elements of the system in the process.

In many current situations, the requirements in design are changing from a detailed "design to the component level" to the design of systems using a "black-box-integration" approach. Given the need to reduce acquisition times, while responding to an ever-changing set of requirements on a continuing basis and with many more suppliers involved, the system *architecture* must allow for the "ease of upgrade and/or modification." In other words, the system *structure* must be such as to facilitate design on an *evolutionary* basis, and with minimum cost. This can be enhanced through a good and comprehensive functional definition of the system in the early conceptual design phase of the life cycle.

Figure 2.19 illustrates a manufacturing system where there are many suppliers (from various locations throughout the world) who produce components for a consumer product that must be effectively integrated and tested. There are fabrication functions, subassembly functions, assembly functions, and test functions. Where, in many instances in the past, the manufacturing activity involved a bottom-up "build" approach, the challenges today relate to the *integration* of the various components into the end product. Without a good early definition and specification of the functional interfaces, the final integration and test activity may result in a costly "trial-by-error" process. In the figure, the example reflects a factory where the subprocesses were being accomplished effectively and efficiently; however, there were considerable problems associated with the "integration" activities; that is, the four critical integration points. The functional interfaces were not well defined from the beginning, causing a great deal of modification and rework downstream.

In completing a functional analysis, care should be taken to ensure that the required resources are properly identified for each function. A timeline analysis may be accomplished to determine whether the functions are to be accomplished in series or in parallel. It may be possible to share resources in some instances; that is, the

readily available spares and repair parts, and to reduce costs from a life-cycle perspective. A good reference is the technical report, *American Defense Preparedness Association Commercial-off-the-Shelf (COTS) Supportability Study,* ADPA, Arlington, Virginia, 1994.

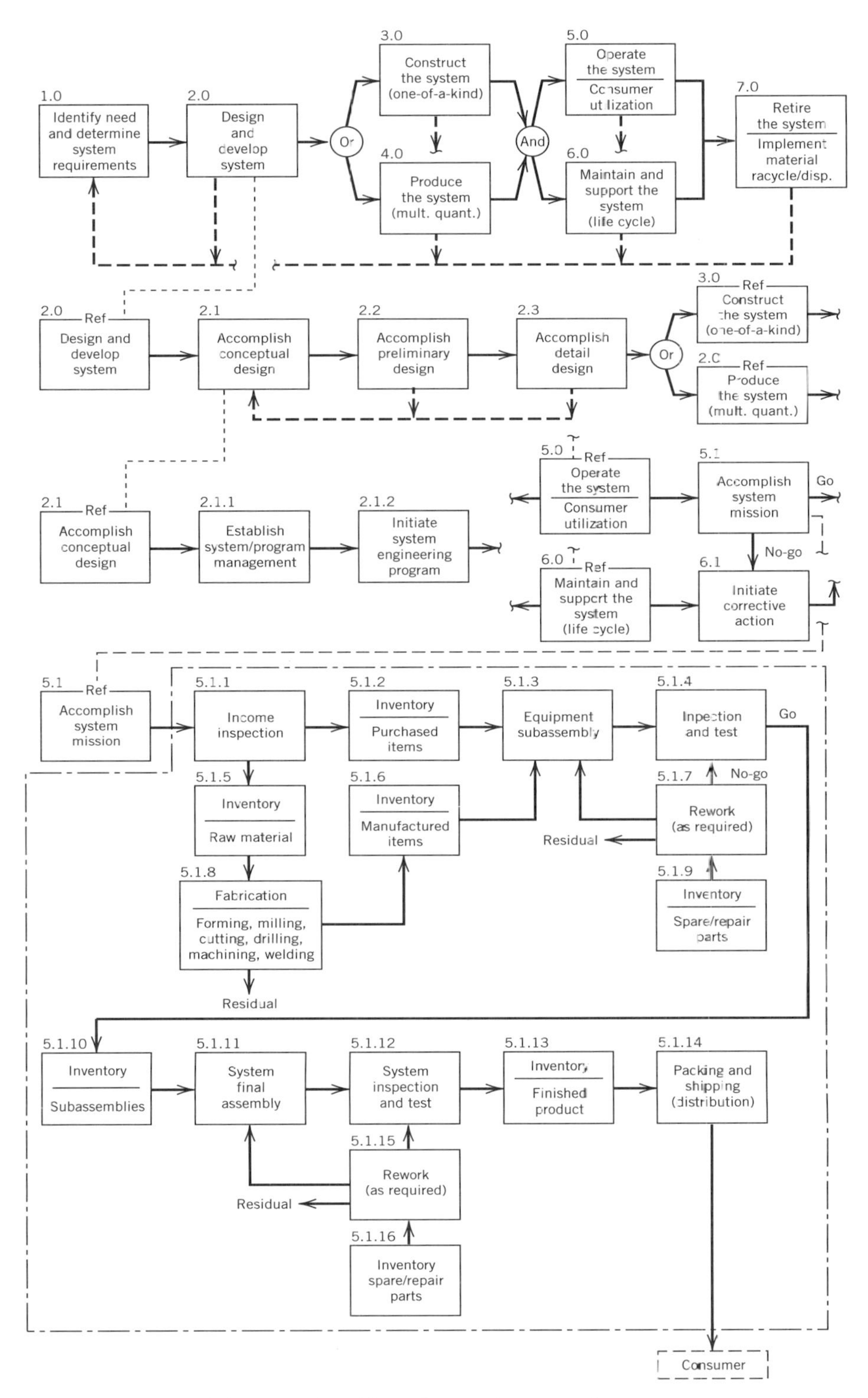

Figure 2.16 Functional flow diagram for a manufacturing system.

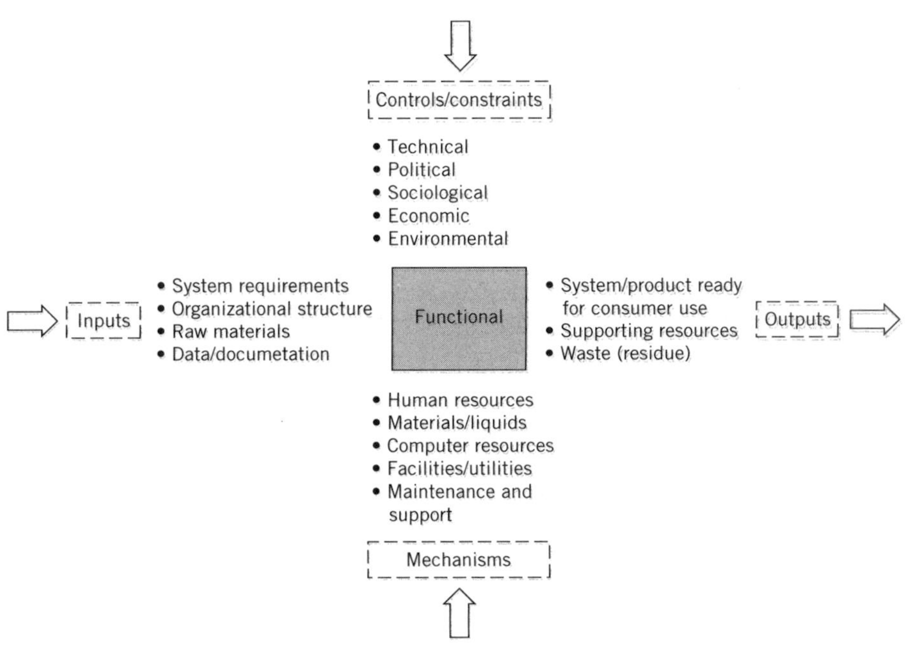

Figure 2.17 Identification of resource requirements (i.e., "mechanisms").

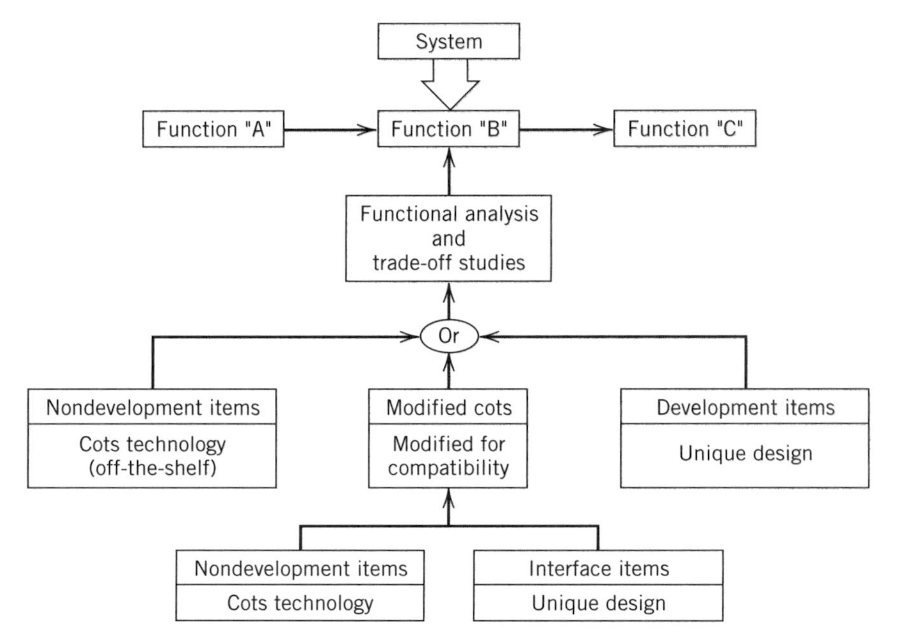

Figure 2.18 Identification of commercial off-the-shelf (COTS) items from functional analysis.

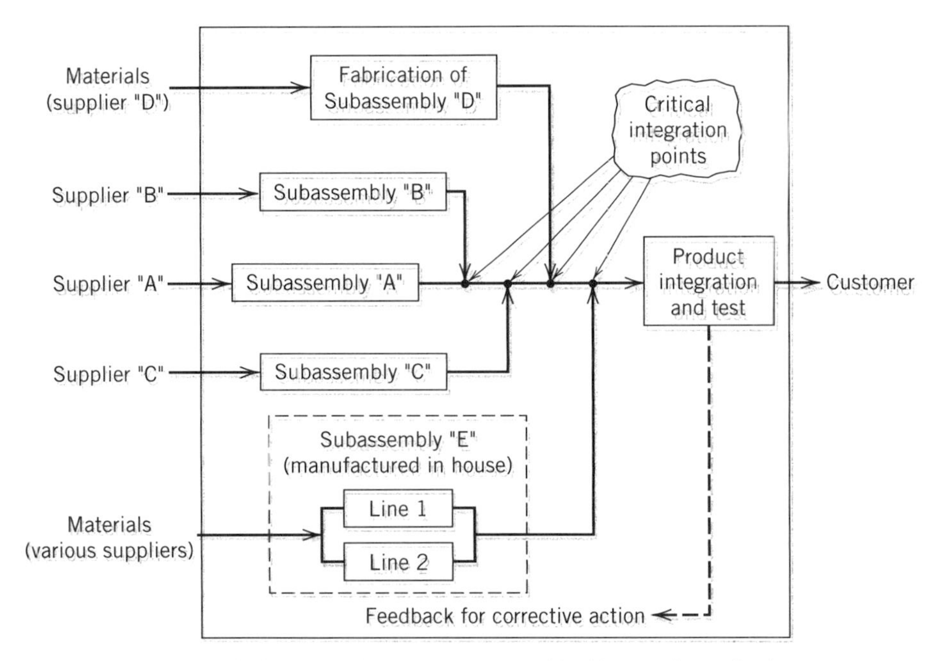

Figure 2.19 Manufacturing system (critical integration points).

same resources may be utilized to accomplish more than one function. The identified resources may be combined and integrated to the extent possible. Every effort should be made to avoid the specification of resources that are not necessary. Figure 2.20 illustrates a documentation format that can be applied to formalize the identification of such resources.

In summary, the functional analysis constitutes a critical step in the early system design and development effort, and it forms a baseline for many activities that are conducted subsequently. For instance, it serves as a basis in the development of the following:

1. Electrical and mechanical design for functional packaging, condition monitoring, and diagnostic provisions.
2. Reliability models and block diagrams.
3. Failure mode, effect, and criticality analysis (FMECA).
4. Fault tree analysis (FTA).
5. Reliability-centered maintenance (RCM) analysis.
6. System safety/hazard analysis.
7. Maintainability analysis.
8. Level-of-repair analysis.
9. Maintenance task analysis (MTA).
10. Operator task analysis (OTA).

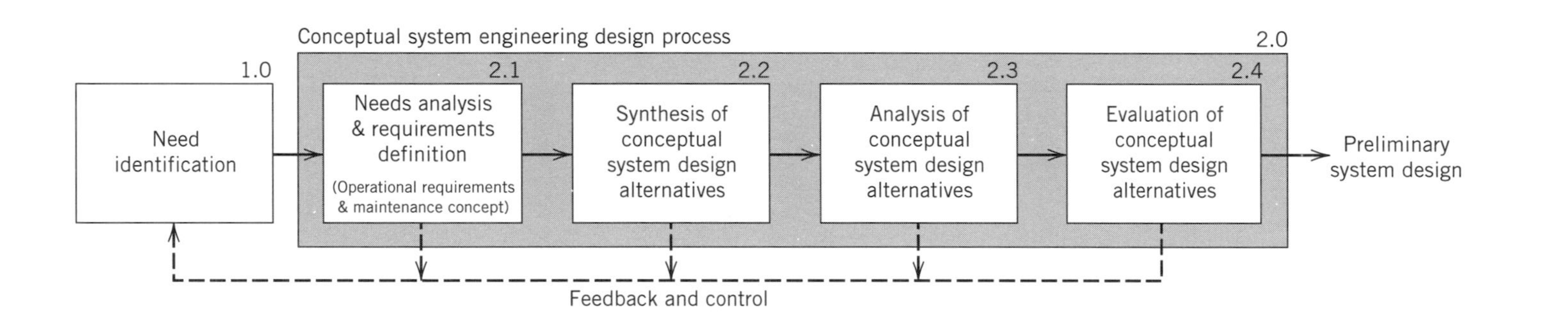

Activity number	Activity description	Required inputs	Expected outputs	Resource requirements (activities/techniques)
1.0	Need identification	Customer surveys; marketing inputs; shipping and servicing department logs; market niche studies; competitive product research	A specific qualitative and quantitative needs statement responding to a current deficiency. Care must be taken to state this need in functional terms.	Benchmarking; statistical analyses of data (i.e., data collected as a result of surveys and consolidated from shipping and servicing logs, etc.)
2.1	Needs analysis and requirements definition	A specific qualitative and quantitaive needs statement expressed in functional terms.	Qualitative and quantitative factors pertaining to system performance levels, geographical distribution of products, expected utilization profiles, user/consumer environment; operational life cycle, effectiveness requirements, the levels of maintenance and support, consideration of the applicable elements of logistic support, the support environment, and so on.	Quality Function Deployment (QFD); input-output matrix; checklists; value engineering; statistical data analysis; trend analysis; matrix analysis; parametric analysis; various categories of analytical models and tools for simulation studies, trade-offs, etc.
2.2	Synthesis of conceptual system design alternatives	Results from needs analysis and requirements definition process; technology research studies; supplier information	Identification and description of candidate conceptual system design alternatives and technology applications.	Pugh's concept generation approach; brainstorming; analogy; checklists.
2.3	Analysis of conceptual system design alternatives	Candidate conceptual solutions and technologies; results from the needs analysis and requirements definition process	Approximation of the "goodness" of each feasible conceptual solution relative to the pertinent parameters, both direct and indirect. This goodness could be expressed as a numeric rating, probabilistic measure, or fuzzy measure.	Indirect system experimentation (e.g., mathematical modeling and simulation); parametric analyses; risk analyses.
2.4	Evaluationof conceptual system design alternatives	Results from the analysis task in the form of a set of feasible conceptual system design alternatives.	A specific qualitative and quantitative needs statement responding to a current deficiency. Care must be taken to state this need in functional terms.	Design-dependant parameter approach; generation of hybrid numbers to represent candidate solution "goodness"; conceptual system design evaluation display.

Figure 2.20 Document format for resource requirements.

11. Operational sequence diagrams (OSDs).
12. Logistic support analysis (LSA).
13. Operating and maintenance procedures.
14. Producibility and disposability analyses.

In the past, the functional analysis has not always been completed in a timely manner, if completed at all! As a result, the various design disciplines assigned to a given program have had to generate their own analyses in order to comply with program requirements. In many instances, these efforts were accomplished independently, and many design decisions were made without the benefit of a *common* baseline to follow. This, of course, resulted in design discrepancies and costly modifications occurring later in the system life cycle.

The functional analysis provides an excellent and very necessary baseline, and all applicable design activities must "track" the same data source in order to meet the objectives for system engineering, as stated in Chapter 1. For this reason, the functional analysis is considered as being a key activity in the system engineering process.

2.8 REQUIREMENTS ALLOCATION

Given a top-level definition of the system through the functional analysis, the next step is to break the system down into components by *partitioning*.[7] This involves a breakdown of the system into subsystems and lower-level elements. The challenge is to identify and group closely related functions into packages, employing a common resource (e.g., equipment, software) to accomplish multiple functions to the extent possible. Although it may be relatively easy to identify individual functional requirements and associated resources on an independent basis, this may turn out to be rather costly when it comes to system packaging, weight, size, and so on. The questions are: What hardware or software can be selected that will perform multiple functions? How can new functions be added without adding any new physical elements to the system structure?

The partitioning of the system into elements is evolutionary in nature. Common functions may be grouped or combined in such a way as to provide a system packaging scheme with the following objectives in mind:

1. System elements may be grouped by geographical location, a common environment, or by similar types of equipment.
2. Individual system "packages" should be as independent as possible with a minimum of "interaction effects" with other packages. A design objective is to be able to remove and replace a given package without having to remove

[7] The concepts of system *architecture* and *partitioning* are presented in E. Rechtin, *Systems Architecting: Creating and Building Complex Systems,* Prentice Hall, Upper Saddle River, NJ, 1991.

and replace other packages in the process, or requiring an extensive amount of alignment and adjustment in the process.

3. In breaking down a system into subsystems, select a configuration where the communications between the subsystems is minimized. In other words, whereas the subsystem *internal* complexity may be high, the *external* complexity should be low. Breaking down the system into packages where there are high rates of information exchange between these packages should be avoided.

An overall design objective is to break down the system into elements such that only a very few critical events can influence or change the inner workings of the various packages that make up the system architecture.

As a result of partitioning, a system may be broken down into components such as shown in Figure 2.21. The process utilized in the identification and packaging of elements is illustrated in Figure 2.22. System functions are identified, broken down into subfunctions, and grouped into three equipment units; that is, Unit A, Unit B, and Unit C. The design should be such that any one of the three units can be removed and replaced without impacting the other units. In other words, there should be a minimum of interaction effects between the three units.

Given the identification of system elements, the next step is to *allocate* or *appor-*

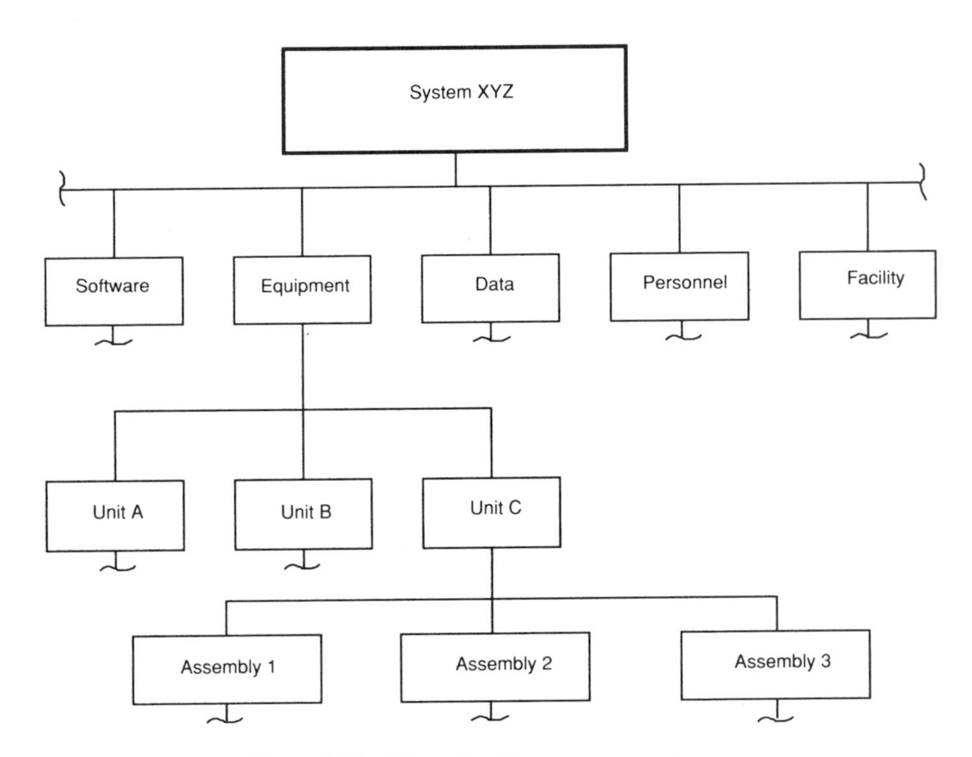

Figure 2.21 Hierarchy of system components.

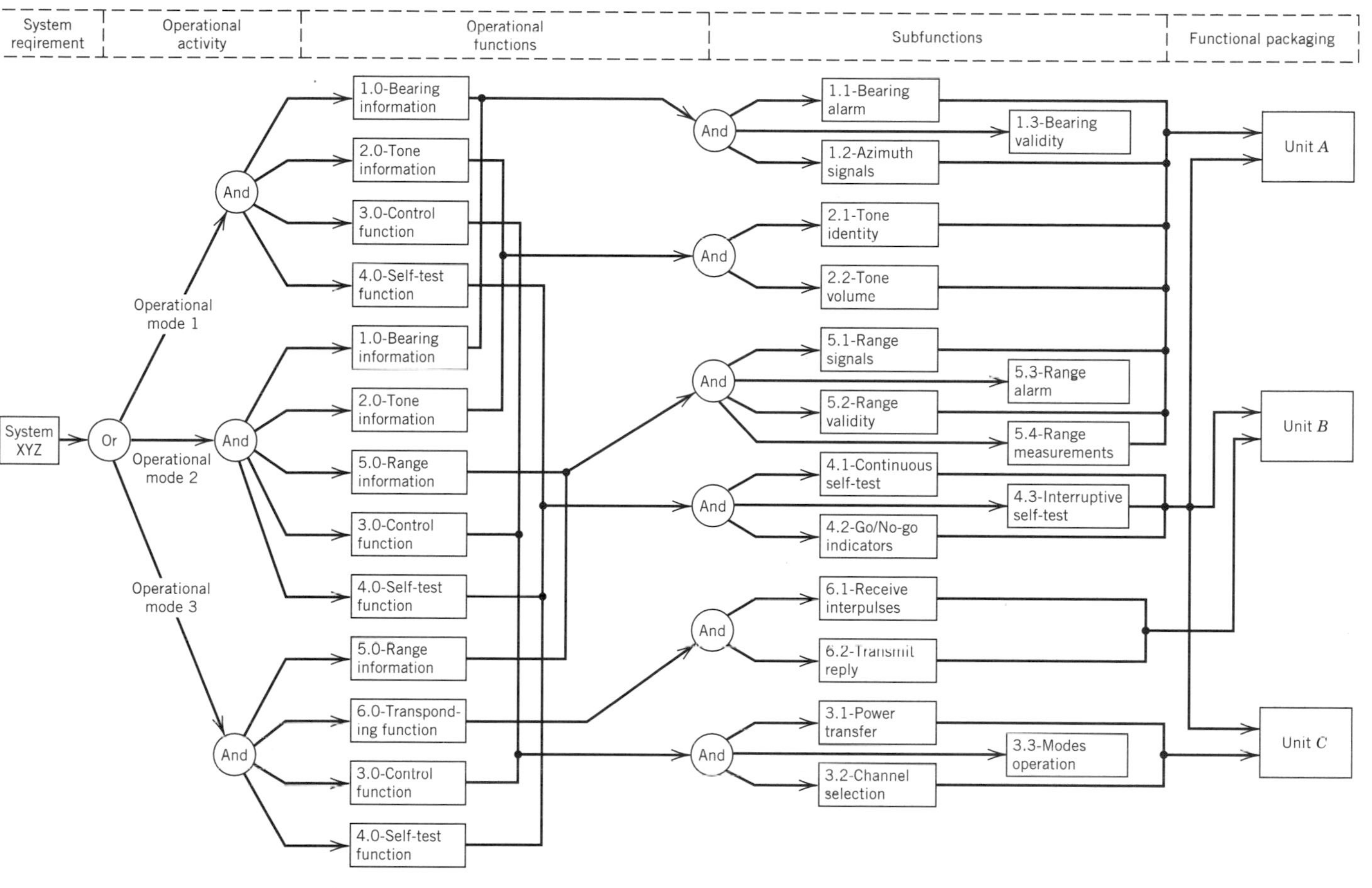

Figure 2.22 Abbreviated functional analysis leading to system packaging.

tion the requirements specified for the system down to the level desired to provide a meaningful *input* to design. This involves a top-down distribution of the quantitative and qualitative criteria developed in Section 2.6. From the prioritized technical performance measures (TPMs) in Figure 2.10, what should be specified for the unit level in Figure 2.22 in order to meet the system-level requirements in Figure 2.10?

Figure 2.23 shows the results of an allocation (in this instance to 4 units). Utilizing an "objectives-tree" approach, as illustrated in Figure 2.7, the designer should establish the appropriate metrics for the system, then the metrics at the next lower level, and so on. There should be a "traceability" of requirements from the top down. Although there may be different measures at each level, those identified at the lower levels must directly support the requirements for the overall system. Further, the depth to which requirements are specified is somewhat dependent on the priorities (i.e., "importance" factors) identified in Figure 2.10. If there is a highly critical requirement from the perspective of the consumer, allocation may be accomplished down to the assembly level in Figure 2.23. On the other hand, if the allocation is accomplished unnecessarily to a very detailed level, the designer may be overly constrained relative to what can be accomplish through the trade-off analysis and evaluation process.

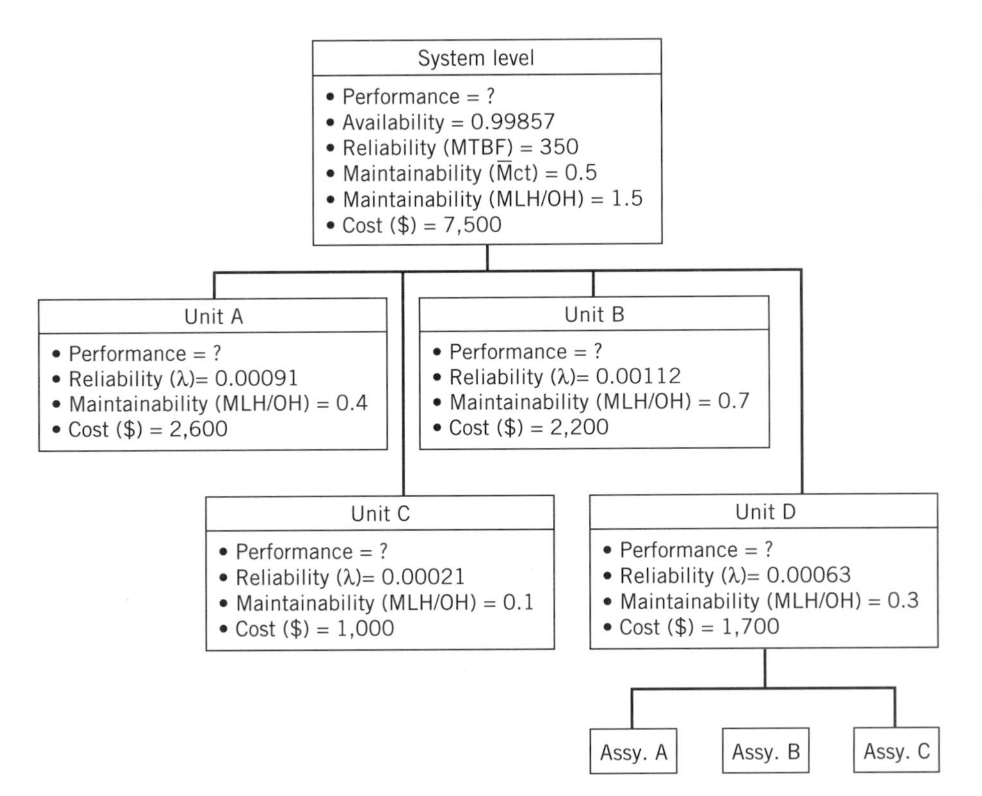

Figure 2.23 Allocation of system requirements.

The allocation process constitutes a top-down specification of design requirements, to the depth necessary to provide input criteria for the design of various system elements. Highly complex new design will require a greater degree of coverage than would be necessary when utilizing commercial-off-the-shelf (COTS) items. The results of the allocation should be incorporated in the applicable development, product, process, and/or material specifications identified in Figure 1.12. By not properly specifying the requirements from the top down, the results can be costly in terms of *overdesign, underdesign,* or both. The risks may be high if the requirements are not addressed from the beginning![8]

2.9 SYSTEM SYNTHESIS, ANALYSIS, AND DESIGN OPTIMIZATION

Synthesis refers to the combining and structuring of components in such a way as to represent a feasible system configuration. The requirements for a system have been established, some preliminary trade-off studies have been completed, and a baseline configuration needs to be developed to demonstrate the concepts discussed earlier. Synthesis is *design.* Initially, synthesis is employed to develop preliminary concepts and to establish basic relationships among the various components of the system. Later, when sufficient functional definition and decomposition have occurred, synthesis is used to further define the "HOWs," in response to the "WHAT" requirements. Synthesis involves the selection of a configuration that could be representative of the form that the system will ultimately take, although a final configuration is certainly not to be assumed at this point.[9]

The synthesis process usually leads to the definition of several possible alternative design approaches, which will be the subject of further analysis, evaluation, refinement, and optimization. As these alternatives are initially structured, it is essential that the appropriate technical performance parameters be properly aligned to applicable components of the system. For instance, technical performance parameters may include factors such as weight, size, speed, capacity, accuracy, volume, range, processing time, along with the reliability and maintainability factors presented in Figure 2.24. These parameters, or measures, must be prioritized and aligned to the appropriate elements of the system (e.g., an equipment, unit or assembly, item of software).

When defining the initial requirements for the system, technical performance measures (TPMs) are established based on their relationship and criticality to the accomplishment of the mission; that is, the impact that a given factor has on cost-effectiveness, or system effectiveness, or performance. These applicable TPMs are prioritized, and their relative relationships are presented in the form of design considerations presented in a hierarchical tree, as illustrated in Figure 2.24. The ranking

[8] The allocation of reliability, maintainability, human, economic, and related factors is discussed further in Chapter 3.

[9] Synthesis is covered further in J. Lacy, *Systems Engineering Management,* McGraw-Hill, New York, 1992.

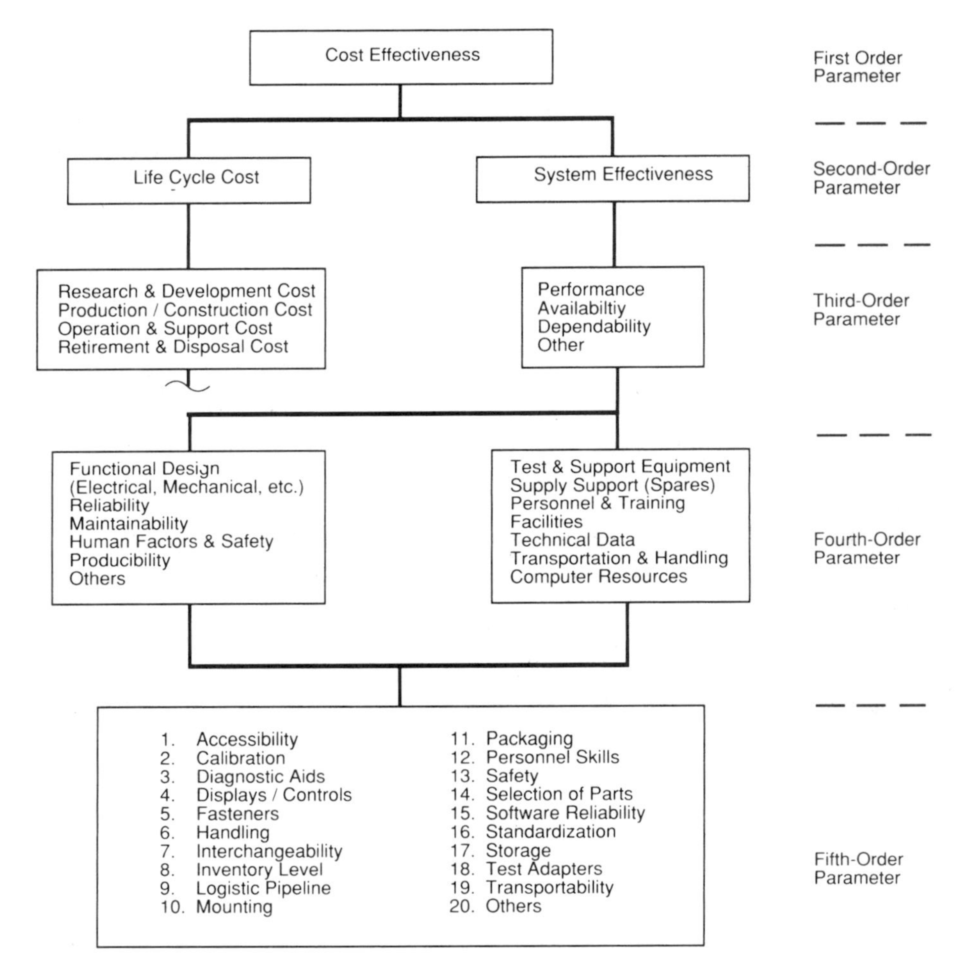

Figure 2.24 Order of evaluation parameters.

of TPMs (and supporting design considerations), which will be built into the program management and review structure, will likely vary from one system to the next. A top-level measure for one system may be "reliability," whereas "availability" may be of greater importance in another example. In any event, the appropriate measures need to be established, prioritized, and included in the specifications. As the design process evolves, these measures will be used for the purposes of analysis and evaluation.

Given a number of alternatives, the evaluation procedure progresses through the general steps illustrated in Figure 2.25 and described as follows:

1. *Definition of analysis goals:* an initial step requires the clarification of objectives, the identification of possible alternative solutions to the problem at hand, and

a description of the analysis approach to be employed. Relative to alternatives, all possible candidates must be initially considered; however, the more alternatives considered, the more complex the analysis process becomes. Thus, it is desirable to first list *all* possible candidates to ensure against inadvertent omissions, and then eliminate those candidates that are clearly unattractive, leaving only a few for evaluation. Those few candidates are then evaluated with the intent of selecting a preferred approach.

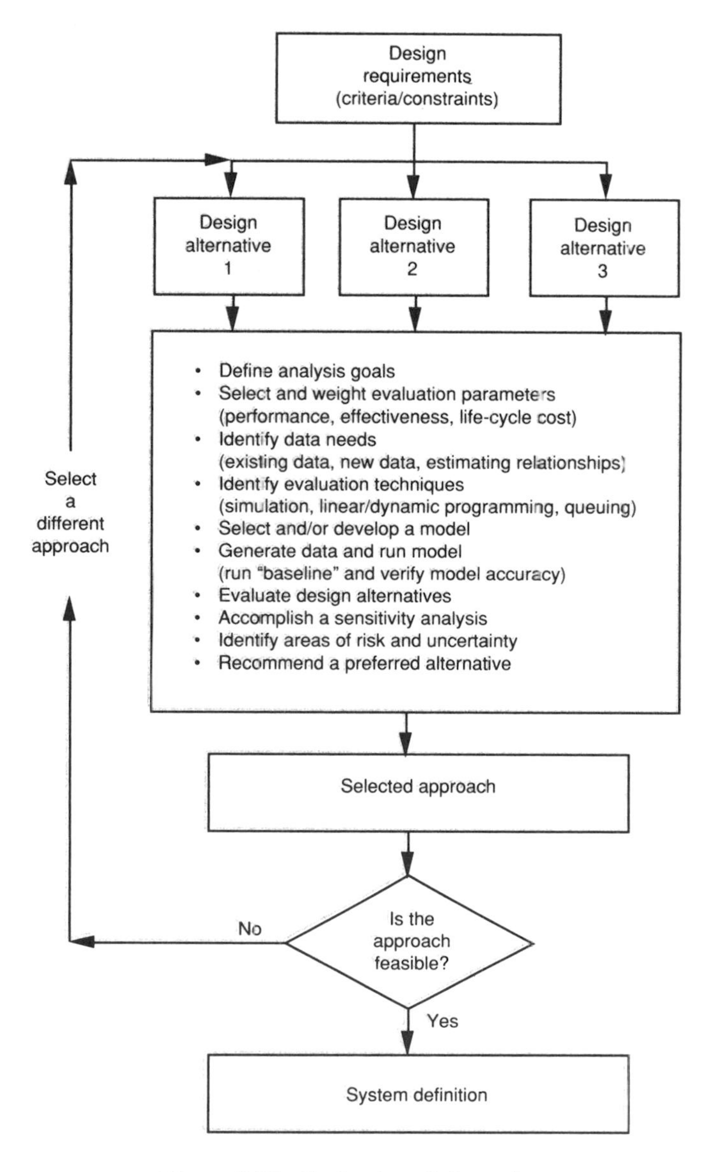

Figure 2.25 Evaluation of alternatives.

2. *Selection and weighting of evaluation parameters:* the criteria used in the evaluation process may vary considerably depending on the stated problem, the system being evaluated, and the depth and complexity of the analysis. From Figure 2.24, parameters of primary significance include cost, effectiveness, performance, availability, and so on. At the detail level, the order of parameters will be different. In any event, parameters are selected, weighted in terms of priority of importance, and are tailored to the system in a meaningful manner.

3. *Identification of data needs:* when evaluating a particular system configuration, it is necessary to consider operational requirements, the maintenance concept, major design features, production and/or construction plans, and anticipated system utilization and product support requirements. Fulfilling this need requires a variety of data, the scope of which depends on the type of evaluation being performed and the program phase during which the evaluation is accomplished. In the early stages of system development, available data are limited; thus, the analyst must depend on the use of various estimating relationships, projections based on past experience covering similar system configurations, and intuition. As the system development progresses, improved data are available (through analyses and predictions) and are used as an input to the evaluation effort. At this point, it is important to initially determine the specific needs for data (i.e., type, quantity, and the time of need), and to identify possible data sources. The nature and validity of the data input for a given analysis could have a significant impact on the risks associated with the decisions made based on the analysis results. Thus, one needs to accurately assess the situation as early as practicable.

4. *Identification of evaluation techniques:* given a specific problem, it is necessary to determine the analytical approach to be used and the techniques that can be applied to facilitate the problem-solving process. Techniques may include the use of Monte Carlo simulation in the prediction of random events downstream in the life cycle, the use of linear programming in determining transportation resource requirements, the use of queuing theory in determining production and/or maintenance shop requirements, the use of networking in establishing distribution needs, the use of accounting methods for life-cycle costing purposes, and so on. Assessing the problem itself and identifying the available tools that can possibly be used in attacking the problem are necessary prerequisites to the selection of a model.

5. *Selection and/or the development of a model:* the next step requires the combining of various analytical techniques into the form of a model, or a series of models, as illustrated in Figure 2.26.[10] A model, as a tool used in problem solving, aids in the development of a simplified representation of the real world as it applies to the problem being solved. The model should (a) represent the dynamics of the system configuration being evaluated; (b) highlight those factors that are most rele-

[10]There are many types of models including physical models, symbolic models, abstract models, mathematical models, and so on. Model, as defined here, refers primarily to a mathematical (or analytical) model.

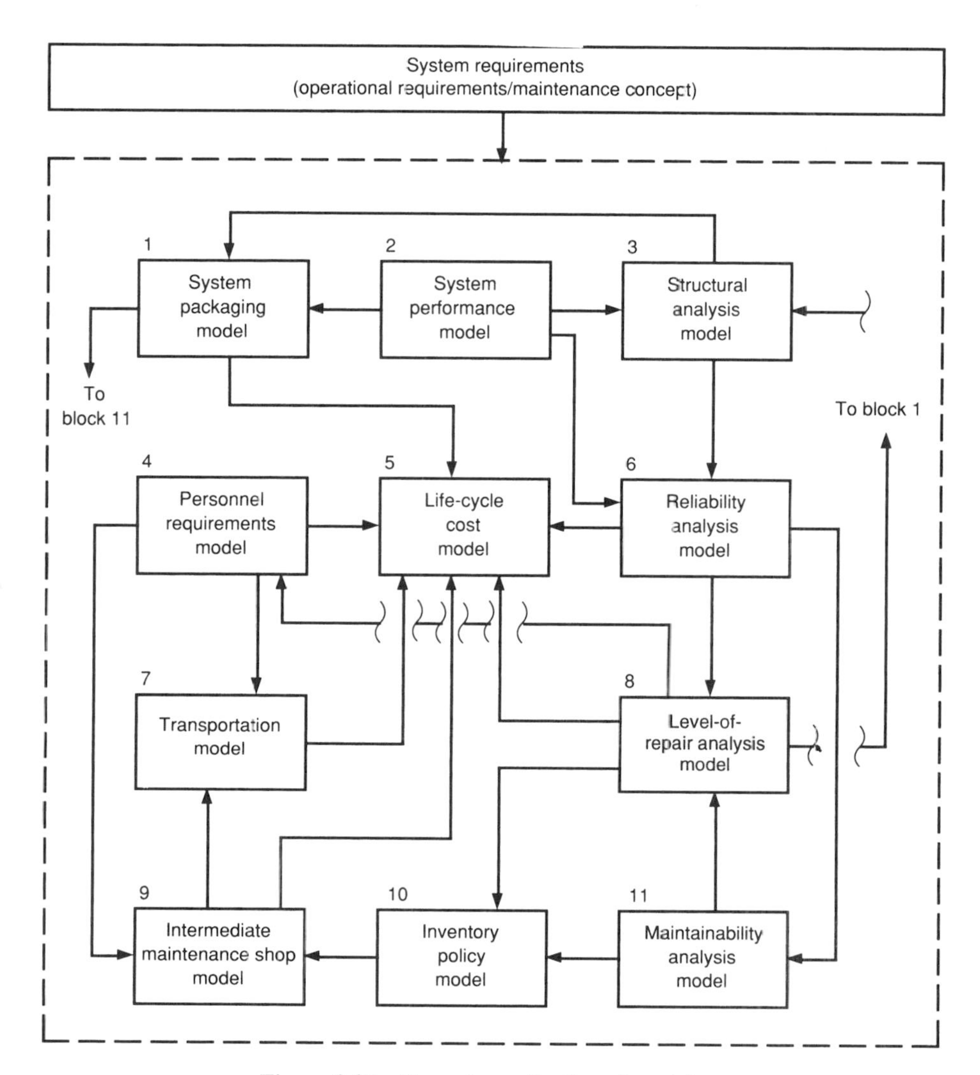

Figure 2.26 Example application of models.

vant to the problem at hand; (c) be comprehensive by including *all* relevant factors and be reliable in terms of repeatability of results; (d) be simple enough in structure so as to enable its timely implementation in problem solving; (e) be designed such that the analyst can evaluate the applicable system configuration as an entity, analyze different components of the system on an individual basis, and then integrate the results into the whole; and (f) be designed to incorporate provisions for easy modification and/or expansion to permit the evaluation of additional factors as required. An important objective is to select and/or develop a tool that will help to evaluate the *overall* system configuration, as well as the *interrelations* of its various components. Models (and their applications) are discussed further in Chapter 4.[11]

6. *Generation of data and model application:* with the identification of analytical techniques and the model selection task accomplished, the next step is to "verify" or "test" the model to ensure that it is responsive to the analysis requirement. Does the model meet the stated objectives? Is it sensitive to the major parameters of the system configuration(s) being evaluated? Evaluation of the model can be accomplished through the selection of a *known* system entity, and the subsequent comparison of analysis results with historical experience. Input parameters may be varied to ensure that the model design characteristics are sensitive to these variations and will ultimately reflect an accurate output as a result.

7. *Evaluation of design alternatives:* each of the alternatives being considered is then evaluated using the techniques and the model selected. The required data are collected from various sources such as existing data banks, predictions based on current design data, and/or gross projections using analogous and parametric estimating relationships. The required data, which may be taken from a wide variety of sources, must be applied in a consistent manner. The results are then evaluated in terms of the initially specified requirements for the system. Feasible alternatives are considered further. Figure 2.27 illustrates some considerations where possible feasible solutions fall within the desired shaded areas.

8. *Accomplishment of a sensitivity analysis:* in the performance of an analysis, there may be a few key system parameters about which the analyst is uncertain because of inadequate data input, poor prediction procedures, "pushing" the state of the art, and so on. There are several questions that need to be addressed: How sensitive are the results of the analysis to possible variations of these uncertain input parameters? To what extent can certain input parameters be varied before the choice of alternatives shifts away from the initially selected approach? From experience, there are certain key input parameters in a life-cycle cost analysis, such as the reliability MTBF and the maintainability $\bar{M}ct$, that are considered to be critical in determining system maintenance and support costs. With good historical field data being very limited, there is a great deal of dependence placed on current prediction and estimating methods. Thus, with the objective of minimizing the risks associated with making an incorrect decision, the analyst may wish to vary the input MTBF and $\bar{M}ct$ factors over a designated range of values (or a distribution) to see what impact this variation has on the output results. Does a relatively *small* variation of an input factor have a *large* impact on the results of the analysis? If so, then these parameters might be classified as being critical TPMs in the overall design review and evaluation process, monitored closely as design progresses, and an additional effort might be generated to modify the design for improvement and to improve the reliability and maintainability prediction methods. In essence, a sensitivity analysis

[11] The development and application of various analytical methods are covered further in most texts on operations research. Two excellent references are (1) F. S. Hillier and G. J. Lieberman, *Introduction to Operations Research,* 6th Ed., McGraw-Hill, New York, 1995; and (2) H. A. Taha, *Operations Research: An Introduction,* 5th Ed., Macmillan, New York, 1992.

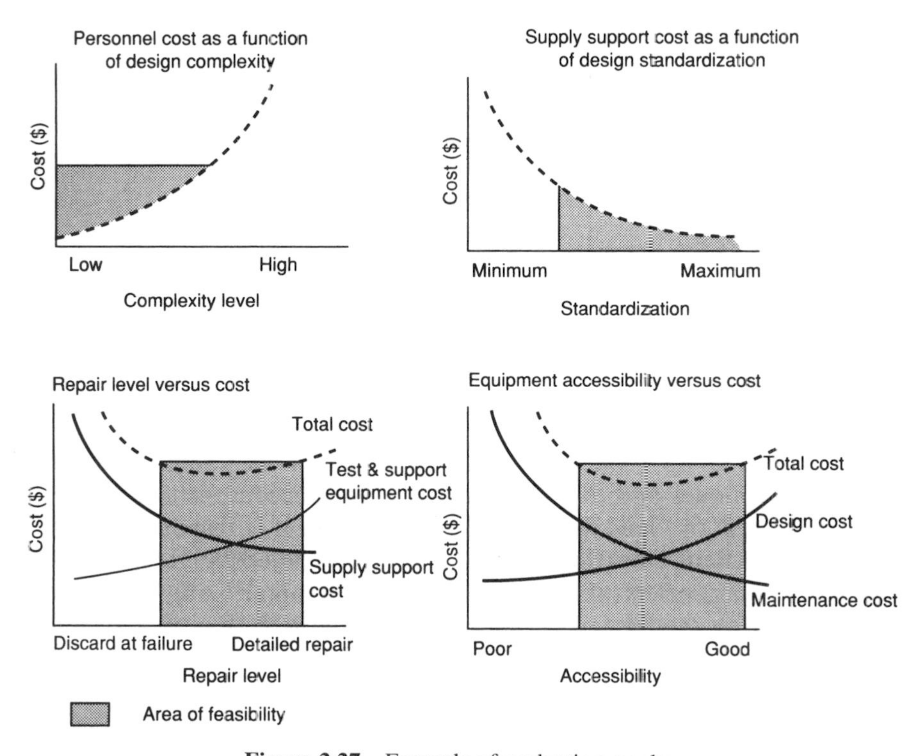

Figure 2.27 Example of evaluation results.

is directed toward determining the relationships between design decisions and output results.

9. *Identification of risk and uncertainty:* the process of design evaluation leads to decisions having a significant impact on the future. The selection of evaluation criteria, the weighting of factors, the selection of the life cycle, the use of certain data sources and prediction methods, and the assumptions made in interpreting analysis results will obviously influence these decisions. Inherent within this process are the aspects of "risk" and "uncertainty" because the future is, of course, unknown. Although these terms are often used jointly, *risk* actually implies the availability of discrete data in the form of a probability distribution around a certain parameter. *Uncertainty* implies a situation that may be probabilistic in nature, but one that is not supported by discrete data. Certain factors may be measurable in terms of risk, or may be stated under conditions of uncertainty. The aspects of risk and uncertainty, as they apply to the system design and development process, must be integrated into the program risk management plan described in Chapter 6, Section 6.6.

10. *Recommendation of preferred approach:* the final step in the evaluation process is the recommendation of a preferred alternative. The results of the analysis should be fully documented and made available to all applicable project design personnel. A statement of assumptions, a description of the evaluation procedure

that was followed, a description of the various alternatives that were considered, and an identification of potential areas of risk and uncertainty should be included in this analysis report.

In Figure 1.12, requirements for the system are established in conceptual design, functional analysis and allocation are accomplished either late in conceptual design or at the start of preliminary system design, and detail design is accomplished on a progressive basis from thereon. Throughout this overall series of steps, there is an ongoing effort involving synthesis, analysis, and design optimization. In the early stages of design, trade-off studies may entail the evaluation of alternative operational profiles, alternative technology applications, distribution schemes, or maintenance concepts. During early preliminary design, alternative methods for accomplishing a given function or alternative equipment packaging schemes may be the focus of analysis. In detail design, the problems will be at a lower level in the overall hierarchical structure of the system.

In any event, the process discussed in Section 2.9 (and illustrated in Figure 2.25) is applicable throughout the system design and development effort. The only difference lies in the depth of analysis, the type of data required, and the model used in accomplishing the analysis. For instance, one can perform a life-cycle cost analysis early in conceptual design or late in detail design. The process is the same in either case; however, the depth of analysis and the data requirements are different. The synthesis, analysis, and optimization process must be tailored to the problem at hand. Too little effort will result in greater risks associated with decision making in design, and too much analysis effort will be expensive.

2.10 DESIGN INTEGRATION

Design integration activities commence during the early stages of conceptual design and extend through system development, production and/or construction, distribution, operational use and sustaining support, and ultimate retirement and the disposal (or recycling) of materials. As the requirements for a new system are established, the *design team* is formed, initially performing system-level design functions, as indicated in Figure 1.12. At this stage, the design team may include only a small number of selected qualified individuals, with the objective of developing a comprehensive System Specification (Type "A"; refer to Figure 1.12). It is important that personnel with the appropriate backgrounds and experience be selected, and these individuals must be able to work together and effectively communicate on a day-to-day basis. The assignment of a large number of individual domain specialists, whose expertise lie in given technical fields, is not appropriate at this stage. The organization of design teams is discussed further in Chapter 7.

As system development progresses, the appropriate design specialists are added to the team. The objective, from a system engineering perspective, is to ensure that the right specialists are available at the time required and that their individual contributions are properly integrated into the whole! The selection of domain specialists is highly dependent on the requirements developed through the functional analysis and allocation process (refer to Sections 2.7 and 2.8). As the criteria for design will vary with the type of system and its mission, the emphasis in assigning the proper level of expertise to the team will be different from one project to the next. Figure 2.28 identifies some of the considerations that must be addressed through the design integration effort.

During the latter phases of the life cycle (i.e., production/construction, system utilization and support), the role of system engineering continues, but in the form of evaluation/validation and the introduction and processing of design changes as necessary. The requirement(s) for *change* may stem from some identified deficiency (i.e., the failure to meet an initially specified requirement), or for the purposes of *continuous process improvement.* Each "engineering change proposal (ECP)" must be evaluated in terms of not just performance issues alone, but in terms of reliability, maintainability, supportability or serviceability, producibility, disposability, and life-cycle cost as well. The design change and modification process is described further in Section 5.4.

Inherent within the established design team activity is the requirement for good communications on a day-to-day basis. Although the colocation of personnel in one geographical area (and the "eyeball-to-eyeball" contact) is preferred, the trends toward "outsourcing" and decentralization often result in the introduction of many

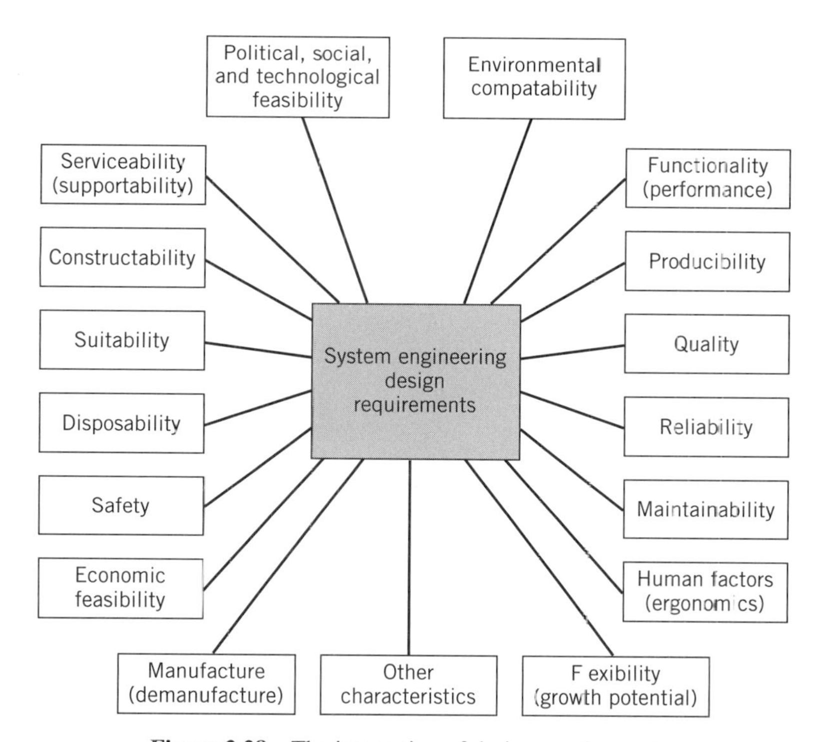

Figure 2.28 The integration of design requirements.

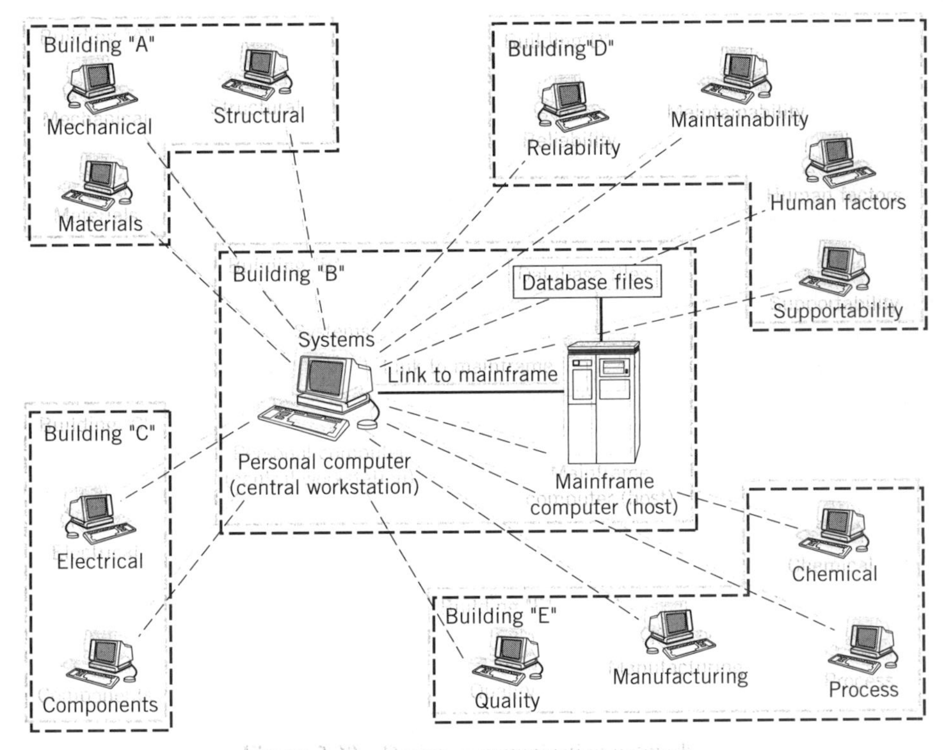

Figure 2.29 Design communication network.

different suppliers located throughout the world.[12] Further, there are design activities being conducted at remote locations that are accomplished concurrently. Thus, the design team becomes heavily dependent on the utilization of computer-aided tools, operating in a network such as illustrated in Figure 2.29.[13]

Successful implementation of the integrated computer-based network shown in Figure 2.29 is highly dependent on the structure of the design database. Such a database may include design drawings and layouts, the presentation of three-dimensional visual models, parts and material lists, prediction and analysis results, supplier data, and whatever else is necessary to describe the system configuration as designed. The designer must be able to gain access to the database and provide input easily, and the results must be transmitted to other members of the design

[12]The term "outsourcing" refers to the practice of soliciting the support of component suppliers to accomplish selected packages of work externally from the producer or prime contractor. Experience indicates that there is a greater use of external suppliers today than in the past. This, in turn, provides some additional challenges relative to maintaining the proper level of *communications* across the project organization.

[13]Included in this network are the appropriate computer-aided design (CAD), computer-aided manufacturing (CAM) or computer-integrated manufacturing (CIM), and computer-aided logistic support (CALS) tools. This area is discussed further in Chapter 4.

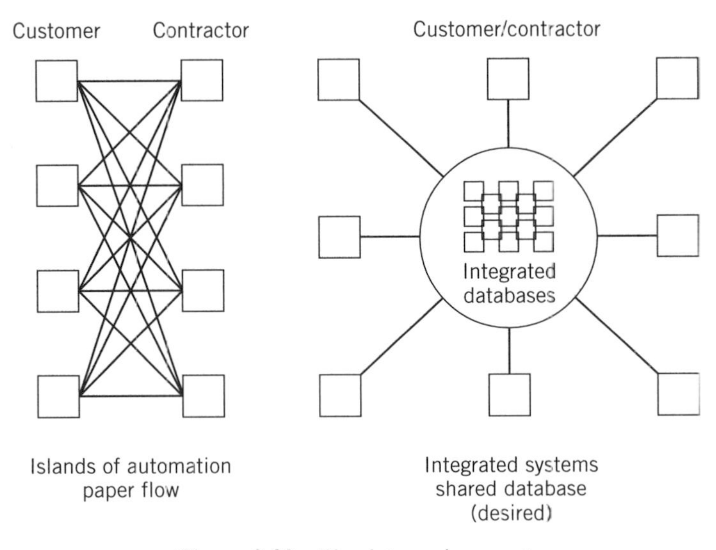

Figure 2.30 The data environment.

team accurately and in a timely manner. The data, usually presented in a digital format, must be available to all members of the design team concurrently. Instead of many different data items "flowing" back and forth between different members of the design team, between the producer (contractor) and consumer (customer), and so on, an integrated shared database structure is necessary, as illustrated in Figure 2.30. This, of course, should facilitate the process of communications, with every member of the design team having access to the same system description.[14]

2.11 TEST AND EVALUATION

As the system design and development activity progresses, there needs to be an ongoing measurement and evaluation (or validation) effort, as indicated in Figure 1.22. In the true sense, a complete evaluation of the system, in terms of meeting the initially specified consumer requirements, cannot be accomplished until the system is produced and functioning in an operational environment. However, if problems occur and system modifications are necessary, the accomplishment of this so far downstream in the life cycle may turn out to be quite costly. In essence, the earlier that problems are detected and corrected, the better off one is in terms of both incorporating the required changes and the costs thereof!

When addressing the subject of evaluation, the objective is to acquire a high

[14] With the advent of new technologies on an almost continuing basis, it is anticipated that the nature of the *data environment* will be changing almost constantly. The objective here is to emphasize the need for good communications through the integration and transfer of design data among members of the design team, supporting organizations, and management.

degree of confidence, as early in the life cycle as possible, that the system will ultimately perform as intended. The realization of this, through the accomplishment of laboratory and field testing involving a physical replica of the system (and/or its components), can be quite expensive. The resources required for testing are often quite extensive, and the necessary facilities, test equipment, personnel, and so on, may be difficult to schedule. Yet we know that a certain amount of formal testing is required in order to properly verify that system requirements have been met.

On the other hand, with a more comprehensive analysis effort and the use of prototyping, it may be possible to verify certain design concepts during the early stages of preliminary and detail design. With the advent of three-dimensional databases and the application of simulation techniques, the designer can now accomplish a great deal relative to the evaluation of system layouts, component relationships and interferences, human–machine interfaces, and so on. There are many functions that can now be accomplished with computerized simulation that formerly required a physical mockup of the system, a preproduction prototype model, or both. The availability of computer-aided design (CAD), computer-aided manufacturing (CAM), computer-aided logistic support (CALS) methods, and related technologies has made it possible to accomplish much in the area of system evaluation, relatively early in the system life cycle when the incorporation of changes can be accomplished with minimum cost.

In determining the needs for test and evaluation, one commences with the initial specification of system requirements in conceptual design. As specific technical performance measures (TPMs) are established, it is necessary to determine the methods by which compliance with these factors will be verified. How will these TPMs be measured and what resources are necessary to accomplish such? Response to this question may be in the form of using simulation and related analytical methods, using an engineering model for test and evaluation purposes, testing a production model, evaluating an operational configuration in the consumer's environment, or a combination of these. In essence, one needs to review the requirements for the system, determine the methods that can be used in the evaluation effort and the anticipated effectiveness of these methods, and develop a comprehensive plan for an overall integrated test and evaluation effort (i.e., Test and Evaluation Master Plan; refer to Figure 1.21). As a point of reference, Figure 2.31 is presented to illustrate suggested categories of testing as they may apply in system evaluation.

2.11.1 Categories of Test and Evaluation[15]

In Figure 2.31, the first category is "analytical," which pertains to certain design evaluations that can be conducted early in the system life cycle using computerized techniques to include CAD, CAM, CALS, simulation, rapid prototyping, and related approaches. With the availability of a wide variety of models, three-dimensional

[15] The categories of test and evaluation may vary by type of system or by functional organization. These categories have been selected as a point of reference for discussions throughout this text.

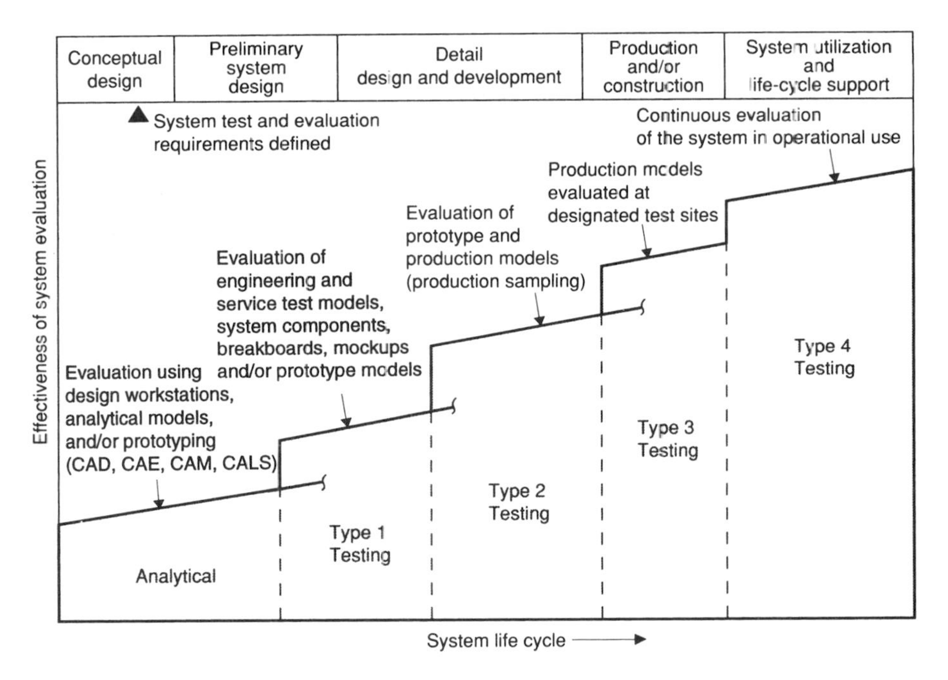

Figure 2.31 Stages of system evaluation during the life cycle.

databases, and so on, the design engineer is now able to simulate human–equipment interfaces, equipment packaging schemes, the hierarchical structures of systems, and activity/task sequences. In addition, through the utilization of these technologies, the design engineer is able to do a better job of predicting, forecasting, and the accomplishment of sensitivity/contingency analyses with the objective of reducing future risks. In other words, a great deal can be now accomplished in system evaluation that, in the past, could not be realized until equipment became available in the latter phases of detail design and development.

"Type 1 testing" refers primarily to the evaluation of system components in the laboratory using engineering breadboards, bench test models, service test models, rapid prototyping, and the like. These tests are designed primarily with the intent of verifying certain performance and physical characteristics, and are developmental by nature. The test models used operate functionally, but do not by any means represent production equipment or software. Such testing is usually performed in the producer/supplier's laboratory facility by engineering technicians using "jury-rigged" test fixtures and engineering notes for procedures. It is during this initial phase of testing that design concepts and technology applications are validated, and changes can be initiated on a minimum-cost basis.

"Type 2 testing" includes formal tests and demonstrations accomplished during the latter stages of the detail design and development phase when preproduction prototype equipment and software are available. Prototype equipment is similar to production equipment (that which will be delivered for operational use), but is not necessarily fully qualified at this point in time.[16] A test program in this area may constitute a series of individual tests, tailored to the need, including the following:

1. *Environmental qualification:* temperature cycling, shock and vibration, humidity, sand and dust, salt spray, acoustic noise, explosion proofing, and electromagnetic interference.
2. *Reliability qualification:* sequential testing, life testing, environmental stress screening (ESS), and test, analyze, and fix (TAAF).
3. *Maintainability demonstration:* verification of maintenance tasks, task times and sequences, maintenance personnel quantities and skill levels, degree of testability and diagnostic provisions, prime equipment—test equipment interfaces, maintenance procedures, and maintenance facilities.
4. *Support equipment compatibility:* verification of the compatibility among the prime equipment, test and support equipment, and ground handling equipment.
5. *Technical data verification:* the verification (and validation) of operating procedures, maintenance procedures, and supporting data.
6. *Personnel test and evaluation:* verification to ensure the compatibility among the human and equipment, the personnel quantities and skill levels required, and training needs.
7. *Software compatibility:* verification that software meets the system requirements, the compatibility between software and hardware, and that the appropriate quality provisions have been incorporated. This includes computer software unit (CSU) and computer software configuration item (CSCI) testing, as reflected in Figure 1.13.

Another facet of testing in this category is production sampling tests, used when multiple quantities of an item are being produced. Although the system (and its components) may have successfully passed the initial qualification tests, there needs to be some assurance that the *same* level of quality has been maintained throughout the production process. The process is usually dynamic by nature, conditions change, and there is no guarantee that the characteristics that have been built into the design will be retained throughout production. Thus, sample systems/components may be selected (based on a percentage of the total produced), and qualification tests may be conducted on a recurring basis. The results are measured and evaluated in terms of whether improvement or degradation has occurred.

"Type 3 testing" includes the completion of formal tests at designated field test sites by user personnel over an extended period of time. These tests are usually conducted after initial system qualification and prior to the completion of the pro-

[16] "Qualified" equipment refers to the production configuration that has been verified through the *successful completion* of environmental qualification tests (e.g., temperature cycling, shock and vibration), reliability qualification, maintainability demonstration, and supportability compatibility tests. Type 2 testing primarily refers that activity associated with the qualification of a system.

duction/construction phase. Operating personnel, operational test and support equipment, operational spares, applicable computer software, and validated operating and maintenance procedures are used. This is the first time that *all* elements of the system (i.e., prime equipment, software, and the elements of support) are operated and evaluated on an integrated basis. A series of simulated operational exercises are usually conducted, and the system is evaluated in terms of performance, effectiveness, the compatibility between the prime mission-oriented segments of the system and the elements of support, and so on. Although Type 3 testing does not completely represent a fully operational situation, the tests can be designed to provide a close approximation.

"Type 4 testing," conducted during the system operational use and life-cycle support phase, includes formal tests that are sometimes conducted to acquire specific information relative to some area of operation or support. The purpose is to gain further insight of the system in the user environment, or of user operations in the field. It may be desirable to vary the mission profile or the system utilization rate to determine the impact on total system effectiveness, or it may be feasible to evaluate several alternative maintenance support policies to see whether system operational availability can be improved. Type 4 testing is accomplished at one or more user operational sites, in a realistic environment, by operator and maintenance personnel, and is supported through the normal maintenance and logistics capability. This is actually the first time that we will really know the true capability of the system.

2.11.2 Integrated Test Planning

Test planning starts in the conceptual design phase when system requirements are initially established. If a requirement is to be specified, there needs to be a way to evaluate and validate the system at a later point in time to ensure that the requirement has been met. Thus, considerations for test and evaluation are intuitive from the beginning.

In Figure 1.21, initial test planning is included in a Test and Evaluation Master Plan (TEMP), prepared in the conceptual design phase. The document includes the requirements for test and evaluation, the categories of test, the procedures for accomplishing testing, the resources required, and associated planning information (i.e., tasks, schedules, organizational responsibilities, and cost).[17]

One of the key objectives of this plan, and of particular significance for system engineering, is the *complete integration* of the various test requirements for the overall system. By referring to the content of Type 2 testing (Section 2.11.1), individual requirements may be specified for environmental qualification, reliability qualification, maintainability demonstration, software functionality, and so on.

[17] In the defense sector, the TEMP is required for most large programs and includes the planning and implementation of procedures for Development Test and Evaluation (DT&E) and Operational Test and Evaluation (OT&E). DT&E basically equates to the Analytical, Type 1, and Type 2 testing described in Section 2.11.1, and OT&E is equivalent to Type 3 and Type 4 testing.

These requirements, stemming from a series of "stand-alone" specifications, may be overlapping in some instances, and conflicting in other cases. Further, not all system configurations should be subjected to the same test requirements. In situations where there are new design technology applications, more up-front evaluation may be desirable, and the requirements for Type 1 testing may be different than for a situation involving the use of well-known state-of-the-art design methods. In other words, in areas where the potential technical risks are high, the requirement for a more extensive evaluation effort early in the system life cycle may be feasible.

In any event, the TEMP represents a significant input relative to meeting the objectives of system engineering. Not only must one understand the system requirements overall, but knowledge of the functional relationships among the various components of the system is necessary. Also, those involved in test planning must be familiar with the objectives of each specific test requirement such as reliability qualification, maintainability demonstration, and so on.[18] A total integrated approach to test and evaluation is essential, particularly when considering the costs associated with testing activities.

2.11.3 Preparation for System Test and Evaluation

Prior to the start of formal testing, an appropriate period of time is designated for the purposes of test preparation. During this time, the proper conditions must be established to ensure effective results. These conditions will, of course, vary depending on the category of testing being undertaken.

During the early phases of design and development, as analytical evaluations and Type 1 testing are accomplished, the extent of test preparation is minimal. On the other hand, the accomplishment of Type 2 and Type 3 testing, where the conditions are designed to simulate realistic consumer operations to the maximum extent possible, will likely require a rather extensive preparation effort. In order to promote a realistic environment, the following factors need to be addressed:

1. *Selection of test item:* the system (and its components) selected for test should represent the most up-to-date design or production configuration, incorporating all of the latest approved engineering changes.
2. *Selection of test site:* the system should be tested in an environment that will be characteristic for user operations; that is, arctic or tropics, flat or mountainous terrain, airborne or ground. The test site selected should simulate these conditions to the maximum extent possible.
3. *Testing procedures:* the fulfillment of test objectives usually involves the accomplishment of both operator and maintenance tasks, and the completion of these tasks should follow formal approved procedures (e.g., validated technical manuals). The recommended task sequences must be followed to ensure proper system operation.

[18] The detailed requirements for reliability qualification testing, maintainability demonstration, and other specialized tests are covered further in Chapter 3.

4. *Test personnel:* this includes (a) the individuals who will actually operate and maintain the system throughout the test, and (b) supporting engineers, technicians, data recorders, analysts, and administrators who provide assistance in conducting the overall test program. Personnel selected for the first category should be representative of user (or consumer) requirements in terms of the recommended quantities, skill levels, and supporting training needs.
5. *Test and support equipment/software:* the accomplishment of system operational and maintenance tasks may require the use of ground handling equipment, test equipment, software, and/or a combination thereof. Only those items that have been approved for operation should be used.
6. *Supply support:* this includes all spares, repair parts, consumables, and supporting inventories that are necessary for the completion of system test and evaluation. Again, a realistic configuration, projected in a real-world environment, is desired.
7. *Test facilities and resources:* the conductance of system testing may require the use of special facilities, test chambers, capital equipment, environmental controls, special instrumentation, and associated resources (e.g., heat, water, air conditioning, power, telephone). These facilities and resources must be properly identified and scheduled.

In summary, the nature of the test preparation function is highly dependent on the overall objectives of the test and evaluation effort. Whatever the requirements may dictate, these considerations are important to the successful completion of these objectives.

2.11.4 Test Performance and Evaluation

With the necessary preparations in place, the next step is to commence with the formal test and evaluation of the system. The system (or elements thereof) is operated and supported in a designated manner, as defined in the TEMP. Throughout this process, data are collected and analyzed, and the results are compared with the initially specified requirements. With the system in operational status (either "real" or "simulated"), the following questions arise:

1. How well did the system actually perform and did it accomplish its mission objective?
2. What is the *true* effectiveness of the system
3. What is the *true* effectiveness of the system support capability?
4. Does the system meet all of the requirements as covered through the specified technical performance measures (TPMs)?
5. Does the system meet all consumer requirements?

A response to these questions requires a formalized data-information feedback capability with the appropriate output in a timely manner. A data subsystem must be developed and implemented with the goal of achieving certain objectives, and these objectives must relate to these questions.[19]

The process associated with formal testing, data collection, analysis, and evaluation is presented in Figure 2.32. Testing is conducted, data are collected and evaluated, and decisions are made as to whether the system configuration (at this stage) meets the requirements. If not, problem areas are identified, and recommendations are initiated for corrective action.

The final step in this overall evaluation effort is the preparation of a final test report. The report should reference the initial test planning document (i.e., the TEMP), describe all test conditions and the procedures followed in conducting the test, identify data sources and the results of the analysis, and include any recommendations for corrective action and/or improvement. Because this phase of activity is rather extensive and represents a critical milestone in the life cycle, the generation of a good comprehensive test report is essential from the historical standpoint.

2.11.5 System Modifications

The introduction of a change in an item of equipment, a software program, a procedure, or an element of support will likely affect many different components of the system. Equipment changes will likely affect software, spare parts, test equipment, technical data, and possibly certain production processes. Procedural changes will affect personnel and training requirements. Software changes may impact hardware and technical data. A change in any given component of the system will likely have an impact (of some kind) on most, if not all, of the other major components of that system.

Recommendations for changes, evolving from test and evaluation, must be dealt with on an individual basis. Each proposed change must be evaluated in terms of its impact on the other elements of the system, and on life-cycle cost, prior to a decision on whether or not to incorporate the change. The feasibility of incorporating the change will depend on the extensiveness of the change, its impact on the system in terms of its ability to perform the designated mission, and the cost of change implementation.

If a change is to be incorporated, the necessary change control procedures described in Chapter 5 must be implemented. This includes consideration of the time when the change is to be incorporated, the appropriate serial-numbered item(s) affected in a given production quantity, the requirements for retrofitting on earlier serial-numbered items, the development and "proofing" of the change modification kits, the geographic location where the modification kits are to be installed, and the requirements for system checkout and verification following the incorporation of the change. A plan should be developed for each approved change being implemented.

[19] Data requirements and the development of a data subsystem are discussed further in B. S. Blanchard and W. J. Fabrycky, *Systems Engineering and Analysis,* 2nd Ed., Prentice Hall, Upper Saddle River, NJ, 1990.

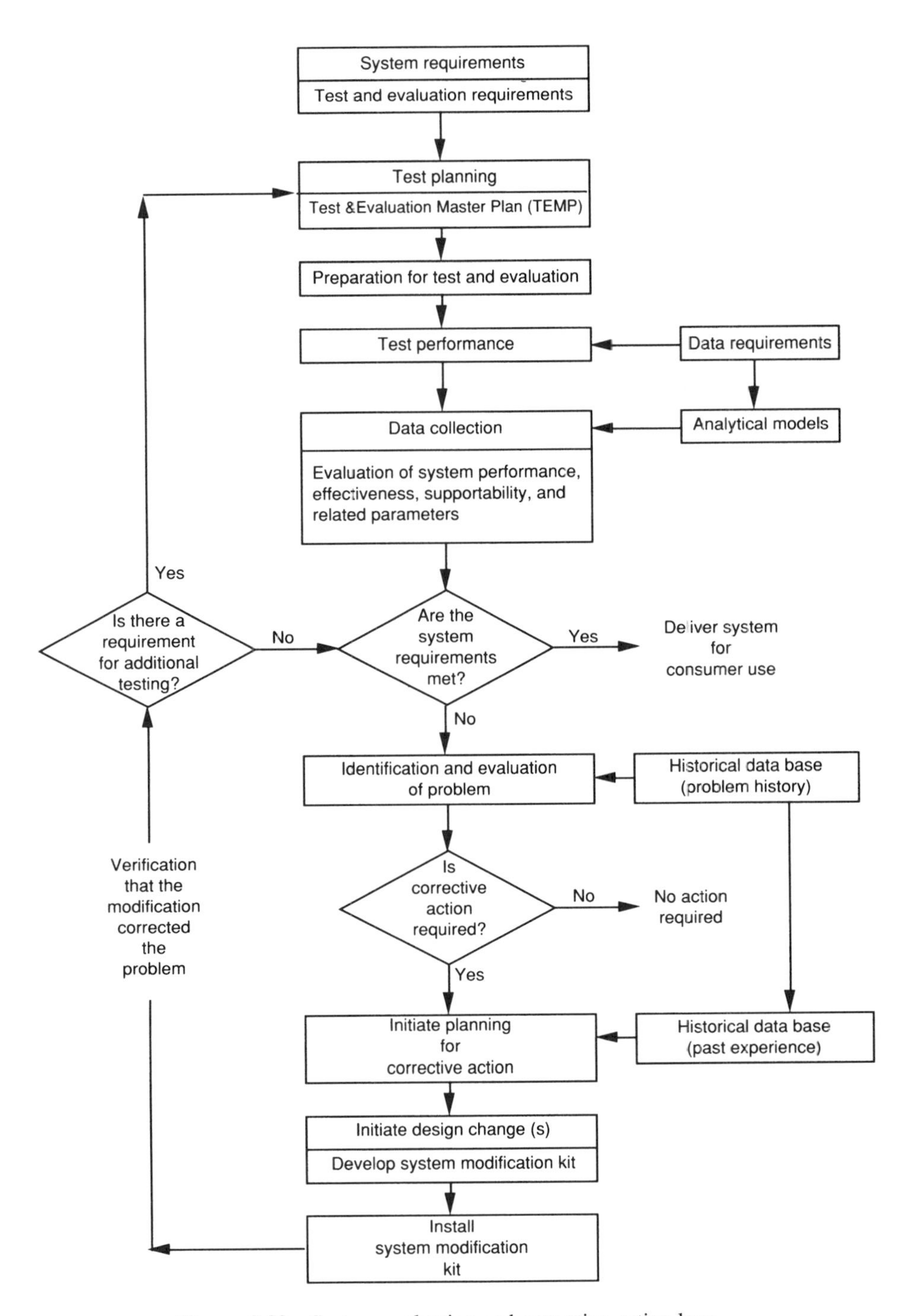

Figure 2.32 System evaluation and corrective-action loop.

2.12 PRODUCTION AND/OR CONSTRUCTION

Depending on the system developed and the nature of its mission, there may be a requirement for follow-on production, or there may be a requirement for construction (refer to Figure 1.10, Examples "A" and "B," respectively). In each case, there are certain unique challenges that must be addressed to ensure that the system configuration, which has been initially designed and verified through evaluation, retains the same high-quality characteristics as it progresses through the production or construction processes. A system may demonstrate a certain degree of effectiveness during formal test and evaluation. However, follow-on replicas of that system configuration, being produced in multiple quantities, may not exhibit the same characteristics. Degradation may occur as a result of a combination of allowable variances in design and/or variances in the different manufacturing processes used in production. A similar effect may take place if the quality of workmanship is not maintained throughout the construction of a one-of-a-kind system configuration.

Current experiences indicate that basic design, given that a *good* robust design exists initially, can be influenced significantly through the follow-on processes employed in production and/or construction. In view of this, there is a need for the continuation of system engineering emphasis and activities throughout the production/construction phase of the life cycle. These activities are discussed further in Chapter 6.[20]

2.13 SYSTEM OPERATIONAL USE AND SUSTAINING SUPPORT

In the system design and development process, consideration must be directed toward (1) the design of the prime mission-oriented segments of the system for supportability, and (2) the design of the system maintenance and support capability itself (refer to Figure 1.10, Examples "A" and "B"). The first item addresses the inherent abilities of the system to be supported in an effective and efficient manner, and the second item covers the resources required to ensure high-quality system support on a sustaining basis. Although a system may be designed and produced with the required effectiveness characteristics incorporated, these characteristics need to be retained for the duration of the life cycle through the accomplishment of *good* maintenance and support practices.

System engineering objectives include not only the initial acquisition of a system with the required characteristics, but the continued maintenance of that system at the required level of effectiveness. Degradation should not take place as a result of inadequate maintenance and support practices. Thus, system engineering emphasis needs to continue throughout the system operational use and sustaining support

[20] In the defense sector, the Department of Defense has recognized these design–production interfaces, and there is currently a great deal of emphasis being placed on "concurrent engineering," "simultaneous engineering," and the like. These concepts include some of the same objectives as defined for system engineering in Chapter 1.

phase of the life cycle. This is basically accomplished through the maintenance and logistics activities described in Chapter 3.

2.14 SYSTEM RETIREMENT AND MATERIAL DISPOSAL

With the concerns for environmental impacts as they exist today, consideration must be directed not only to the acquisition and utilization of the system throughout its intended life cycle, but to the requirements associated with system retirement and the appropriate disposal of its components. There are many systems in use today that, when they become obsolete, will be *costly* to phase out of the inventory. Although some system components can be appropriately recycled, with the resulting materials made available for other uses, there are a number of other components that cannot be consumed without creating a detrimental impact on the environment.

Relative to the role of system engineering, program objectives need to address the retirement and disposal phase of the life cycle, as well as the earlier phases. The design for "disposability," or "design for the environment," should be included in the criteria for analyses and early design decisions.

2.15 SUMMARY

Although some terms and definitions are introduced in Chapter 1, the purpose of this chapter is to relate these to the system life cycle. Further, a baseline needs to be established in order to provide a frame of reference for the discussion of individual design disciplines, design methods, and the activities associated with system engineering.

The *system engineering process,* discussed throughout this chapter, is presented in the form of an "overview"! As one proceeds through the subsequent chapters of this text, the concepts introduced here are amplified to a much greater degree. However, the material in this chapter is a necessary prerequisite to the information presented later.

QUESTIONS AND PROBLEMS

1. Identify the basic steps in the system engineering process, and describe some of the *inputs* and *outputs* associated with each step.
2. What is the purpose of *feasibility analysis?* What information is desired from such an analysis?
3. Why is the definition of system *operational requirements* important? What is included?
4. Why is the definition of the system *maintenance concept* important? What is included? How does the maintenance concept relate to the *maintenance plan?*

5. Identify a specific *problem* that you wish to solve through the design and development of a new system. For your system:

   (a) Describe the current deficiency and identify the *need* for the new system.
   (b) Accomplish an abbreviated feasibility analysis and discuss the various alternative technical approaches that you may wish to consider in designing the new system.
   (c) Define the basic operational requirements for the new system.
   (d) Define the maintenance concept for the new system.
   (e) Identify the critical technical performance measures (TPMs) based on the defined operational requirements and maintenance concept. Describe the process leading from the identification of TPMs to the identification of specific design characteristics.

6. What is meant by "Quality Function Deployment (QFD)?" What are some of the benefits that can be derived from its application?

7. Identify a new system requirement and apply the QFD process (or something of an equivalent nature) in defining the specific characteristics that should be included in the design (demonstrate the application by applying QFD to a real situation).

8. Describe how the QFD process can be beneficially applied in fulfilling the objectives of system engineering.

9. What is meant by *functional analysis?* When should it be accomplished (if at all)? Why is it important in system engineering? What purpose(s) does it serve?

10. For the system selected in Problem 5, accomplish a functional analysis. Construct a functional block diagram showing three levels of *operational* functions. From one of the blocks in the operational functional flow diagram, show two levels of *maintenance* functions. Show how the operational functions and the maintenance functions relate.

11. Select one block from the operational functional diagram and one block from the maintenance functional diagram in Problem 10, and show inputs-outputs and how specific resource requirements are identified (e.g., hardware, software, people, facilities, data, etc.). Show an example by documenting the resource requirements using a format similar to what is presented in Figure 2.20.

12. Why is the identification and description of system-level *functional interfaces* important? What can happen if these interfaces are not well defined?

13. Identify some applications of functional analysis.

14. Describe what is meant by *allocation* or *partitioning?* What is its purpose? To what depth should it be applied? How can the process of allocation influence system design?

15. For the system configuration described in Problem 5, show a breakdown of the system into its subsystems and lower-level elements. Accomplish an allocation

of requirements specified through the TPMs at the system level to the next level below.

16. What are the basic steps involved in *system analysis?* Construct a basic flow diagram illustrating the process, showing the steps, and including feedback provisions.

17. Describe what is meant by *synthesis.* How do the functions of *analysis, synthesis,* and *evaluation* relate to each other?

18. What is a *model?* Identify some of the basic characteristics of a model. List some of the benefits associated with the usage of mathematical models in system analysis. What are some of the problems/concerns?

19. What is meant by *sensitivity analysis?* What are some of the objectives of performing a sensitivity analysis? Benefits?

20. In your opinion, what are some of the major problems in implementing the process described in Figure 2.25? Identify at least three areas of concern.

21. What are some of the "challenges" associated with the day-to-day design process that must be addressed for successful implementation of the system engineering process?

22. How is a system *validated* in terms of compliance with the initially specified requirements?

23. How are test requirements determined?

24. Select of a system of your choice and develop a comprehensive outline for a test and evaluation plan. Identify the categories of test, and describe the *inputs* and *outputs* for each category.

25. Describe some of the considerations associated with the initiation of design changes resulting from test and evaluation.

26. Describe the process associated with the initiation and implementation of design changes. What considerations must be incorporated to enhance the implementation of the system engineering process?

27. Why is system engineering important in the production/construction phase? Operational use and maintenance and support phase? Retirement and disposal phase?

3 System Design Requirements

System design requirements evolve from the identification of a consumer need and are developed through the accomplishment of a feasibility analysis, the definition of system operational requirements and the maintenance concept, the development and prioritization of technical performance measures (TPMs), and the completion of a top-level functional analysis and allocation; that is, those activities that are accomplished early in the system engineering process described in Sections 2.1 through 2.8 in Chapter 2. Initially, these requirements are developed to provide a complete definition of the system in *functional* terms, along with the appropriate metrics. The results are presented in the form of a System Specification (Type "A"; refer to Figure 1.12).

As system development progresses, these requirements are broken down in sufficient detail as to describe the performance, effectiveness, and related characteristics for each major component of the system. "Design-to" criteria are identified, applied to each indenture level in the system hierarchy, and are included in a series of lower-level specifications; that is, development, product, process, material, and supporting specifications.

Given the basic *input* guidelines for design, the process then becomes a series of investigations, trade-off studies, evaluations, and the selection of ways in which the specified design requirements can be met. The iterative process of *synthesis, analysis,* and *evaluation* prevails as one progresses from the definition of system-level requirements during conceptual design through the design of lower-level components during the preliminary and detail design stages, as conveyed in Figure 3.1, which is a simplified evolution from Figure 1.12. Design activities are inherent within each of the blocks in the figure.

The ongoing *design process* occurs as a result of a *team* effort, combining the necessary expertise from various technical specialities. Depending on the system in question and its mission, there may be electrical requirements, mechanical requirements, structural requirements, material requirements, hydraulic requirements, reliability requirements, environmental requirements, maintainability requirements, quality requirements, and so on. These design requirements, which are initially identified through the top-down process illustrated in Figure 1.14 and later supported by the prioritized TPMs (refer to Section 2.6), will vary somewhat as one progresses through the steps in Figure 3.1. In the early stages, the design team may include only a very few selected individuals with the appropriate system-level design experience. Later, additional personnel may be brought into the process as the needs dictate. Thus, the "makeup" of the design team will likely vary as the overall system development process evolves. One of the objectives in system engineering is to

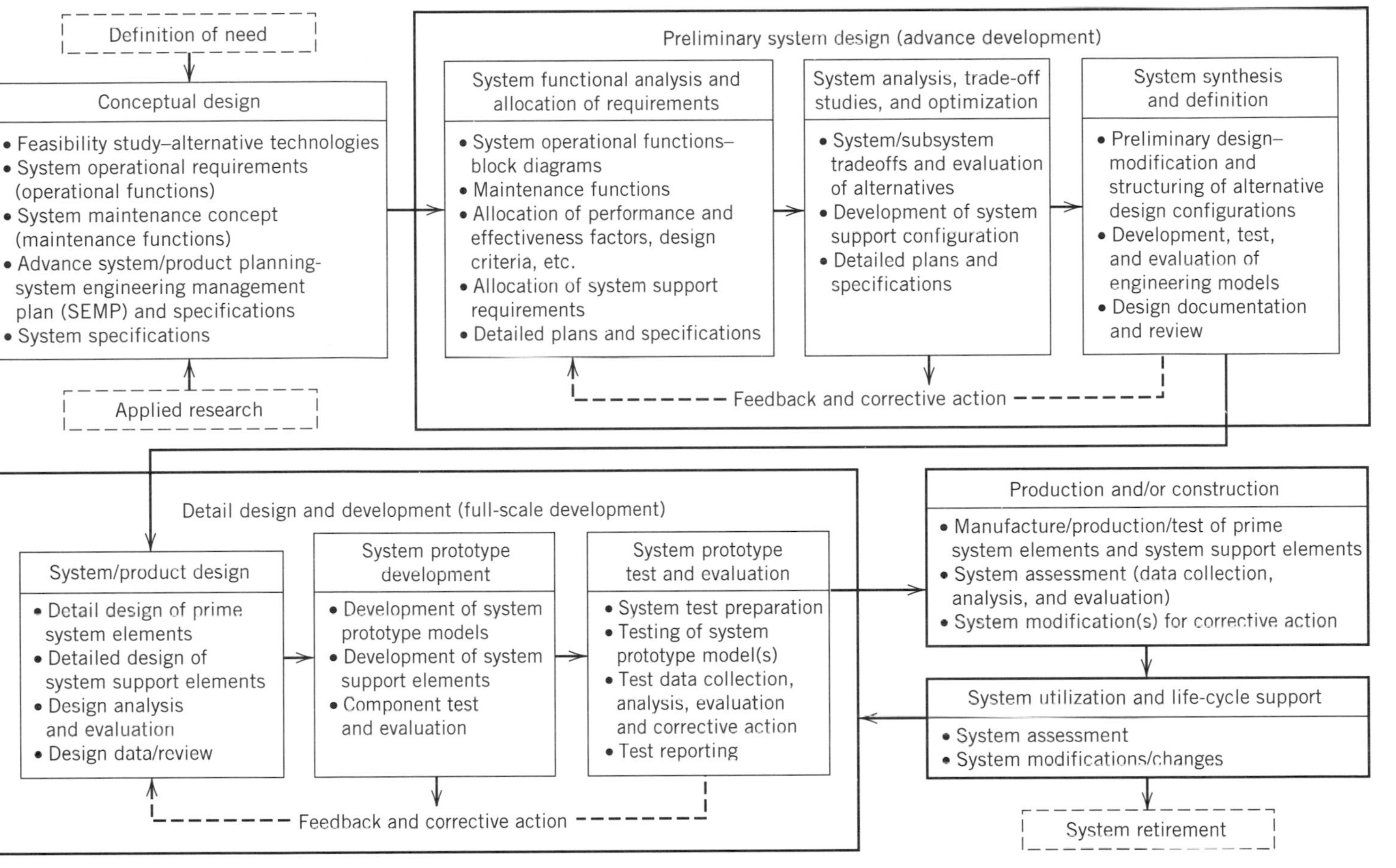
Definition of need
Conceptual design
• Feasibility study–alternative technologies
• System operational requirements (operational functions)
• System maintenance concept (maintenance functions)
• Advance system/product planning-system engineering management plan (SEMP) and specifications
• System specifications
Applied research
Preliminary system design (advance development)
System functional analysis and allocation of requirements
• System operational functions–block diagrams
• Maintenance functions
• Allocation of performance and effectiveness factors, design criteria, etc.
• Allocation of system support requirements
• Detailed plans and specifications
System analysis, trade-off studies, and optimization
• System/subsystem tradeoffs and evaluation of alternatives
• Development of system support configuration
• Detailed plans and specifications
System synthesis and definition
• Preliminary design–modification and structuring of alternative design configurations
• Development, test, and evaluation of engineering models
• Design documentation and review
Feedback and corrective action
Detail design and development (full-scale development)
System/product design
• Detail design of prime system elements
• Detailed design of system support elements
• Design analysis and evaluation
• Design data/review
System prototype development
• Development of system prototype models
• Development of system support elements
• Component test and evaluation
System prototype test and evaluation
• System test preparation
• Testing of system prototype model(s)
• Test data collection, analysis, evaluation and corrective action
• Test reporting
Feedback and corrective action
Production and/or construction
• Manufacture/production/test of prime system elements and system support elements
• System assessment (data collection, analysis, and evaluation)
• System modification(s) for corrective action
System utilization and life-cycle support
• System assessment
• System modifications/changes
System retirement

Figure 3.1 The major steps of system design and development.

ensure the proper and timely selection and integration of the required design disciplines, combined in such a manner as to enable the development of a system that will respond to the consumer need in a cost-effective way. Not only will a typical program include designers representing different disciplines and with a wide variety of backgrounds and experiences, but the specific requirements for these areas of expertise will shift from one program phase to the next.

Chapter 2 sets the stage in defining the overall development process. It is now appropriate to address some of the *specifics* relative to design requirements. The purpose of this chapter is to cover these requirements through *specifications,* and to review some of the details as they pertain to individual design disciplines. An introduction to a select sample of disciplines is included, some commonalities are noted, and the importance of design integration through the application of system engineering methods is highlighted.[1]

3.1 DEVELOPMENT OF SPECIFICATIONS AND DESIGN CRITERIA

The initial definition of system requirements is projected through a combination of formal specifications and planning documentation. Specifications basically cover the *technical* requirements for system design, and planning documentation includes all *management* requirements necessary to fulfill program objectives. The combination of specifications and plans is considered as the basis for all future program engineering and management decisions.

The scope and depth of such documentation depend on the nature, size, and complexity of the system. In addition, the extent to which new design is feasible (where extra guidance and controls are desired), versus the selection of an "off-the-shelf" capability, will dictate the amount of documentation necessary. For small and relatively simple items, the technical specification and program planning requirements may be included in a single document. On the other hand, for large-scale systems there may be a significant assemblage of documentation. In either case, the amount of documentation must be tailored to the need as dictated by the degree of technical and management controls necessary to accomplish program objectives.

In dealing with large systems, there are numerous elements that must be covered by specifications. Some components of the system may require an extensive amount of research-and-development effort, whereas other components are procured directly from existing supplier inventories. For new items, some are developed by the major producer of the system, and others are developed by suppliers remotely located in various parts of the world. In manufacturing, certain components may be produced in multiple quantities using conventional methods, whereas a special process may be required to produce other items. There may be a variety of specifications necessary to provide the guidance and controls associated with the development of the system and its components.

[1] It should be emphasized that no attempt has been made to cover all of the disciplines that may be required in the design of a system. Only a few have been identified, with the intent of highlighting those areas that are not always properly addressed as part of a typical project activity.

When preparing and applying specifications, there are different classifications, as noted and illustrated in Figure 3.2:[2]

1. *System specification* (Type "A"): includes the technical, performance, operational, and support characteristics for the system as an entity. It includes the allocation of requirements to functional areas, and it defines the various functional-area interfaces. The information derived from the feasibility analysis, operational requirements, maintenance concept, and the functional analysis is covered (refer to Sections 2.3 to 2.8)

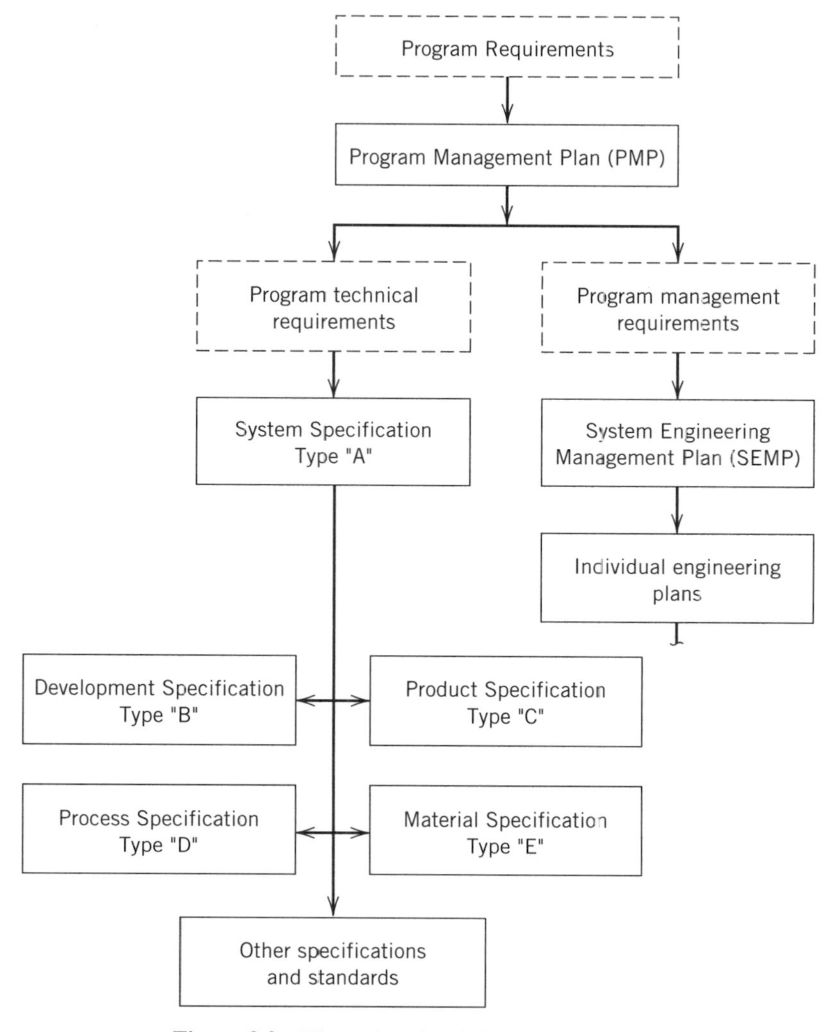

Figure 3.2 Hierarchy of technical specifications.

[2] These specification classifications were taken from MIL-STD-490, Military Standard, "Specification Practices," Department of Defense, Washington, DC, latest revision.

2. *Development specification* (Type "B"): includes the technical requirements for any item below the system level where research, design, and development are accomplished. This may cover an equipment item, assembly, computer program, facility, critical item of support, and so on. Each specification must include the performance, effectiveness, and support characteristics that are required in the evolving of design from the system level and down.
3. *Product specification* (Type "C"): includes the technical requirements for any item below the top system level that is currently in the inventory and can be procured "off the shelf." This may cover standard system components (equipment, assemblies, units, cables), a specific computer program, a spare part, a tool, and so on.
4. *Process specification* (Type "D"): includes the technical requirements that cover a service that is performed on any component of the system (e.g., machining, bending, welding, plating, heat treating, sanding, marking, packing, and processing).
5. *Material specification* (Type "E"): includes the technical requirements that pertain to raw materials, mixtures (e.g., paints, chemical compounds), and/or semifabricated materials (e.g., electrical cable, piping) that are used in the fabrication of a product.

The preparation of specifications is a key engineering activity. The system specification is prepared during the conceptual design phase. Development and product specifications are based on the results of "make-or-buy" decisions and are generally prepared during preliminary design.[3] Process and material specifications are more oriented to production and/or construction activities, and are generally prepared during the detail design and development phase. The relative timing of these specifications, in terms of program scheduling, is illustrated in Figures 1.12 and 1.21.

For large-scale systems, involving a wide mix of component suppliers, it is likely that many specifications will be generated and applied at varying stages in the system design and development process. In reviewing past experiences associated with different programs, the generation/application of many different specifications on an independent basis has resulted in conflicts (i.e., contradictions relative to design criteria), as well as questions pertaining to which specification takes precedence. Additionally, there has been a tendency to not only specify the "WHATs" but the "HOWs!" This, in turn, can lead to a costly result. Specifications should be prepared to cover "performance" requirements, or the "*what*-is-required" case.

To help resolve the precedence problem, a "documentation tree" (or "specification tree") should be prepared, showing the hierarchy of specifications (and plans) from the system specification and down. Referring to the process of requirements allocation in Section 2.8, it is necessary to establish requirements at the system level

[3]The results of "make-or-buy" decisions determine whether an item is to be manufactured within the producer's facility or purchased from an outside source. Economic factors and scheduling requirements, combined with the availability of sources of supply, are prime considerations in the decision-making process. Make-or-buy decisions are covered further in Chapter 8.

first, and then allocate these requirements down to the various components of the system. When developing specifications, which dictate the design requirements for the various system components, it is essential that a good comprehensive *system* specification be developed first, and then supplement this specification with the generation of good development, product, process, and/or material specifications as applicable.

In Figure 2.21 (Chapter 2), a preliminary hierarchy of system components is shown as a basis for the allocation of requirements. Figure 3.3 shows a variation of this hierarchy, converted into the form of a specification tree. Basically, the system specification is the *top* technical document for design. Other specifications supplement the system specification to varying degrees. Further, an order of precedence must be established to provide guidance as to which specification governs in the event of possible conflicts.

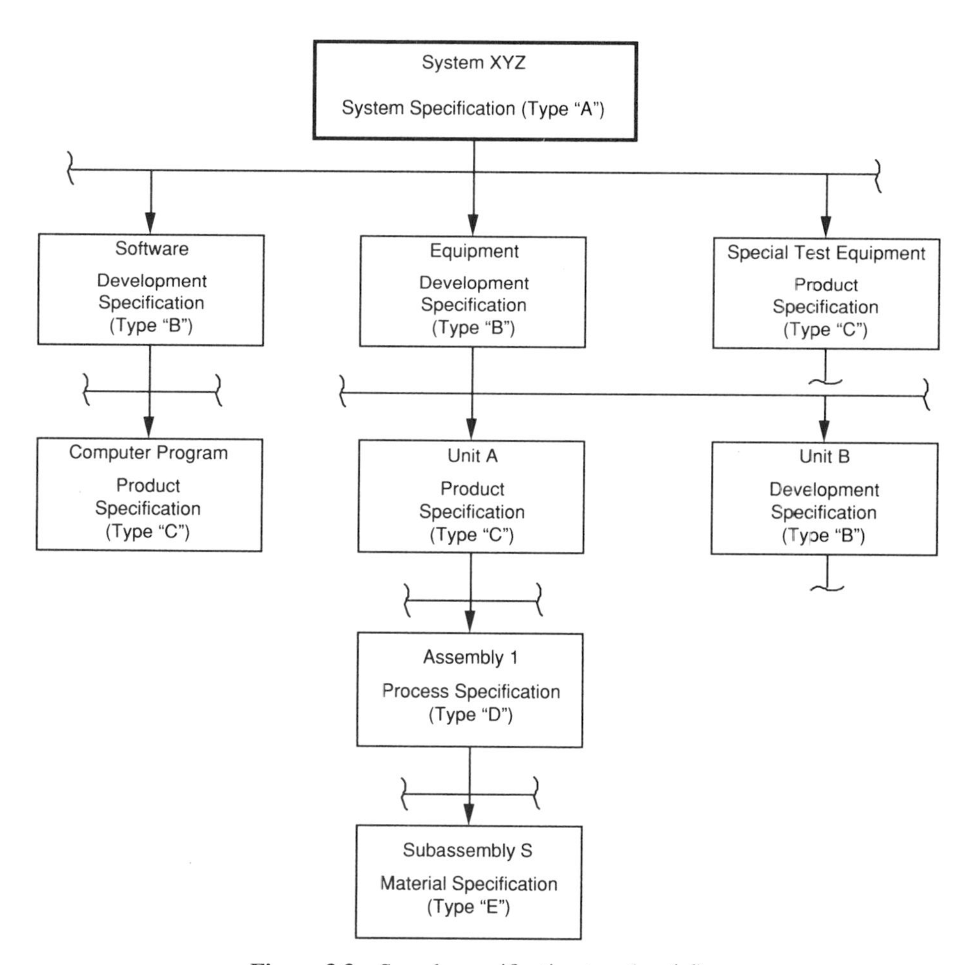

Figure 3.3 Sample specification tree (partial).

With the System Specification (Type "A") being the prime document for technical guidance, it is appropriate that the responsibility for its preparation and implementation be assigned as a system engineering task. Care must be taken to ensure that *all* significant design characteristics, applicable at the system level, are included. The requirements must be properly *integrated,* and meaningful technical performance measures (TPMs) must be identified.[4] TPMs include those quantitative characteristics of the system that are initially specified, then reflected in the follow-on design, and later used as measures against which the system is evaluated (e.g., speed, range, accuracy, size, capacity, MTBF, MLH/OH, $\bar{M}ct$, and cost).

A sample format for a system specification is presented in Figure 3.4. The specification includes a description of the system, its major characteristics, some general criteria for design and assembly, major data requirements, logistics and producibility considerations, test and evaluation requirements, and quality assurance provisions. The intent is to provide a description of the functional baseline for the system. It constitutes the framework used in the preparation of subordinate specifications by the responsible designer for the numerous components of the system, and it serves as the major technical reference for all program planning documentation. The system specification covers the results of the feasibility analysis, the definition of system operational requirements and the maintenance concept, and a "functional" description of the system based on the functional analysis. The allocated factors, discussed in Section 2.8, are derived directly from system-level TPMs, and are included in the development, product, process, and/or material specifications as appropriate.

As a final comment, the preparation of a good system specification is highly dependent on the abilities of those accomplishing the task relative to their thorough understanding of the system in total, its intended mission, its components and their interrelationships, the various design disciplines required and their interfaces, and so on. It is not sufficient to merely prepare a series of individual writeups covering each discipline, stapled together, and submitted as a specification. This type of an output usually results in contradictions, confusion, and inefficiencies throughout all subsequent phases of system design and development. Without a good technical baseline, many of the design decisions made later will be "suspect"! Thus, the realization of a good comprehensive and highly integrated specification is critical from the beginning.

3.2 THE DESIGN PROCESS AND DESIGN OBJECTIVES

Based on the system specification, there may be a variety of "design-to" requirements, such as those illustrated in Figure 3.5. These requirements may be mutually supportive by nature, or there may be some inherent conflicts in goals. These goals

[4] Technical performance measures (TPMs) in the system specification should include those measures of effectiveness (and supporting factors) that are identified as system-level parameters described in Section 2.6 and highlighted in Figure 2.10.

System Specification

1.0 Scope

2.0 Applicable Documents

3.0 Requirements

- 3.1 System Definition
  - 3.1.1 General Description
  - 3.1.2 Operational Requirements (Need, Mission, Utilization Profile, Distribution, Life Cycle)
  - 3.1.3 Maintenance Concept (Levels of Repair)
  - 3.1.4 Functional Analysis and System Definition
  - 3.1.5 Allocation Requirements
  - 3.1.6 Functional Interfaces and Criteria
  - 3.1.7 Environmental Conditions
- 3.2 System Characteristics
  - 3.2.1 Performance Characteristics
  - 3.2.2 Physical Characteristics
  - 3.2.3 Effectiveness Requirements
  - 3.2.4 Reliability
  - 3.2.5 Maintainability
  - 3.2.6 Usability (Human factors)
  - 3.2.7 Supportability
  - 3.2.8 Transportability/Mobility
  - 3.2.9 Flexibility
  - 3.2.10 Other
- 3.3 Design and Construction
  - 3.3.1 CAD/CAM Requirements
  - 3.3.2 Materials, Processes, and Parts
  - 3.3.3 Mounting and Labeling
  - 3.3.4 Electromagnetic Radiation
  - 3.3.5 Safety
  - 3.3.6 Interchangeability
  - 3.3.7 Workmanship
  - 3.3.8 Testability
  - 3.3.9 Economic Feasibility
- 3.4 Documentation/Data
- 3.5 Logistics
  - 3.5.1 Maintenance Requirements
  - 3.5.2 Supply Support
  - 3.5.3 Test and Support Equipment
  - 3.5.4 Personnel and Training
  - 3.5.5 Facilities and Equipment
  - 3.5.6 Packaging, Handling, Storage, and Transportation
  - 3.5.7 Computer Resources (Software)
  - 3.5.8 Technical Data
  - 3.5.9 Customer Services
- 3.6 Producibility
- 3.7 Disposability
- 3.8 Affordability

4.0 Test and Evaluation

5.0 Quality Assurance Provisions

6.0 Distribution and Customer Service

Figure 3.4 Example Type "A" System Specification format.

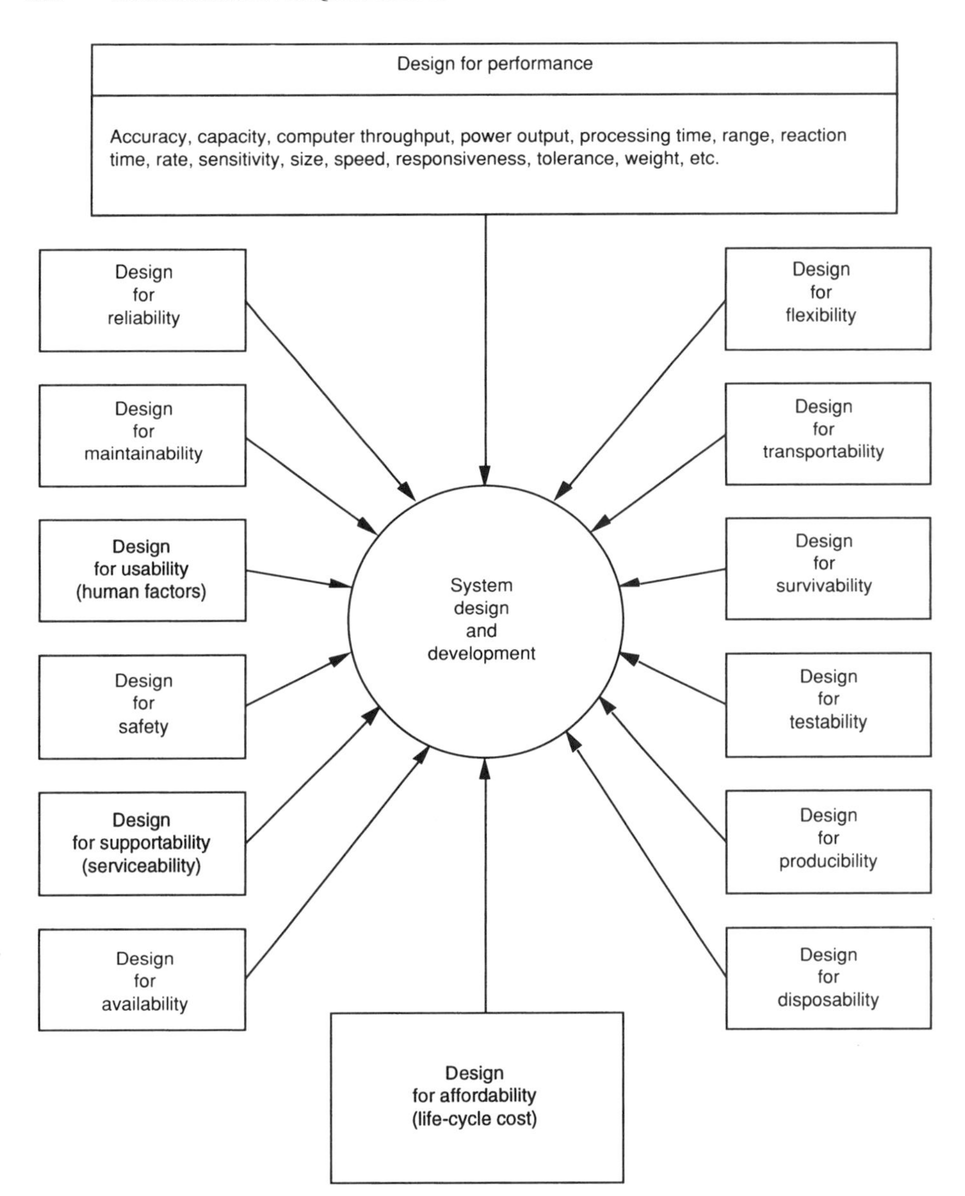

Figure 3.5 System design requirements.

are viewed in terms of relative importance (refer to Figure 2.10), and design optimization is accomplished through trade-off studies with the objective of establishing a mutually satisfactory approach.

In response to the specification and the established goals, certain categories of engineering expertise are identified as being necessary for the design and development of the system in question. These categories, and associated levels of effort, depend on the nature and complexity of the system and on the size of the project.

For relatively small systems/products such as a radio, an electrical household appliance, or an automobile, the quantity and variety of engineering expertise may be limited. On the other hand, there are many large-scale systems that require the combined input of specialists representing a wide variety of engineering disciplines. Two examples of relatively large projects are noted:

1. *Commercial aircraft system:* Aeronautical engineers determine aircraft performance requirements and design the overall airframe structure. Electrical engineers design the aircraft power distribution system and ground power requirements. Electronic engineers of various types are responsible for the development of subsystems such as radar, navigation, communications, and data recording and handling. Mechanical engineers are called on in the areas of mechanical structures, linkages, pneumatics, and hydraulics. Metallurgists are needed in the selection and application of materials for the aircraft structure. Reliability and maintainability engineers are concerned with availability, mean time between failure (MTBF), mean corrective maintenance time ($\bar{M}ct$), maintenance labor hours per operating hour (MLH/OH), and system logistic support. Human factors engineers are interested in the man-machine functions, cockpit and cabin layout, and design of the various operator control panels. System engineering is concerned with the overall development of the airplane as a system and ensuring the proper integration of the numerous aircraft subsystems. Industrial engineers of various types are directly involved in the production of the aircraft itself and its many components. Test engineers are required to evaluate the system to ensure conformance to consumer requirements. Other engineering specialties are employed on a task-by-task basis.
2. *Ground mass-transit system:* Civil engineers are required for the layout and/or design of railroad tracks, tunnels, bridges, cables, and facilities. Electrical engineers are involved in the design of automatic train control provisions, traction power, substations and power distribution, automatic fare collection, digital data systems, and so on. Mechanical engineers are necessary in the design of passenger vehicles and related mechanical equipment. Architectural engineers may provide support in the construction of passenger terminals. Reliability and maintainability engineers would likely be involved in the design for system availability and the incorporation of supportability characteristics. Human factors engineers are involved in the aspects of lighting, comfort ventilation, boarding access, handicapped accommodations, audio boarding instructions, and comfort stations. Industrial engineers will deal with the production aspects of passenger vehicles and vehicle components. Test engineers will evaluate the system to ensure that all performance, effectiveness, and system support requirements are met. Engineers in the planning and marketing areas will be required to keep the public informed and to promote the technical aspects of the system (i.e., keep the politicians and local citizens happy). Additional engineering specialists of various categories will be required to perform specific project-related tasks on an as-required basis.

Although these examples may not necessarily be all inclusive, it is apparent that many different engineering disciplines are directly involved. Some of the more traditional disciplines such as electrical engineering and mechanical engineering are fractionated and broken down into specific job-oriented classifications. Engineering requirements for many large projects may include hundreds of individuals (or more in the case of an aircraft development project or equivalent) with varied backgrounds assigned to perform engineering functions. These engineers, forming a part of a larger organization, must not only be able to communicate with each other, but must be conversant with such supplemental activities as purchasing, accounting, manufacturing, and legal.

When considering personnel loading for large projects, there are likely to be some fluctuations. Depending on the functions to be performed, some engineers will be assigned to a given project until system development is completed, some will be assigned through the completion of production, and others will be brought in for a short term to perform specific tasks. The requirements, in terms of needed engineering expertise, will also vary from phase to phase because the areas of emphasis change as system development progresses. During the early phases of advance planning and conceptual design, individuals with a broad systems-oriented background are needed in greater quantities than detailed design specialists, whereas the reverse may be true during the detail design and development phase. In any event, the needs for engineering will change as the system evolves through its planned life cycle.

An additional characteristic associated with large projects pertains to the split in design engineering workload between the major system producer and the suppliers of system components. A great deal of development, evaluation, production, and support of system components is accomplished at supplier facilities located throughout the world (the percentage of supplier activity in terms of total acquisition cost often ranges up to 75%; refer to Chapter 8). In other words, the major producer who is ultimately responsible for the development, integration, production, and support of the total system as an entity is greatly dependent on the results of engineering activities being accomplished at numerous disperse locations.

The project environment for the design and development of a large number of systems today is highly "dynamic"! There are many individuals with different specialties and backgrounds, rotating "on" and "off" of the program at varying times. The need for good communications is essential, as well as having a good understanding of the numerous interfaces that exist. The electrical design engineer needs to understand his (or her) interface relationships with the mechanical designer, the structural engineer, the reliability engineer, and/or the human factors specialist. The logistics engineer needs to understand the design process and the responsibilities of the electronics engineer. To acquire the necessary design integration, within the context of system engineering, requires this understanding and appreciation, along with good communications.

To provide a better understanding of design requirements overall (with the objective of further promoting the integration process), a few design disciplines have been selected for the purposes of additional emphasis. The disciplines of reliability, maintainability, human factors, safety, software development, logistics engineering, producibility, disposability, quality engineering, and value/cost engineering are discussed in the following sections. These areas, by themselves, certainly do not represent the total spectrum of design activity. However, in the past, some of these particular requirements have not been adequately reflected in many of the systems developed, perhaps due to a lack of understanding and appreciation for these areas as they apply to design. As a result, these disciplines (and others) have been addressed independently through specifications and standards, and have not been well integrated into the mainstream design effort. Further, when addressing the individual requirements for each of the disciplines presented, one will find some commonalities! Through a review of these requirements, it is hoped that an even better appreciation of the need for total design integration will take place.

3.3 SELECTED DESIGN ENGINEERING DISCIPLINES

3.3.1 Reliability Engineering[5]

Reliability, in a generic sense, can be defined as "the probability that a system or product will perform in a satisfactory manner for a given period of time when used under specified operating conditions." The *probability* factor relates to the number of times that one can expect an event to occur in a total number of trials. A probability of 95%, for example, means that (on the average) a system will perform properly 95 out of 100 times, or that 95 of 100 items will perform properly.

The aspect of *satisfactory performance* relates to the system's ability to perform its mission. A combination of qualitative and quantitative factors defining the functions that the system is to accomplish, usually presented in the context of the system specification, is included. These factors are defined under system operational requirements described in Section 2.4.

The element of *time* is most significant because it represents the measure against which the degree of performance can be related. A system may be designed to perform under certain conditions, but for how long? Of particular interest is the ability to predict the probability of a system surviving for a designated period of time without failure. Other time-related measures are mean time between failure (MTBF), mean time to failure (MTTF), mean cycles between failure (MCBF), and failure rate (λ).

The fourth key element in the reliability definition, *specified operating conditions,* pertains to the environment in which the system will operate. Environmental requirements are based on the anticipated mission scenarios (or profiles), and appropriate considerations for reliability must include temperature cycling, humidity, vibration and shock, sand and dust, salt spray, and so on. Such considerations must not only address the conditions when the system is operating and in a "dynamic"

[5]The intent herein is to provide an introductory overview of reliability engineering, both in terms of definitions and program requirements, and not to cover the subject in depth! However, it is highly recommended that the subject area be pursued further. Three good references are (1) D. Kececiogly, *Reliability Engineering Handbook,* 2 vols., Prentice Hall, Upper Saddle River, NJ, 1991; (2) W. G. Ireson, and C. F. Coombs (Eds.), *Handbook of Reliability Engineering and Management,* McGraw-Hill, New York, 1988; and (3) J. Knezevic, *Reliability, Maintainability, and Supportability,* McGraw-Hill, New York, 1993. Additional references on reliability are included in the bibliography in Appendix F.

state, but the conditions of the system during the accomplishment of maintenance activities, when the system (or components thereof) is being transported from one location to another, and/or when the system is in the storage mode. Experience indicates that the transportation, handling, maintenance, and storage modes are often more critical from a reliability standpoint than the environmental conditions during the periods of actual system utilization.

This definition of reliability is rather basic, and it can be applied to almost any type of system. However, there are instances in which it may be more appropriate to define reliability in terms of some specific mission scenario. In such cases, reliability can be defined as "the probability that a system will perform a designated mission in a satisfactory manner." This definition may, of course, imply the accomplishment of maintenance activities, as long as it does not interfere with the successful completion of the mission. The aspect of maintenance is covered more extensively in subsequent sections of this text.

In applying reliability requirements to a specific system, one needs to relate these requirements in terms of some quantitative measure (or a combination of several figures of merit). The basic reliability function, $R(t)$, may be stated as

$$R(t) = 1 - F(t) \tag{3.1}$$

where $R(t)$ is the probability of success, and $F(t)$ is the probability that the system will fail by time $t.F(t)$ represents the failure distribution function.

When dealing with failure distributions, one often assumes average failure rates and attempts to predict the expected (or average) number of failures in a given period of time. To assist in this prediction, the Poisson distribution (which is somewhat analogous to the binomial distribution) can be applied. This distribution is generally expressed as

$$P(x,t) = \frac{(\lambda t)^x e^{-\lambda t}}{x!} \tag{3.2}$$

where λ represents the average failure rate, t is the operating time, and x is the observed number of failures.

This distribution states that if an average failure rate (λ) is known for an item, then it is possible to calculate the probability, $P(x,t)$, of observing 0, 1, 2, 3, . . . , n number of failures when the item is operating for a designated period of time, t. With this in mind, the Poisson expression may be broken down into a number of terms:

$$1 = e^{-\lambda t} + (\lambda t)e^{-\lambda t} + \frac{(\lambda t)^2 e^{-\lambda t}}{2!} + \frac{(\lambda t)^3 e^{-\lambda t}}{3!} + \ldots + \frac{(\lambda t)^n e^{-\lambda t}}{n!} \tag{3.3}$$

where $e^{-\lambda t}$ represents the probability of zero failures occurring in time, t, $(\lambda t)e^{-\lambda t}$ is the probability that one (1) failure will occur, and so on.

In addressing the reliability objective, dealing with the probability of success, the first term in the Poisson expression is of significance! This term, representing the "exponential" distribution, is often assumed as the basis for specifying, predicting, and later measuring the reliability for a system.[6] In other words,

$$R = e^{-\lambda t} = e^{-t/M} \tag{3.4}$$

where M is the MTBF. If an item has a constant failure rate, the reliability of that item at its mean life is approximately 0.37, or there is a 37% chance that the item will survive its mean life without failure.

Figure 3.6 presents the traditional exponential reliability curve. The basic underlying assumption is that the failure rate is constant. When dealing with failure rates, it is necessary to view such in terms of both time and life-cycle activity. Figure 3.7 presents some typical failure-rate curve relationships. Although somewhat "puristic" in nature, the illustrations are included to support additional discussion of reliability.

In Figure 3.7, the "bathtub" curve will vary somewhat depending on the type of equipment (whether electronic or mechanical), the degree of system/equipment maturity (new design or production versus state-of-the-art), and so on. Usually, there is an initial "break-in" or "infant mortality" period in which a certain amount of "debugging" or "burn-in" is required in order to reach a stabilized condition. Design and/or manufacturing defects often occur, maintenance is accomplished, and corrective action is taken to resolve any outstanding problems. Subsequently, when stability is acquired, the failure rate is relatively constant until such point in time when components begin to wear out, causing an ever-increasing failure rate as time evolves.

The curves presented in Figure 3.7 may also be highly influenced by individual program activities. For example, it is not uncommon for the customer to demand that a system (or components thereof) be delivered earlier than initially scheduled. With the objective to respond, the producer may eliminate certain essential quality checks in the production process in order to "get the equipment/software out the door"! This usually leads to more initial defects, the consumption of more resources for maintenance and support than initially anticipated, and a higher degree of customer dissatisfaction in the long term. In Figure 3.7, the system becomes operational at the early stages of the bathtub curve before stabilization is attained.

In the world of software, failures may be related to calendar time, processor time, the number of transactions per period of time, the number of faults per module of code, and so on. Expectations are usually based on an operational profile and criticality to the mission. Thus, an accurate description of the mission scenario(s) is required. As the system evolves from the design and development stage to the operational utilization phase, the ongoing maintenance of software often becomes a major issue. Whereas the failure rate of equipment generally assumes the profiles in Figure

[6] It should be noted that many of the assumptions herein are based on an average, or *constant,* failure rate. Although this assumption sometimes simplifies the reliability calculation process, there are instances in which failure rates are constantly changing. In these situations, it may be more appropriate to assume a Weibull distribution (or equivalent) in lieu of the negative exponential.

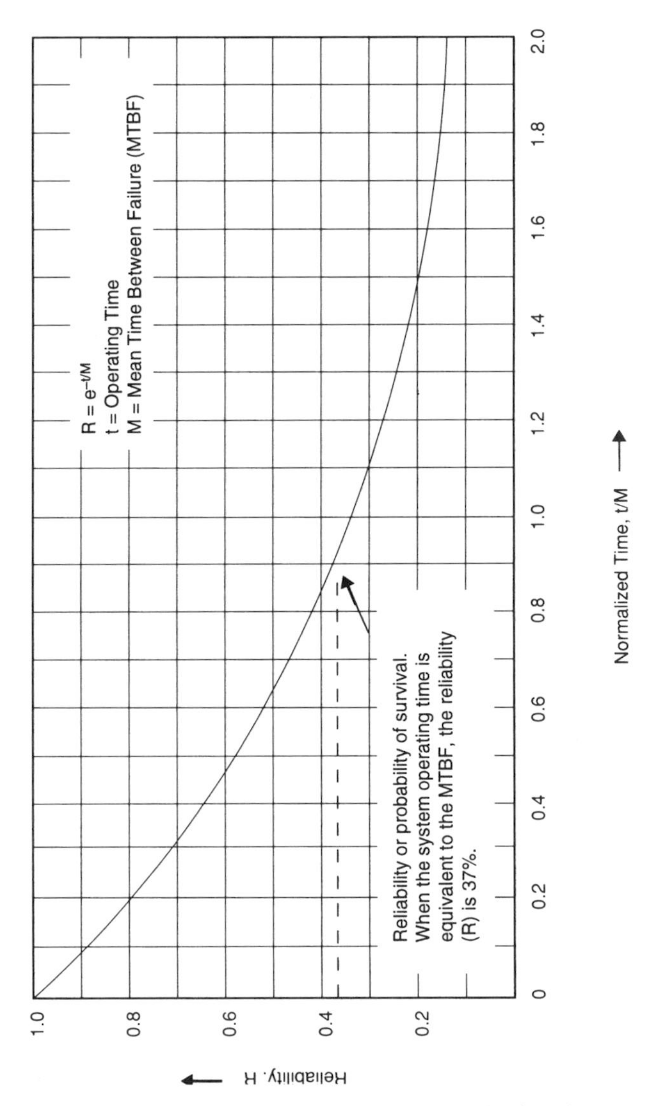

Figure 3.6 Traditional reliability exponential function.

3.7, the maintenance of software often has a negative effect on the overall system reliability. The performance of software maintenance on a continuing basis, along with the incorporation of system changes in general, usually impacts the overall failure rate, as shown in Figure 3.8. When a change or modification is incorporated, "bugs" are usually introduced and it takes a while for these to be "worked out" of the system.

From Figure 3.7, and Equations (3.2) through (3.4), the failure rate constitutes the number of failures occurring during a specified interval of time, or

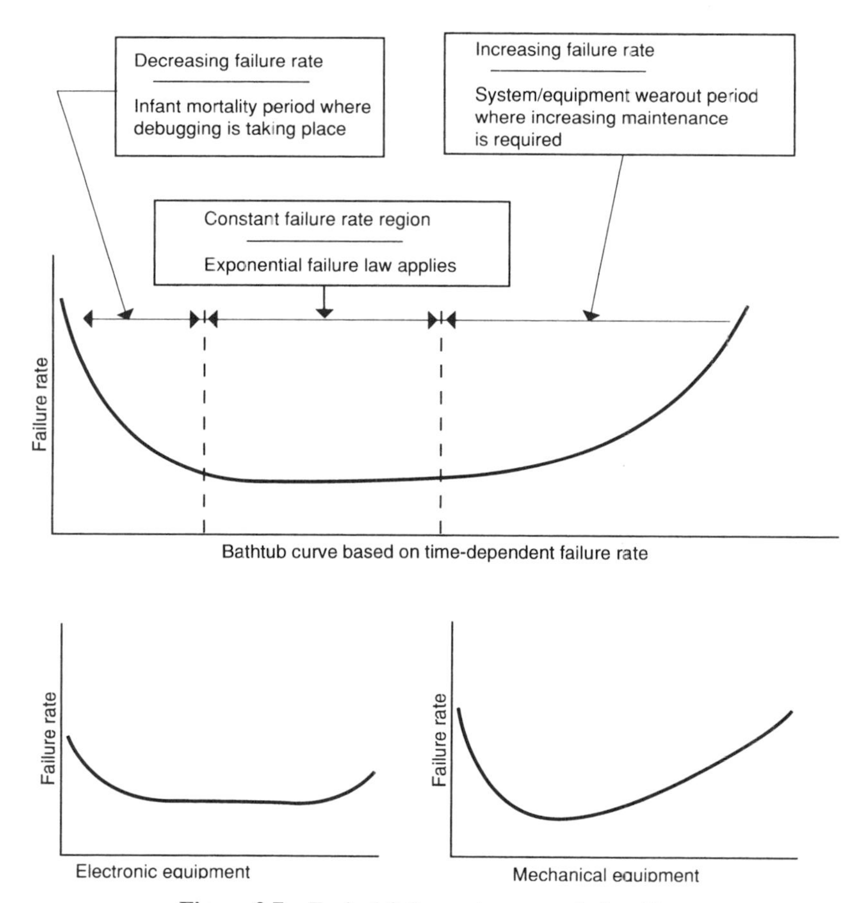

Figure 3.7 Typical failure-rate curve relationships.

$$\lambda = \frac{\text{number of failures}}{\text{total operating hours}} \tag{3.5}$$

More specifically, the failure rate may be expressed in terms of failures per hour, failures per million hours, or percent failures per 1000 hours.[7]

Additionally, when defining failures in a pure reliability sense, this refers to "primary" or "catastrophic" failures; that is, instances when the system is not operating in accordance with specification requirements due to an actual component failure stemming from an overstressed condition. A component failure may, in turn, cause other components to fail through a chain reaction of events. Thus, we need to con-

[7] This definition applies primarily to operating equipment. Failure rates may also be expressed in terms of failures per cycle of operation, errors per page of documentation, errors per operator or maintenance task, failures per module of software, and so on.

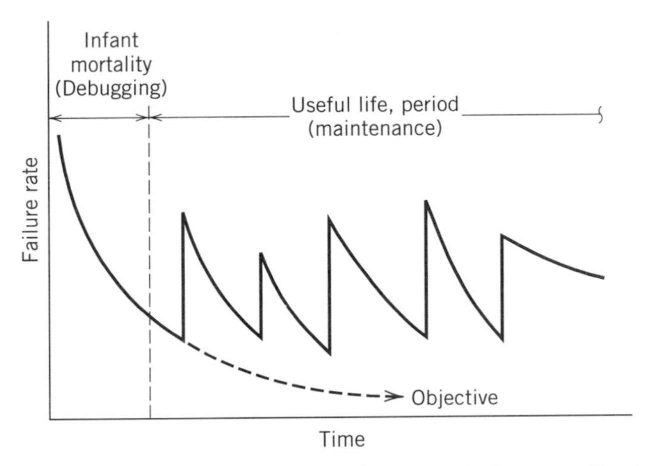

Figure 3.8 Failure-rate curve with maintenance (software application).

sider both primary catastrophic failures and secondary failures, sometimes known as dependent failures.[8]

With the identification of reliability measures, it is now appropriate to show some applications. System components are functionally related to each other through a series relationship, a parallel relationship, or a combination of these. Figure 3.9 illustrates some examples.

Figure 3.9(A) shows a series network. All components must operate in a satisfactory manner if the system is to function properly. The reliability, or the probability of success, of the system is the product of the reliabilities for the individual components and is expressed as

$$R_s = (R_A)(R_B)(R_C) \tag{3.6}$$

If the system operation is to be related to a specific time period, by substituting Equation (3.4) into Equation (3.6), the overall reliability of the series network is

$$R_s = e^{-(\lambda_A + \lambda_B + \lambda_C)t} \tag{3.7}$$

Figure 3.9(B) illustrates a parallel redundant network with two components. The system will function if either “A” or “B” or both, are working. The reliability expression for this network is

$$R_s = R_A + R_B - (R_A)(R_B) \tag{3.8}$$

[8] The overall frequency of unscheduled maintenance considers primary failures, secondary failures, manufacturing defects, operator-induced failures, maintenance-induced failures, defects due to handling, and so on. From a systems engineering perspective, this overall frequency factor needs to be addressed, and is discussed further in Section 3.3.2.

Now, consider a network with three components in parallel, as shown in Figure 3.8(C). For the system to fail, all three components must fail individually. The reliability of the network is

$$R_s = 1 - (1 - R_A)(1 - R_B)(1 - R_C) \tag{3.9}$$

In the event that all three components are identical, the reliability expression in Equation (3.9) can be simplified to

$$R_s = 1 - (1 - R)^3$$

For a system with n components, the expression becomes

$$R_s = 1 - (1 - R)^n \tag{3.10}$$

Incorporating redundancy in design helps to improve system reliability. The effects of redundancy on design, presented in a simple generic sense, is illustrated in Figure 3.10. One can also determine the degree of reliability improvement through redundancy by developing some mathematical examples using Equations (3.8) and (3.9).

Redundancy can be applied in design at different hierarchical indenture levels of

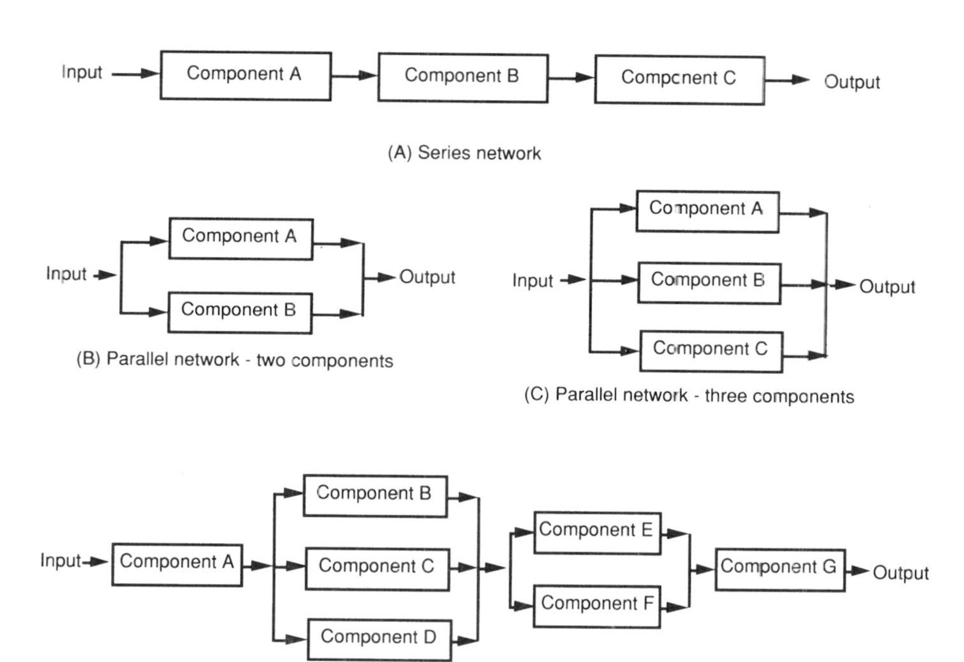

Figure 3.9 Reliability component relationships.

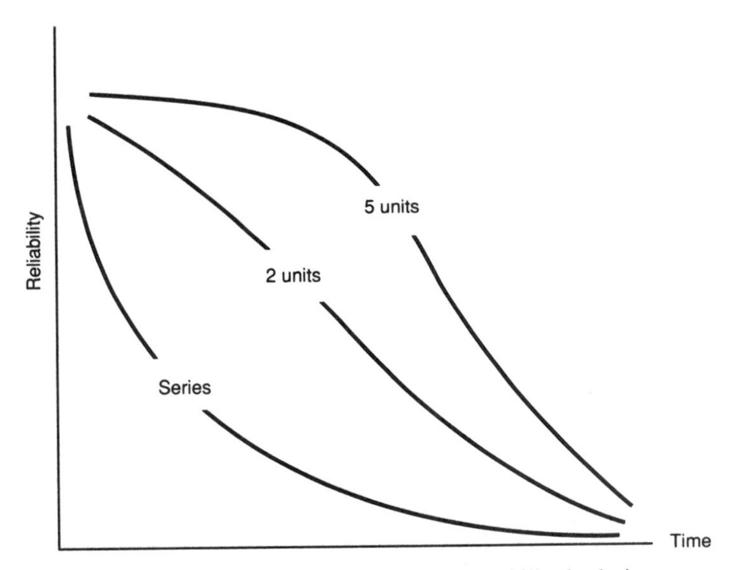

Figure 3.10 Effects of redundancy on reliability in design.

the system. At the subsystem level, it may be appropriate to incorporate parallel functional capabilities, where the system will continue to operate if one path fails to function properly. The flight control capability (incorporating electronic, digital, and mechanical alternatives) in an aircraft is an example where there are alternate paths in case of a failure in any one! At the detailed piece-part level, redundancy may be incorporated to improve the reliability of critical functions, particularly in areas where the accomplishment of maintenance is not feasible. For example, in the design of many electronic circuit boards, redundancy is often built-in for the purposes of improving reliability, while the accomplishment of maintenance is not practical.

The application of redundancy in design is a key area for evaluation. Although redundancy per se does improve reliability, the incorporation of extra components in the design requires additional space, and the costs are higher. This leads to a number of questions: Is redundancy really required in terms of criticality relative to system operation and the accomplishment of the mission? At what level should redundancy be incorporated? What type of redundancy should be considered ("active" or "standby")? Should maintainability provisions be considered? Are there any alternative methods for improving reliability (e.g., improved part selection, part derating)? In essence, there are many interesting and related concerns that require further investigation.

Given an introduction to series and parallel networks, the next step is to combine these, as shown in Figure 3.9(D). The reliability expression for this network can be derived by applying Equations (3.6), (3.8), and (3.9). Thus,

$$R_s = (R_A)[1 - (1 - R_B)(1 - R_C)(1 - R_D)][R_E + R_F - (R_E)(R_F)](R_G) \qquad (3.11)$$

In evaluating combined series-parallel networks, like the one illustrated in Figure 3.9D, the analyst should first evaluate the parallel redundant elements to obtain the unit reliability, and then combine the unit(s) with other elements of the system in a series format. Overall system reliability is determined by calculating the product of all series reliabilities.

Through various applications of series-parallel networks, a system reliability block diagram can be developed for use in reliability allocation, modeling and analyses, predictions, and so on. The reliability block diagram is derived directly from the system functional analysis described in Section 2.7 (refer to Figures 2.11 to 2.16), and is expanded downward, as illustrated in Figure 3.11. Figure 3.12 shows an expanded block diagram. The block diagram describes the system reliability in terms of various component relationships.

The material presented in this section is obviously not intended to be a comprehensive text on the subject of reliability, but enough information is included to provide the reader with some overall knowledge of key terms, definitions, and the prime objectives associated with the discipline. Basically, the subject of reliability is being presented as one of the many disciplines requiring consideration within the overall context of system engineering. A general familiarization of the subject area is necessary, as well as an understanding of some of the activities that are usually undertaken in the performance of a typical reliability program. By having covered some key terms and definitions, it is now appropriate to describe related program activities.[9]

When implementing a reliability program for a typical large-scale system, the tasks identified in Figure 3.13 are generally applicable. Although there are variations from one program to the next, the performance of these tasks in terms of overall program phasing is assumed to be in accordance with Figure 3.14. The major program phases, and system-level activities, are derived from the baseline presented in Figure 1.12 (Chapter 1).

From Figure 3.13, the reliability tasks listed can be categorized under three basic areas: (1) program planning, management, and control (Tasks 1–5); (2) design and analysis (Tasks 6–16); and (3) test and evaluation (Tasks 17–20). The first category of tasks must be closely integrated with system engineering activities and reflected in the System Engineering Management Plan (SEMP). The second group of tasks constitutes tools used in support of the mainstream design engineering effort, in response to the reliability requirements included in the system specification and the program plan. The third group, involving reliability testing, must be integrated with system-level testing activities and covered in the test and evaluation master plan (TEMP). Although these tasks are primarily in response to reliability program requirements, there are many interfaces with basic design functions and with other supporting disciplines such as maintainability and logistic support.

Although brief task descriptions are included in Figure 3.13, some additional comments covering a select few are noted for the purposes of emphasis.

[9] Although specific reliability tasks should be tailored to the system and the associated program needs, the tasks listed in Figure 3.13 are assumed as being typical for the purposes of discussion.

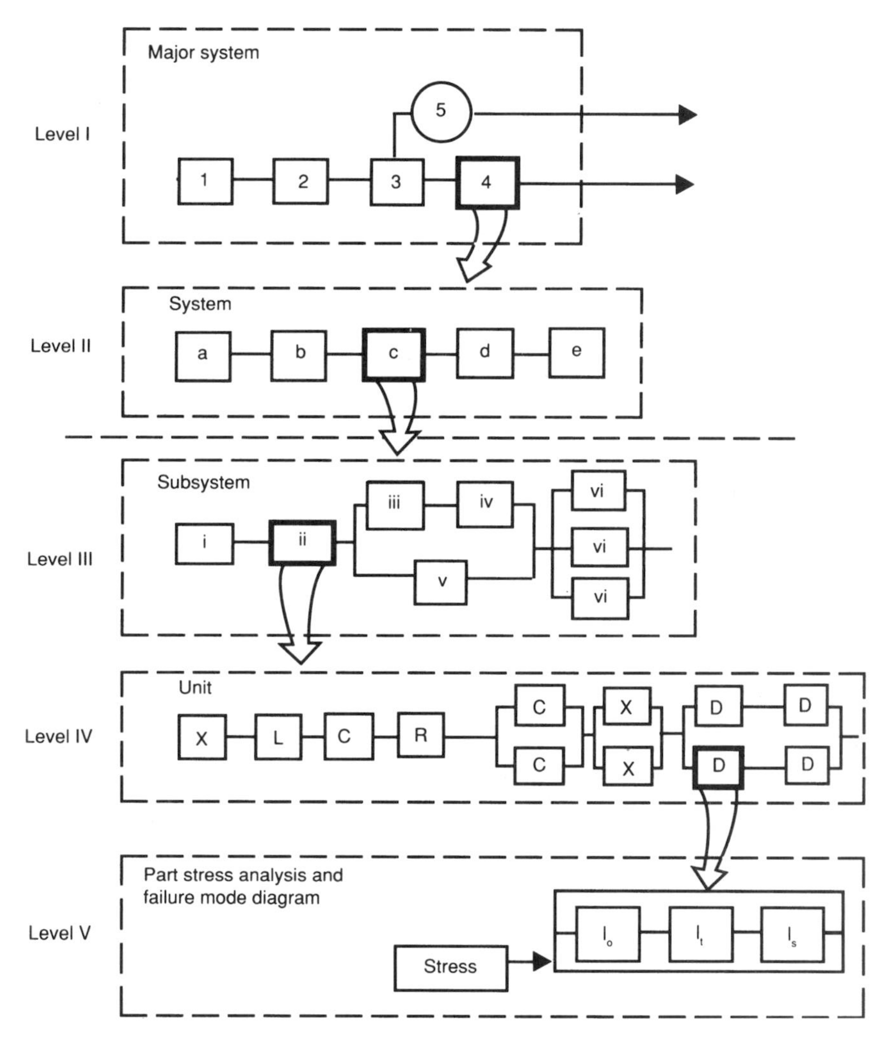

Figure 3.11 Progressive expansion of the reliability block diagram. (*Source:* MIL-HDBK-338, Military Handbook, *Electronic Reliability Design Hanbook,* Department of Defense, Washington, DC.)

1. *Reliability program plan:* Although the requirements for a reliability program may specify a separate and independent effort, it is *essential* that the program plan be developed as part of, or in conjunction with, the System Engineering Management Plan (SEMP). Organizational interfaces, task inputs-outputs, schedules, and so on, must be directly supportive of system engineering activities. Also, reliability activities must be closely integrated with maintainability and logistic support functions, and must be included in the respective plans for these program areas (which also should be tied directly into the SEMP). The SEMP is introduced in Section 1.3 (refer to Figure 1.21), and is described further in Chapter 6.

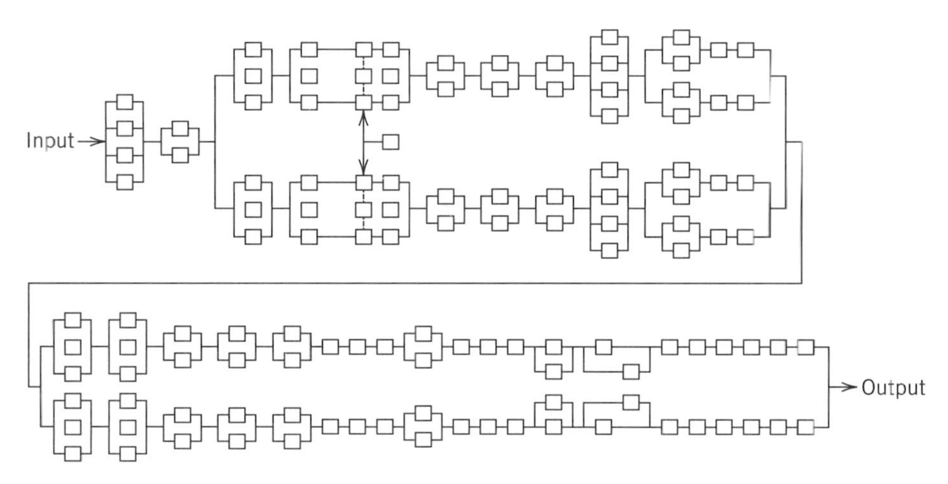

Figure 3.12 Expanded reliability block diagram of system.

2. *Reliability modeling:* This task, along with several others (e.g., allocation, prediction, stress/strength analysis, tolerance analysis), depends on the development of a good reliability block diagram (refer to Figure 3.12). The block diagram should evolve directly from, and support, the system functional analysis and associated functional flow diagrams (refer to Section 2.7). Further, the reliability block diagram is used for analyses and predictions, the results of which are provided as a major input to maintainability, human factors, logistics, and safety analyses. The reliability block diagram represents a major link in a long series of events, and must be developed in conjunction with these other activities.

3. *Failure mode, effect, and criticality analysis* (FMECA): The FMECA is a tool that has many different applications. Not only is it an excellent design tool for determining cause-and-effect relationships and identifying weak links, but it is useful in maintainability for the development of diagnostic routines. It is also required in the accomplishment of logistic support analysis (LSA) relative to the identification of both corrective and preventive maintenance requirements. The FMECA constitutes a major input to the reliability-centered maintenance (RCM) program. It is used to supplement both the fault-tree analysis and the hazard analysis accomplished in a system safety program. The FMECA is a critical activity, must be accomplished in a timely manner (early during preliminary design and subsequently updated on an iterative basis), and must be directly tied into these other activities. Figure 3.15 points to the application of the FMECA to a package handling system, and Case Study A.1, Appendix A, describes the FMECA process.

4. *Fault-tree analysis* (FTA): The FTA is a deductive approach involving the graphical enumeration and analysis of different ways in which a particular system failure can occur, and the probability of its occurrence. A separate fault tree may be developed for every critical failure mode, or undesired top-level event. Attention is focused on this top-level event and the first-tier causes associated with it. Each of these causes is next investigated for its causes, and so on. The FTA is narrower in

Program Task	Task Description and Application
1. Reliability Program Plan	To develop a reliability program that identifies, integrates, and assists in the implementation of all management tasks applicable in fulfilling reliability program requirements. This plan includes a description of the reliability organization, organizational interfaces, a listing of tasks, task schedules and milestones, applicable policies and procedures, and projected resource requirements. This plan must tie directly into the System Engineering Management Plan (SEMP).
2. Review and control of suppliers of subcontractors	To establish initial reliability requirements and to accomplish the necessary program review, evaluation, feedback, and control of component supplier/subcontractor program activities. Supplier program plans are developed in response to the requirements of the overall Reliability Program Plan for the system.
3. Reliability program reviews	To conduct periodic program and design reviews at designated milestones; e.g., conceptual design review, system design reviews, equipment/software design reviews, and critical design review. The objective is to ensure that reliability requirements will be achieved.
4. Failure reporting, analysis, and corrective-action system (FRACAS)	To establish a closed-loop failure reporting system, procedures for analysis and for determining the cause(s) of failures, and documentation for recording the corrective action initiated.
5. Failure review board (FRB)	To establish a formal review board to review significant or critical failures, failure trends, corrective-action status, and to assure that adequate actions are being taken in a timely manner to resolve any outstanding problems.
6. Reliability modeling	To develop a reliability model for making initial numerical allocations, and for subsequent estimates to evaluate system/component reliability. As design progresses, a reliability block diagram is developed and used as a basis for accomplishing periodic reliability predictions. The reliability block diagram should evolve directly from the system functional flow block diagram.
7. Reliability allocation	To allocate, or apportion, top system-level requirements to lower indenture levels of the system (e.g., subsystem, unit, assembly). This is accoplished to the depth necessary to provide specific criteria as an input to design.
8. Reliability prediction	To estimate the reliability of a system (or components thereof) based on a given design configuration. This is accomplished periodically throughout the system design and development processs to determine whether the initially specified system requiremtns are likely to be met given the proposed design at that time.
9. Failure mode, effect, and criticality analysis (FMECA)	To identify potential design weaknesses through a systematic analysis approach considering all possible ways in which a component can fail (the modes of failure), the possible causes for each failure, the likely frequency of occurance, the criticality of failure, the effects of each failure on system operation (and on various system components), and any corrective action that should be initiated to prevent (or reduce the probability of) the potential problem from occuring in the future. Refer to Case Study A.1, Appendix A.
10. Fault-tree analysis (FTA)	To determine system design weaknesses using a deductive approach involving the graphical enumeration and analysis of different ways in which a system failure can occur. Refer to Case Study A.2, Appendix A.
11. Reliability-centered maintenance (RCM)	To identify alternatives and determine the best overall program for preventive maintenance using life-cycle criteria. Refer to Case Study A.3, Appendix A.
12. Sneak circuit analysis (SCA)	To identify possible latent paths that could cause the occurance of unwanted functions, assuming that all components are functioning properly in the beginning.
13. Electronic parts/circuits tolerance analysis	To examine the effects of parts/circuits electrical tolerances, specified over a range of operations (performance, temperature, etc.), on system reliability. The objective is to asses part drift characteristics, possible tolerance buildup, and identify design weaknesses.
14. Parts program	To establish a procedure for controlling the selection and use of standard and nonstandard parts.
15. Reliability critical items	To identify components requiring "special attention" because of their complexity, their relatively short life, and/or their use in new state-of-the-art technology application. Critical items usually require special maintenance/logistic support provisions.
16. Effects of testing, storage, handling, packaging, transportation, and maintenance	To determine the effects of these activities (i.e., handling, transportation, etc.) on system, or component, reliability.
17. Environmental stress screening	To plan and implement a program where the system (or components thereof) is tested using various environmental stresses; e.g., thermal or temperature cycling, vibration and shock, burn-in, X-ray, etc. The objective is to stimulate potential relevant failures early in the life cycle.
18. Reliability development/ growth testing	To plan and implement a "test-analyze-and-fix" procedure whereby system/component weaknesses can be identified, modifications can be incorporated, and reliability growth can be realized as the system development process evolves. This is an iterative activity, and involves performance testing, environmental testing, accelerated testing, and so on.
19. Reliability qualification test	To plan and implement a program where sequential testing is accomplished, using a preproduction prototype and considering statistical "accept" and "reject" criteria, to measure the relability MTBF of the system. This occurs prior to entering production.
20. Production reliability acceptance test	To plan and implement a program where testing is accoplished, on a sampling basis, throughout the production process to ensure that degradation has not occured as a result of that process

Figure 3.13 Reliability engineering program tasks.

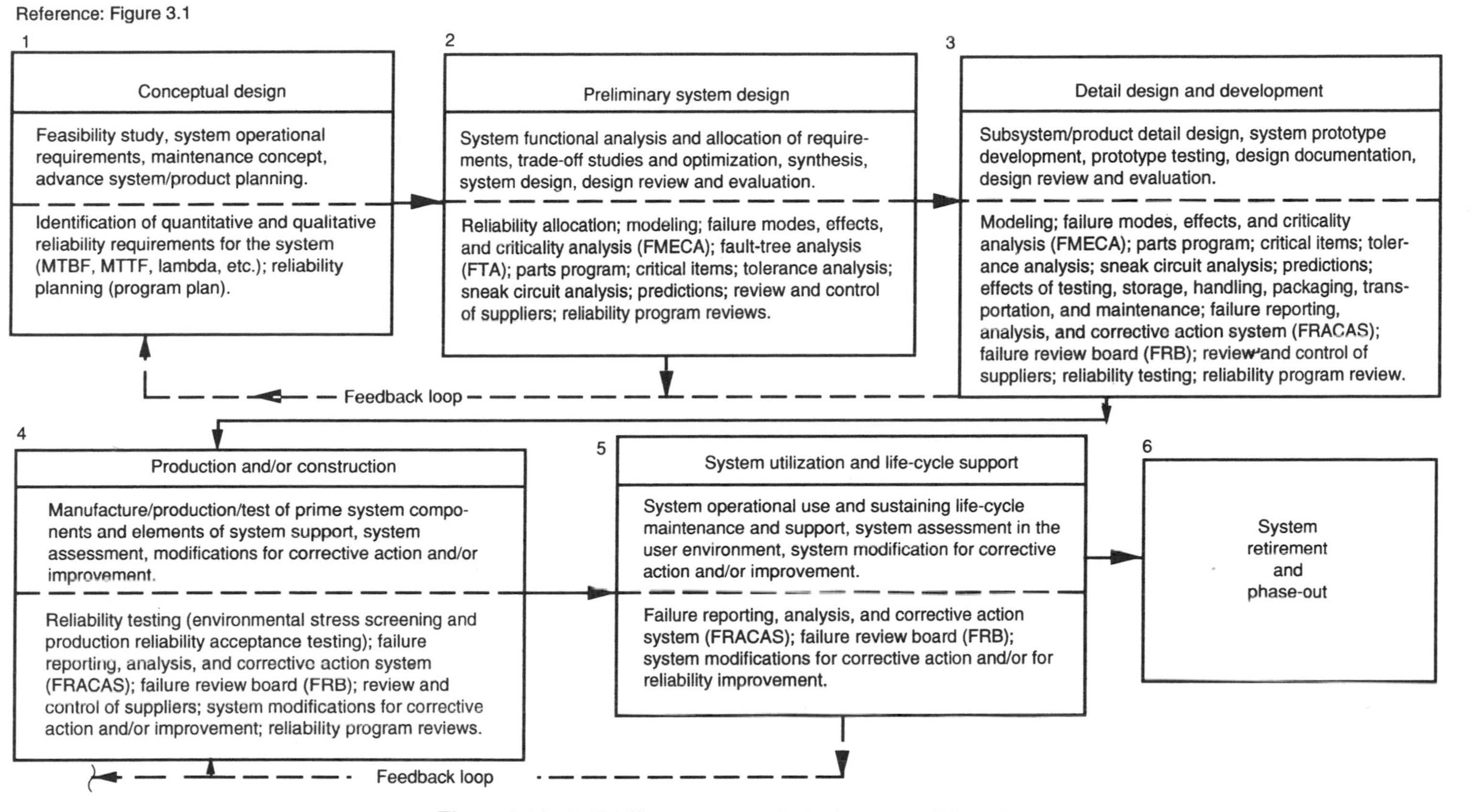

Reference: Figure 3.1
1
Conceptual design
Feasibility study, system operational requirements, maintenance concept, advance system/product planning.
Identification of quantitative and qualitative reliability requirements for the system (MTBF, MTTF, lambda, etc.); reliability planning (program plan).
2
Preliminary system design
System functional analysis and allocation of requirements, trade-off studies and optimization, synthesis, system design, design review and evaluation.
Reliability allocation; modeling; failure modes, effects, and criticality analysis (FMECA); fault-tree analysis (FTA); parts program; critical items; tolerance analysis; sneak circuit analysis; predictions; review and control of suppliers; reliability program reviews.
3
Detail design and development
Subsystem/product detail design, system prototype development, prototype testing, design documentation, design review and evaluation.
Modeling; failure modes, effects, and criticality analysis (FMECA); parts program; critical items; tolerance analysis; sneak circuit analysis; predictions; effects of testing, storage, handling, packaging, transportation, and maintenance; failure reporting, analysis, and corrective action system (FRACAS); failure review board (FRB); review and control of suppliers; reliability testing; reliability program review.
Feedback loop
4
Production and/or construction
Manufacture/production/test of prime system components and elements of system support, system assessment, modifications for corrective action and/or improvement.
Reliability testing (environmental stress screening and production reliability acceptance testing); failure reporting, analysis, and corrective action system (FRACAS); failure review board (FRB); review and control of suppliers; system modifications for corrective action and/or improvement; reliability program reviews.
5
System utilization and life-cycle support
System operational use and sustaining life-cycle maintenance and support, system assessment in the user environment, system modification for corrective action and/or improvement.
Failure reporting, analysis, and corrective action system (FRACAS); failure review board (FRB); system modifications for corrective action and/or for reliability improvement.
6
System retirement and phase-out
Feedback loop

Figure 3.14 Reliability program tasks in the system life cycle.

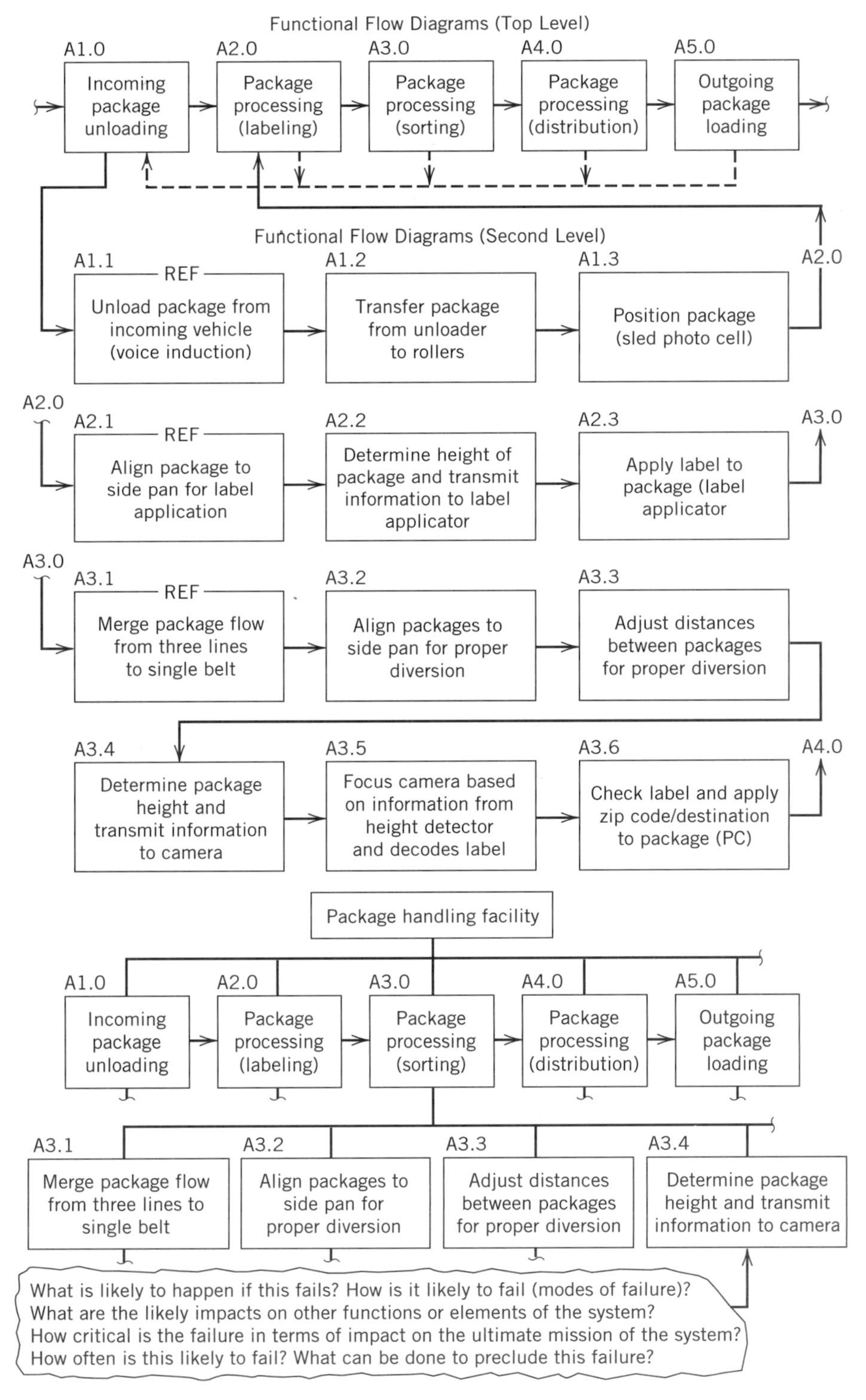

Figure 3.15 Application of FMECA to a package handling system.

focus than the FMECA and does not require as much input data. Case Study A.2, Appendix A, describes the FTA and its application.

5. *Reliability-centered maintenance* (RCM) analysis: The RCM analysis includes an evaluation of the system/process, in terms of the life cycle, to determine the best overall program for preventive (scheduled) maintenance. Emphasis is on the establishment of a cost-effective preventive maintenance program based on reliability information derived from the FMECA (i.e., failure modes, effects, frequency, criticality, and compensation through preventive maintenance). Case Study A.3, Appendix A, describes the RCM analysis and its application.

6. *Failure reporting, analysis, and corrective-action system* (FRACAS): Although this is identified as a reliability program task designed to address recommendations for corrective action as a result of catastrophic failures, the overall task objective relates closely with the system engineering feedback and control loop. Often, as problems arise and corrective action is initiated, the events that take place and the results are not adequately documented. Although it is important to respond to the "short-term" needs (i.e., correct outstanding problems in an expeditious manner), it is also important to provide some "long-term memory" through good reporting and documentation. This task should be tied directly with the system engineering reporting, feedback, and control process.

7. *Reliability qualification testing:* This task, usually accomplished as part of Type 2 testing, should be defined in the context of the *total* system test and evaluation effort (refer to Figure 2.31). The specific requirements will depend on the system complexity, the degree of design definition, the nature of the mission that the system is expected to accomplish, and the TPMs (and their priorities) established for the system. Additionally, for this and any other individual test, there are certain expectations and opportunities for gathering information. For instance, the objective of environmental qualification testing is to determine whether the system will perform in a specified environment. In accomplishing this test, it may be possible to gather some reliability information by observing system operating times, failures, and so on. This, in turn, may permit a reduction in subsequent reliability testing. A second example pertains to the gathering of maintainability data during the performance of formal reliability testing. As failures occur during the test, maintenance actions can be evaluated in terms of elapsed times and resource requirements. This, in turn, may allow for some reductions in both maintainability and supportability test and evaluation efforts. In other words, there are numerous possibilities for reducing costs (while still gathering the necessary information) through the accomplishment of an integrated testing approach. Thus, reliability testing must be viewed in the context of the overall system test effort, and the requirements for this must be covered in the TEMP.

In summary, the tasks identified in Figure 3.13 are generally accomplished in response to some detailed specification or program requirement. For many programs, these are completed on a relatively independent basis. Yet, the interfaces are many, and there are some excellent possibilities for task integration, resulting in reduced costs. As one progresses through this text, opportunities for integration are discussed further. The intent of this section is to provide an introduction to the requirements associated with most reliability programs.

3.3.2 Maintainability Engineering[10]

Maintainability is an inherent characteristic of system design that pertains to the ease, accuracy, safety, and economy in the performance of maintenance actions. It deals with component packaging, diagnostics, part standardization, accessibility, interchangeability, mounting and labeling, and so on. A system should be designed such that it can be maintained without large investments of time and resources (e.g., personnel, materials, test equipment, facilities, data), and at minimum cost, while still fulfilling its designated mission. Maintainability is the *ability* of an item to be maintained, whereas maintenance constitutes those actions taken to restore an item to (or retain an item in) a specified operating condition. Maintainability is a design parameter, whereas maintenance is the result of design.

Maintainability, defined in the broadest sense, can be measured in terms of a combination of maintenance times, personnel labor hours, maintenance frequency factors, maintenance cost, and related logistic support factors. There is no single measure that will address *all* issues. For instance, an objective may be to shorten the elapsed time for accomplishing maintenance by adding more personnel (and possibly with greater skills). Although such an action may reduce the time requirement, it may cause an increase in personnel requirements and a resultant increase in life-cycle cost. Further, it may be desirable to reduce the frequency of unscheduled maintenance by adding the requirements for more scheduled maintenance. In doing such, there may be an increase in the overall frequency of maintenance, and the life-cycle cost may increase as well. In essence, these factors (as applicable) must be addressed on a collective basis, as well as being considered in conjunction with the reliability measures discussed in Section 3.3.1.

One of the most commonly used measures of maintainability is the aspect of "time." In Figure 3.16, the overall time spectrum can be broken down into different applications. "Uptime" pertains to the elapsed time applicable to the system when in operational use, or when in a standby or ready state awaiting for use. On the other hand, "downtime" refers to the total elapsed time required, when the system is not operational, to accomplish corrective maintenance and/or preventive maintenance. These categories of maintenance are defined as follows:

1. *Corrective maintenance:* The unscheduled actions, initiated as a result of failure (or a perceived failure), that are necessary to *restore* a system to its required level of performance. Such activities may include troubleshooting, dis-

[10]The objective is to provide an introductory overview of maintainability engineering, including definitions and program requirements, and not to cover the subject in depth. However, for more information, two good references are (1) J. D. Patton, *Maintainability and Maintenance Management,* Instrument Society of America, Research Triangle Park, NC, 1980; and (2) B. S. Blanchard, D. Verma, and E. Peterson, *Maintainability: A Key to Effective Serviceability and Maintenance Management,* John Wiley, New York, 1995. Additional references are included in Appendix F.

assembly, repair, remove and replace, reassembly, alignment and adjustment, checkout, and so on. Additionally, this includes all software maintenance that is not initially planned; e.g., *adaptive* maintenance, *perfective* maintenance, and so on.

2. *Preventive maintenance:* The scheduled actions necessary to *retain* a system at a specified level of performance. This may include periodic inspections, servicing, calibration, condition monitoring, and/or the replacement of designated critical items.

In Figure 3.16, total maintenance downtime (MDT) is the elapsed time required to repair and restore a system to full operational status, and/or to retain a system in that condition. MDT can be broken down into the following components:

1. *Active maintenance time* ($\bar{M}$): That portion of downtime when corrective and/or preventive maintenance activities are being accomplished. This factor is often expressed as

$$\bar{M} = \frac{(\lambda)(\bar{M}ct) + (fpt)(\bar{M}pt)}{\lambda + fpt} \tag{3.12}$$

   where $\bar{M}$ is the mean active maintenance time, $\bar{M}ct$ is the mean corrective maintenance of time, $\bar{M}pt$ is the mean preventive maintenance time, fpt is the frequency of preventive maintenance, and λ is the failure rate (or frequency of corrective maintenance).

2. *Logistics delay time* (LDT): That portion of downtime when the system is not operational because of delays associated with the support capability; for example, waiting for a spare part, waiting for the availability of test equipment, waiting for the use of a special facility.
3. *Administrative delay time* (ADT): That portion of downtime when the necessary maintenance is delayed for reasons of an administrative nature; for example, the unavailability of personnel because of other priorities, organizational constraints, labor strikes.

When looking at these elements of downtime from the design engineer's perspective, it is quite common to address only the *active* maintenance segment (i.e., $\bar{M}$). This is because of being able to directly relate system characteristics such as diagnostic capability, accessibility, and interchangeability to downtime. The producer (i.e., contractor) is responsible for, and usually can control, this element, whereas the LDT and ADT factors are primarily influenced by the consumer (i.e., customer). From the perspective of system engineering, one needs to deal with the *entire* downtime spectrum. There is little point in constraining the design of prime equipment (i.e., an item must be designed such that it can be repaired in 30 minutes) if the support capability is such that it takes 3 months to acquire the necessary spare part. In essence, the entire spectrum must be considered as reflected in Figure 2.4, and each of these time elements represents an important measure.

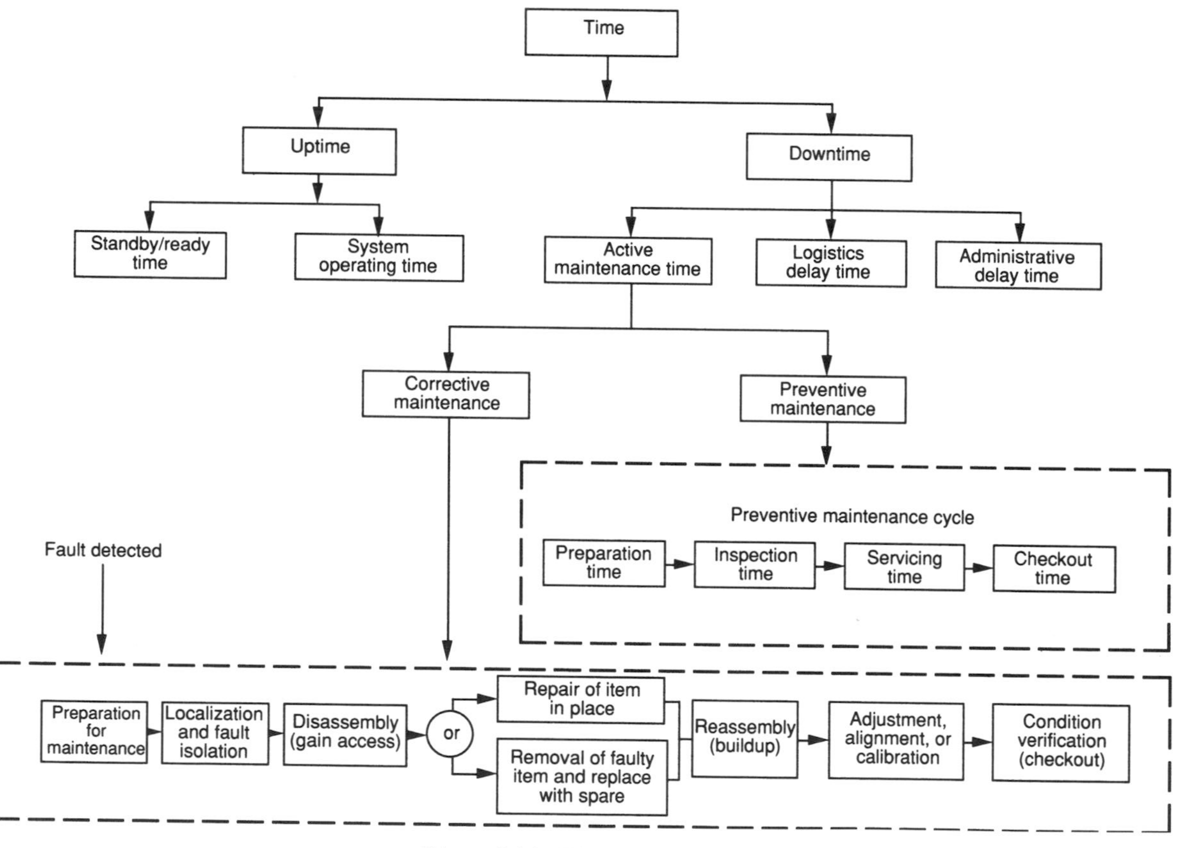

Figure 3.16 Time relationships.

By referring to the time relationships presented in Figure 3.16, as well as the factors in Equation (3.12), active maintenance time ($\bar{M}$) can be broken down into corrective maintenance and preventive maintenance times. The mean corrective maintenance time ($\bar{M}ct$) is expressed as

$$\bar{M}ct = \frac{\Sigma(\lambda_i)Mct_i}{\Sigma(\lambda_i)} \tag{3.13}$$

where Mct_i represents the time that it takes to progress through the corrective maintenance cycle illustrated in Figure 3.16 (for the ith item), and λ_i is the corresponding failure rate. In the event of a fixed number of maintenance actions, n, then

$$\bar{M}ct = \frac{\sum_{i=1}^{n} Mct_i}{n} \tag{3.14}$$

$\bar{M}ct$, which is a weighted average of repair times using reliability factors, is equivalent to the mean time to repair (MTTR), a measure that is commonly used for maintainability.

The time-dependency relationship between the probability of corrective maintenance and the time allocated for accomplishing corrective maintenance can be expected to produce a probability density function in one of three common forms, as illustrated in Figure 3.17(A):

1. *The normal distribution:* applies to relatively simple and common maintenance actions where times are fixed with very little variation.
2. *The exponential distribution:* applies to maintenance actions involving part substitution methods of fault isolation in large systems that result in a constant failure rate.
3. *The log-normal distribution:* applies to most maintenance actions involving detailed tasks with unequal frequency and time durations.

Experience has indicated that in most instances, the distribution of maintenance times for complex systems follows the log-normal approximation. From Figure 3.17(B), the key maintainability parameters are the mean time to repair (Point 1), the median time to repair (Point 2), and the maximum time to repair (Point 3). Whereas the "mean" value constitutes the measure that is most commonly used, the median and maximum time values are appropriate measures used in certain applications.

The median active corrective maintenance time ($\tilde{M}ct$) is that value that divides all of the repair-time values such that 50% are less than the median and 50% are greater than the median. For the normal distribution, the median is the same as the mean, and the median in the log-normal distribution is the same as the geometric mean ($MTTR_g$) illustrated in Figure 3.17(B). The median, represented by Point 2, is calculated as

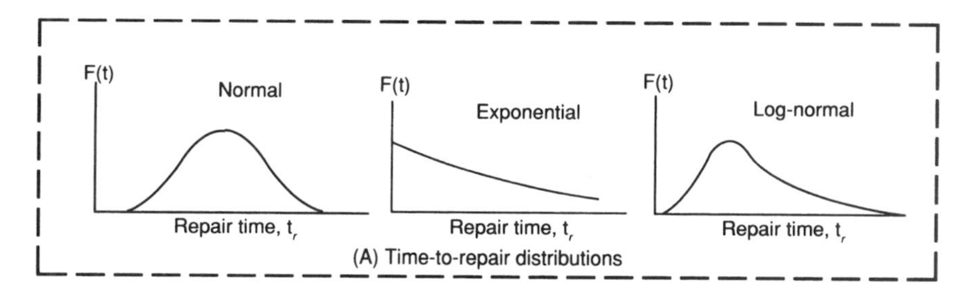

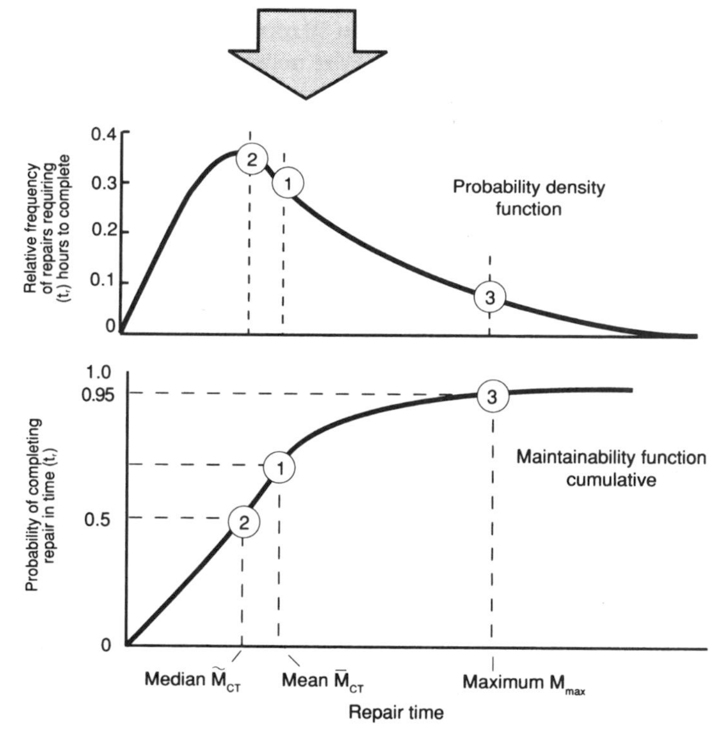

Figure 3.17 Maintainability distributions.

$$\tilde{M}ct = \text{antilog} \frac{\Sigma(\lambda_i)(\log Mct_i)}{\Sigma(\lambda_i)} = \text{antilog} \frac{\sum_{i=1}^{n} \log Mct_i}{n} \tag{3.15}$$

The maximum active corrective maintenance time (M_{max}) can be defined as that value of downtime below which a designated percent of all maintenance actions can be expected to be completed. This is represented by Point 3 in Figure 3.17(B). Selected points, in the log-normal distribution, at the 90th or 95th percentile are generally used. The maximum corrective maintenance time is expressed as

$$\mathrm{M_{max}} = \text{antilog } [\overline{\log \mathrm{Mct}} + Z\sigma_{\log}\mathrm{Mct}_i] \tag{3.16}$$

where $\overline{\log \mathrm{Mct}}$ is the mean of the logarithms of Mct_i, Z is the standard variate at the point where $\mathrm{M_{max}}$ is defined (1.65 at 95%, 1.28 at 90%, 1.04 at 85%, and so on; refer to the normal distribution tables in any text on statistics), and σ is the standard deviation of the sample logarithms of average repair times, Mct_i.

In the area of preventive maintenance, both the mean and the median measures are used. The mean preventive maintenance time ($\bar{\mathrm{M}}\mathrm{pt}$) can be determined by

$$\bar{\mathrm{M}}\mathrm{pt} = \frac{\Sigma(\mathrm{fpt})(\mathrm{Mpt}_i)}{\Sigma(\mathrm{fpt}_i)} = \frac{\sum_{i=1}^{n} \mathrm{Mpt}_i}{n} \tag{3.17}$$

where fpt_i is the frequency of the individual (ith) preventive maintenance action, and Mpt_i is the associated elapsed time to perform the preventive maintenance required.

The median value for preventive maintenance, like the requirement for corrective maintenance specified in Equation (3.15), is determined from

$$\tilde{\mathrm{M}}\mathrm{pt} = \text{antilog } \frac{\Sigma(\mathrm{fpt}_i)(\log \mathrm{Mpt}_i)}{\Sigma(\mathrm{fpt}_i)} \tag{3.18}$$

Preventive maintenance may be accomplished while the system is in full operation, or the requirements for such could result in downtime. In this instance (and in the case for corrective maintenance), only those actions that are accomplished and result in downtime are considered. Maintenance actions that do not result in system downtime are basically accounted for through the personnel labor-hour and maintenance cost measures of maintainability.[11]

Although the various measures of elapsed time are extremely important, one must also consider the maintenance labor hours expended in the process. When dealing with the ease and economy in the performance of maintenance, an objective is to obtain the proper balance between elapsed time, labor hours, and personnel skills at minimum maintenance cost. Personnel time may be expressed in terms of maintenance labor hours per system operating hour (MLH/OH), maintenance labor hours per cycle of system operation (MLH/cycle), maintenance labor hours per maintenance action (MLH/MA), or maintenance labor hours per month (MLH/month). Any of these factors can be presented in terms of mean values, such as mean corrective maintenance labor hours ($\overline{\mathrm{MLH}}_c$), which can be expressed as

$$\overline{\mathrm{MLH}}_c = \frac{\Sigma(\lambda_i)(\mathrm{MLH}_i)}{\Sigma(\lambda_i)} \tag{3.19}$$

[11] Although maintainability has already been defined in the broadest context, there are additional definitions that relate to a specific measure. With regard to "time," it can be defined as "the measure of the ability of an item to be retained in or restored to a specified condition when maintenance is performed by personnel having specified skills, using prescribed procedures and resources, at each prescribed level of maintenance and repair."

where λ_i is the failure rate of the ith item, and MLH_i maintenance labor hours necessary to accomplish the related corrective maintenance actions.

Having covered the aspect of corrective maintenance, the values for mean preventive maintenance labor hours and mean total maintenance labor hours (to include *all* corrective and preventive maintenance actions) can be determined in a similar manner. These factors, predicted for each level of maintenance identified in the system maintenance concept, can be utilized in determining specific maintenance and logistic support requirements and associated costs.

A third measure of maintainability (in addition to the time and labor-hour factors) is maintenance frequency. From Section 3.3.1, the frequency factors associated with primary and secondary failures are basically reflected through the reliability MTBF and λ measures. These measures are certainly important for determining the overall frequency of unscheduled maintenance; however, there are additional considerations such as manufacturing defects, operator-induced failures, maintenance-induced failures, and defects due to handling that may be relevant (refer to footnote 8 in this chapter). Also, one must consider the aspect of preventive maintenance. With this in mind, it is appropriate to look at the total spectrum of maintenance and the measure of mean time between maintenance (MTBM). This can be calculated as

$$\mathrm{MTBM} = \frac{1}{1/\mathrm{MTBM_u} + 1/\mathrm{MTBM_s}} \tag{3.20}$$

where $\mathrm{MTBM_u}$ is the mean interval of unscheduled (or corrective) maintenance, and $\mathrm{MTBM_s}$ is the mean interval of scheduled (preventive) maintenance. The reciprocals of $\mathrm{MTBM_u}$ and $\mathrm{MTBM_s}$ are equivalent to the maintenance rates, or the maintenance actions per hour of system operation. $\mathrm{MTBM_u}$ should be equivalent to MTBF assuming that the possibilities of operator-induced defects, maintenance-induced defects, and so on, have been "designed out" of the system.

Within the overall spectrum of activity represented by the MTBM factor, there are some maintenance actions that result in the removal and replacement of components, and the requirement for spare parts. These actions, in response to both corrective and preventive maintenance requirements, can be measured in terms of mean time between replacement (MTBR), a factor of MTBM. In essence, the MTBM factor reflects *all* maintenance actions, some of which result in item replacements.

Figure 3.18 shows a given system where there were 100 unscheduled maintenance actions recorded over a specific segment of time. In all instances, some organizational-level maintenance was accomplished relative to diagnostics and checkout. In 25 cases, it was impossible to verify that a problem existed as the system appeared to be operating properly when checked. Therefore, no items were removed and replaced. In the other 75 instances, a given component was suspect, resulting in a removal and replacement action. Of the components removed for higher-level maintenance (i.e., intermediate level), a problem was verified in 45 instances and repair was accomplished on-site, 12 components were sent to the factory for higher-level repair, 3 components were condemned (determined to be beyond economic repair), and there were 15 components where no defect was noted.

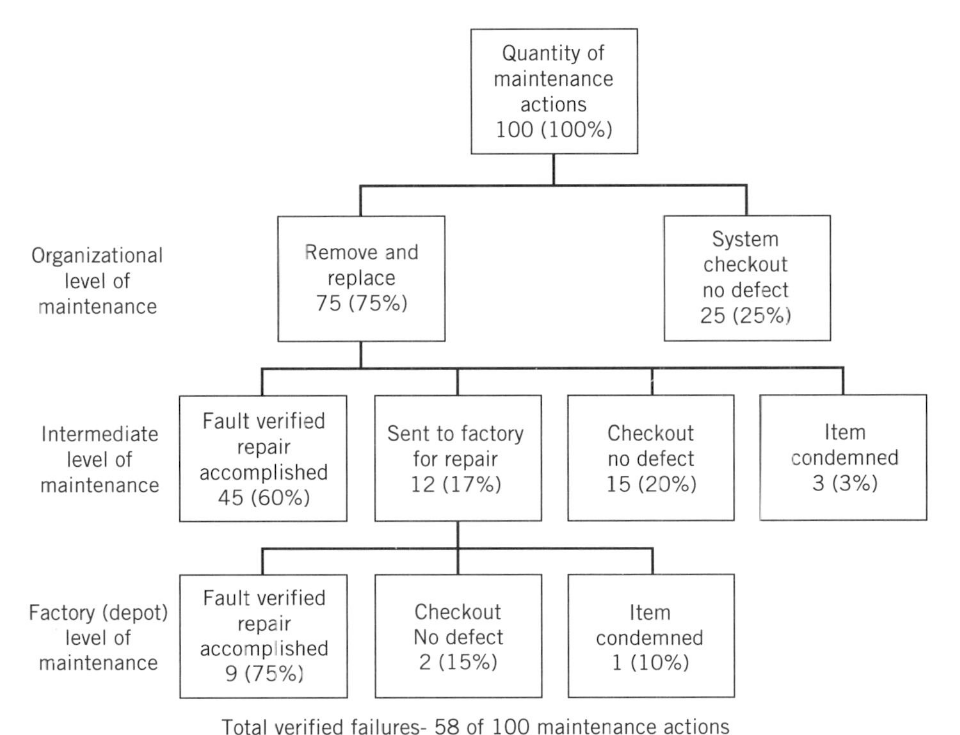

Figure 3.18 System XYZ unscheduled maintenance actions.

Of the 12 components sent to the factory, 10 were considered as being faulty. Through a review of these factors, it can be seen that the MTBM figure must consider all of the 100 maintenance actions, the MTBR figure can be related to the 75 replacements (at the organizational level), and the MTBF measure (as defined in a puristic reliability sense) pertains to the 58 components where actual catastrophic failures were confirmed. From, a *systems* perspective, however, there were 100 failures in total, whether they be charged to an element of equipment, a module of software, or to a human being.

Given the definitions associated with MTBM, MTBR, MTBF, MDT, $\bar{M}ct$, $\bar{M}pt$, $\bar{M}$, and so on, it is important to relate some of these figures of merit to a higher-order system parameter. From Figure 2.24, reliability and maintainability factors are, for example, key inputs in determining system availability which, in turn, is a major element of system effectiveness. Although the specific measures may vary significantly from one system application to the next, the term "availability" is used quite often as a system measure. Availability can be expressed as follows:

$$A_o = \frac{\text{MTBM}}{\text{MTBM} + \text{MDT}} = \frac{\text{uptime}}{\text{uptime} + \text{downtime}} \tag{3.21}$$

where A_o is operational availability. This definition of availability relates to the consumer's operational environment where MTBM reflects *all* maintenance requirements and MDT represents *all* downtime considerations. In instances where a producer is responsible for designing a system to meet a certain availability requirement, and where the producer has no influence or control of the consumer's support structure, it may be appropriate to define availability as

$$A_a = \frac{\text{MTBM}}{\text{MTBM} + \bar{\text{M}}} \tag{3.22}$$

where A_a is achieved availability. It should be noted that the LDT and ADT factors are not considered here. Progressing one step further, there are instances in which availability is defined as

$$A_i = \frac{\text{MTBF}}{\text{MTBF} + \bar{\text{M}}\text{ct}} \tag{3.23}$$

where A_i represents inherent availability. Note that preventive maintenance is not included here. Employing this figure of merit as a system measure may be appropriate from a contractual standpoint where the producer is somewhat isolated from the consumer environment. However, in dealing with system engineering requirements, the A_o factor is more relevant than either the A_a or A_i factors.

Figures 1.3 and 2.24 show two sides of the balance. The reliability and maintainability factors described herein are significant contributors (along with performance) in measuring the *technical* effectiveness of the system. Reliability and maintainability parameters are combined to determine availability, and system availability constitutes a major input in determining system effectiveness. At the other end of the balance is life-cycle cost (LCC). LCC is a function of research-and-development cost, production/construction cost, operation and support cost, and retirement and disposal cost. The consequences of reliability and maintainability have a direct impact on each of these major cost categories. However, the greatest impact of these design characteristics is on operational and support costs, where the frequency of maintenance and downtime factors are significant in determining the overall support capability for the system. If these characteristics are not appropriately considered in system design, the "iceberg" effect illustrated in Figure 1.4 will likely prevail!

The material presented to this point is intended to provide a familiarization with the terms and definitions associated with maintainability. Maintainability is one of the many disciplines requiring consideration within the overall context of system engineering. A general understanding of the subject is necessary, as well as some familiarity with the activities that are usually undertaken in the performance of a typical maintainability program. Having covered some key terms and definitions, it is now appropriate to describe related program activities.[12]

[12] Although specific maintainability tasks should be tailored to the system and associated program needs, the tasks listed in Figure 3.19 are assumed as being typical for the purposes of discussion.

When implementing a maintainability program for a typical large-scale system, the tasks identified in Figure 3.19 are generally applicable. Although there are variations from one situation to the next, the performance of these tasks in terms of overall program phasing is assumed to be in accordance with Figure 3.20. The major program phases, and system-level activities, are derived from the baseline presented in Figure 1.12 (Chapter 1).

In Figure 3.19, the maintainability program tasks listed can be categorized under (1) program planning, management, and control (Tasks 1–4); (2) design and analysis (Tasks 5–12); and (3) test and evaluation (Task 13). The first category of tasks must be closely integrated with system engineering activities and reflected in the SEMP. The second group of tasks constitutes tools used in support of the mainstream design engineering effort, in response to maintainability program requirements included in the system specification and the program plan. The third area of activity, maintainability demonstration, must be integrated with system-level testing activities and covered in the TEMP. Although these tasks are primarily in response to maintainability program requirements, there are many interfaces with basic design functions and with other supporting disciplines such as reliability and logistic support.

Although brief task descriptions are included in Figure 3.19, some additional comments, as they pertain to a select few, are noted for the purposes of emphasis.

1. *Maintainability program plan:* Although the requirements for a maintainability program may specify a separate and independent effort, it is *essential* that the program plan be developed as part of, or in conjunction with, both the Reliability Program Plan (refer to Figure 3.13 Task 1) and the SEMP. Organizational interfaces, task input-output requirements, schedules, and so on, must be integrated with reliability program requirements and must be directly supportive of system engineering activities. Also, maintainability activities must be closely integrated with human factors and logistic support functions, and must be included in the respective plans for these program areas. The SEMP is introduced in Section 1.3 (refer to Figure 1.21), and is described further in Chapter 6.

2. *Maintainability modeling:* The completion of this task, along with several others (e.g., allocation, prediction, FMECA, maintainability analysis), depends on the development of functional-level diagrams, similar to the one presented in Figure 3.21. These diagrams should evolve directly from, and must support, the system functional analysis and associated functional flow diagrams described in Section 2.7 (refer to Figures 2.11 to 2.16). The objective is to illustrate system packaging concepts, diagnostic capabilities (depths of localization and fault isolation), items that are repaired in place or removed for maintenance, and so on. The results of this task constitute a major input to the maintenance task analysis (MTA) and the logistic support analysis (LSA) and must be provided in a timely manner.

3. *Failure mode, effect, and criticality analysis* (FMECA): FMECA, as it applies to maintainability, is primarily used as an aid in the development of system packaging schemes and diagnostic routines, and is employed to assist in determining critical preventive maintenance requirements. This task should be closely integrated

Program Task	Task Description and Application
1. Maintainability Program Plan	To develop a maintainability program plan that identifies, integrates, and assists in the implementation of all management tasks applicable in fulfilling maintainability program requirements. This plan includes a description of the maintainability organization, organizational interfaces, a listing of tasks, task schedules and milestones, applicable policies and procedures, and projected resource requirements. This plan must tie directly into the System Engineering Management Plan (SEMP).
2. Review and control of suppliers or subcontractors	To establish initial maintainability requirements and to accomplish the necessary program review, evaluation, feedback, and control component supplier/subcontractor program activities. Supplier program plans are developed in response to the requirements of the overall Maintainability Program Plan for the system.
3. Maintainability program reviews	To conduct periodic program and design reviews at designated milestones; (e.g., conceptual design review, system design reviews, equipemnt/software design reviews, and critical design review). The objective is ensure that maintainability requirements will be achieved.
4. Data collection, analysis, and corrective-action system	To establish a closed-loop system for data collection, analysis, and the initiation of recommendations for corrective action: The objective is to identify potential maintainability design problems.
5. Maintainability modeling	To develop a maintainability model for making initial numerical allocations, and for subsequent estimates to evaluate system/component maintainability. As design progresses, maintainability top-down functional block diagrams, logic troubleshooting flow diagrams, and so on, are developed and are used as a basis for accomplishing periodic predictions, logistic support analysis, and testability analysis. These should evolve directly from the system-level maintenance functional flow block diagrams.
6. Maintainability allocation	To allocate, or apportion, top system-level requirements to lower indenture levels of the system (e.g., subsystem, unit, assembly). This is accomplished to the depth necessary to provide specific criteria as an input to design.
7. Maintainability prediction	To estimate the maintainability of a system (or components thereof) based on a given design configuration. This is accomplished periodically throughout the system design and development process to determine whether the initially specified system requirements are likely to be met given the proposed design at that time.
8. Failure mode, effect, and criticality analysis (FMECA)—maintainability information	To identify potential design weaknesses through a systematic analysis approach considering all possible ways in which a component can fail (the modes of failure), the possible causes for each failure, the likely frequency of occurance, the criticality of failure, the effects of each failure on system operation (and on various system components), and any corrective action that should be initiated to prevent (or reduce the probability of) the potential problem from occuring in the future. The objective is to determine maintainability design requirements as a result of anticipated corrective and/or preventive maintenance needs. Refer to Case Study A.1, Appendix A.
9. Maintainability analysis	To accomplish various design-related studies pertaining to equipment packaging schemes, fault-isolation and diagnostic provisions, built-in test versus external test equipment, levels of repair, component standardization, producibility considerations, and so on. Maintainability mathematical models, level-of-repair analysis models, and life-cycle cost analysis models are utilized as required.
10. Maintenance task analysis (MTA)	To evaluate design data and determine weaknesses relative to the maintainability characteristics incorporated in the design, and to determine the maintenance and support resources required for the system. Refer to Case Study A.4, Appendix A.
11. Level-of-repair analysis (LORA)	To evaluate system components to determine whether it is more economical to repair the item or to discard it in the event of failure. Refer to Case Study A.5, Appendix A.
12. Maintainability data for the detailed maintenance plan and the logistic support analysis (LSA)	To identify and prepare maintainability data as they apply to the various elements of logistic support—spare and repair parts, test and support equipment, personnel quantities and skill levels, training, facilities, technical manuals, and software.
13. Maintainability demonstration	To plan and implement a program where testing is accomplished (either sequential testing or a "fixed" sample size), using a preproduction prototype and considering statistical "accept" and "reject" criteria, to measure the maintainability characteristics of the system. These characteristics may include $\overline{Mct}$, MLH/OH, $\overline{Mpt}$, or equivalent. This test is accomplished prior to entering production.

Figure 3.19 Maintainability engineering program tasks.

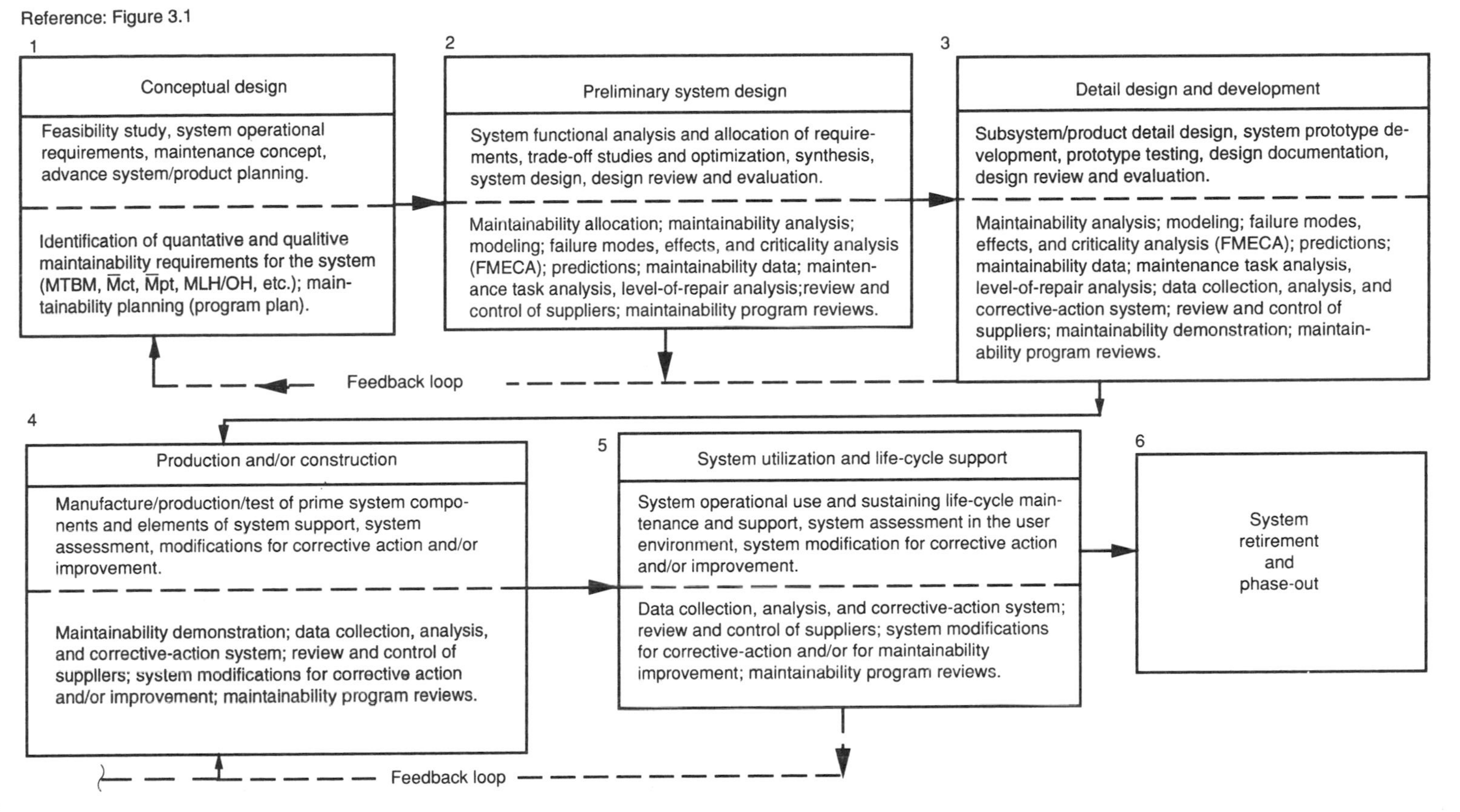

Figure 3.20 Maintainability program tasks in the system life cycle.

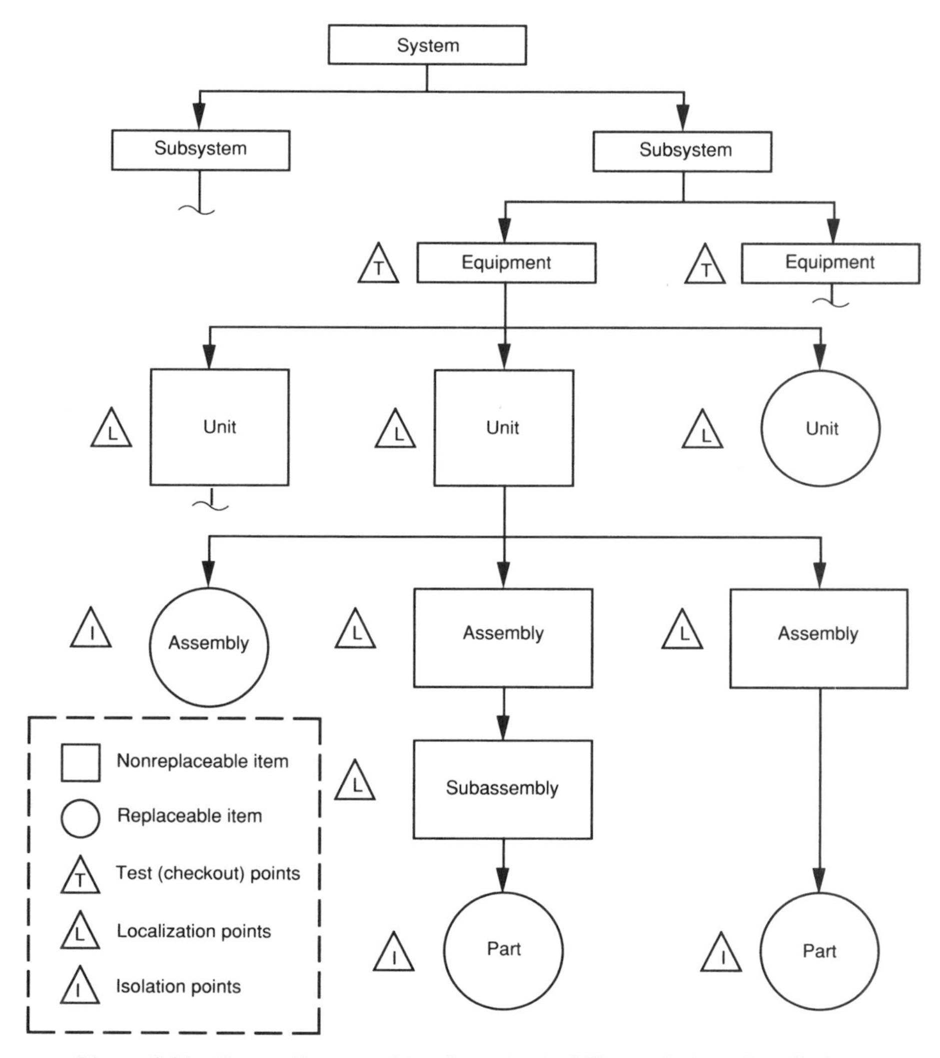

Figure 3.21 System/decomposition for maintainability analysis and prediction.

with reliability and logistics activities because the FMECA is also a required task in these program areas. Case Study A.1, Appendix A, describes the FMECA process.

4. *Maintainability analysis:* This includes the accomplishment of many different design-related studies dealing with system functional packaging concepts, levels of diagnostics, levels of repair, built-in versus external test, and so on. It must be accomplished in conjunction with the FMECA and maintainability modeling, and it must be coordinated with logistic support analysis (LSA) requirements. The LSA also requires the level-of-repair analysis and life-cycle cost analysis in fulfilling the requirements related to the design for supportability. Case Study A.6, Appendix A,

describes an evaluation of alternative design configurations accomplished in support of a maintainability analysis effort.

5. *Maintenance task analysis* (MTA): This includes a detailed analysis and evaluation of the system to (a) assess a given configuration relative to the degree of incorporation of maintainability characteristics in design and compliance with the initially specified requirements, and (b) to determine the maintenance and logistic support resources required in order to support the system throughout its planned life cycle. Such resources may include maintenance personnel quantities and skill levels, spares and repair parts and associated inventory requirements, tools and test equipment, transportation and handling requirements, facilities, technical data, computer software, and training requirements. Such an evaluation may be accomplished during the preliminary and detail design phases utilizing available design data as the source of information and/or through a review and assessment of an existing item using checklists as an aid. A MTA may be conducted on a commercial off-the-shelf (COTS) item in the event that the maintenance resource requirements have not already been identified. This task should be closely coordinated with human-factors activities (i.e., the operator task analysis and the development of operational sequence diagrams) and with logistics activities (i.e., the MTA is an integral part of the logistic support analysis effort). Case Study A.4, Appendix A, includes an abbreviated example of the results of a MTA.[13]

6. *Level-of-repair analysis* (LORA): This includes an evaluation of various system components to determine whether it is economically feasible to accomplish repair of the item or to discard it in the event of failure. If repair is to be accomplished, should the component be repaired at the intermediate level or at the factory (i.e., depot)? A LORA may be accomplished initially, in the development of the system maintenance concept, to provide design guidelines for packaging, diagnostics, and so on, and later in the evaluation of a given design configuration to determine maintenance resource requirements. The LORA should be accomplished in conjunction with the MTA and as part of the logistic support analysis effort. Case Study A.5, Appendix A, includes an example of the LORA process.

7. *Maintainability demonstration:* This task, usually accomplished as part of Type 2 testing, should be defined in the context of the *total* system test and evaluation effort. The objective of maintainability demonstration is to simulate different maintenance task sequences, record the associated maintenance times, and verify the adequacy of the resources required to support the demonstrated maintenance activities (e.g., spare/repair parts, support equipment, software, personnel quantities and skills, and data). The results from this activity should not only determine whether maintainability requirements have been met, but should help to determine whether the supportability objectives have been met in response to logistic support requirements. Maintainability demonstration requirements must be covered in the TEMP.

[13] A more in-depth presentation of the MTA, its content, and the procedure for accomplishing such is included in B. S. Blanchard, *Logistics Engineering and Management,* 4th Ed., Prentice Hall, Upper Saddle River, NJ, 1992.

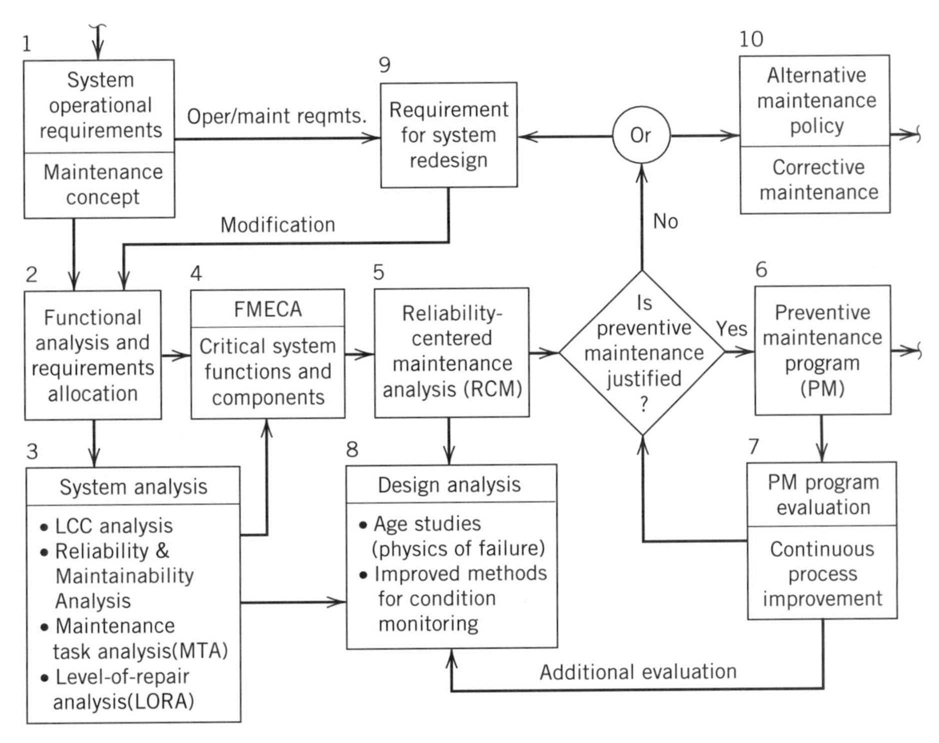

Figure 3.22 Example of the relationships between selected reliability and maintainability tools.

In summary, the tasks identified in Figure 3.19 are generally accomplished in response to some detailed specification or program requirement. Like reliability, these tasks are completed on a relatively independent basis for many programs. Yet, the interfaces are numerous, and there are some excellent opportunities for task integration, resulting in reduced program costs. Figure 3.22 conveys an example of the relationships between selected reliability and maintainability tools. As one progresses further through this text, the opportunities for integration will become even more apparent. The intent of this section is to provide an introduction to the requirements associated with most maintainability programs.

3.3.3 Human-Factors Engineering[14]

Quite often in the design and development of a system, emphasis is placed on hardware and software, and the human element tends to be ignored. For the system to

[14] The objective is to provide an introduction to human factors (or human engineering), and not to cover the subject in depth. However, for more information, three good references are (1) D. Meister, *Behavioral Analysis and Measurement Methods,* John Wiley, New York, 1985; (2) M. S. Sanders and E. J.

be complete, the human being and the interfaces between the human and the other elements of the system (e.g., equipment, software, facilities, data, elements of logistic support) must be considered. Optimum hardware, or software, design alone will not guarantee effective results.

The field of *human factors,* sometimes known as "ergonomics" or "human engineering," refers to the *design of a system or product for human use.* Considerations in design must include the following factors:

1. *Anthropometric factors:* Anthropometry deals with the measurement of the dimensions and the physical characteristics of the human body (e.g., standing height, sitting height, arm reach, breadth, buttock–knee length, hand size, and weight). When establishing basic design requirements involving the human being (for work-space application, work-surface design, control panel layout), one obviously must take into consideration the physical dimensions of the human body. Both "structural" dimensions (when the body is fixed and in a static state) and "functional" dimensions (when the body is engaged in some physical activity and in a dynamic state) must be measured and used in designing for the performance of operational functions and maintenance functions. Further, the design engineer must consider both *male* and *female* dimensions, with the appropriate ranges of variability (usually from the 5th to the 95th percentiles). For instance, the height of a male may range from 63.6 inches (5th percentile), 68.3 inches (50th percentile), and 72.8 inches (95th percentile), and the height of the female is 59.0 inches (5th percentile), 62.9 inches (50th percentile), and 67.1 inches (95th percentile). Although the average values may be used, the design of work spaces, surfaces, and so on, must consider possible variations for both male and female operators and maintainers; for example, from the 5th percentile female to the 95th percentile male. For specific design criteria, the reader should refer to additional references.[15]

2. *Human sensory factors:* This category relates to the human sensory capacities, particularly sight or vision, hearing, feel or touch, smell, and so on. In the design of workstations, surfaces, operator consoles, and panels, the engineer must be cognizant of the human's capability relative to sight as it pertains to vertical and horizontal fields of view, angular fields of view, the detection of certain objects from different angles, the detection of certain colors and varying degrees of brightness from different angles, and so on. The placement of panel displays and controls as a function of use and the employment of different color combinations to facilitate the accomplishment of manual tasks require knowledge of the human being's capability for seeing. Additionally, the designer needs to understand the human's capacity for hearing in terms of both frequency and intensity (or amplitude). The design of work

McCormick, *Human Factors in Engineering Design,* 7th Ed., McGraw-Hill, New York, 1992; and (3) W. E. Woodson, B. Tillman, and P. Tillman, *Human Factors Design,* 2nd Ed., McGraw-Hill, New York, 1991. Additional references are included in Appendix F.

[15] Anthropometry data are included in National Aeronautics and Space Administration (NASA): Volume 1, *Anthropometric Source Book;* Volume 2, *A Handbook of Anthropometric Data;* and Volume 3, *Annotated Bibliography;* NASA Reference Publication 1024, 1978. Also, refer to Kroemer, Meister, Sanders, Van Cott, and Woodson in Appendix F.

areas for oral communications and/or the use of auditory displays requires knowledge relative to the effects of noise on the performance of work. For instance, as the noise level increases, a human begins to experience discomfort and both productivity and efficiency decrease. If the noise level approaches 120 to 130 dB, then a physical sensation in some form, or pain, will likely occur. In essence, the system designer needs to integrate the capabilities of the human into the final product.[16]

3. *Physiological factors:* Although the study of physiology is obviously well beyond the scope of this text, it is appropriate to recognize the effects of environmental stresses on the human body during the performance of manual tasks. "Stress" refers to any type of external activity, or environment, that acts on an individual in such a manner as to cause a degrading impact. Some typical causes of stress are (1) high and low temperatures, or temperature extremes, (2) high humidity, (3) high levels of vibration, (4) high levels of noise, and/or (5) large amounts of radiation ortoxic substances in the air. To varying degrees, these environmental effects will cause an impact on human performance in a negative manner; that is, physical fatigue will occur, motor response is slower, mental processes slow down, and the likelihood of error increases. These externally related stress factors will normally result in individual human "strain." Strain may, in turn, have an impact on any one or more of the human body biological functions (e.g., the circulatory system, digestive system, nervous system, and respiratory system). Measures of strain may include parameters such as blood pressure, body temperature, pulse rate, and oxygen consumption. These factors of strain, caused by external stresses, will definitely have an impact on the performance of human operator and maintenance functions if the design fails to consider the physiological effects on the human.

4. *Psychological factors:* This category relates to the factors that pertain to the human mind; that is, the emotions, traits, attitudinal responses, and behavioral patterns as they relate to job performance. All other conditions may be perfect relative to completing a task in an effective manner. However, if the individual operator (or maintenance technician) lacks the proper motivation, initiative, dependability, self-confidence, communication skills, and so on, the likelihood of performing the task in an effective manner is extremely low. Generally, one's attitude, initiative, motivation, and so on, are based on the needs and expectations of the individual. This, in turn, is a function of system design and the organizational environment within which the individual performs. If the tasks to be completed are perceived as being too complex, the individual may become frustrated, a poor attitude may develop, and errors will occur. On the other hand, if the tasks are overly simple and routine, there is little challenge, boredom prevails, and errors will occur as a result of attitude. Further, as an external factor, the management style of the supervisor may cause an attitudinal problem. In any event, it is appropriate to consider the possible psycho-

[16]Human sensory factors are covered further in H. P. Van Cott, and R. G. Kinkade (Eds.), *Human Engineering Guide to Equipment Design,* U.S. Government Printing Office, Washington, DC, 1972; and M. S. Sanders and E. J. McCormick, *Human Factors in Engineering Design,* 7th Ed., McGraw-Hill, New York, 1992.

logical effects on the human when involved in the design and development of a system.[17]

Human-factors requirements stem from the definition of operational requirements, the maintenance concept, and the functional analysis, as illustrated in Figure 3.23. From Sections 2.4 and 2.5, operational and maintenance functions are identified, trade-off analyses are conducted in response to the "HOWs" relative to accomplishing the function, and those functions that involve human activity (i.e., tasks that are completed manually or semiautomatically) are noted. Functions that are allocated to the human are further broken down into job operations, duties, tasks, subtasks, and task elements. A "task element" may be classified as the smallest logically definable facet of activity that requires an individual behavioral response (e.g., the identification of a signal on a display, the actuation of a switch on a panel). Progressing upward, task elements are combined in logical sequences to make up subtasks, which are combined to make up tasks, and so on. These activity sequences are defined through various analyses (i.e., operator task analysis, timeline analysis, workload analysis, etc.), with the consideration of personnel factors in mind. Ultimately, the specific requirements for operator personnel and maintenance personnel, in terms of quantities and skill levels are identified.[18]

In the implementation of a human-factors program for a typical large-scale system, the tasks identified in Figure 3.24 are generally applicable. There are (1) program planning, management, and control tasks (Tasks 1–3); (2) design and analysis tasks (Tasks 4–13); and (3) test and evaluation tasks (Tasks 14–15). Although brief task descriptions are included in the figure, some additional comments pertaining to a few are noted for the purposes of emphasis.

1. *Human-factors program plan:* Although the requirements for a human-factors program may specify a separate and independent effort, it is *essential* that the program plan be developed as part of, or in conjunction with, the Reliability Program Plan (Figure 3.13, Task 1), the Maintainability Program Plan (Figure 3.19, Task 1), and the System Engineering Management Plan (SEMP). Many of the activities in each of the plans are mutually supportive and require integration in terms of task input-output requirements, schedules, and so on.

2. *Functional analysis:* The purpose of the functional analysis (in this context) is to identify those functions that are to be performed by the human being, and where there is a human–machine interface. This activity should evolve directly from, and must support, the system functional analysis and associated functional flow diagrams described in Section 2.7 (refer to Figures 2.11 to 2.16).

[17] Additional information on human behavioral characteristics, psychological factors, motivation, attitude, leadership characteristics, and so on, may be found in most texts dealing organizational dynamics, behavioral science, and related subjects.

[18] The requirements for personnel (quantities and skill levels) in support of system maintenance activities are also defined through maintainability and logistic support program activities.

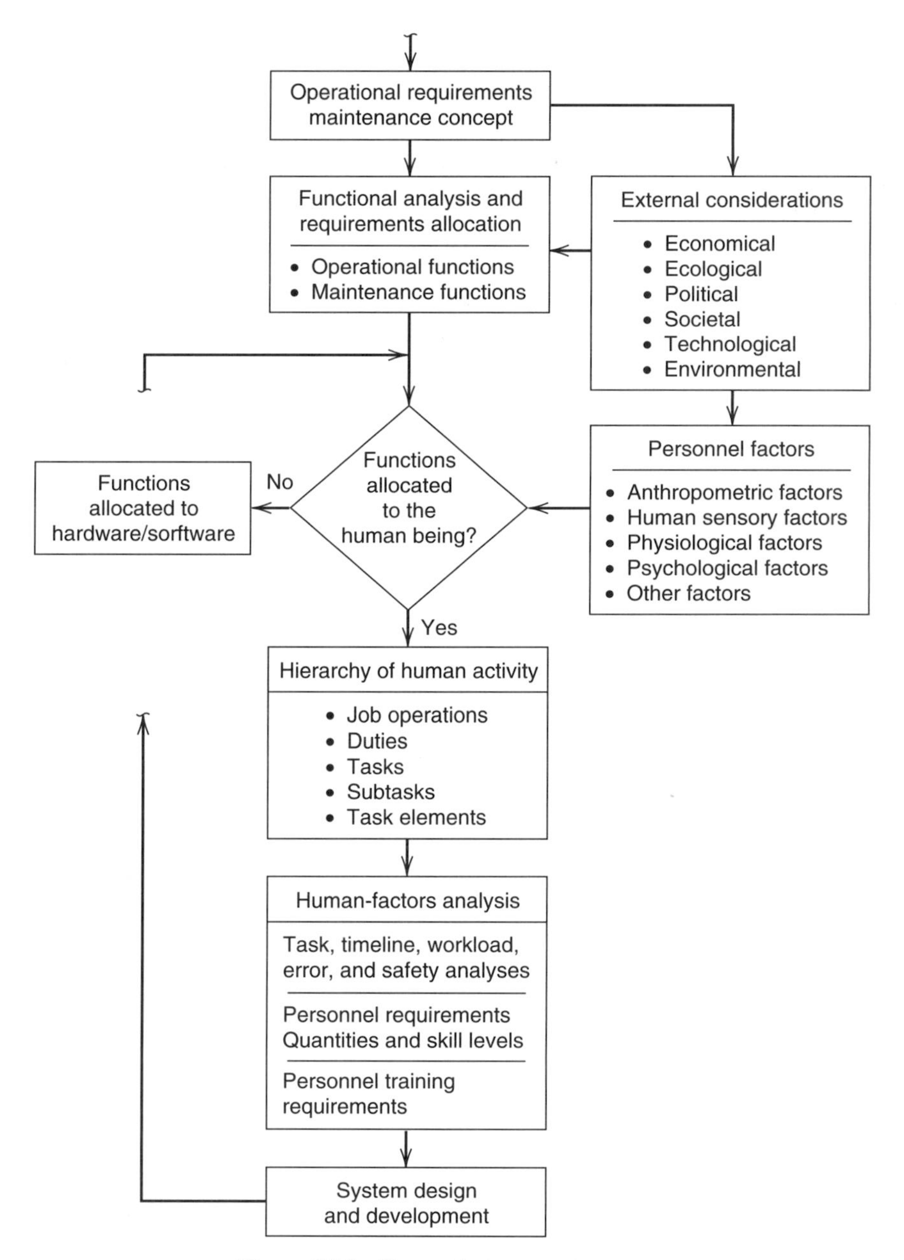

Figure 3.23 Human-factors requirements.

Program Task	Task Description and Application
1. Human Factors Program Plan	To develop a human factors program plan which identifies, integrates, and assists in the implementation of all management tasks applicable in fulfilling human factors engineering requirements. This plan includes a description of the human factors organization, organizational interfaces, a listing of tasks, task schedules and milestones, applicable policies and procedures, and projected resource requirements. This plan must tie directly into the System Engineering Management Plan (SEMP).
2. Review and control of suppliers or subcontractors	To establish initial human factors requirements and to accomplish the necessary program review, evaluation, feedback, and control of component supplier/subcontractor program activities. Supplier program plans are developed in response to the requirements of the overall Human Factors Program Plan for the system.
3. Human factors program reviews	To conduct periodic program and design reviews at designated milestones; e.g., conceptual design review, system design reviews, equipment/software design reviews, and critical design review. The objective is to ensure that human factors requirements will be achieved.
4. System analysis (mission analysis)	To determine the overall capabilities and the performance requirements for the system, and to develop appropriate mission scenarios identifying basic activity sequences. This should be accomplished as part of the system requirements definition process in conceptual design.
5. Functional analysis	To identify the major functions that the system is to perform (based on operational requirements), and to develop functional flow block diagrams defining system design requirements in functional terms. This task must "track" the system-level functional analysis.
6. Function allocation	To conduct trade-off studies, evaluate, and determine the resources required in accomplishing the functions identified through the Functional Analysis activity; i.e., determining the "HOWs" (versus the "WHATs"), particularly in situations where there are human-machine interfaces.
7. Detailed operator task analysis	To evaluate functions that are to be accomplished by the human, and to establish a hierarchical breakdown to the lowest level where human activity exists; i.e., job operation, duty, task, sub-task, and task element. Personnel quantity and skill-level requirements are identified through analysis.
8. Operational sequence diagrams	To identify the human-machine interfaces, and to develop a sequential flow of information, decisions, and actions through the generation of operational sequence diagrams (OSDs).
9. Timeline analysis	To select and evaluate critical task sequences, and to verify that the necessary events can be performed and that they are compatible in terms of allocated time; i.e., can the tasks be performed within the appropriate time allotted for accomplishing the mission?
10. Workload analysis	To evaluate human operator activities throughout a given mission scenario (or through a number of designated scenarios) to determine the workload level; e.g., the relationship between the maximum time allowed and the actual time for task performance.
11. Error analysis	To systematically determine the various ways in which errors can be made by the human, and to make design recommendations to reduce the likelihood of such errors occurring in the future. This task is comparable to the reliability FMECA, except that the system/equipment failures are the result of *human* errors.
12. Safety analysis	To systematically evaluate, through cause-and-effect analysis, the effects of system/equipment failures on safety. Although safety pertains to both personnel and equipment, the aspect of *personnel* safety is emphasized herein. This task ties in directly with the reliability FMECA and the Human Factors Error Analysis.
13. Models and/or mockups	To develop a three-dimensional physical model or a mockup of the system (or a component thereof) to demonstrate human-machine interfaces, spatial relationships, equipment layouts, panel displays, accessibility provisions for maintenance, and so on.
14. Training program requirements	To plan and implement a formal training program. This includes the determination of personnel training requirements (quantity of personnel and the skill levels desired as an output), categories of training, training equipment, training data, training facilities, mockups and models, special training aids, and so on. The plan should include a description of the training organization, a listing of tasks, task schedules and milestones, policies and procedures, and projected resource requirements.
15. Personnel test and evaluation	To plan and implement a program to physically demonstrate human-machine interfaces, task sequences, task times, personnel quantity and skill-level requirements, the adequacy of operating procedures, the adequacy of personnel training, and so on. This test and evaluation activity is accomplished prior to entering production.

Figure 3.24 Human-factors engineering program tasks.

3. *Detailed operator task analysis:* This part of the overall human-factors analysis effort constitutes the expansion of major functions from the system functional analysis into job operations, duties, tasks, and so on. Ultimately, this will lead to the definition of operator and maintenance personnel requirements, in terms of quantities and skill levels, and the subsequent development of training program requirements (Figure 3.24, Task 14). With the identification of personnel and training requirements, close coordination must be established with reliability, maintainability, and logistics program activities as there are common interests in this area.

4. *Operational sequence diagrams:* As part of the human-factors design analysis effort, operational sequence diagrams (OSDs) are developed to show various groups of activities involving the human–machine interface. An example of an OSD is presented in Figure 3.25, where a communications sequence between operators and workstations is illustrated. Through a symbolic presentation, different actions are shown that, in turn, lead to the identification of specific design requirements. Of significance is the requirement that OSDs must evolve from the functional analysis.

5. *Personnel test and evaluation:* The purpose of this task is to demonstrate selected human activity sequences to verify operating/maintenance procedures, and to ensure compatibility between the human and other elements of the system. Demonstrations are conducted using a combination of analytical computer simulations, physical mockups (wooden, metal, and/or cardboard), and preproduction prototype equipment. Computerized simulations may include the insertion of a 5th percentile female or a 95th percentile male into a work space, in a sitting or standing position, in order to evaluate activity sequences and space requirements. A great deal of information can be acquired through use of the appropriate computer graphics employing a three-dimensional database. Type 2 testing, using preproduction prototype equipment, may include the use of personnel, trained as recommended from the results of Task 14, in the performance of selected operator and/or maintenance task sequences accomplished in accordance with approved procedures. The conductance of such tests should not only allow for the evaluation of critical human–machine interfaces, but should provide reliability information pertaining to operator functions, maintainability data when maintenance tasks are performed, verification and validation of information in formal technical manuals/procedures, verification of the adequacy of the training program for operator and maintenance personnel, and so on. Basically, this activity must be coordinated with other testing requirements, and must be covered in the TEMP.

In summary, the tasks identified in Figure 3.24 are generally accomplished in response to some detailed program requirement, often completed on an independent basis. However, the interfaces are numerous, and it is essential that these requirements be appropriately integrated into the overall system engineering process.

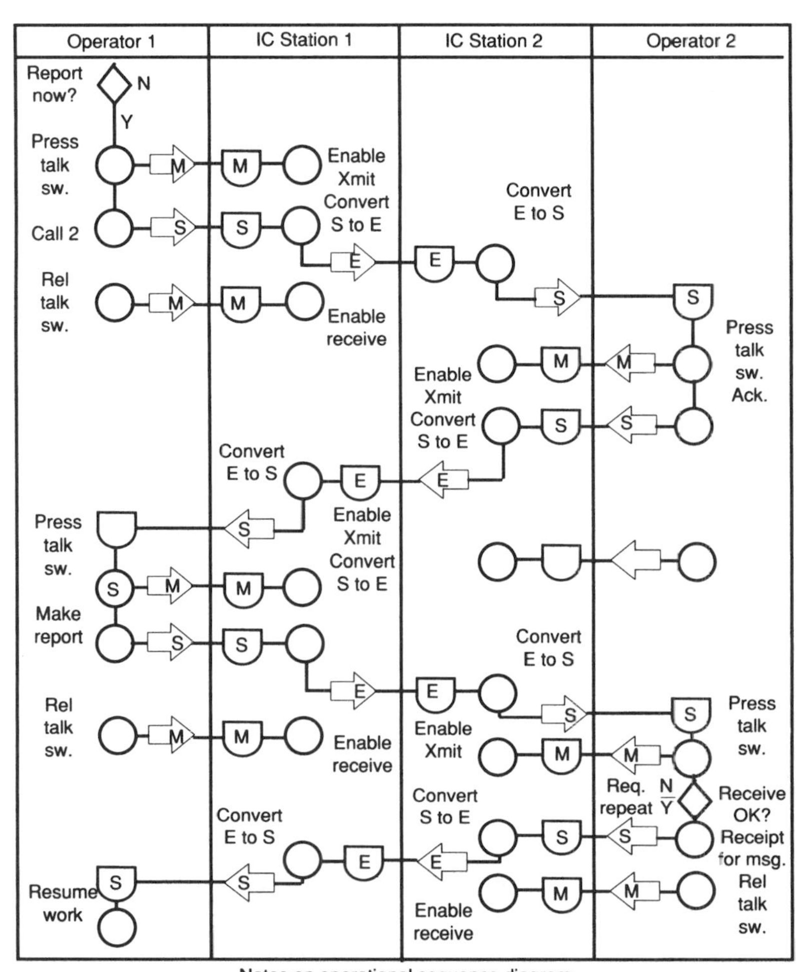

Notes on operational sequence diagram

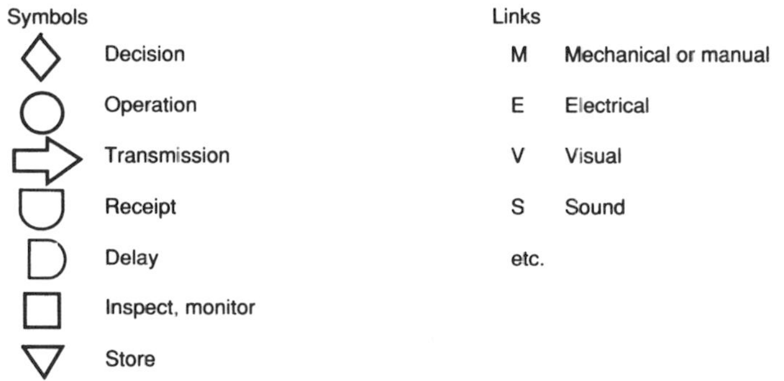

Stations or subsystems are shown by columns; sequential time progresses down the page.

Source: MIL-H-46855, Military Specification, "Human Engineering Requirements For Military Systems, Equipment and Facilities," Department of Defense, Washington, D.C.

Figure 3.25 Example operational sequence diagram. (*Source:* MIL-H-46855, Military Specification, "Human Engineering Requirements for Military Systems, Equipment and Facilities," Department of Defense, Washington, DC.)

3.3.4 Safety Engineering[19]

Safety is a system design characteristic. The selection of certain materials in the design and construction of a system element could produce harmful toxic effects on the human; the placement and mounting of components could cause injuries to the operator and/or the maintainer; the use of certain fuels, hydraulic fluids, and/or cleansing liquids could result in an explosive environment; the location of certain electronic components close together may cause the generation of an electrical hazard; the performance of a series of strenuous tasks during the operation or maintenance of the system could cause personal injury; and so on.

Safety is important, both from the standpoint of the human operator and/or maintainer and from the standpoint of the equipment and other elements of the system. Through faulty design, one can create problems that could result in human injury. Also, problems can be created that result in damage to other elements of the system. In other words, the concerns in design deal with both personal safety and equipment safety.

Relative to the system design and development process, safety engineering requirements are comparable in nature to those described for reliability, maintainability, and human factors (Sections 3.3.1, 3.3.2, and 3.3.3., respectively). Figure 3.26 provides a listing of safety program tasks for a typical large-scale system. There are (1) program planning, management, and control tasks (Tasks 1–3); (2) design and analysis tasks (Tasks 4–7); and (3) test and evaluation tasks (Tasks 8–9). Referring to the figure, there are three basic tasks that require additional comment.

1. *System safety program plan:* Although the requirements for this task may specify a separate and relatively independent effort, it is *essential* that the program plan be developed as part of, or in conjunction with, the Reliability Program Plan (Figure 3.13, Task 1), the Maintainability Program Plan (Figure 3.19, Task 1), the Human-Factors Program Plan (Figure 3.26, Task 1), and the System Engineering Management Plan (SEMP). Tasks 4 and 5 of the safety program (fault-tree analysis and hazard analysis) are closely related to the reliability FMECA, the maintainability analysis (diagnostics and testability analysis), and the human-factors safety analysis. Task 7 should tie in with the reliability FRACAS and the maintainability Task 4 (data collection and analysis). Task 8 (training program) should be related to the human-factors Task 14. Task 9 (testing) should be coordinated with reliability Tasks 18–20, maintainability Task 13, and human-factors Task 15. Many of the activities in each of the plans are mutually supportive and require integration in terms of task input-output requirements, schedules, and so on.

2. *Fault-tree analysis (FTA):* This is an ongoing top-down analytical process, using deductive analysis and Boolean methods, for determining system events that

[19] The objective is to provide an introduction to safety engineering. For a more in-depth coverage, two good references are (1) W. Hammer, *Occupational Safety Management and Engineering,* 4th Ed., Prentice Hall, Upper Saddle River, NJ, 1989; and (2) H. E. Roland and B. Moriarty, *System Safety Engineering and Management,* 2nd Ed., John Wiley, New York, 1990.

Program Task	Task Description and Application
1. System Safety Program Plan	To develop a system safety program plan that identifies, integrates, and assists in the implementation of all management tasks applicable in fulfilling safety engineering requirements. This plan includes a desription of the safety engineering organization, organizational interfaces, a listing of tasks, task schedules and milestones, applicable policies and procedures, and projected resource requirements. This plan must tie directly into the System Engineering Management Plan (SEMP).
2. Review and control of suppliers or subcontractors	To establish initial system safety requirements and to accomplish the necessary program review, evaluation, feedback, and control of component supplier/subcontractor program activities. Supplier program plans are developed in response to the requirements of the overall System Safety Program Plan for the system.
3. System safety program reviews	To conduct periodic program and design reviews at designated milestones, e.g., conceptual design review, system design reviews, equipment/software design reviews, and critical design review. The objective is to ensure that safety engineering requirements will be achieved.
4. Fault-tree analysis (FTA)	To accomplish a fault-tree analysis (FTA) for determining system events that may cause undesireable events (or hazards), and to establish a ranking of these undesireable events. Fault-tree diagrams are developed from early hazard analyses, critical paths are identified, and probable causes are noted (a top-down approach). This task is closely related to the reliability FMECA. Refer to Case Study A.2, Appendix A.
5. Hazard analysis	To accomplish an analysis of the system with the objective of (a) identifying all major hazards and the anticipated probability of occurance, (b) identifying the "cause' factors that will result in a hazard, (c) evaluating the impacts (effects) on the system in the event that hazards occur, and (d) categorizing the identified hazards, i.e., catastrophic, critical, marginal, negligible. This task is closely related to the reliability FMECA and the Human-Factors Safety Analysis.
6. Risk analysis	To initiate a risk management program for the evaluation and control of the probability of occurrance and the consequences of hazardous events. Risk analysis, risk assessment, and risk abatement activities are included.
7. Data collection, analysis, feedback, feedback, and corrective action	To plan and implement a data collection and reporting capability for identifying and evaluating potential areas of risk. Participate in failure analysis activity and in accident investigations as appropriate. Recommendations for corrective action are initiated in areas when potential risk exists.
8. Safety training program	To plan and implement a training program covering the procedures and steps necessary to ensure that operator and maintenance personnel are properly trained in the performance of all system functions. This includes consideration of the requirements for training materials and data, training equipment, training aids, training facilities, and so on.
9. Safety test and evaluation	To plan and implement a program to test the system (and its components) to ensure that it can be safely operated and maintained, and that all necessary safety precautions have been taken. This test and evaluation activity is accomplished prior to entering production.

Figure 3.26 Safety engineering program tasks.

will, in turn, cause undesirable events, or hazards. Further, these events are ranked in terms of their influence in causing the potential hazards. Fault-tree logic diagrams are developed commencing with the top event and proceeding downward through successive levels of causation steps, determining at each level what the next set of events will be! Fault-tree analysis is closely related to both reliability and maintainability analysis, particularly when considering possible symptoms and frequencies of failure, diagnostic and test routines, and so on. Case Study A.2, Appendix A, describes the FTA approach.

3. *Hazard analysis:* The objective of this task is to evaluate the design and determine possible events that could result in hazards at the system level. By simulating failures, critical activities, and so on, at the component level, one can (through a "cause-and-effect" analysis) identify possible hazards, anticipated frequency of occurrence, and classification in terms of criticality. Recommendations for design change are made where appropriate. This task, with regard to methodology and objectives, is very closely related to the reliability FMECA (which also categorizes events in terms of criticality) and the human-factors safety analysis.

In summary, the tasks identified in Figure 3.26 are generally accomplished in response to some detailed program requirement, often completed on an independent basis. However, the interfaces are numerous, and it is essential that these requirements be appropriately integrated into the overall system engineering process.

3.3.5 Software Engineering[20]

With today's trends and the continuing development of computer technology, *software* is becoming (if not already) a significant factor in the configuration of many systems. There have been estimates that software elements are inherent within most of the large system design and development efforts today; that is, from 50 to 75% of the makeup for many large systems constitutes software. System engineers can no longer make hardware design decisions without considering the software implications. At the same time, software engineers must develop their software within the context of the requirements for the overall *system* and not as an independent entity.

Software is *not* a system in itself, but a major element of a system along with the hardware, people, facilities, data, and so on (refer to Figure 2.21). Software requirements evolve directly from system requirements through the functional analysis and allocation process described in Sections 2.7 and 2.8. In Figure 1.13, software requirements at the system level evolve from the functional analysis in block 0.2, and from blocks 1.1 and 1.2 in Figure 1.12 as the major subsystems and system elements become defined. Operating and maintenance functions are defined (representing the "WHATs"), trade-off analyses are accomplished to determine a preferred design approach (identifying the "HOWs"), and those requirements specifying software are identified. Subsequently, the software development process follows the top-down approach conveyed in Figure 1.14.[21]

Figure 3.27, which represents a simplification of Figure 1.13, emphasizes the software-hardware interface requirements. The software, hardware, and other elements of the system must be developed in a highly *integrated* and *concurrent* manner, with the necessary coordination and communications taking place on a daily

[20] The object here is to present an overview of software, and to relate its importance in the context of the system engineering process. Four good references are (1) B. W. Boehm, *Software Engineering Economics,* Prentice Hall, Inc., Upper Saddle River, NJ, 1981; (2) R. S. Pressmen, *Software Engineering: A Practitioner's Approach,* McGraw-Hill, New York, 1992; (3) A. P. Sage and J. D. Palmer, *Software Systems Engineering,* John Wiley, New York, 1990; and (4) K. D. Shere, *Software Engineering and Management,* Prentice Hall, Upper Saddle River, NJ, 1988. Refer to Appendix F for additional references.

[21] It should be noted that software development often assumes an *object-oriented* approach where "objects" are identified from the "bottom-up" and software modules are developed around these objects. Although this approach may be highly beneficial in many respects, problems often occur when the software developer fails to address the aspects of *functionality* from the top down! Software modules are developed, but when combined and integrated with other elements of the system, there is often a "mismatch!" Thus, although the object-oriented approach is feasible, the results must be compatible with the functions defined through the process described in Section 2.7 and the various software modules should fit within some functional block (refer to Figures 2.11 to 2.17).

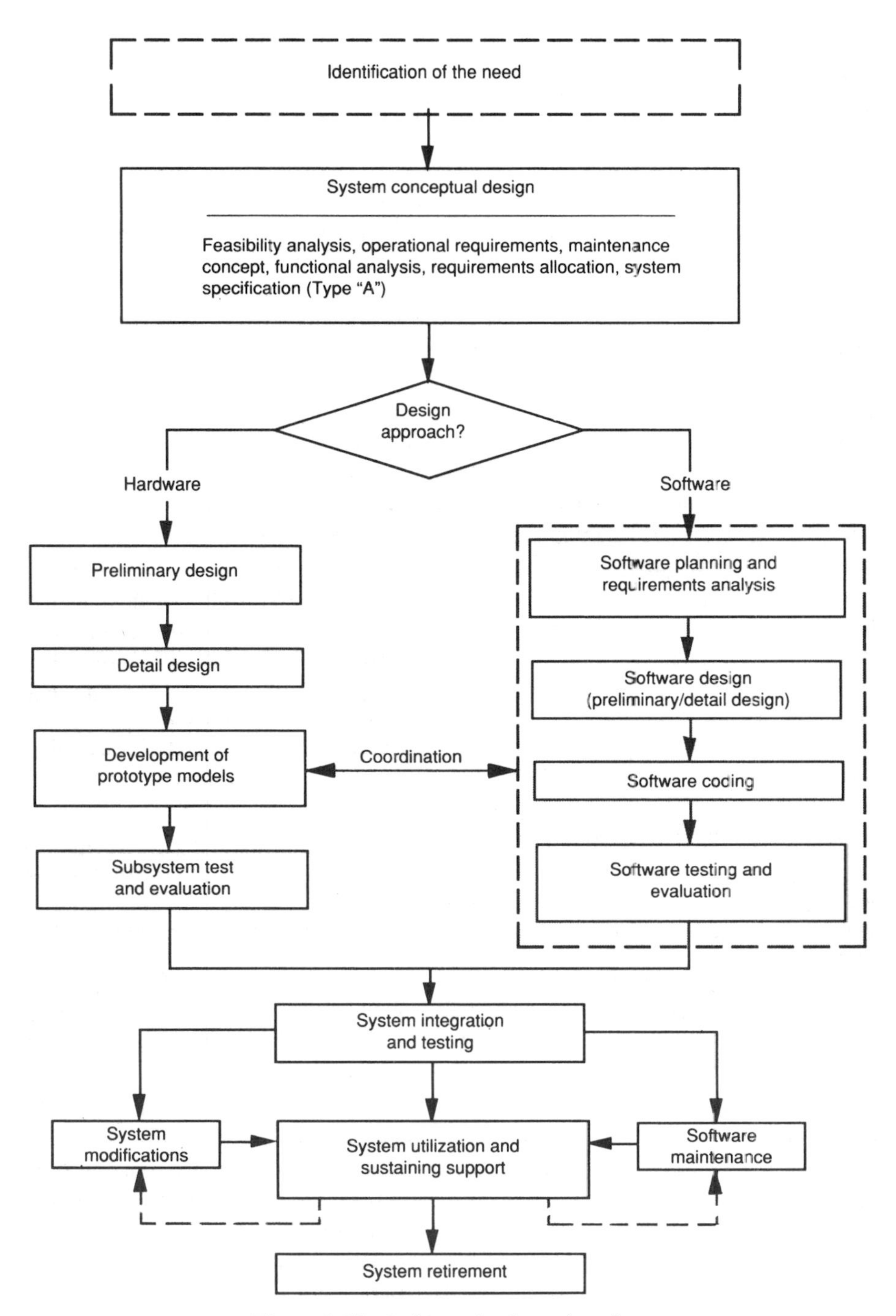
Identification of the need
System conceptual design
Feasibility analysis, operational requirements, maintenance concept, functional analysis, requirements allocation, system specification (Type "A")
Design approach?
Hardware
Software
Preliminary design
Detail design
Development of prototype models
Subsystem test and evaluation
Coordination
Software planning and requirements analysis
Software design (preliminary/detail design)
Software coding
Software testing and evaluation
System integration and testing
System modifications
System utilization and sustaining support
Software maintenance
System retirement

Figure 3.27 Software-hardware interface.

basis. The software development cycle can be expanded to reflect the detailed steps of software requirements analysis, detailed design, coding, unit test, and validation/verification. In Section 1.2.5, there are a number of recognized models for software development to include the "Waterfall Model" (Figure 1.16), the "Spiral Model" (Figure 1.17), and the "Vee Model" (Figure 1.18). Although the nomenclature may vary from one application to the next, the basic overall objectives are similar.

In support of the various software models, the following basic steps are noted:

1. *Software planning:* a definition of the scope, the user's operating environment, problem statement, development and support strategy, basic functional characteristics, interfaces, procurement philosophy, and so on, is included in the *software plan* (sometimes known as the computer resources plan). Additionally, the plan must include program schedules, cost data, and an identification of the resources necessary to implement the proposed program. Resource requirements may include system designers, programmers, technical writers, computer hardware for development testing, existing software that may be used during development, and so on. The software plan must be closely integrated with the SEMP.

2. *Requirements analysis:* A definition of system requirements (i.e., system functional analysis, operator functions, maintenance functions, allocations), the identification of software functions, performance factors, specific software design constraints/criteria, top-level software functional block diagrams or information/data flow diagrams, and so on, are included. Software requirements are derived from system-level requirements, and information/data flow diagrams should evolve directly from the functional block diagrams describing the system, as discussed in Section 2.7. This information is generally included in a software requirements specification; that is, a Type "B" specification.

3. *Preliminary design:* The development of a software hierarchical structure is accomplished to establish the basic relationships between the functional elements of the software, modules, and coupling interface requirements. Information/data flow diagrams are developed further, and database requirements are defined.

4. *Detail design:* The development of detailed flow charts (from the functional flow diagrams) is accomplished, and the requirements for program design language(s) and the application of the appropriate design tools are identified.

5. *Software coding:* The preparation and writing of programs using structured code, the appropriate code format and code documentation, debugging and error-check procedures, and so on, are accomplished.

6. *Software testing and evaluation:* A "verification" of software design is accomplished to ensure that each of the various individual software products fulfills its specific purpose. The verification process often constitutes a step-by-step approach to testing one program, followed by the testing of a second program given the results of the first, and so on. Verification is iterative by nature and is generally oriented to specific software items. On the other hand, "validation" is directed toward the system level. The goal is to ensure that the overall software element(s) of the system complies with the requirements of the system specification. This is accomplished

through system test and evaluation where software is integrated with hardware, operating personnel, facilities, elements of support, and so on.

7. *Software maintenance:* This refers to the follow-on upgrading and maintenance of software for the purposes of improvement; i.e., *adaptive* and/or *perfective* maintenance. Care must be taken to ensure that such maintenance does not cause the rippling effect illustrated in Figure 3.8.

With software requirements growing in terms of system applications, and software costs increasing at a rapid rate, it is essential that these requirements be appropriately *integrated* into and *controlled* through the system engineering process.

3.3.6 Producibility Engineering

Producibility is a measure of the relative ease and economy of producing a system or a product. The characteristics of design must be such that an item can be produced easily and economically, using conventional and flexible manufacturing methods and processes without sacrificing function, performance, effectiveness, or quality. Some major objectives in designing a system for producibility are noted as follows:

1. The quantity and variety of components utilized in system design should be held to a minimum. Common and standard items should be selected where possible, and there should be a number of different supplier sources available throughout the planned life cycle of the system.
2. The materials selected for constructing the system should be standard, available in the quantities desired and at the appropriate times, and should possess the characteristics for easy fabrication and processing. The design should preclude the specification of peculiar shapes requiring extensive machining and/or the application of special manufacturing methods.
3. The design configuration should allow for the easy assembly (and disassembly as required) of system elements; that is, equipment, units, assemblies, and modules. Assembly methods should be simple, repeatable, economical, and should not require the utilization of special tools and devices or high personnel skill levels.
4. The design configuration should be simplistic to the extent that the system (or product) can be produced by more than one supplier, using a given data package and conventional manufacturing methods/processes. The design should be compatible with the application of CAD/CAM technology where appropriate.

The basic underlying objectives are "simplicity" and "flexibility" in design. More specifically, it is the goal to minimize the use of critical materials and critical processes, the use of proprietary items, the use of special production tooling, the application of unrealistic tolerances in fabrication and assembly, the use of special test systems, the use of high personnel skills in manufacturing, and the production/procurement lead times.

Producibility engineering is the effort necessary to ensure a smooth transition from system design and development to its production (or construction). It requires the coordinated activities of design engineering and manufacturing, and generally includes accomplishment of the following program tasks:

1. *Producibility plan:* A plan is developed to ensure the adequate implementation of producibility requirements throughout a given program. This plan may be included as a section in SEMP, or as a separate independent document.

2. *Producibility requirements:* Specific system-level design requirements are developed during the conceptual design and advanced planning phase, and are included in the System Type "A" Specification (refer to Figure 3.3). Additionally, requirements at the subsystem, equipment, unit, assembly, and component level must be included in the appropriate development, process, product, and material specifications (Types "B," "C," "D," and "E").

3. *System analysis and trade-off studies:* Producibility must be considered as a system/product design characteristic, and each design alternative should be evaluated using the general approach illustrated in Figure 3.28.

4. *Design reviews:* Both informal and formal design reviews must cover producibility requirements in the evaluation process. Given a definitive design configuration, the questions in Figure 3.29 (and in Appendix C) may be utilized for assessment purposes. Recommendations for improvement can be processed, hopefully, before the design is assumed as being fixed!

In summary, producibility engineering pertains to that part of system design that addresses the ability of an item to be reproduced in multiple quantities, based on a given design configuration. It is not uncommon for a product to be developed as a result of an advanced research-and-development effort, using rough engineering laboratory practices, and then to have an immediate request for 300 more (wanted yesterday). Although the results of the laboratory experience proved to be successful, the engineering model developed will likely include the use of unapproved nonstandard components available from laboratory stocks. In other words, the laboratory model developed may have successfully demonstrated the accomplishment of certain performance-related goals, but it does not at this stage represent a production configuration. The task yet to be completed is the conversion of the laboratory model into a production prototype, built to manufacturing data/drawings and incorporating the appropriate approved standard components. The objective is to convert the results of research and development efforts into a producible entity.

3.3.7 Quality Engineering[22]

In recent years, United States industry, the Department of Defense, and others have placed a great deal of emphasis on the subject of *quality.* The prime motivator is

[22] Selected references covering various facets of quality, quality assurance, quality control, and so on, are identified in the bibliography in Appendix F.

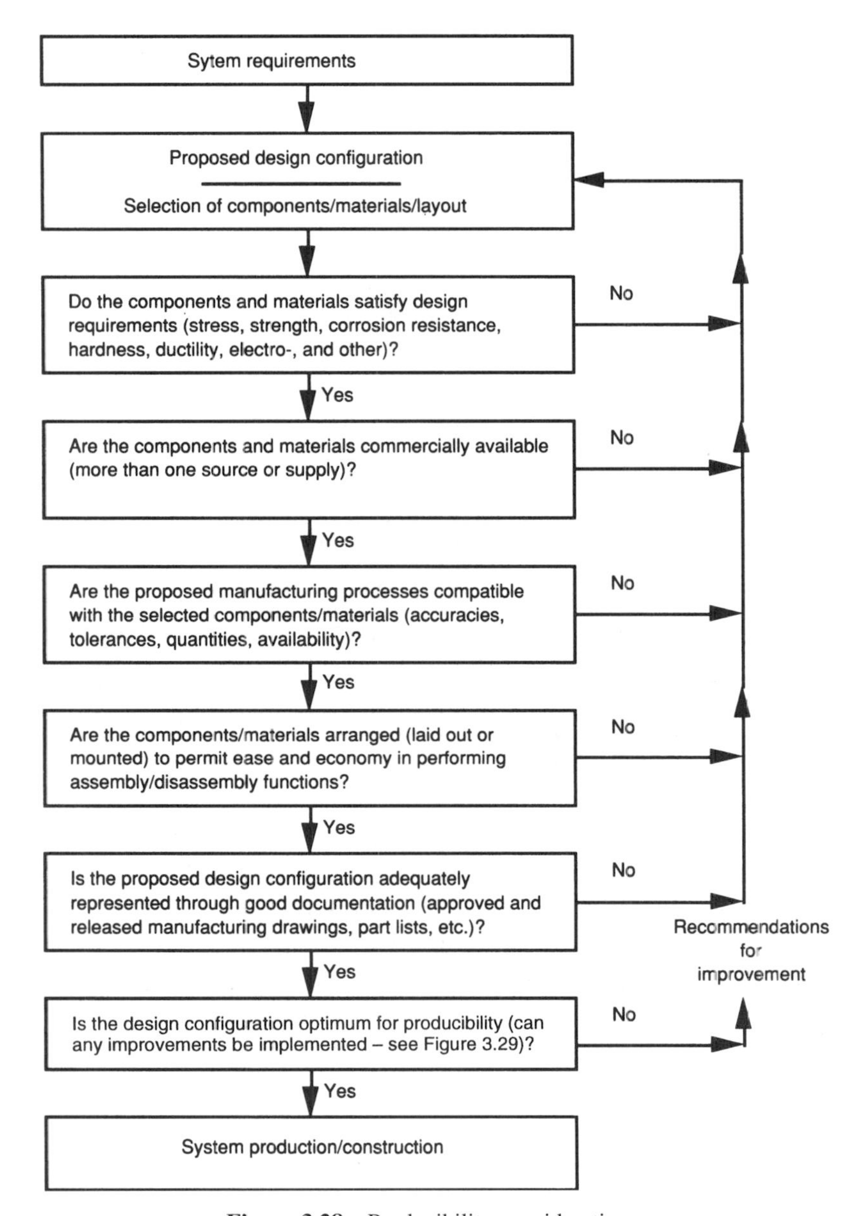

Figure 3.28 Producibility considerations.

that of "survival" in a highly competitive international environment. In general, the availability of cost-effective, high-quality systems/products from international sources has been increasing, and competition is encouraging industries to do a better job in the design and production of systems. As a result, the field of "quality," although not new, is undergoing a change in emphasis.

Producibility Checklist
1. Can the design be further simplified?
2. Can two or more components be combined into one?
3. Can components with slight differences be made identical?
4. Can the design be standardized to a greater degree; i.e., a greater use of standard components and materials?
5. Can the design be modified to avoid the use of critical and/or proprietary components and materials?
6. Can a standard off-the-shelf component/material item be used to replace a manufactured item?
7. Are the manufacturing methods consistent and standard relative to forming, bending, stressing, forging, machining, types of castings, fits, hole sizes, fillet and edge radii, countersinks, clearances, etc.?
8. Can a different material be used which is easier to machine?
9. Are the proposed manufacturing methods compatible to CAD/CAM methods where applicable?
10. Are the accuracies/tolerances reasonable and consistent with multiple manufacturing process capabilities?
11. Can the cost be reduced through the use of a less costly component, material, or simpler manufacturing process? Have the most economical manufacturing processes been specified?
12. Can the weight be reduced through component/material substitution? Can a lighter gauge material be used?
13. Are the specifications/standards consistent with the planned production environment?
14. Have component/material coating requirements been minimized?
15. Have inspection and test requirements been minimized? Have nondestructive testing methods been developed where required? Is the approach economical?
16. Can residual and/or waste materials be recycled or disposed of without causing any environmental degradation?

Figure 3.29 Selected producibility checklist

In the past, the fulfillment of quality objectives has primarily been accomplished in the Production and/or Construction Phase of the life cycle through the implementation of formal quality control (QC) or quality assurance (QA) programs. Statistical process control (SPC) techniques, incoming and in-process inspection activities, closely monitored supplier control programs, periodic audits, and selected problem-solving methods have been implemented with the objective of attaining a designated level of system quality. These efforts have basically been accomplished "after the fact," and the overall results have been questionable!

Recently, the aspect of "quality" has been viewed more from a top-down life-cycle perspective, and the concept of "total quality management" (TQM) has evolved.[23] TQM can be described as a total integrated management approach that addresses system/product quality during all phases of the life cycle and at each level

[23] The Department of Defense, in particular, has been advocating TQM, and there are many references covering the subject from different perspectives. Three such references are (1) T. R. Stuelpnagel, "Total Quality Management," *National Defense,* Journal of the American Defense Preparedness Association (ADPA), Arlington, VA, November 1988; (2) DOD 5000.51G, "Total Quality Management—A Guide for Implementation," Department of Defense, Washington, DC; and (3) SOAR-7, "A Guide for Implementing Total Quality Management," Reliability Analysis Center, Rome Air Development Center, Griffiss AFB, New York, 1990.

in the overall system hierarchy. It provides a before-the-fact orientation to quality, and it focuses on system design and development activities, as well as production, manufacturing, assembly, construction, product support, and related functions. TQM is a unification mechanism linking human capabilities to engineering, production, and support processes. It provides a balance between the "technical system" and the "social system." Some specific characteristics of TQM are as follows:[24]

1. Total customer satisfaction is the primary objective, as compared to the practice of accomplishing as little as possible in conforming to the minimum requirements. The customer orientation is important (versus the "What can I get away with?" approach).
2. Emphasis is placed on the iterative practice of "continuous improvement" as applied to engineering, production and support processes, functions, and the like. The objective is to seek improvement on a day-to-day basis, as compared to the often-imposed last-minute single thrust initiated to force compliance with some standard. The Japanese practice this approach through the implementation of a process known as *kaizen.*
3. In support of item 2, an individual understanding of processes, the effects of variation, the application of process control methods, and so on, is required. If individual employees are to be productive relative to continuous improvement, they must be knowledgeable of various processes and their inherent characteristics. Variability must be minimized (if not eliminated).
4. TQM emphasizes a total organizational approach, involving every group in the organization and not just the quality control function. Individual employees must be motivated from within, and should be recognized as being key contributors to meeting quality objectives.

Included within the broad spectrum of TQM are the very important aspects of engineering and the "design for quality;" that is, quality engineering. From Section 1.2.3, the projected life cycles, as illustrated in Figure 1.10, must be considered in *total!* A system is conceived, designed, produced, utilized, and supported throughout its planned life cycle. As part of the initial system design effort, consideration must be given toward (1) the design of the process that will be utilized to produce that system, and (2) the design of the support configuration that will be utilized to provide the necessary ongoing maintenance and support for that system. As the interactions between these various facets of program activity are numerous, it is important that these areas be addressed on an integrated basis from the start!

These program relationships have been recognized by the Department of Defense through initiation of the concept of "concurrent engineering," which is defined as

[24]The reader is encouraged to review the literature covering Japanese activities in the overall field of "quality." Also, the writings of Crosby, Deming, and Juran will provide greater insight into the subject area and are highly recommended (refer to Appendix F). This section discusses some of the principles of TQM; that is, customer satisfaction, individual participation, continuous improvement, robust design, variability, management responsibility, and supplier integration.

"a systematic approach to the integrated, concurrent design of products and their related processes, including manufacture and support. This approach is intended to cause the developers, from the outset, to consider all elements of the product life cycle from conception through disposal, including quality, cost, schedule, and user requirements."[25] The objectives of concurrent engineering include (1) improving the quality and effectiveness of systems/products through a better integration of requirements, and (2) reducing the system/product development cycle time through a better integration of activities and processes. This, in turn, should result in a reduction in the total life-cycle cost for a given system.

From the perspective of this text, the primary thrust is *quality engineering* and its role as a part of the system engineering process. In a relationship similar to that expressed for logistics (refer to Section 3.3.8), there is the larger concern for TQM and there are some specific concerns associated with quality as it pertains to engineering design. With this in mind, the following activities are considered to appropriate with regard to system engineering:

1. *Quality planning:* The development of a TQM plan (or equivalent) must be accomplished during conceptual design and updated during preliminary and detail design as required. Inherent within this overall plan are quality engineering activities including the (a) determination of engineering design requirements using a QFD, "house of quality," or equivalent approach (refer to Section 2.6); (b) evaluation and design of manufacturing and assembly processes in response to design technology decisions; (c) participation in the evaluation and selection of system components and supplier sources; (d) preparation of product, process, and material specifications as required (Types "C," "D," and "E"); (e) participation in on-site supplier reviews; and (f) participation in formal design reviews. These, and related, activities should also be included in the SEMP.[26]

2. *Quality in design:* This area of activity, viewed in the broad context, pertains to many of the issues discussed throughout the earlier sections of this chapter. Emphasis is directed toward design simplicity, flexibility, standardization, and so on. Of a more specific nature are the concerns for "variability," where a reduction in the variation of the dimensions for specific component designs, or tolerances in process designs, will likely result in an overall improvement. Taguchi's general approach to "robust design" is to provide a design that is insensitive to the variations normally encountered in production and/or in operational use. The more robust the design, the less the support requirements, the lower the life-cycle cost, and the higher the degree of effectiveness. Overall design improvement is anticipated through a combination of careful component evaluation and selection, the appropriate use of statisti-

[25] IDA Report R-338, "The Role of Concurrent Engineering in Weapons System Acquisition," Institute for Defense Analysis, Alexandria, VA, 1988.

[26] "House of quality" refers to a basic methodology used to implement a "quality function deployment (QFD)" program. QFD focuses on planning and communications using a cross-functional team approach. It provides a framework for assessing product attributes and for transforming them into engineering design requirements. Refer to J. R. Hauser and D. Clausing, "The House of Quality," *Harvard Business Review,* May-June 1988, pp. 63–73.

cal process control (SPC) methods, and application of experimental testing approaches, applied on a continuous basis.[27]

The subject of quality pertains to both the technical characteristics of design and the humanistic aspects in the accomplishment of design activities. Not only is there a concern relative to the selection and application of components, but the successful fulfillment of quality objectives is highly dependent on the behavioral characteristics of those involved in the design process. A thorough understanding of customer requirements, good communications, a "team" approach, the willingness to accept the basic principles of TQM, and so on, are all necessary. In this respect, the objectives of quality engineering are inherent within the scope of system engineering.

3.3.8 Logistics, Serviceability, and Supportability Engineering[28]

Logistics, as a discipline, includes a wide variety of activities conducted throughout the system life cycle. Such activities deal with the overall flow of materials from the supplier or producer to the consumer, the distribution of products by the producer for consumer use, the transportation and handling of materials between two or more points, and the sustaining customer service and maintenance support of a system (or product) throughout its planned life cycle.

The activities within the broad spectrum of logistics are often subdivided into two basic areas as noted:

1. In the *commercial* sector, logistics can be defined as the "process of planning, implementing, and controlling the efficient, cost-effective flow and storage of raw materials, in-process inventory, finished goods, and related information from the point of origin to the point of consumption for the purpose of conforming to customer requirements."[29] The emphasis in this area is on the procurement and distribution of relatively small *consumable* items, and the primary activities are highlighted in Figure 3.30(A).
2. In the *defense* sector, logistics can be defined as "a disciplined, unified, and iterative approach to the management and technical activities necessary to (a) integrate support considerations into system and equipment design; (b) develop support requirements that are related consistently to readiness objec-

[27] Genichi Taguchi has developed mathematical techniques for application relative to the evaluation of design variables. Refer to P. J. Rose, *Taguchi Techniques for Quality Engineering,* McGraw-Hill, New York, 1988.

[28] To gain a complete perspective of the field of logistics, it is recommended that additional study in this area be pursued. Three good references are (1) B. S. Blanchard, *Logistics Engineering and Management,* 4th Ed., Prentice Hall, Upper Saddle River, NJ, 1992 (this conveys an *engineering* approach); (2) N. A. Glaskowsky, D. R. Hudson, and R. M. Ivie, *Business Logistics,* 3rd Ed., Dryden Press, Orlando, FL, 1991 (this conveys the industrial *business* approach); and (3) DSMC, *Integrated Logistics Support Guide,* Defense Systems Management College, Fort Belvoir, VA. Additional references are included in Appendix F.

[29] Council of Logistics Management (CLM), Oak Brook, Illinois.

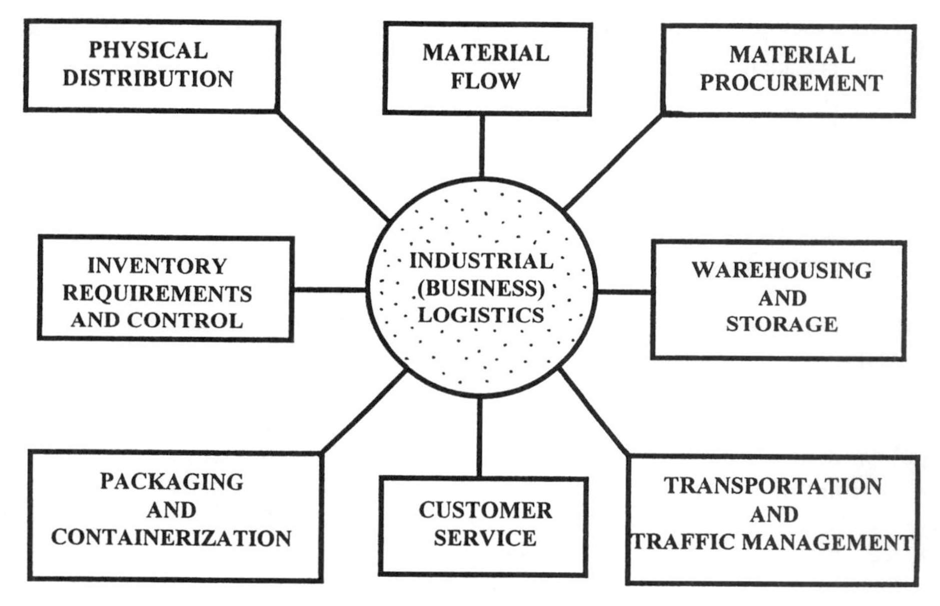

(A) The elements of industrial business logistics (consumables)

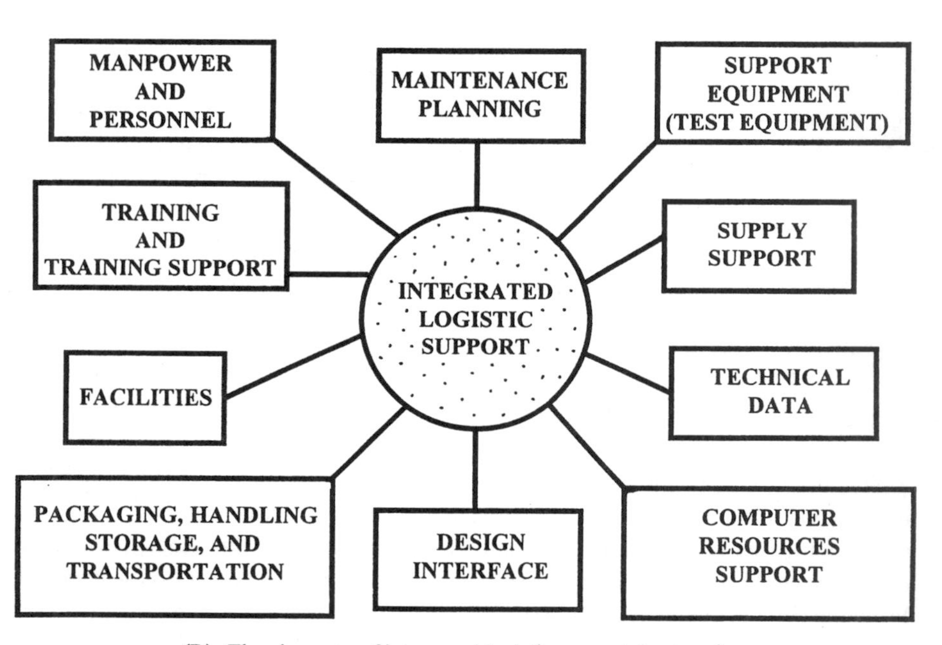

(B) The elements of integrated logistic support (systems)

Figure 3.30 The elements of logistics, serviceability, and supportability.

tives, to design, and to each other; (c) acquire the required support; and (d) provide the required support during the operational phase at minimum cost."[30] The emphasis here is on the procurement, acquisition, distribution, and sustaining maintenance support of *systems,* and the primary activities are identified in Figure 3.30(B).

In evaluating the two basic approaches to logistics (i.e., the first dealing with *consumables* and the second addressing *systems*), the activities represented in Figure 3.30(A) are also conducted within the context of those system-level activities presented in Figure 3.30(B). Further, the activities reflected in Figure 3.30(B) are applicable to any category of systems, whether one is involved in defense systems, manufacturing systems, commercial transportation systems, and the like (see Section 1.2.2). The approach here not only addresses the procurement and acquisition of *systems,* but the maintenance and support of those systems throughout their planned life cycle. In essence, one must address the overall maintenance and support *infrastructure,* as being a significant element of the system, when defining consumer requirements and throughout the subsequent design and development process (refer to Figure 1.7).

Figure 3.31, which is an extension of Figure 1.10, illustrates an example of the system life cycle and the activities for an automative vehicle that must be addressed on a concurrent basis. Because of the costs associated with past experiences and the practice of not considering system support issues until "after the fact," it has become essential that *supportability* and *serviceability* considerations be included in the early system design and development process. The appropriate characteristics must be "built into" the basic design such that the ultimate configuration can be maintained, serviced, and supported in an effective and efficient manner throughout its planned life cycle.[31]

When addressing the entire maintenance and support infrastructure, one must consider the flow of materials and products from the suppliers, through the process of construction and/or production, and for the subsequent distribution and installation of components at the desired consumer locations. The basic objective is to provide an operating system for consumer use. At the same time, there is a *reverse* flow of items from the consumer site to various supporting organizations for the purposes of providing maintenance. This infrastructure, which can be considered as a major subsystem, is depicted in Figure 3.32.

[30] This definition initially evolved from Department of Defense Instruction 5000.2, "Defense Acquisition Management Policies and Procedures," Part 7, 1991. A second definition of logistics, developed by the International Society of Logistics (SOLE), is "the art and science of management, engineering, and technical activities concerned with requirements, design, and supplying and maintaining resources to support objectives, plans, and operations." This definition is *conceptual* in nature, and it does convey the broad aspects of the field.

[31] The objectives of *serviceability* and *supportability* are similar to those specified for *maintainability* (described in Section 3.3.2). These terms are often used interchangeably, depending on whether one is dealing with defense systems, automotive systems, or depending on one's individual background and experience.

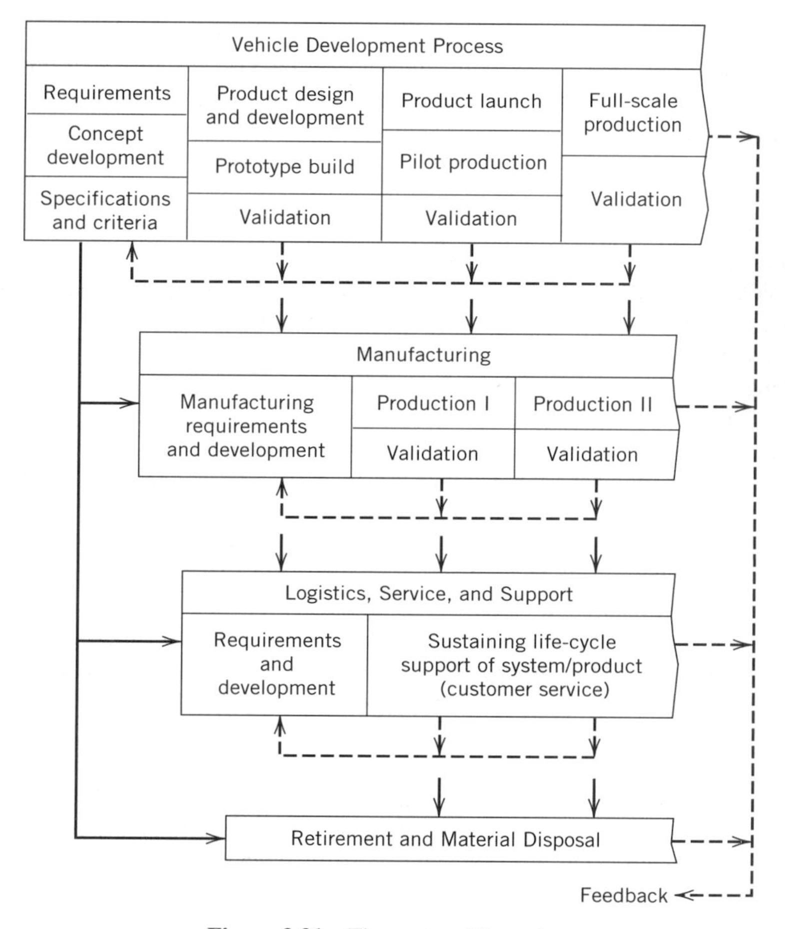

Figure 3.31 The system life cycle.

Figure 3.33 (an extension of Figure 3.32) shows the activities and some of the resources that may be applicable to the sustaining life-cycle maintenance and support of a typical system. This illustrated flow, which is derived from the initial maintenance concept (refer to Section 2.5), provides a basis for the identification of specific maintenance resource requirements, which may include all, or a combination, of the following elements:[32]

1. *Manpower and personnel:* includes all personnel required in the installation, checkout, operation, handling, and sustaining maintenance of the system throughout its planned life cycle. Maintenance personnel considerations cover activities at all levels of maintenance, operation of test equipment, operation of facilities, and so on.

[32] B. S. Blanchard, *Logistics Engineering and Management,* 4th Ed., Prentice Hall, Upper Saddle River, NJ, 1992.

2. *Training, training equipment, and devices:* includes the "initial" training of all system operator and maintenance personnel, and the follow-on replenishment training to cover attrition and replacement personnel. Training equipment, training simulators, mockups, training data and manuals, special facilities, special devices and aids, and software to support personnel training operations are also included.

3. *Supply support:* includes all spares (units, assemblies, modules, etc.), repair parts, consumables, special supplies, and related inventories needed to support prime mission-oriented equipment, software, test and support equipment, transportation and handling equipment, training equipment, and facilities. Provisioning documentation, procurement functions, warehousing, distribution of material, and personnel associated with the acquisition and maintenance of spare/repair part inventories at all support locations are also included in this category.

4. *Test and support equipment:* includes all tools, special condition monitoring equipment, diagnostic and checkout equipment, metrology and calibration equipment, maintenance stands, and servicing and handling equipment required to support operation, transportation, and scheduled and unscheduled maintenance actions associated with the system or product. Both "peculiar" (newly developed) and common "standard" (existing and already in the inventory) items must be covered.

5. *Packaging, handling, storage, and transportation:* includes all special provisions, materials, containers (reusable and disposable), and supplies necessary to support packaging, preservation, storage, handling, and/or transportation of prime

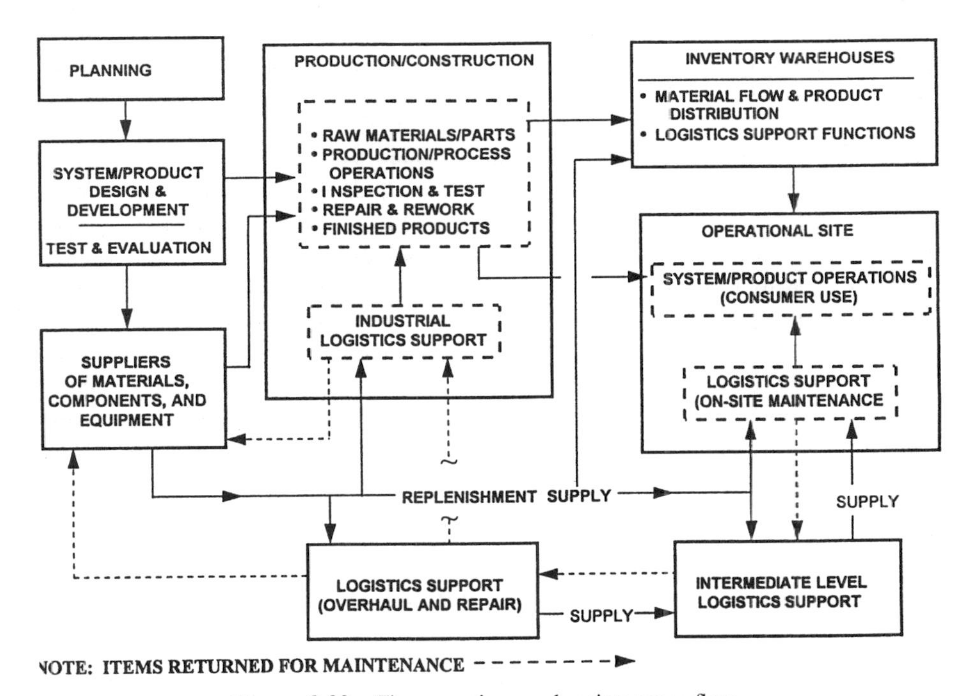

Figure 3.32 The operations and maintenance flow.

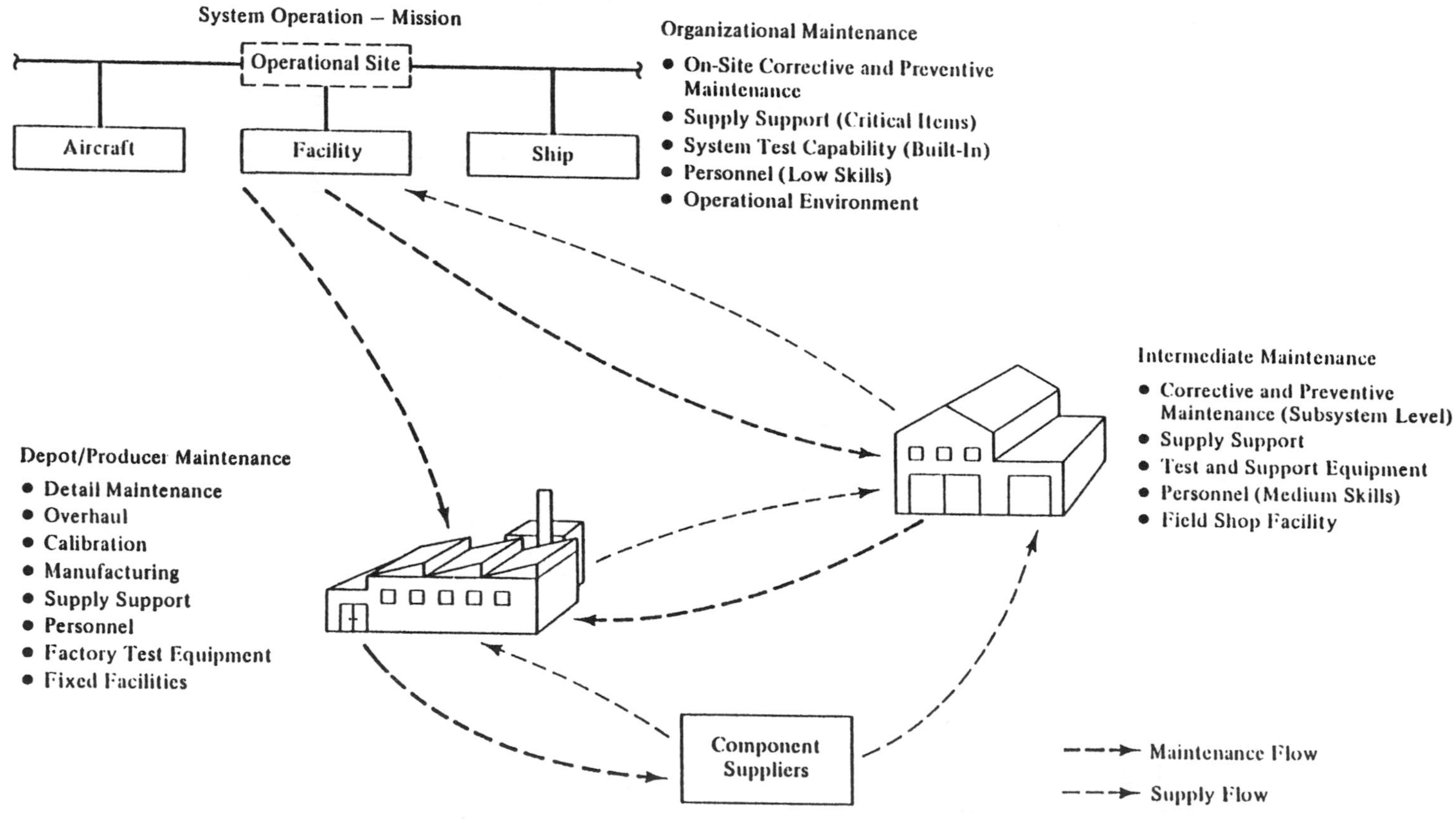
System Operation – Mission
Operational Site
Aircraft
Facility
Ship
Organizational Maintenance
• On-Site Corrective and Preventive Maintenance
• Supply Support (Critical Items)
• System Test Capability (Built-In)
• Personnel (Low Skills)
• Operational Environment
Intermediate Maintenance
• Corrective and Preventive Maintenance (Subsystem Level)
• Supply Support
• Test and Support Equipment
• Personnel (Medium Skills)
• Field Shop Facility
Depot/Producer Maintenance
• Detail Maintenance
• Overhaul
• Calibration
• Manufacturing
• Supply Support
• Personnel
• Factory Test Equipment
• Fixed Facilities
Component Suppliers
Maintenance Flow
Supply Flow

Figure 3.33 The system support infrastructure.

mission-oriented equipment, test and support equipment, spares and repair parts, personnel, technical data, and mobile facilities. In essence, this category covers the initial distribution of products and the transportation of personnel and materials for maintenance purposes.

6. *Facilities:* includes all special facilities needed for system operation and the performance of maintenance functions at each level. Physical plant, real estate, portable buildings, housing for personnel, intermediate maintenance shops, calibration laboratories, and special depot or overhaul facilities must be considered. Capital equipment and utilities (heat, power, energy requirements, environmental controls, communications, etc.) are generally included as part of facilities.

7. *Technical data:* includes system installation and checkout procedures, operating and maintenance instructions, inspection and calibration procedures, overhaul procedures, modification instructions, facilities information, drawings and specifications, and associated databases that are necessary the performance of system operation and maintenance functions. Information processing requirements (networks and equipment) are also included in this category.

8. *Computer resources:* includes all software, computer equipment, tapes/disks, databases, and accessories necessary in the performance of system maintenance functions at each level. This covers condition monitoring requirements and maintenance diagnostic aids.

These basic elements of support represent significant segments of the system. They enable the performance of system maintenance activities, or sustaining customer service. There are distinct "measures" (i.e., technical performance measures) associated with each of these segments, and these factors must be considered in the overall system evaluation process. For example, what is the probability of having a spare part available when required? What is the probability of mission success, given a designated level of inventory at a specific location? What economic order quantity (EOQ) factors should be assumed in the provisioning and procurement of spare parts? What is the process time, or turnaround time (TAT), for items at the intermediate level of maintenance? What is the reliability of the test equipment? What are the personnel quantity and skill-level requirements for system maintenance? What are the transportation times between the various levels of support? What are the reliability requirements for ground handling equipment? These and related factors need to be considered in the early systems modeling and analysis activity (along with MTBM, MTBF, MDT, $\bar{M}ct$, MLH/OH, etc.).

The elements of system support must be addressed as an inherent part of the *maintenance planning* activity accomplished in the development of the maintenance concept during conceptual design (refer to Section 2.5). Maintenance planning then continues throughout design, is expanded with the aid of the logistic support analysis, and the various elements of support are properly integrated into the overall system configuration.

As stated in Section 1.1, experience associated with many large-scale systems has indicated that a significant proportion of the life-cycle cost for a given system is attributed to the maintenance and support requirements for that system. Further, in evaluating "cause-and-effect" relationships, much of this cost is the direct result of decisions made during the early phases of system/product design and development. Thus, logistics (and particularly the aspects dealing with system supportability) must be addressed in terms of the entire system life cycle. In other words, there are logistics *planning* activities conducted at the inception of a program, there are logistics *engineering* activities conducted throughout all phases of system design and development, there are *provisioning* and *acquisition* (or procurement) functions in preparation for and throughout the consumer utilization phase, and there are logistics *evaluation* and *assessment* activities conducted throughout the period when the system is being utilized in the field.

The primary focus in this text is on those aspects of logistics that relate to engineering design and are inherent within the system engineering process. With the objective of "designing for supportability and serviceability," requirements are initially established, functional analyses and allocations are accomplished, system analyses and trade-off studies are conducted to ensure a preferred design approach, design reviews are conducted, and the evaluation of supportability is accomplished as part of the overall system test and evaluation effort. Supportability refers to those characteristics of design that relate to the ease and economy of system support (refer to Appendix C).

Relative to logistics engineering, the basic objectives are to initially *influence design* and subsequently *determine logistic support resource requirements* based on an assumed design configuration. In Figure 3.34, these objectives are best accomplished through an ongoing iterative process involving the integration and application of various techniques and functions to ensure that supportability requirements are considered in the design process. This process, known as the logistic support analysis (LSA), is an inherent part of the system engineering process, and responds to a variety of specific objectives. Basically:

1. LSA aids in the initial establishment of supportability requirements during conceptual design through the evaluation of system operational requirements, alternative technology applications, and alternative maintenance and support concepts.
2. LSA aids in the evaluation of alternative system, or equipment, design configurations; for example, alternative repair policies, packaging schemes, diagnostic routines, and selection of components.
3. LSA aids in the evaluation of a given design configuration ("fixed" or "assumed") to determine specific logistic resource requirements, for example, personnel and training, spare/repair parts, test equipment, transportation and handling, data, and facilities. This is usually accomplished through a detailed maintenance task analysis, a maintenance engineering analysis, or the equivalent (refer to Case Study A.4).
4. LSA aids in the measurement and evaluation of an operating system in the field, in terms of its effectiveness and supportability in the user's environment.

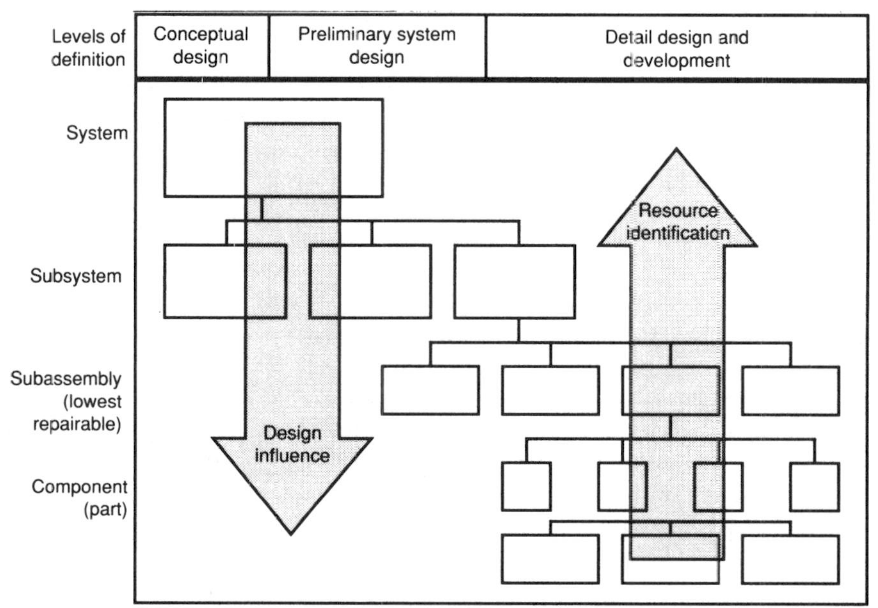

Figure 3.34 Logistics support analysis emphasis. (*Source:* Adapted from AMC, *Logistics Support Analysis; LSA Primer,* Pamphlet 700-22, USAMC Material Readiness Support Activity, Lexington, KY.)

> Field data are collected, evaluated, and utilized to update the LSA initially developed through the earlier design evaluation effort.[33]

The LSA process includes the evaluation of many different alternatives, following the steps illustrated in Figure 2.25. Inherent within this activity is the accomplishment of life-cycle cost (LCC) analysis; level-of-repair analysis (LORA); reliability-centered maintenance (RCM) analysis; failure mode, effect, and criticality analysis (FMECA); testability and diagnostic analysis; and so on. Analytical techniques such as simulation, linear programming, dynamic programming queuing analysis, and control theory may be employed in solving a wide variety of problems.

[33] Much of the LSA material discussed herein was initially derived from MIL-STD-1388-1A, Military Standard, "Logistic Support Analysis," Department of Defense, Washington, DC, 1983. While this particular reference is no longer being imposed on many DOD/military-related programs, the concepts and methods described within are valid and can be applied to any category of system. In this context, it serves as a good reference. Additionally, these same concepts/methods have been applied earlier under such titles as *maintenance engineering analysis* (MEA), *maintenance level analysis* (MLA), *maintenance engineering analysis record* (MEAR), *maintenance analysis data* (MAD), and other titles of a similar nature. Independent of what is is called, the concepts/methods/techniques are appropriate in the design of a system for supportability.

In the implementation of a logistics program for a typical large-scale system, the tasks identified in Figure 3.35 are generally applicable. Figure 3.36 provides a "generic" overview of these tasks in terms of a program timeframe. There are (1) program planning, management, and control tasks (Tasks 1–4); (2) logistics engineering tasks (Tasks 5–13, 16); and (3) system/product support tasks (Tasks 14–15, 17–20). Although brief task descriptions are included in Figure 3.35, some additional comments are appropriate.

1. *Integrated Logistic Support Plan* (ILSP): ILSP, usually initiated during conceptual design and updated in preliminary design, covers *all* program activities dealing with system logistic support; that is, planning activities, design activities, procurement and acquisition activities, and sustaining support activities. The ILSP may include a LSA Plan (LSAP), an Integrated Support Plan (ISP), and the like, depending on the nature of a given program. As such, its scope is rather broad in content. Those activities that deal with engineering and system design are critical in meeting the objectives of system engineering; thus, these activities must be appropriately integrated into the SEMP.

2. *Logistics engineering:* This general category (as defined herein) covers all activities to include the initial definition of system requirements, the design for system supportability, and the follow-on test and evaluation effort. Engineering activities are directed toward (1) the design of the prime mission-oriented segments of the system for effective and economical support, and (2) the design of the overall support capability such that it will be responsive to system requirements.[34] Meeting these objectives can be accomplished, to a large extent, through the incorporation of reliability, maintainability, and human-factors principles in system design. There is a great deal of commonality between the objectives of logistics engineering and those described for reliability, maintainability, and human factors. Thus, there are many tasks, which are similar in nature, that are accomplished in response to the requirements specified for a given program. These tasks must be properly integrated through the SEMP.

3. *System/product support:* Given that a system has been initially designed and produced (or constructed), there are a series of activities that are necessary to ensure that the system is maintained in such a manner that it will *continue* to meet consumer needs. These activities deal with the procurement, distribution, and sustaining maintenance and support of the system in the field and are critical if the system is to fulfill it overall objectives. The system engineering role here is that of ongoing assessment, evaluation, data feedback, and system modification for corrective action as required.

[34] Through the years, much progress has been made in terms of responding to the first objective; that is, design of the prime elements of the system for supportability. However, in the opinion of the author, very little has been accomplished relative to the design of the support capability.

Program Task	Task Description and Application
1. Integrated Logistic Support Plan (ILSP)	To develop an ILS Plan which identifies, integrates, and assists in the implementation of all logistic support tasks for a given program. The ILS Plan should cover a number of individual logistic element sub-plans to include: a. Detailed Maintenance Plan (initially the maintenance concept) b. Reliability and Maintainability Plan (interface requirements) c. Logistic Support Analysis Plan (LSAP) d. Supply Support Plan e. Test and Support Equipment Plan f. Personnel Training Plan g. Technical Data Plan h. Packaging, Handling, Storage, and Transportation Plan i. Facilities Plan j. Computer Resources Plan k. Distribution and Consumer Support Plan (user operations) l. Post-Production Support Plan m. System Retirement Plan The ILS Plan includes a description of logistics concepts and acquisition strategy, logistics organization, organizational interfaces, a listing of program tasks, task schedules and milestones, applicable policies and procedures, and projected resource requirements. The ILS Plan must tie directly into the System Engineering Management Plan (SEMP), particularly those elements dealing with logistics engineering.
2. Review and control of suppliers or subcontractors	To establish initial logistic support requirements and to accomplish the necessary program review, evaluation, feedback,and control of supplier/ subcontractor program activities. Each supplier should develop an Integrated Support Plan (ISP) in response to the overall ILS Plan for the system.
3. Logistic Support Analysis (LSA) Program	To plan and implement a logistics engineering program applicable throughout system design and development.
4. Logistics program reviews	To conduct periodic program and design reviews at designated milestones; e.g., conceptual design review, system design reviews, equipment/software design reviews, and critical design review. The objective is to ensure that logistics requirements will be achieved.
5. Operational use study	To determine system operational requirements and the maintenance concept as a basis for defining system support requirements.
6. Mission hardware, software, and support system standardization	To define supportability and related design constraints for the system based on existing and planned logistic support resources. To what existing environment should the support capability be designed?
7. Comparative analysis	To develop a "baseline comparison system," representing the characteristics of the proposed new system, for projecting supportability quantitative and qualitative requirements, and to use as a basis against which alternative configurations are evaluated.
8. Technological opportunities	To identify and establish new design technology approaches, technology advancements, and so on, to achieve possible supportability improvements in the new system. This activity stems from the results of logistics research.
9. Supportability-related design factors	To establish qualitative and quantitative supportability characteristics resulting from the evaluation of alternative operating concepts, maintenance and support policies, and technology applications.

Figure 3.35 Logistics support program tasks.

Program Task	Task Description and Application
10. Functional requirements identification	To identify the operational and support functions that must be accomplished by the system, and to develop operational and maintenance functional flow block diagrams. From the functional diagrams, FMECA and Reliability-Centered Maintenance (RCM) data must be generated to define corrective and preventive maintenance requirements at the system level.
11. Support system alternatives	To identify and describe feasible support alternatives for the new system configuration being developed.
12. Evaluation of alternatives and trade-off analyses	To accomplish trade-off studies and to define a preferred support configuration as a result. The accomplishment of level of repair analyses, the evaluation of alternative diagnostic approaches, the evaluation of different equipment packaging schemes and mounting provisions, and so on, are included.
13. Task analysis	To identify and analyze required operational and maintenance tasks (which are derived from functions) in order to define specific logistic support resource requirements; i.e., spare and repair parts, test and support equipment, personnel quantities and skill levels, facilities, data, software, and so on.
14. Early fielding analysis	To assess the impact of introducing new systems/components on existing systems and on the environment. To identify the logistic support resources required during the transition from production to full-scale consumer operations.
15. Post-production support analysis	To analyze the life-cycle support requirements for the new system (prior to discontinuing production) to ensure that adequate logistic support resources will be available throughout the remaining life of the system; i.e., can the system be adequately supported after the production capability has been discontinued?
16. Supportability test and evaluation	To verify that the initially specified supportability requirements for the system have been met, and to establish a procedure whereby recommendations for corrective action and/or improvement can be processed.
17. Provisioning and acquisition of logistics elements	To plan and implement a program for the provisioning, procurement, and acquisition of spare and repair parts, special items of test equipment, transportation and handling equipment, facilities, properly trained personnel, technical data, and software. This may include the production of various elements of support, the procurement of off-the-shelf items of equipment, the development of technical publications, and/or the construction of a new facility.
18. Consumer service or field service engineering	To establish a sustaining capability where the producer/contractor provides on-site engineering services to the consumer in support of system operation and maintenance activities.
19. Field data collection, analysis, feedback, and corrective-action system	To plan and implement a data collection and reporting capability in order to provide an assessment of system operations and support in the field. Recommendations for corrective action and/or improvement are initiated as appropriate.
20. System modifications	To establish a procedure whereby system modifications can be implemented to correct a known deficiency. This includes the initial planning for modifications, the development and "proofing" of modification kits, installation of modifications in the field, and verification of system operation after the changes have been incorporated.

Figure 3.35 *(Continued)*

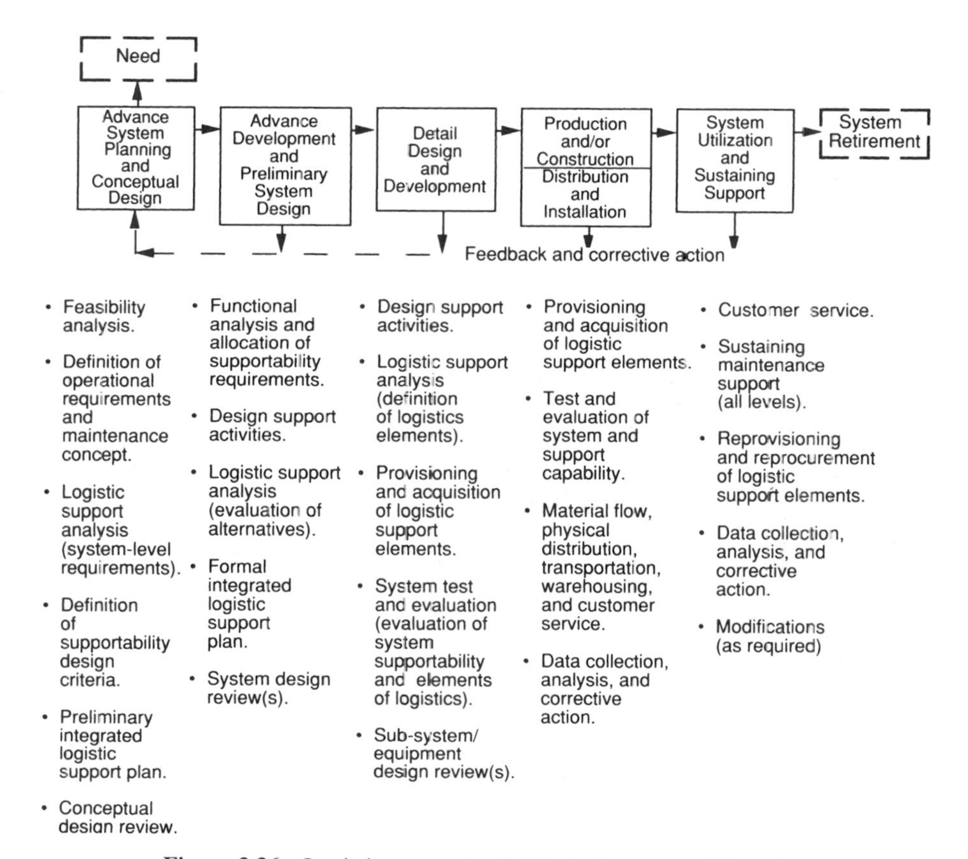

Figure 3.36 Logistics support activities in the system life cycle.

3.3.9 Value/Cost Engineering[35]

The material presented thus far has primarily emphasized the *technical factors* associated with the system, as referenced in Figure 1.23. These factors, which include performance, reliability, maintainability, human factors, supportability, and quality, represent only one side of the overall spectrum. The other side of the spectrum pertains to *economic factors,* and a proper balance between the two must be attained.

In the system evaluation process, these technical and economic factors are often combined in such a manner as to provide a measure of effectiveness (MOE) for a given system. Although these effectiveness figures of merit (FOMs) will vary from one application to the next, a few examples are noted:

[35]Economics and life-cycle costing are covered in detail in W. J. Fabrycky and B. S. Blanchard, *Life-Cycle Cost and Economic Analysis,* Prentice Hall, Upper Saddle River, NJ, 1991; G. J. Thuesen and W. J. Fabrycky, *Engineering Economy,* 8th Ed., Prentice Hall, Upper Saddle River, NJ, 1994; and R. D. Stewart, *Cost Estimating,* 2nd Ed., John Wiley, New York, 1990. Additional references are noted in Appendix F.

$$\text{Effectiveness FOM} = \frac{\text{performance} \times \text{availability}}{\text{life-cycle cost}} \tag{3.24}$$

$$\text{Effectiveness FOM} = \frac{\text{system capacity}}{\text{revenues} - \text{cost}} \tag{3.25}$$

$$\text{Effectiveness FOM} = \frac{\text{life-cycle cost}}{\text{facility space}} \tag{3.26}$$

$$\text{Effectiveness FOM} = \frac{\text{supportability}}{\text{life-cycle cost}} \tag{3.27}$$

Or other

With regard to the *economic* side of the balance, both *revenues* and *costs* must be considered, as conveyed in Figure 3.37, particularly in the commercial sector

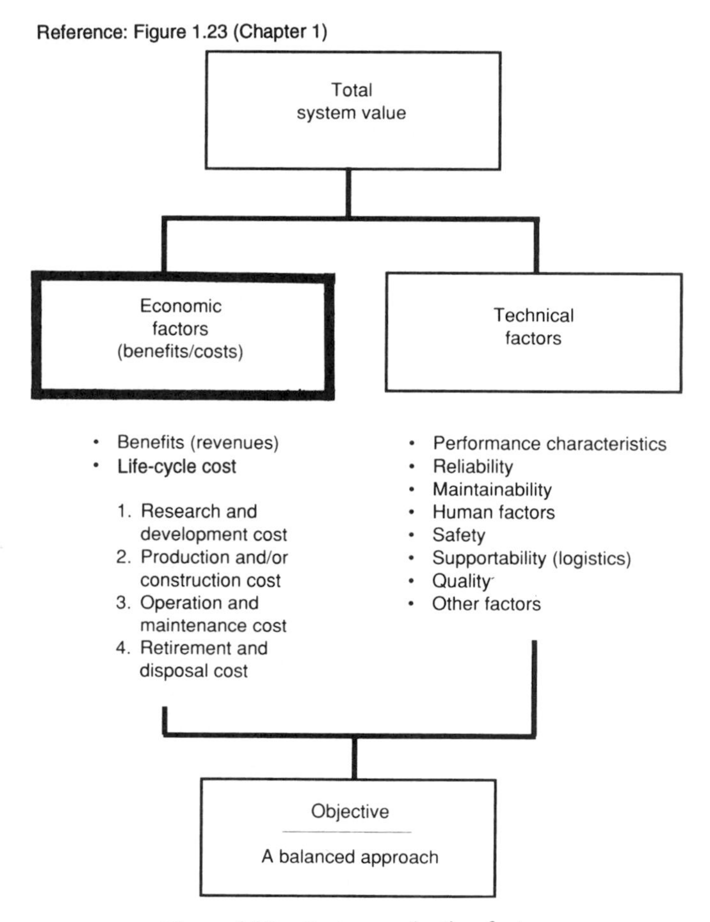

Figure 3.37 System evaluation factors.

where the loss of revenues often represents a major segment of cost. However, the emphasis in this section is on "cost;" that is, the total cost of all activities throughout the system life cycle. *Life-cycle cost* (LCC) includes the consideration of *all* future costs associated with research and development (i.e., design), construction and/or production, distribution, system operation, maintenance and support, retirement, and material disposal and/or recycling. It involves the costs of all technical and management activities throughout the system life cycle; that is, producer activities, contractor and supplier activities, and consumer or user activities. Additionally, costs are often related to "functions" accomplished over the long term, as compared to the rather short-term perspective conveyed through the traditional accounting structure for most organizations. With this in mind, one may pose the following questions:

1. For your company or organization, do you know the costs associated with each of the *functions* being accomplished?
2. Do you know what functions constitute the *high-cost contributors* over the long term? For a given system, what are the high-cost elements?
3. Are you aware of the *cause-and-effect relationships* and the *criticalities* as they relate to the accomplishment of a given mission (or operational scenario)?
4. What are the *high-risk* areas or elements?

The answers to these and related questions are not easily attained. Yet, individual design and management decisions are often based on some smaller aspect of cost (e.g., initial purchase price or acquisition cost) without first assessing the consequences of these decisions in terms of of *total* cost. As conveyed in Section 1.2 (Chapter 1), many of the decisions made in the early stages of system design will have a large impact on the costs of downstream activities such as production, operations, maintenance and support, and retirement and material disposal. Although some of these early decisions may be necessary, one is remiss unless they are made in the context of total life-cycle cost. *Full-cost visibility* is essential if the risks associated with the decision-making process are to be properly assessed.

Life-cycle cost analyses, in one form or another, are accomplished throughout system design and development, during construction/production, and/or for the purposes of assessment while the system is being utilized in the field. The completion of such an effort generally requires that one follow certain steps, such as those presented in Figure 3.38.[36]

In Figure 3.38, one of the first steps in the process is to describe the system in *functional* terms, and then to construct a functional flow diagram covering all of the activities in the system life cycle, evolving from the identification of need through retirement and material disposal (refer to Section 2.7). Given this, it is necessary to develop a *cost breakdown structure* (CBS), such as shown in Figure 3.39. The CBS

[36] The life-cycle cost analysis process is covered in detail in Appendix B.

1. Describe the system configuration being evaluated in functional terms, and identify the appropriate technical performance measures (TPMs) or applicable "metrics" for the system.
2. Describe the system life cycle and identify the major activities in each phase as applicable (system design and development, construction and/or production, utilization, maintenance and support. retirement and disposal).
3. Develop a work breakdown structure (WBS), or cost breakdown structure (CBS), covering all activities and work packages throughout the life cycle.
4. Estimate the appropriate costs for each category in the WBS (or CBS), using activity-based costing (ABC) methods, or equivalent.
5. Develop a computer-based model to facilitate the life-cycle cost analysis process.
6. Develop a cost profile for the "baseline" system configuration being evaluated.
7. Develop a cost summary, identifying the high-cost contributors (i.e., high cost "drivers").
8. Determine the "cause-and-effect" relationships, and identify the "causes" for the high-cost areas.
9. Conduct a sensitivity analysis to determine the effects of input factors on the analysis results. and identify the high-risk areas
10. Construct a Pareto diagram and rank the high-cost areas in terms of relative importance and requiring immediate management attention.
11. Identify feasible alternatives (potential areas for the improvement), construct a life-cycle cost profile for each, and construct a break-even analysis showing the point in time when a given alternative assumes a point in preference.
12. Recommend a preferred approach, and develop a plan for system modification and improvement (this may entail a modification of equipment or software, a facility change, and/or a change in some process). This constitutes an ongoing iterative approach for continuous process improvement.

Figure 3.38 The basic steps in a life-cycle cost analysis.

constitutes a vehicle for including all costs, and is broken down to the depth required to provide the appropriate level of visibility for determining the costs of various functions, processes, and/or elements of the system over time. The CBS serves as a structure that will allow for the initial allocation of cost targets in a "design-to-cost" application (refer to Figure 2.23), and for the subsequent collection of costs in a "life-cycle cost analysis." Costs are estimated for each year in the system life cycle, inflationary and other influencing factors are included, costs profiles are developed, and costs are summarized by category in the CBS. The high-cost contributors are noted, cause-and-effect relationships are established, a sensitivity analysis is accomplished, feasible alternatives are evaluated, and recommendations are made based on the results.

A life-cycle cost analysis may serve many purposes, and the possible applications are varied, as conveyed in Figure 3.40. Of particular note is the use of LCC analysis in the evaluation of different design configurations in the early stages of system development, the evaluation of different commercial off-the-shelf (COTS) alternatives, and in the evaluation of an existing system configuration with the objective of identifying the high-cost contributors leading to possible recommendations for product/process improvement. In each application, the steps identified in Figure 3.38 and the process illustrated in Figure 3.41 are followed.

Figure 3.42 provides an example of LCC analysis applications in the system

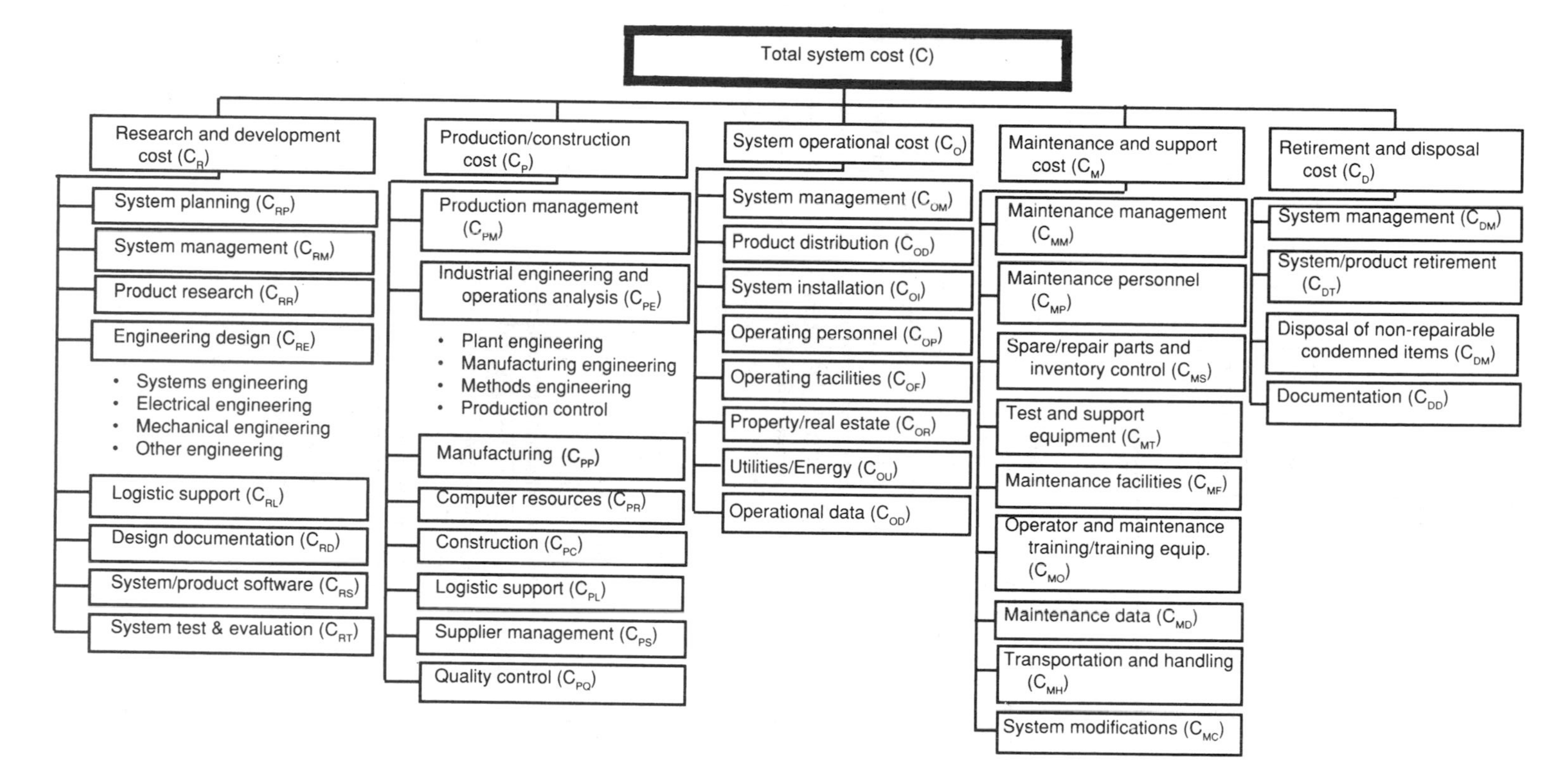

Figure 3.39 Sample cost breakdown structure (CBS).

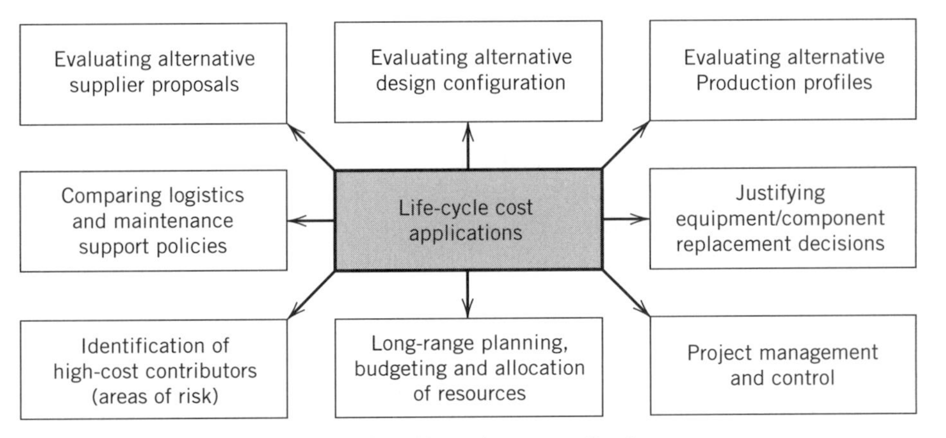

Figure 3.40 Life-cycle cost applications.

design and development process. Cost targets may be established initially in conceptual design through the development of TPMs (refer to Section 2.6). Trade-off studies are accomplished during the preliminary and detail design phases to support design and procurement decisions. During the latter stages of detail design and throughout the construction/production and system utilization phases, LCC analyses may be conducted for assessment of the overall cost-effectiveness of the system. Computer-based models are used to facilitate the analysis process (as shown in Figure 2.26). Figure 3.43 shows the LCC analysis as it may be applied throughout the system life cycle. For a more in-depth discussion of life-cycle costing, the analysis process, and its benefits, refer to Appendix B.

3.3.10 Environmental Engineering

Although the previous sections in this chapter dealt primarily with some of the more tangible considerations in design, it is essential that one also consider the aspect of "design for the environment!" *Environment,* in this context, refers to the numerous external factors that must be addressed in the overall system development process. In addition to the *technological* and *economic* factors discussed previously, one must deal with *ecological, political,* and *social* considerations as well. The system being developed must be compatible with, acceptable, and ultimately must exist within an environment that addresses the many factors illustrated in Figure 3.44. A requirement within the spectrum of systems engineering is to ensure that the system being developed will be socially acceptable, is compatible with the political structure, is technically and economically feasible, and will not cause a degradation to the environment overall.

Of particular interest here are the ecological considerations. Ecology generally pertains to the study of the relationships between various organisms and their environment. This includes consideration of plant, animal, and human populations in terms of rate of population growth, food habits, reproductive habits, and ultimate

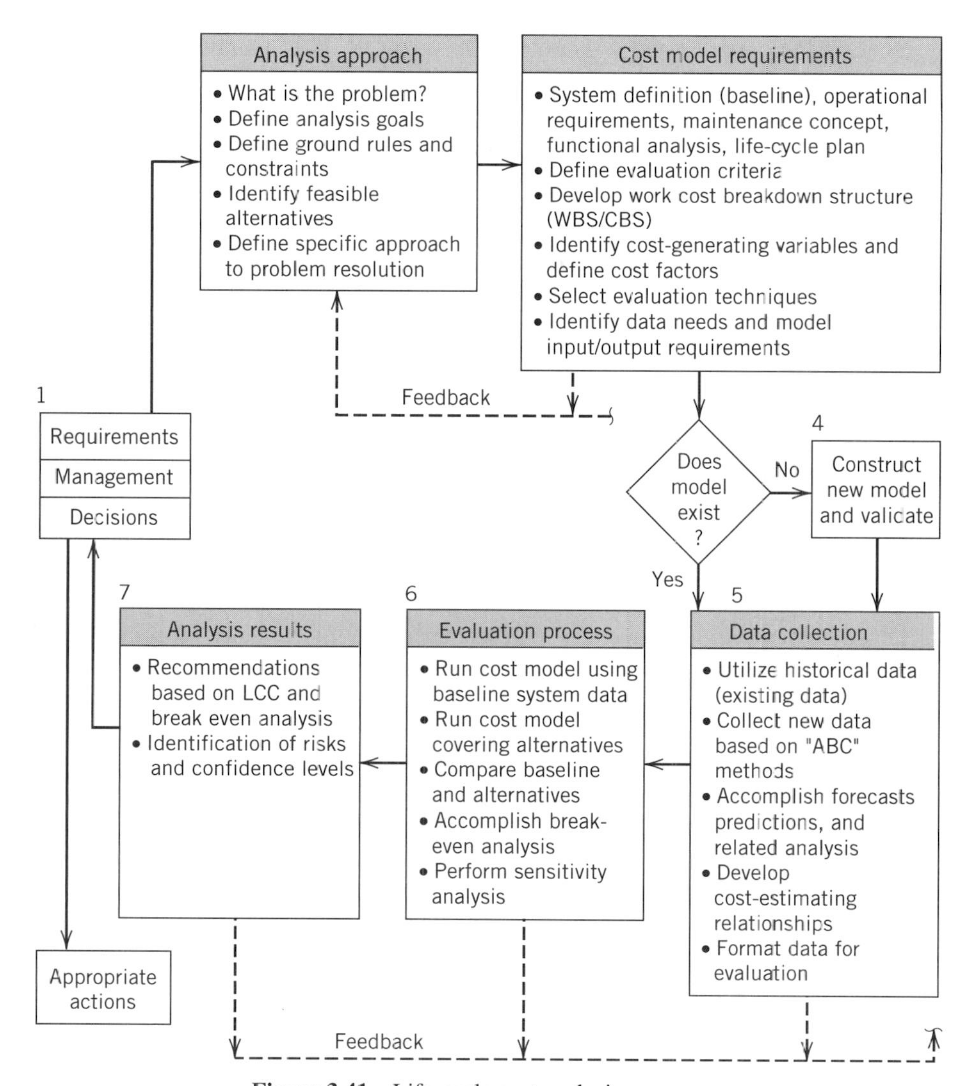

Figure 3.41 Life-cycle cost analysis process.

death. In other words, one is addressing the conventional biological process as viewed in the broad context.

In recent decades, the world population growth, combined with the technological changes associated with our living standards, has created a greater consumption of our resources, resulting in potential shortages, which in turn has stimulated shifts toward establishing other means for accomplishing objectives. Concurrently, the amount of waste has increased significantly. The net effects of this have caused alterations to the basic biological process, and to some extent these alterations have been harmful. Of particular concern are those problems dealing with following:

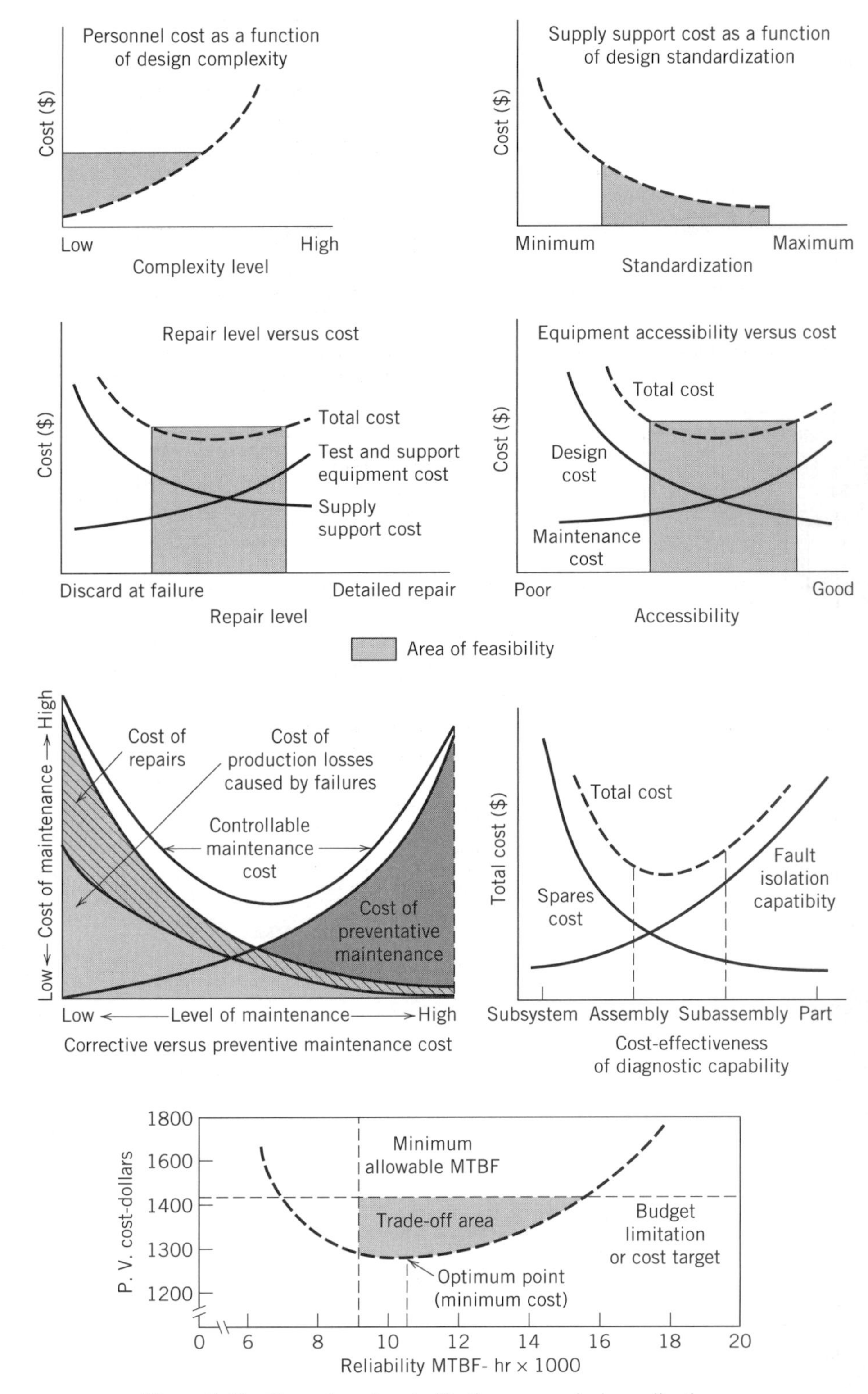

Figure 3.42 Examples of cost-effectiveness analysis applications.

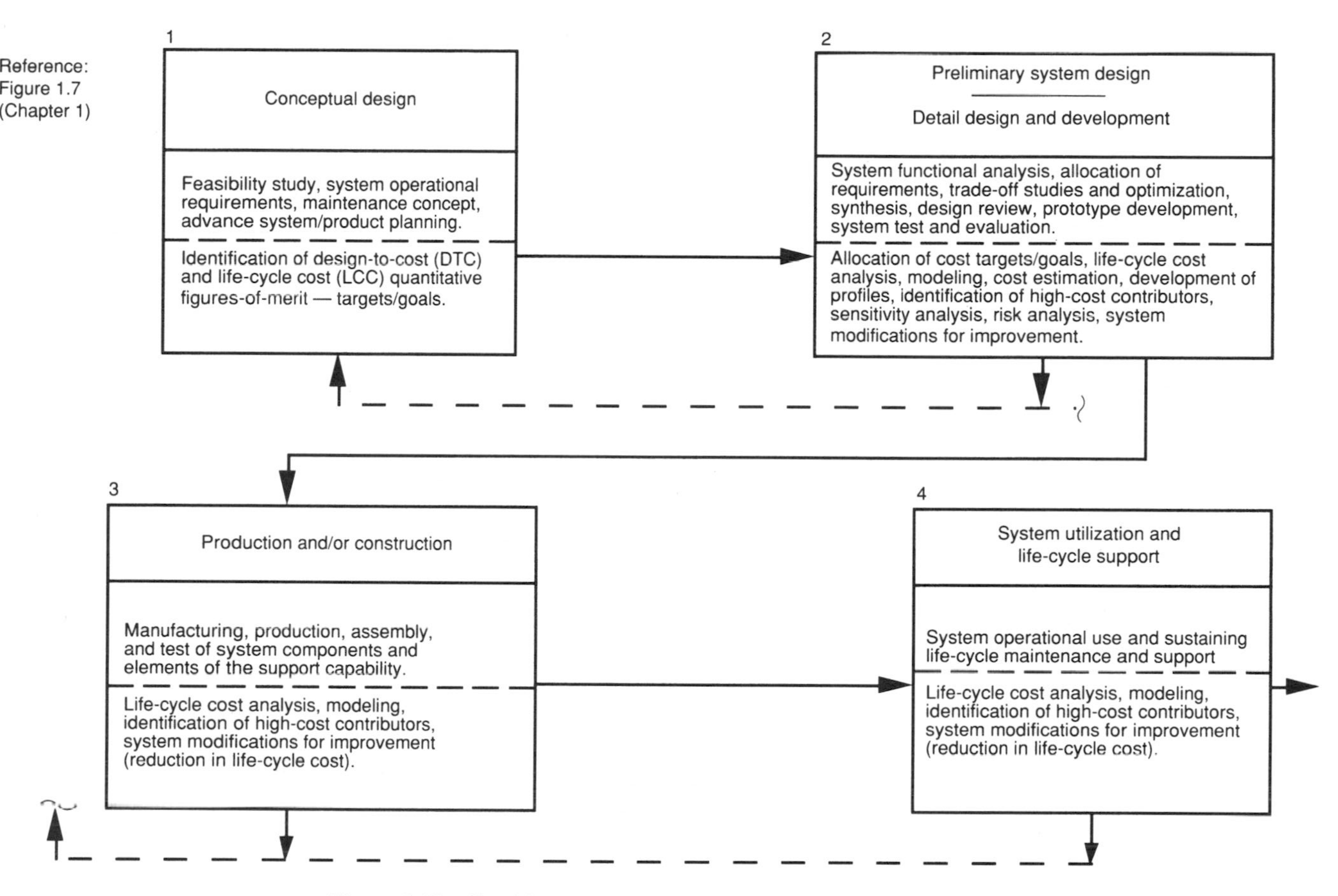
Reference:
Figure 1.7
(Chapter 1)
1
Conceptual design
Feasibility study, system operational requirements, maintenance concept, advance system/product planning.
Identification of design-to-cost (DTC) and life-cycle cost (LCC) quantitative figures-of-merit — targets/goals.
2
Preliminary system design
Detail design and development
System functional analysis, allocation of requirements, trade-off studies and optimization, synthesis, design review, prototype development, system test and evaluation.
Allocation of cost targets/goals, life-cycle cost analysis, modeling, cost estimation, development of profiles, identification of high-cost contributors, sensitivity analysis, risk analysis, system modifications for improvement.
3
Production and/or construction
Manufacturing, production, assembly, and test of system components and elements of the support capability.
Life-cycle cost analysis, modeling, identification of high-cost contributors, system modifications for improvement (reduction in life-cycle cost).
4
System utilization and life-cycle support
System operational use and sustaining life-cycle maintenance and support
Life-cycle cost analysis, modeling, identification of high-cost contributors, system modifications for improvement (reduction in life-cycle cost).

Figure 3.43 Considerations of value/cost in the system life cycle.

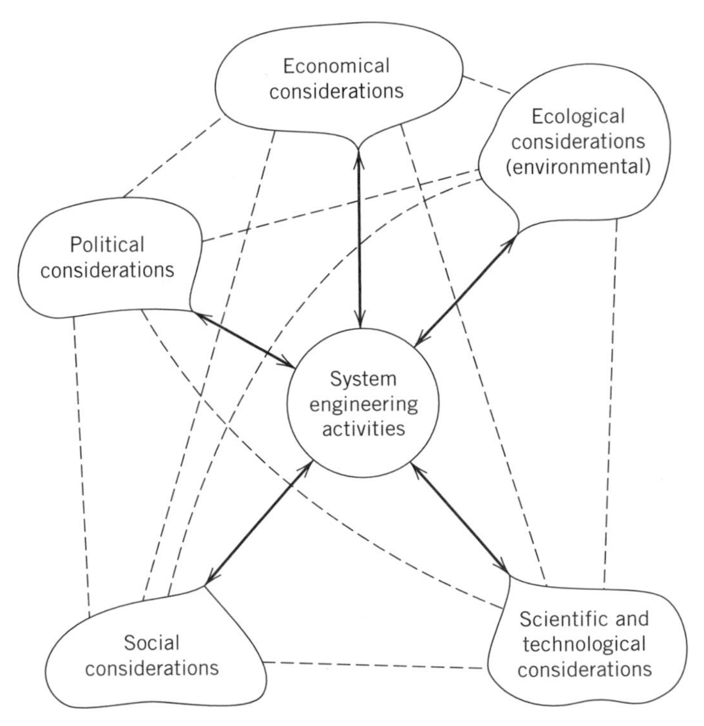

Figure 3.44 Environmental influences on system design and development.

1. *Air pollution and control:* any gaseous, liquid, or solid material suspended in air that could result in health hazards to humans. Air pollutants may fit into categories to include particulate matter (small substances in air resulting from fuel combustion, incineration of waste materials, or industrial processes), sulfur oxides, carbon monoxide, nitrogen oxides, and hydrocarbons.
2. *Water pollution and control:* any contaminating influence on a body of water brought about by the introduction of materials that will adversely affect the organisms living in that body of water (measure of dissolved oxygen content).
3. *Noise pollution and control:* the introduction of industrial noise, community noise, and/or domestic noise that will result in harmful affects on the humans (e.g., loss of hearing).
4. *Radiation:* any "natural" or "human-made" energy transmitted through space that will result in harmful affects on the humans.
5. *Solid waste:* any garbage and/or refuse (e.g., paper, wood, cloth, metals, plastics, etc.) that will result in a health hazard. Roadside dumps, piles of industrial debris, junk car yards, and so on, are good examples of solid waste. Improper solid-waste disposal may be a significant problem in view of the

fact that flies, rats, and other disease-carrying rodents are attracted to areas where there are solid wastes. In addition, there may be a significant impact on air pollution if windy conditions prevail or on water pollution if the solid waste is located near a lake, river, or stream.

In the design of systems, all phases of the life cycle must be addressed, including the "retirement and material disposal" block in Figure 3.31. When the system and its components are retired from the inventory, either because of obsolescence and there is no longer a need or for the purposes of maintenance when items are removed in order to accomplish repair, those items must be of such a makeup that they can be disposed of without causing any negative impacts on the environment. More specifically, a prime objective is to design components such that they can be *reused* in other similar applications. If there are no opportunities for "reuse," then the component should be designed such that it can be decomposed, with the residual elements being *recycled* and converted into materials that can be remanufactured for other purposes. Further, the *recycling process* itself should not create any detrimental affects on the environment.

Thus, in the development of systems and in the selection of components, the designer needs to be sure that the materials selected can be reused if possible, will not cause any toxicity problems, and can be decomposed without adding to the solid-waste inventory that currently exists in many areas. Care must also be taken to ensure that the product characteristics do not generate the need for a nonreusable container or packing materials for transportation that will cause problems. Figure 3.45 conveys a decision-making logic approach that may be applied and that will be helpful in the design and development of systems.

3.4 SUMMARY

Inherent within the system engineering process described in Chapter 2 are the requirements for reliability, maintainability, supportability, quality, and the like. A few design disciplines such as these have been selected for discussion in Section 3.3 of this chapter. In each instance, there are certain steps that are followed in order to meet the objectives as specified. Initially, the requirements for reliability, maintainability, and so on, must be established in defining operational requirements and the maintenance concept for the system. Functional analyses and the allocation of these requirements are necessary to identify input criteria for design. Analyses and trade-off studies are accomplished in the design optimization process. Finally, the initially specified requirements are verified through system test and evaluation. These steps, which are characteristic in each instance, are illustrated in Figure 3.46.

Although the design disciplines in this chapter have been introduced as separate individual requirements, there is a certain degree of interdependence among them. Maintainability requirements are based on reliability, supportability requirements are dependent on reliability and maintainability data, safety factors are based on

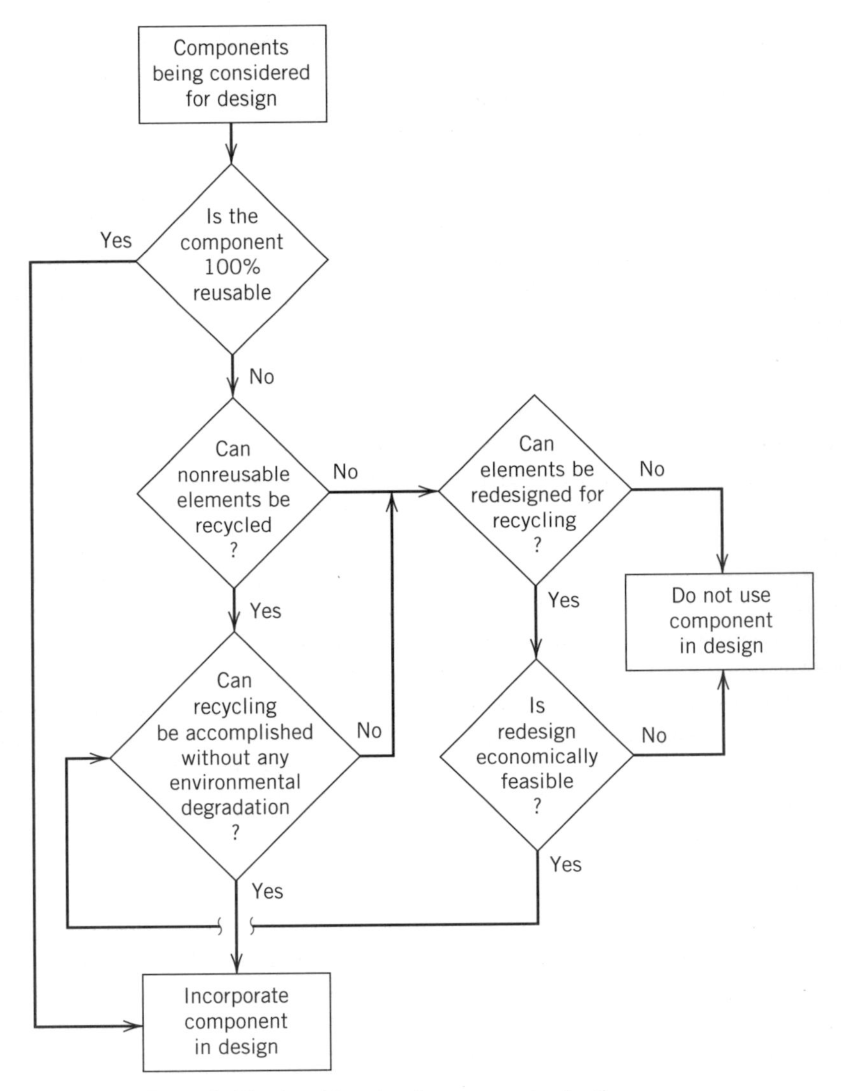

Figure 3.45 An abbreviated component-selection process.

human factors, and so on. These disciplines not only build on the basic design (i.e., electrical design, mechanical design, etc.), but they build on each other. An attempt is made to show these relationships through the order of material presentation in Section 3.3.

Finally, with the objectives of system engineering in mind, it is essential that the appropriate level of communications be established among these disciplines. This communication must be reflected throughout the individual respective program plans, and there must be a free exchange of design-related data in order to fulfill the various analyses and design support functions. The necessity to integrate these

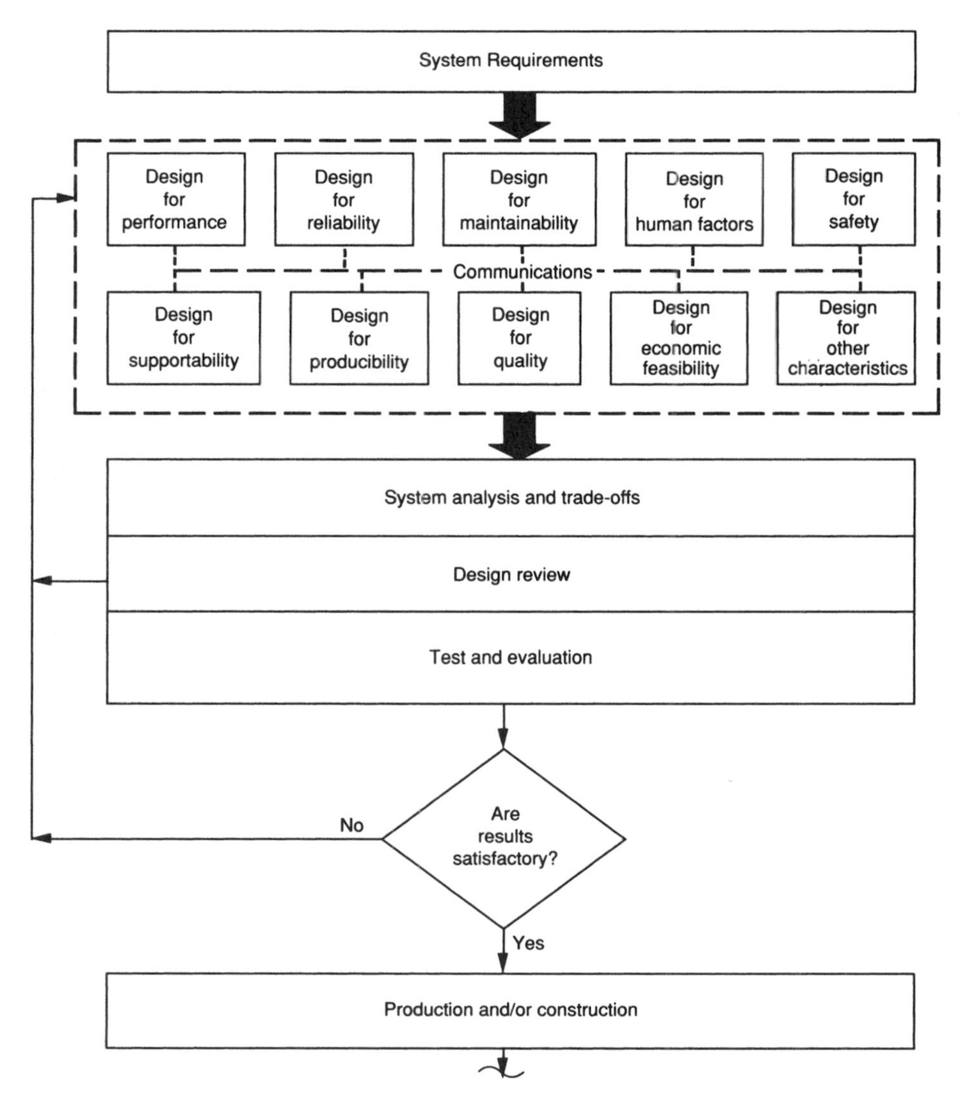

Figure 3.46 The design process.

activities into a total effective engineering design effort is a major aspect of system engineering.

QUESTIONS AND PROBLEMS

1. What is a "documentation tree"? Why is it important to generate one for a program?

2. Select a system of your choice and develop a detailed outline of a Type "A" Specification for that system.

3. Define "reliability." What are the measures of reliability?

4. One hundred (100) parts are tested for 10 hours and 10 failures occurred during the test. The times when the failures occurred are 1, 3, 6, 2, 3, 6, 8, 9, 2, and 1 hour, respectively. What is the failure rate?

5. Field data have indicated that Unit "A" has a failure rate of 0.0004 failure per hour. Calculate the reliability of the unit for a 150-hour mission.

6. A system consists of four subassemblies connected in series. The individual subassembly reliabilities are A = 0.98, B = 0.85, C = 0.90, and D = 0.88. Determine the overall system reliability.

7. A system consists of three subsystems in parallel. Subsystem "A" has a reliability of 0.98, Subsystem "B" has a reliability of 0.85, and Subsystem "C" has a reliability of 0.88. Calculate the overall system reliability.

8. In Figure 3.47, the component reliabilities are A = 0.95, B = 0.97, C = 0.92, D = 0.94, E = 0.90, and F = 0.88. Determine the overall network reliability.

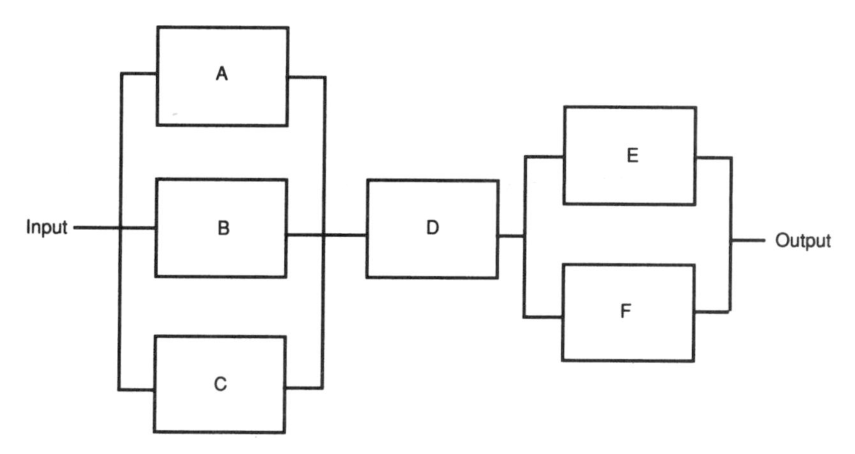

Figure 3.47 Problem 8 network.

9. Develop the overall reliability expression (R_n) for the network shown in Figure 3.48.

10. There are a variety of aids (or tools) that can be effectively utilized in the design process to help meet the objectives of reliability engineering. Briefly describe the objective and the application of each of the following (What is it? How can it be applied? What are the anticipated results?): reliability modeling, reliability allocation, reliability prediction, FMECA, SCA, and FRACAS.

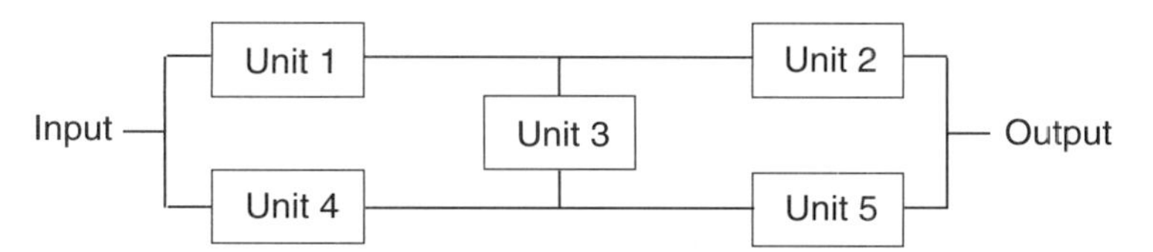

Figure 3.48 Problem 9 network.

11. Define maintainability. What are the measures of maintainability?

12. The corrective maintenance task times in Figure 3.49 were observed.

(a) What is the range of observations?
(b) Using a class interval width of 4, determine the number of class intervals. Plot the data and construct a curve. What type of distribution is indicated by the curve?
(c) What is the $\bar{M}ct$?
(d) What is the geometric mean of the repair times?
(e) What is the standard deviation of the sample data?
(f) What is the $\bar{M}_{max}$ value (assume 90%)?

Task time (min)	*Frequency*	*Task time (min)*	*Frequency*
41	2	37	4
39	3	25	10
47	2	35	5
36	5	31	7
23	13	13	3
27	10	11	2
33	6	15	8
17	12	29	8
19	12	21	14

Figure 3.49 Problem 12 data.

13. Describe the objective and application of each of the following (What is it? How can it be applied? What are the anticipated results?): maintenance concept, maintainability allocation, maintainability analysis, maintainability prediction, and FMECA (as it applies to maintainability).

14. Calculate as many of the following parameters as you can with the given information: Determine

A_i	MTBM
A_a	MTBF
A_o	$\bar{M}$
$\bar{M}ct$	$MTTR_g$
M_{max}	

Given:
$\lambda = 0.004$
Total operation time = 10,000 hours
Mean downtime = 50 hours
Total number of maintenance actions = 50
Mean preventive maintenance time = 6 hours
Mean logistics plus administrative time = 30 hours

15. Define human factors. What are some of the measures of human factors?

16. Identify (by classification) some of the characteristics that must be considered in the design for human factors.

17. Describe the objective and the application of each of the following (What is it? How can it be applied? What are the anticipated results?): functional analysis, detailed task analysis, error analysis, safety/hazard analysis, OSD, and FTA.

18. What is included in the definition of training requirements? How are these requirements determined?

19. Define logistics. Identify some of the logistics activities as they apply to the system life cycle. What is meant by "logistics engineering"? Define "supportability" and "serviceability."

20. Describe some of the measures of logistic support (as defined herein).

21. What is the LSA? What activities are included? How does the LSAP relate to the ILSP? How do these relate to the SEMP?

22. How do reliability, maintainability, human factors, safety, and logistic support requirements relate to the system-level functional analysis described in Chapter 2?

23. Define "software." Identify some of the measures of software.

24. How are the initial requirements for software determined?

25. Describe the process for software design and development (identify the steps). How does this process relate to the system engineering process described in Chapter 2.

26. How would you define "software reliability"? "Software maintainability"?

27. Define TQM. Describe what is meant by "quality engineering." What is meant by "SPC"? How does quality relate to the system engineering process? What are some of the "barriers" to the successful implementation of TQM?

28. Describe what is meant by "concurrent engineering." How does it relate to system engineering?

29. Define "life-cycle cost." How does LCC relate to value? How are economic factors considered in the system design process?

30. What is meant by "CBS"? What does it include? Define "cost profile," "break-even analysis," "sensitivity analysis," and "risk analysis."

31. Select a system of your choice and accomplish a life-cycle cost analysis in accordance with the steps identified in Figure 3.38 and the process described in Appendix B.

32. When one "designs for the environment," what factors should be addressed?

33. How can engineering design impact ecological conditions? What design objectives should you impose in this area?

34. Identify an industry of your choice and assume that your community has been selected as a possible site for a new plant. Develop a checklist of the factors that you would use in assessing the environmental impact of the new plant.

35. Describe, in your own words, the commonalities (or similarities) that exist among the design disciplines covered in this chapter.

4 Engineering Design Methods and Tools

Throughout the system engineering process described in Chapter 2, there are a series of design activities accomplished with the objective of providing a system that will fulfill a designated consumer need. The successful completion of these activities, amplified to some extent through the specific requirements covered in Chapter 3, is dependent on the appropriate application of selected design methods and practices. This, in turn, is strongly influenced by technology and the tools available to and used by the responsible design engineer.

For years, the basic design process had involved a series of activities, accomplished on an individual basis using step-by-step manual procedures. Ideas were generated, conceptually oriented layout or arrangement drawings were prepared and approved, system components were evaluated and selected from design standards documentation, detail drawings and parts lists were developed and reviewed, mock-ups and models were constructed, and so on. In essence, the design process involved a long series of activities, often requiring a great deal of time and often not very well coordinated.

With the advent of computer technology, the design process has been undergoing significant change. Through the introduction of computer graphics in the late 1950s and early 1960s, user-input devices in the late 1960s and early 1970s (keyboards, light pens, joysticks), and the current development of sophisticated design workstations, system/product design includes the application of many new innovations. Computer technology is available to facilitate the generation of graphics material, the accomplishment of analyses, and the fulfillment of data management activities. It is within this context (or environment) that the subject of system engineering needs to be addressed. The purpose of this chapter is to briefly highlight some of these relatively recent concepts in design (e.g., the application of computer-aided design methods), and to discuss system engineering objectives as they relate to current design methods.

4.1 CONVENTIONAL DESIGN PRACTICES

For most projects dealing with small and large-scale systems, the major steps in system development include conceptual design, preliminary design, and detail design, as illustrated in the flow process presented in Figure 1.12. This flow process should, of course, be "tailored" to the specific requirement. Although the basic steps are applicable in the development of all systems, the level of effort and the duration of the project will vary from one situation to the next.

Inherent within the fundamental steps identified in Figure 1.12 are many different activities, all related to the prime objective of designing a system to meet a consumer need. For relatively large projects, such as the commercial aircraft system and the ground mass-transit system referenced in Section 3.2, there may be a requirement for technical expertise representing many different disciplines; for example, electrical engineers, mechanical engineers, structural engineers, materials engineers, aerospace engineers, civil engineers, and reliability engineers. In support of the responsible design engineers in the various fields, there is a need for draftsmen, technical illustrators, component part specialists, laboratory technicians, computer programmers, test technicians, purchasing and contracts specialists, legal experts, and so on. There are many different and varied levels of specialization required for most projects, and each assigned specialty contributes to the design.

To review the steps in design further, Figure 4.1 is presented as an amplification of Figure 3.1. As design progresses, actual definition is accomplished through documentation in the form of plans and specifications (already discussed), procedures, drawings, material and part lists, reports and analyses, computerized data bases, and so on. The design configuration may be the best possible in the eyes of the designer; however, the results are practically useless unless properly documented, so that others can first understand what is being conveyed and then be able to translate the output into a producible entity.

When addressing the aspect of documentation, emphasized in Figure 4.1 to define the various levels of system development, the results of design have generally been conveyed through a combination of the following:

1. *Design drawings:* assembly drawings, specification control drawings, construction drawings, installation drawings, logic diagrams, piping diagrams, schematic diagrams, interconnection diagrams, wiring and cable harness diagrams.
2. *Material and part lists:* part lists, material lists, long-lead-item lists, bulk-item lists, provisioning lists.
3. *Analysis reports:* trade-off study reports supporting design decisions, finite-element analysis reports, reliability and maintainability analyses and predictions, safety reports, logistic support analysis records, configuration identification reports, computer software documentation.

From Figure 4.1 and the steps in design definition, ideas are generated and converted to drawings, drawings are reviewed by various interested disciplines and/or organizations, recommended changes are initiated and incorporated as appropriate, and approved drawings (designated by drawing "sign-off") are released for production. For the most part, the steps in this informal day-to-day process have been completed in series and often require a great deal of time. For instance, an electrical engineer may start the process with a proposed layout of components on a circuit

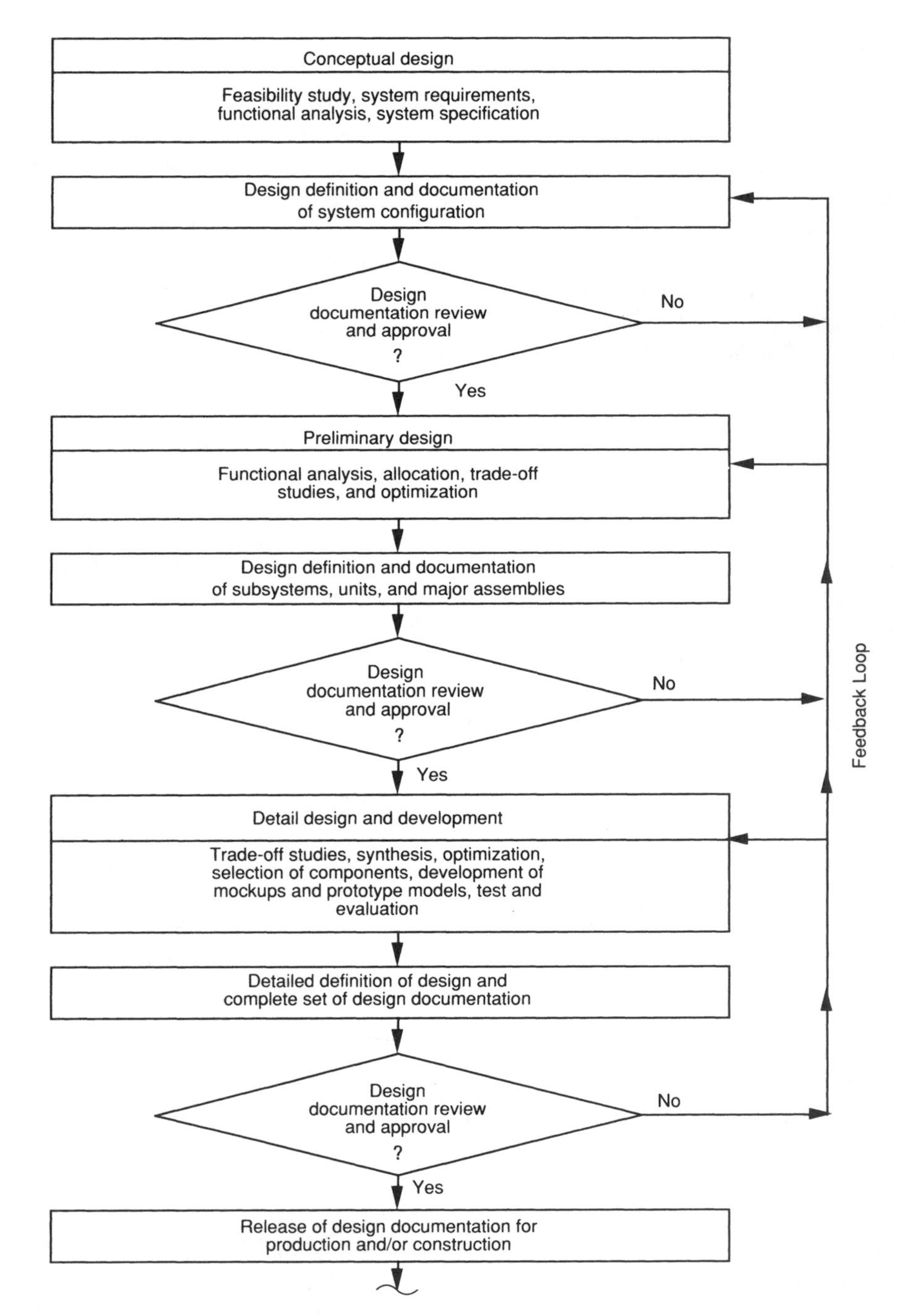
Conceptual design
Feasibility study, system requirements, functional analysis, system specification
Design definition and documentation of system configuration
Design documentation review and approval ?
No
Yes
Preliminary design
Functional analysis, allocation, trade-off studies, and optimization
Design definition and documentation of subsystems, units, and major assemblies
Design documentation review and approval ?
No
Yes
Feedback Loop
Detail design and development
Trade-off studies, synthesis, optimization, selection of components, development of mockups and prototype models, test and evaluation
Detailed definition of design and complete set of design documentation
Design documentation review and approval ?
No
Yes
Release of design documentation for production and/or construction

Figure 4.1 Basic design sequence.

board, a mechanical engineer may then provide the necessary structural and cooling requirements, a reliability engineer may follow with a prediction and evaluation of the selected components, and so on. These responsible individuals may be located in different buildings, and the documentation processing and communications often become quite lengthy. Further, this procedure takes on an additional degree of complexity as the number of drawings and drawing change notices increases.

In essence, this day-to-day design activity, particularly for large-scale systems, may include the generation and processing of hundreds of drawings and drawing change notices (DCNs). Many different design disciplines may be represented, with engineering personnel located throughout the producer's organization and, in some instances, at various remote supplier facilities. The communications, both in terms of verbal discussions pertaining to design approaches and the handling of design documentation, are marginal and the elapsed times are often extensive. In some cases, it may require a month, or more, to route a single drawing through the review steps for approval.

These somewhat conventional practices of the past have created some major challenges! In numerous instances, procedures have been bypassed, changes have been implemented without the proper approvals and necessary coordination, and appropriate configuration controls have not been practiced. In essence, the basic objectives of system engineering have not been followed.

4.2 DESIGN TECHNOLOGIES AND TOOLS

With the advancements in computer technology through the past three decades, many new tools and techniques have been developed and adapted. Not only have the capabilities of mainframe computers increased significantly, but the availability of personal computers (PCs) has literally changed our ways of doing business, particular with regard to engineering design. Additionally, the development of software packages and associated computer models has increased exponentially, and the results have provided the design engineer (and the manager) with a wide variety of tools intended to enhance productivity. Such tools include word processors, graphics packages, spreadsheets, database management packages, and analysis routines.

Given the appropriate computerized design aids, the engineer is able to accomplish a performance analysis employing simulation methods (dynamic versus static, stochastic versus deterministic, continuous versus discrete) in a relatively short time frame. Mathematical programming methods (linear programming, quadratic programming, dynamic programming) can be used in the solving of resource allocation and assignment problems. Statistical tools are available for the plotting of distributions and for determining related characteristics (e.g., mean, standard deviation, range, maximum value). Project management aids are used to plot scheduling networks (e.g., PERT/CPM) and cost projections. Database management models are employed extensively for data acquisition and storage, information processing, and report generation. Finally, there are a wide variety of specialized engineering tools that can be effectively utilized by the designer to help solve specific problems (e.g., the design of a digital filter, the layout of components on a circuit board, and the accomplishment of a reliability analysis for System XYZ).

Relative to application in the design and development process, the availability of these tools offers a number of distinct advantages:

1. The combination of personal computers (PCs), included as part of individual design workstations located throughout the entire project design area, with the connections to a central workstation and a mainframe computer (or equivalent), has created an excellent data communications network. Not only is it possible to transmit many different categories of data, in varying formats, to all applicable workstations on a concurrent basis, but this can be accomplished rapidly and efficiently. Design data packages can be developed by individual designers, transmitted to many others simultaneously, and reviewed with recommended changes submitted to the designer in a very short period of time. This capability minimizes the requirement for completing tasks in series, as described in Section 4.1, and allows for a reduction in the overall system development time. Figure 4.2 illustrates this concept.[1]

2. The versatility and variety of software packages/models provide the designer with many new tools, not readily available in the past. For instance, in systems design, the use of simulation methods early in the conceptual phase enables the designer to do a better job in determining operational requirements and accomplishing a performance analysis. Three-dimensional computer models may be developed in order to evaluate a variety of possible system configurations, to study the interrelationships among system components, to investigate space allocations, to study the performance of human task sequences, and so on. The use of mathematical/statistical models allows the designer to investigate many more alternatives (as compared to what has been possible in the past), involving numerous calculations and the processing of large amounts of data in a short period of time. In essence, the design engineer of today has many more tools available, and the capabilities of these tools allow for a more in-depth analysis and investigation of design alternatives. This capability, applied early in the system life cycle, helps to reduce the risks associated with design decision making by eliminating nonfeasible options in the beginning.

3. The data handling capabilities, provided through computer technology, allow for the acquisition, processing, storage, and retrieval of greater varieties and larger amounts of data than in the past. Data storage and retrieval methods are simplier, and data processing times are much less. Not only is it easy to capture certain design information on drawings, but the generation and distribution of part/material lists can be handled in an expeditious manner. Further, reports and technical publications can be produced automatically by employing a combination of graphics and word processing methods.

[1] One of the objectives through concurrent engineering is to reduce the length of the system acquisition process by appropriately applying new technologies in design and development.

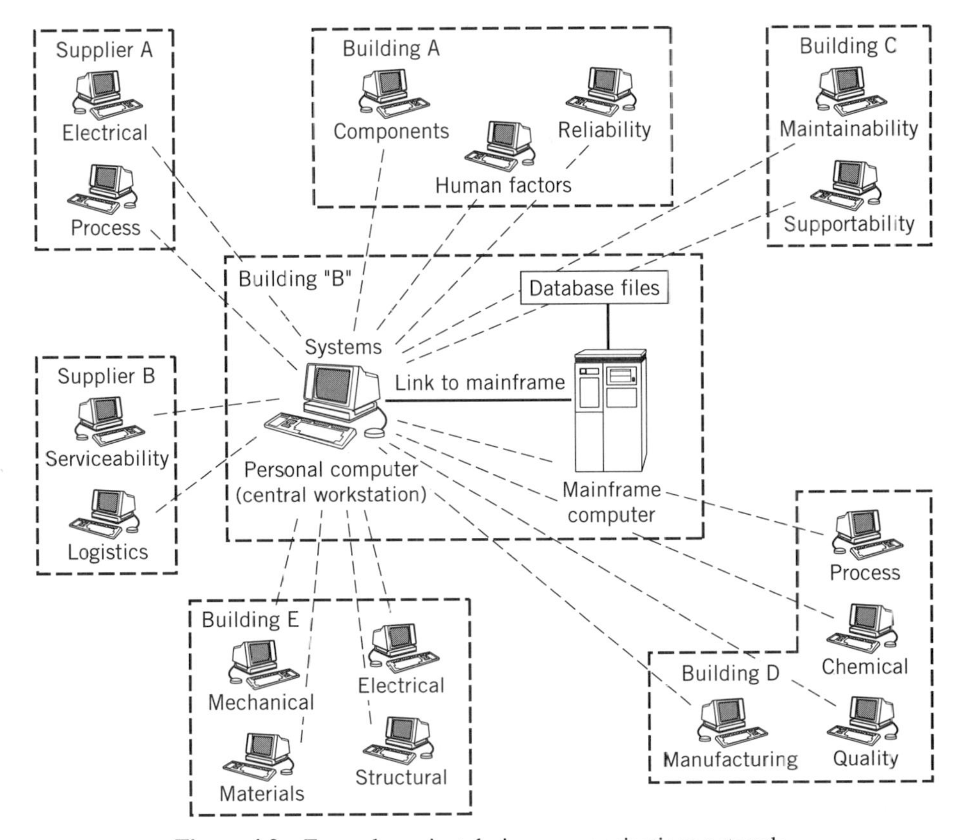

Figure 4.2 Example project design communications network.

From Chapter 1 and the definition of system engineering objectives, the advent of computer technology (along with the many tools that are currently available) can have a very beneficial impact. Specifically:

1. A more in-depth and complete analysis of system requirements can be accomplished during an early stage in design when the life-cycle impacts are the greatest. This, in turn, tends to foster more emphasis on a top-down, life-cycle approach in system development.
2. An improved communications process can evolve as a result of (a) being able to disseminate data rapidly and effectively, (b) being able to transmit information to multiple individuals and/or organizations on a concurrent basis, and (c) being able to incorporate design changes expeditiously. These improvements in communications help to provide the necessary integration of the many different disciplinary areas involved in the design process.

On the other hand, adapting to the new computerized technology is not without some concerns and challenges! First, the development and implementation of a standardized approach for converting design information into a digital format are required. All designers, supporting personnel, producer and consumer organizations, and suppliers must utilize the same language and data format, follow identical practices, and so on.[2]

Second, the methods for linking personal computers (PCs) throughout a company via a local-area network (LAN), between companies via a wide-area network (WAN), and/or the linking of PCs to mainframe computers are critical to the communications process. Figure 4.2, illustrating a "star topology," is one approach, with the objective of allowing for the transmission of design information to multiple workstations on a concurrent basis. Another approach, often utilized, is a circular configuration (i.e., "ring topology"), where data must pass through various workstations in sequence prior to arriving at the centralized computer. This configuration generally does not allow for the transmission to all interested designers simultaneously, and the process assumes a series of actions progressing from one remote workstation to another.

In any event, the design of the protocols and the various networks (e.g., LANs, WANs) must be such as to promote the rapid and effective communication of system/product design information. Further, the hardware and software packages to be utilized must be fully compatible! The various workstations must be able to "talk to each other," and the appropriate levels of communications within a network, and between networks, must be possible. In other words, methods and procedures must be established to allow for the rapid and efficient transmission of design data between the various suppliers and the producer (i.e., the contractor), and between the producer and the consumer (i.e., the customer).[3]

In considering the system engineering objectives highlighted in Chapter 1, these must be viewed in the context of the overall design environment. The objectives relate to the application of a total, top-down, integrated, life-cycle approach in the development of a system that will meet the needs of the consumer, both effectively and efficiently. The successful fulfillment of these objectives is, of course, dependent on the activities accomplished throughout the system development process. This, in turn, is highly influenced by the design capability and environment that exist within the producer/supplier organization. If the appropriate tools are available and properly utilized, the realization of system engineering objectives may be relatively easy. On the other hand, if the environment is not conducive to the accom-

[2] The Department of Defense has recognized this requirement and is attempting to provide some standardization through the application of MIL-D-28000, Military Specification, "Digital Representation for Communication of Product Data: IGES Application Subsets," Department of Defense, Washington, DC. This standard includes the definition of a series of application-specific subsets of the initial graphics exchange specification (IGES), the popular name for American National Standard ANSI Y14.26M, "Digital Representation for Communication of Product Definition Data."

[3] The development of the initial graphics exchange specification (IGES) was initiated to allow for the transfer of design data between networks, CAD/CAM systems, and so on. The capability is being supplemented by the product data exchange specification (PDES), which will encompass the complete set of data elements that defines a system/product for all applications over its expected life cycle.

plishment of *good* design, then additional efforts will be required to ensure that the desired system engineering objectives are met.

Thus, the requirements for system engineering will vary not only as a function of the nature and complexities of the system being developed, but must include consideration of the capability designated to support the design process. The degree to which computerized technology is being utilized may have a significant impact on the system engineering process.

4.2.1 The Use of Simulation in System Engineering[4]

Simulation is the process of designing and utilizing an operational model of a system to conduct experiments for the purpose of either understanding the behavior of the system or for evaluating alternative strategies and/or system design configurations. The objective is to construct a simplified representation of a system or a process in order to facilitate the analysis, synthesis, and/or evaluation of the system/process. The use of simulation is particularly appropriate in the early stages of system development prior to the point when the various physical elements of the system are available for evaluation; that is, during the "analytical" stage specified in Figure 2.31.

The use of simulation methods can be applied in the development of three-dimensional computer-aided design (CAD) models to show the overall system configuration and its components, their location, accesses, interrelationships, and so on. A design engineer can visually see the system configuration, as an entity, during the early stages of preliminary design. This will, of course, enable the designer to evaluate different alternatives, identify potential problems, and accomplish an early synthesis of the design. Some applications include the simulation of an airplane and the layout of its components, an aircraft cockpit with crew, an automobile with driver, a submarine with installed components, a manufacturing facility with capital equipment installed, and so on. Simulation methods can also be used to illustrate process flow rates (e.g., the flow of materials through a factory along with stoppages), different transportation routings, reliability failure patterns, maintenance and support policy alternatives, and so on.

As conveyed in the earlier chapters, one of the objectives of system engineering is to gain as much visibility as possible in the early stages of system development. The intent is to investigate all feasible approaches to design, identify and eliminate potential problems, select a preferred design configuration, and reduce (if not eliminate) the possibility of risk. The use of simulation techniques allows the designer to investigate many different potential design solutions prior to the procurement of equipment, the development of software, and the acquisition of other physical elements of the system. This, in turn, can lead to significant reductions in cost.

[4] A good reference is S. V. Hoover and R. F. Perry, *Simulation: A Problem Solving Approach,* Addison-Wesley, Reading, MA, 1990. The subject of simulation is covered in a number of texts dealing with operations research methods. Refer to Appendix F.

4.2.2 The Use of Rapid Prototyping[5]

In the area of software development in particular, designers are oriented toward the building of "one-of-a-kind" software packages. The issues in software development differ from those in other areas of engineering in that mass production is not the normal objective. Instead, the goal is to develop software that accurately portrays the features that are desired by the user; that is, the interfaces with the customer. For instance, in the design of a complex workstation display, the user may not at first comprehend the implications of the proposed command routines and data format on the screen. When the system is ultimately delivered, problems occur, and the "user interface" is not acceptable for one reason or another. Changes are then recommended, implemented, and the costs of modification and rework are usually high.

The alternative is to develop a "protoype" early in the system design process, design the applicable software, involve the user in the operation of the prototype, identify areas that need improvement, incorporate the necessary changes, involve the user once again, and so on. This iterative and evolutionary process of software development, accomplished throughout the preliminary and detail design phases, is referred to as *rapid prototyping.* Rapid prototyping is a practice that is often implemented and is inherent within the system engineering process, particularly in the development of large software-intensive systems.

4.2.3 The Use of Mockups

Although much information may be acquired through the use of simulation methods in early design, it may be desirable to construct a three-dimensional scale model or physical *mockup* of the system, or an element thereof, during the preliminary or detail design phase to provide a realistic replica of a proposed equipment/facility configuration. These models or mockups can be produced to any desired scale and to varying degrees of detail depending on the level of emphasis required. Mockups may be constructed of heavy cardboard, wood, metal, or a combination of materials. Mockups can be developed on a relatively inexpensive basis and in a short period of time. The utilization of mockups can provide many benefits and their applications are as noted:

1. They provide the design engineer with the opportunity of experimenting with different facility layouts, packaging schemes, panel displays, and so on, prior to the preparation of formal design data.
2. They provide the reliability/maintainability/human-factors engineer with the opportunity to accomplish a more effective review of a proposed design configuration for the incorporation of supportability characteristics. Problem areas readily become evident.

[5]Two references include B. W. Boehm, *Software Engineering Economics,* Prentice Hall, Upper Saddle River, NJ, 1981; and B. Thome (Ed.), *Systems Engineering: Principles and Practice of Computer-Based Systems Engineering,* John Wiley, New York, 1993.

3. They provide the maintainability/human-factors engineer with a tool for use in the accomplishment of predictions and detailed task analyses. It is often possible to simulate operator and maintenance tasks to acquire task sequence and time data.
4. They provide the design engineer with an excellent tool for conveying the final design approach during a formal design review.
5. They serve as an excellent marketing tool.
6. They can be employed to facilitate the training of system operator and maintenance personnel.
7. They are utilized by production and industrial engineering personnel in developing fabrication and assembly procedures and in the design of factory tooling and associated test fixtures.
8. At a later stage in the system life cycle, they may serve as a tool for the verification of a modification kit design prior to the preparation of formal data and the development of kit hardware.

In general, mockups are extremely beneficial. They have been used effectively in facility design, aircraft design, and the design of smaller systems/equipments.

4.3 COMPUTER-AIDED DESIGN (CAD)[6]

Computer-aided design (CAD), defined in a broad sense, refers to the application of computerized technology to the design process.[7] With the availability of computer tools that can be appropriately utilized in the performance of certain design functions, the designer can accomplish more, at a faster pace, and earlier in the life cycle. These tools, including supporting software, incorporate graphics capabilities (vector and raster graphics, line and bar charts, *x–y* plotting, scatter diagrams, three-dimensional displays), analytical capabilities (mathematical and statistical programs for analysis and evaluation), and data management capabilities (data storage and retrieval, data processing, drafting, and reporting). These capabilities are usually combined into integrated packages for application to solve a specific design problem. A few examples of design applications follow:

1. The utilization of graphics, combined with word processing and database management capabilities, enables the designer to:

(a) Lay out components on electrical/electronic circuit boards, design routing paths for logic circuitry, incorporate diagnostic provisions on microelectronic

[6]Two good references are (1) I. Zeid, *CAD/CAM Theory and Practice,* McGraw-Hill, New York, 1991; and (2) J. K. Krouse, *What Every Engineer Should Know About Computer-Aided Design and Computer-Aided Manufacturing,* Marcel Dekker, New York, 1982.

[7]Other terms often utilized to define this area of activity are computer-aided engineering (CAE) and computer-aided systems engineering (CASE).

chips, and so on. CAD capabilities are being used extensively in the design of LSI/VLSI electronic modules and standard packages.

(b) Lay out individual components for sizing, positioning, and space allocation purposes through the use of three-dimensional displays.

(c) Develop solids models enabling assemblies, surfaces, intersections, interferences, and so on, to be clearly delineated via the automatic generation of isometric and exploded views of detailed dimensional and assembly drawings. Component surface areas, volumes, weights, moments of inertia, centers of gravity, and other parameters can be determined automatically. CAD capabilities are being used extensively in the development of solids models for large systems; for example, airplanes, surface ships, submarines, ground vehicles, facilities, bridges, dams, and highways. Many of these models allow the designer to view the system as an entity while providing a top-down hierarchical breakout of system components at different levels. Both two- and three-dimensional displays can be presented, using color graphics as an enhancement.

(d) Develop three-dimensional models of facilities, operator consoles, maintenance work spaces, and the like, for the purposes of evaluating the interface relationships between humans and other elements of the system. CAD tools are being utilized to simulate both operator and maintenance task sequences.

2. The utilization of analytical methods, combined with word processing, spreadsheet, and database management capabilities, enables the designer to:

(a) Accomplish system requirements and performance analyses in support of design trade-off studies; for example, finite-element analysis, structural analysis, stress–strength analysis, thermal analysis, weight/loads analysis, materials analysis.

(b) Perform reliability analyses in support of design; for example, allocations, predictions, FMECA, FTA, sneak circuit analysis, critical useful life analysis, environmental stress analysis.

(c) Perform maintainability analyses in support of design; for example, allocations, predictions, FMECA, diagnostic and test requirements analysis, maintenance task analysis.

(d) Perform human-factors analyses; for example, functional analysis, operator task analysis, OSDs, training requirements analysis.

(e) Perform safety analyses; for example, hazard analysis, fault-tree analysis.

(f) Perform logistic support analyses; for example, maintenance requirements analysis, level-of-repair analysis, spares requirements analysis, transportation requirements analysis, test equipment requirements analysis, facilities requirements analysis.

(g) Perform value/cost analyses; for example, value engineering analysis, life-cycle cost analysis.

3. The utilization of database management, combined with graphics, spreadsheet, and word processor capabilities, enables the designer to:

(a) Develop functional flow diagrams, information/data flow diagrams, dependency diagrams, reliability block diagrams, action diagrams, decision trees and tables.

(b) Develop and maintain a database that includes historical design data, part lists, material lists, supplier information, technical reports. The purpose is to be able to store standard data on common items and to be able to retrieve (or recall) such information quickly and reliably for future applications.

(c) Develop a management information system (MIS) that enables the accomplishment of project review and control functions; for example, data communications, personnel loading projections, cost projections, TPM "tracking," PERT/CPM reporting, project reporting requirements.

Through a review of these areas of application, one can see that the utilization of computer technology is widespread, and is growing at a rapid rate. On the other hand, although there are many uses of computer-aided design technology, they are not very well integrated.

From the overall life cycle illustrated in Figure 4.3, CAD tools are being applied throughout the design and development process, the results of which feed directly into the computer-aided manufacturing (CAM) and computer-aided logistic support (CALS) capabilities.[8] The application of CAD tools has been evolving since the 1960s and 1970s, primarily following a "bottom-up" approach.

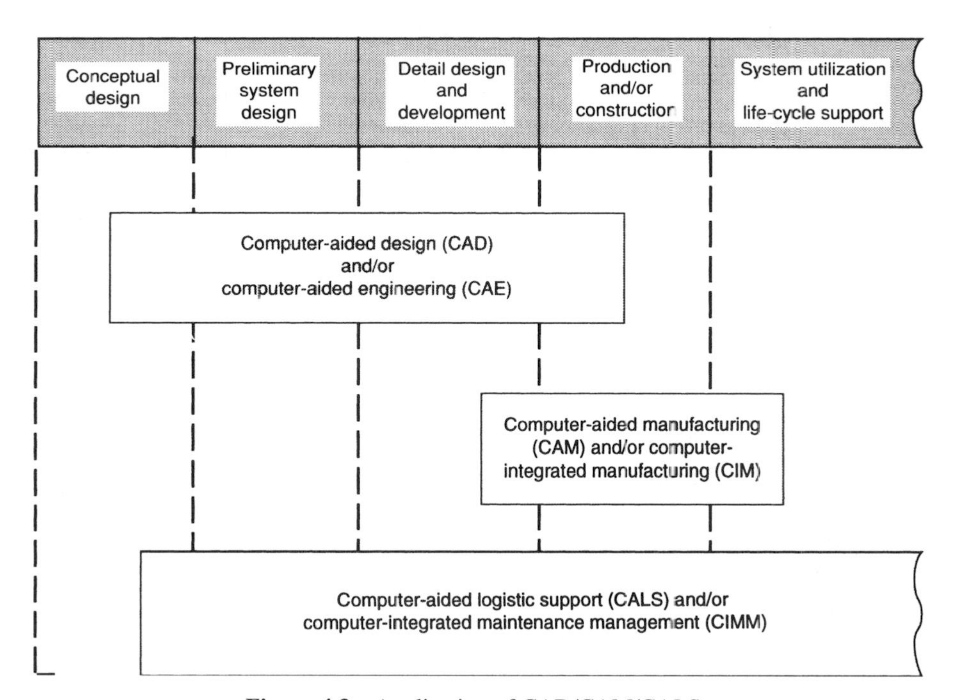

Figure 4.3 Application of CAD/CAM/CALS.

[8] CAM and CALS are discussed further in Sections 4.4 and 4.5, respectively.

Unlike the integrated network concept illustrated in Figure 4.2, individual design workstations have been developed on a somewhat independent basis. A workstation constitutes a grouping or arrangement of equipment (e.g., graphics terminal, computer, keyboard), combined and laid out in such a manner as to help the designer in completing selected tasks. The capabilities of a design workstation will vary depending on (1) the specific design functions to be performed, (2) the nature and complexity of the system being developed (and its components), (3) the education and vision of the responsible designer, in terms of his/her interest and willingness to adapt relative to the utilization of new computer technologies, (4) the degree of management support relative to modernizing the design capability, and/or (5) the availability of the necessary budgetary resources to acquire and support the capital equipment items that are incorporated as part of the design workstation. In many instances, adapting to computerized technologies in design requires a "cultural" change, or reorientation, pertaining to the methods used in task accomplishment. Further, whereas experience has indicated that the use of CAD methods is very cost-effective in the long term, the initial acquisition cost of the required capital assets may be perceived as being too high. In any event, these apparent obstacles must be overcome if progress is to be made in this area.

The capabilities of design workstations have evolved from a configuration where the user was able to design a specific component part, or perform a detailed analysis of some rather isolated function of the system, to a comprehensive configuration integrating many of the diverse technologies discussed earlier. Initially, a graphics terminal was used to design a part, analyze stresses, study and evaluate a mechanical action, and so on. Now, a design workstation can be utilized to effectively accomplish many of the functions noted earlier; for example, the layout of components on circuit boards, the development of three-dimensional solid models, and the utilization of sophisticated analytical methods. Relative to the future, much more can be accomplished in terms of design integration and the incorporation of new capabilities, and the potential for additional growth is great!

The total integrated flow concept illustrated in Figure 4.4 reflects a long-range goal.[9] At the beginning, a comprehensive design workstation can be developed and constructed, along with the appropriate software, to reflect the capabilities inferred through the integrated approach presented in Figure 4.2. In addition to the requirements for electrical design, mechanical design, and structural design, reliability, maintainability, and human factors, supportability and comparable requirements must be integrated into the design process. In other words, all of the design requirements specified in Figure 3.5 (Chapter 3, Section 3.2) must be appropriately integrated, the design communications network illustrated in Figure 4.2 must be available and utilized, and each of the design workstations assigned to a given project must have the capability to deal with these overall requirements.

In addition to providing a capability that will promote a completely integrated design approach, a provision must be incorporated to allow for the smooth transition

[9] In developing an ideal configuration, the capabilities must, of course, be *tailored* to the specific needs of the design organization.

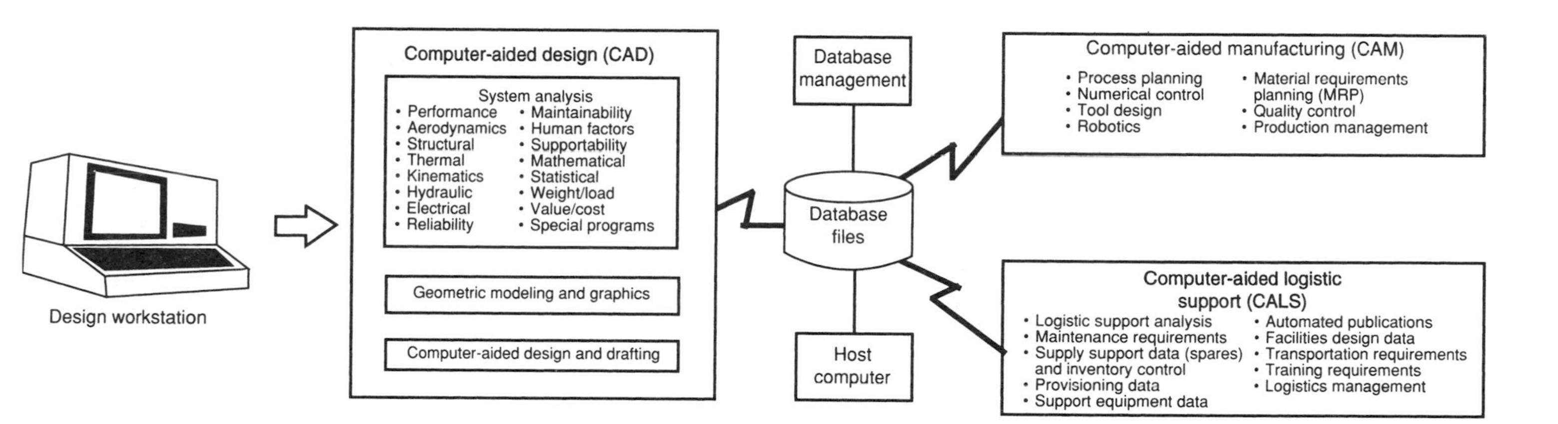
Design workstation
Computer-aided design (CAD)
System analysis
• Performance
• Aerodynamics
• Structural
• Thermal
• Kinematics
• Hydraulic
• Electrical
• Reliability
• Maintainability
• Human factors
• Supportability
• Mathematical
• Statistical
• Weight/load
• Value/cost
• Special programs
Geometric modeling and graphics
Computer-aided design and drafting
Database management
Database files
Host computer
Computer-aided manufacturing (CAM)
• Process planning
• Numerical control
• Tool design
• Robotics
• Material requirements planning (MRP)
• Quality control
• Production management
Computer-aided logistic support (CALS)
• Logistic support analysis
• Maintenance requirements
• Supply support data (spares) and inventory control
• Provisioning data
• Support equipment data
• Automated publications
• Facilities design data
• Transportation requirements
• Training requirements
• Logistics management

Figure 4.4 Major CAD/CAM/CALS interfaces.

of information from the design process to the manufacturing process and to the system support structure. Perhaps, in the not too distant future, the design engineer (with the assistance of a team of discipline-oriented experts) will be able to provide the appropriate design input at the workstation that will, in turn, automatically flow into both the CAM and CALS processes, through the application of effective data communications methods. This provision, of course, will require complete compatibility (language, data format, etc.) among CAD, CAM, and CALS, at both the producer (contractor) and the supplier levels.

Relative to the current status of CAD tools and their applications, many industrial firms have developed and installed CAD capabilities, integrated design workstations, supporting software, and the like. These capabilities vary significantly from one installation to the next. However, in most instances (and particularly for large-scale systems), the use of graphics technology, analytical methods, and database management capabilities has been effectively employed in accomplishing some of the more traditional design activities; for example, electrical circuit design, mechanical packaging design, and structural design. On the other hand, very little has been accomplished relative to the integration of reliability, maintainability, human factors, and supportability into this process; that is, the design disciplines discussed throughout Chapter 3.

Although computer technology has been successfully applied in the development of reliability models, maintainability models, logistic support models, and so on, these tools have been utilized on a "stand-alone" basis. In general, the design activities associated with these disciplines have been accomplished independently, not properly integrated into the basic engineering design process, and the tools that have been developed in these areas have not been integrated with each other or into the more traditional design workstations.[10]

In response to the need to become more effective from the standpoint of contributing to the design process in a timely manner, there has been a great deal of effort expended during the past few years in developing reliability, maintainability, and supportability models. If one were to survey the literature in this area, a listing of over 350 models of varying types could be identified, and this number is growing at a rapid rate. A brief description of a few of these models, which are being utilized in support of different design applications today, follows to provide some insight as to the type of tools that have been available:

1. *Reliability Prediction Program* (RPP). RPP constitutes a series of computer software routines designed to aid in the accomplishment of reliability predictions based on component part stress factors. A system may contain any number of assemblies, subassemblies, and component parts (limited only by available disk space); component failure-rate information can be introduced through a series of menu prompts; part application data and stress factors can be addressed (temperature, power, etc.); and the prediction of MTBF, or λ, values can be determined for the

[10]The term "islands of automation" is often used when referring to analytical tools that are not integrated and are employed on an independent basis. There are many areas where this condition exists today.

system in question. Implementation of RPP can be accomplished using a personal computer and is available for all current versions of MIL-HDBK-217, *Reliability Prediction of Electronic Equipment,* Department of Defense, Washington, DC. (Powertronic Systems, Inc., 13700 Chef Menteur Highway, P.O. Box 29109, New Orleans, LA 70189.)

2. *PC Availability.* This model utilizes Markov Analysis to study the influence of failure rates, repair rates, and logistic support on system availability. The objective is to provide assistance in the development of optimum system configuration design and repair policies. (Management Sciences, Inc., 6022 Constitution Ave., N.E., Albuquerque, NM 87110.)

3. *PC Predictor.* This is a reliability model that automatically applies the part stress analysis or part count methods of MIL-HDBK-217E to produce equipment failure-rate estimates and to accomplish reliability predictions. (Management Sciences, Inc., 6022 Constitution Ave., N.E., Albuquerque, NM 87110.)

4. *Tiger Computer Program.* This is a family of computer programs that can be used to evaluate, by Monte Carlo simulation, an equipment or a large-scale complex system in order to establish various reliability, readiness, and availability measures. Key features include the ranking of equipment by degree of unreliability and unavailability, evaluating a mission with multiphase types, and performing sensitivity analyses on a complex system by downgrading or upgrading the characteristics of each equipment. (Reliability Engineering, Naval Sea Systems Command, Department of the Navy, Washington, DC 20362.)

5. *Mechanical Reliability Prediction Program* (MRP). MRP can be used in accomplishing a reliability prediction of mechanical equipment. The program calculates the reliability of components such as static and dynamic seals, springs, solenoids, valve assemblies, bearings, gears and splines, pump assemblies, filters, brakes and clutches, and actuators. (Powertronic Systems, Inc., 13700 Chef Menteur Highway, P.O. Box 29109, New Orleans, LA 70189.)

6. *Mechanical Reliability Prediction Program* (MECHREL). This program is utilized in the performance of reliability prediction of mechanical systems. Component parts (e.g., gearboxes, valves, pumps, bearings filters) are defined in terms of stresses, failure rates, failure modes, and so on, and combined to predict system MTBF. (Eagle Technology, Inc., 2300 S. Ninth St., Arlington, VA 22204.)

7. *Maintainability Effectiveness Analysis Program* (MEAP). This model is used to compute maintenance times for electronic and electromechanical components, and to accomplish maintainability predictions for systems in accordance with MIL-HDBK-472, *Maintainability Prediction.* Predicted values of MTTR, M_{max}, and MDT for various indenture levels of a system can be determined. (Systems Effectiveness Associates, Inc., 20 Vernon St., Norwood, MA 02062.)

8. *Maintainability Prediction Program* (MPP). MPP can be used to predict the maintainability of electrical, electronic, electromechanical, and mechanical systems. Predictions of MTTR ($\bar{M}ct$), M_{max} (60th to 90th percentile), $\overline{MMH/OH}$, and $\overline{MMH/}$ MA factors associated with maintenance activities at the organizational, intermedi-

ate, and depot/supplier levels can be accomplished. Repair-time factors include localization, fault isolation, disassembly, interchange, reassembly, alignment, and checkout. The quantity of replaceable items, which may be a mixture of assemblies, subassemblies, and component parts, is limited only by disk storage space. MPP, using a personal computer, can be utilized to accomplish a maintainability prediction in accordance with MIL-HDBK-472 (Notice 1, Procedure V, Method B), *Maintainability Prediction,* Department of Defense, Washington, DC. (Powertronic Systems, Inc., 13700 Chef Menteur Highway, P.O. Box 29109, New Orleans, LA 70189.)

9. *Failure Modes, Effects, and Criticality Analysis* (FME). FME can be used in the analysis and development of a FMECA for a system, and can be applied to a user-defined hierarchy, including up to 25 levels of assemblies, subassemblies, components, and so on. FME allows for the determination of failure modes, percent contributions, local effects and effects on the system, frequencies of occurrence, severity classifications, failure detection methods, and recommended compensating maintenance provisions. The data structure is presented in "tree" form, and operations may be accomplished by highlighting individual tree elements. FME can be implemented using a personal computer, and the input-output requirements need to be integrated with reliability and maintainability prediction programs (e.g., the results from the RPP and MPP models). This program includes all features necessary for developing the FMECA in accordance with MIL-STD-1629A (Notice 1), "Procedures for Performing a Failure Mode, Effects and Criticality Analysis," Department of Defense, Washington, DC. (Powertronic Systems, Inc., 13700 Chef Menteur Highway, P.O. Box 29109, New Orleans, LA 70189.)

10. *Equipment Designer's Cost Analysis System* (EDCAS). EDCAS is a design tool that can be used in the accomplishment of a level-of-repair analysis (LORA). It includes a capability for the evaluation of repair versus discard-at-failure decisions, and it can handle up to 3500 unique items concurrently (i.e., 1500 line replaceable units, 2000 shop replaceable units). Repair-level analysis can be accomplished at two indenture levels of the system, and the results can be used in determining optimum spare/repair part requirements and in the accomplishment of life-cycle cost analysis (LCCA). EDCAS is available through a simplified personal computer-oriented University Edition (UE) and a full-scale laboratory model edition, and is compatible with the requirements of MIL-STD-1390B, "Level of Repair," Department of Defense, Washington, DC. (Systems Exchange, Monterey, CA.)

11. *Optimum Repair Level Analysis* (ORLA) *Model.* A level-of-repair analysis (LORA) model used to examine the economic feasibility of maintenance and support alternatives. Up to four alternatives can be evaluated, with life-cycle cost broken down into 13 distinct logistics areas. (U.S. Army-MICOM, Code AMSMI-LC-TA-L, Redstone Arsenal, AL 35898.)

12. *Level of Repair* (NAVY). This model is used in the performance of a level-of-repair (LOR) analysis during the design and development of new systems and equipment. One objective is to establish a least-cost maintenance policy, and to

influence design in order to minimize logistic support costs. (MIL-STD-1390B, Military Standard, "Level of Repair," Department of Defense, Washington, DC.)

13. *Network Repair Level Analysis* (NRLA). NRLA is used in establishing equipment and component repair levels, and for making repair versus discard-at-failure decisions in both the design of new equipment and in provisioning. (U.S. Air Force, Air Force Acquisition Logistics Center, Wright-Patterson AFB, OH 45433.)

14. *Optimum Supply and Maintenance Model* (OSAMM). OSAMM is a level-of-repair analysis (LORA) model used to determine the optimum economic maintenance policy for each item that fails, to identify items in terms of repair versus discard, to identify economic screening criteria, and to evaluate support equipment/repairmen options. Four levels of support and four indenture levels of equipment can be evaluated. (U.S. ARMY-CECOM, Fort Monmouth, NJ 07703.)

15. *Computed Optimization of Replenishment and Initial Spares Based on Demand and Availability* (CORIDA). CORIDA computes initial and replenishment spare/repair part requirements, for multiple levels of maintenance, in terms of costs and distribution. It addresses organizational and budgeting constraints, and is utilized in support of provisioning activities. (Thomson-CSF Systems Canada, 350 Sparks St., Suite 406, Ottawa, Ontario KIR 7S8, Canada.)

16. *OPUS Model.* OPUS is a versatile model used primarily for spare/repair parts and inventory optimization. It considers different operational scenarios and system utilization profiles in determining demand patterns for spares, and it aids in the evaluation of various design packaging schemes. Alternative support policies and structures may be evaluated on a cost-effectiveness basis. (Systecon AB, Linnëgatan 5, Box 5205, S-10245, Stockholm, Sweden.)

17. *VMETRIC.* This is a spares model that can be used to optimize system availability by determining the appropriate individual availabilities for system components and the stockage requirements for three indenture levels of equipment (e.g., line replaceable units, shop replaceable units, and subassemblies) at all echelons of maintenance. Outputs include optimum stock levels at each echelon of maintenance, EOQ quantities, and optimal reorder intervals. (Systems Exchange, Monterey, CA.)

18. *Systems and Logistics Integration Capability* (SLIC). This program is an integrated logistic support analysis data management system designed to respond to CALS objectives in producing a mini LSAR in accordance with MIL-STD-1388-2, "Requirements for a Logistic Support Analysis Record." (Integrated Micro Systems, Inc., 306 Pinecliff Dr., P.O. Box 1438, Seneca, SC 29679.)

19. *Distributed Integrated Logistic Support Analysis* (DILSA). This program is a distributed database processor designed to produce a mini logistic support analysis record (LSAR). It incorporates reliability, maintainability, and logistics data, and creates LSAR reports in accordance with MIL-STD-1388-2, "Requirements for a Logistic Support Analysis Record." (Logistic Engineering Associates, 2700 Navajo Road, Suite A, El Cajon, CA 92020.)

20. *System Design Utility* (SDU). The SDU model can be applied during the early stages of system design in the accomplishment of requirements allocation (i.e.,

the top-down apportionment of system-level requirements to lower-level elements of the system), and in the later stages of development in the accomplishment of design trade-off studies. It incorporates the flexibility to allow for the evaluation of different system architectures, as well as providing database management capability for configuration control purposes. The output from the SDU model can be utilized in a variety of situations, including providing an input for EDCAS. (Systems Exchange, Monterey, CA.)

21. *Requirements Driven Development* (RDD-100). RDD is a modeling approach employed to evaluate various design alternatives and to assess system behavior. This approach utilizes functional block diagrams to describe sequential and concurrent activities, input-output requirements, data flow and physical item flow, and to assist in system decomposition. RDD is a system engineering method, supported by an executable graphic modeling language and computer-based tools developed to assist the engineer in designing a complex system through functional analysis and allocation. System platforms include Apollo, Apple Macintosh II, and Sun with 8 Mb RAM. (Ascent Logic Corp., 180 Rose Orchard Way, Suite 200, San Jose, CA 95134; or Omnitech Systems, Inc., 2070 Chain Bridge Road, Suite 320, Vienna, VA 22180.)

22. *Repairable Equipment Population System* (REPS). REPS is designed to evaluate a system capability. It tracks the scenario of a homogeneous population of repairable equipment meeting a demand. These units fail and are repaired through a repair channel. After a designated time period, the old units are replaced by new ones. The model is used to calculate the optimum combination of units to deploy, the number of repair channels to maintain, and the retirement age, based on the annual equivalent cost to maintain the system to meet a certain demand. (Systems Engineering Design Laboratory, ISE Department, Virginia Tech, Blacksburg, VA 24061.)

23. *Life-Cycle Cost Calculator* (LCCC). LCCC aids in the accomplishment of a life-cycle cost analysis, using a cost breakdown structure (CBS) methodology. Objectives and activities are linked to resources, and constitute a logical subdivision of cost by functional activity area, major element of a system, and/or one or more discrete classes of common or like items. It provides a mechanism for the initial allocation of cost, cost categorization, and finally cost summation. It has the capability of identifying cost contributions, both in real and discounted dollars, at any level in the CBS. (Systems Engineering Design Laboratory, ISE Department, Virginia Tech, Blacksburg, VA 24061.)

24. *Cost Analysis Strategy Assessment* (CASA). CASA is utilized to develop life-cycle cost (LCC) estimates for a wide variety of systems and equipment. It incorporates various analysis tools into one functioning unit and allows the analyst to generate data files, perform life-cycle costing, sensitivity analysis, risk analysis, cost summaries, and the evaluation of alternatives. (Defense Systems Management College, DSS Directorate (DRI-S), Fort Belvoir, VA 22060.)

25. *Computer-aided System Engineering* (CORE). CORE supports both static and dynamic analysis of system requirements, functional behavior, and system ar-

chitecture, with automatic generation of customizable reports via a simple scripting language. The model enables the user to develop a system description in a well-defined, structured manner; provides a full database framework and flexible schema to support requirements traceability; and can be easily implemented using a PC with a Microsoft Windows capability (VITECH Corp., 2070 Chain Bridge Road, Suite 320, Vienna, VA 22182).

These models are only a representative few in the total spectrum of tools available throughout the industrial and government sectors today. However, as indicated earlier, they have been developed on an independent basis and are being utilized outside of the mainstream design effort. The primary objectives in the future are to

1. Evaluate and integrate these and other tools, as applicable, such that they can be utilized interactively as conveyed through the illustration in Figure 2.26. The goal is to develop a set of tools that (a) can be effectively utilized in response to a variety of needs, (b) can address the system as an entity, while being utilized to evaluate different components of the system, and (c) can "talk to each other" in terms of data communications.

2. Incorporate these and other tools, as appropriate, into the workstations for all responsible designers. The design engineer, using the ideal workstation shown in Figure 4.4, should not only have available the necessary tools for the accomplishment of a structural analysis, but for the accomplishment of a reliability analysis as well. The same individual may not accomplish both tasks; however, he or she needs to view the results of each (and their impacts on each other) on a concurrent basis in order to make intelligent design decisions.

With regard to system engineering, as it applies to CAD applications, the challenges for the future are directly in line with these objectives. The appropriate tools must be available and well-integrated into the typical design workstation, in such a manner as to promote the proper consideration of all disciplines in the system design process.

4.4 COMPUTER-AIDED MANUFACTURING (CAM)[11]

Computer-aided manufacturing (CAM) refers to the application of computerized technology to the manufacturing or production process. This application, reflected in the context of the system life cycle presented in Figure 4.3, primarily includes the use of automated methods as they pertain to the following activities:

1. *Process planning:* throughout the production process, there may be a series of steps required to fabricate a component, to assemble a group of items, or a combi-

[11] Another term sometimes utilized to define this area of activity is computer-integrated manufacturing (CIM).

nation thereof! Process planning addresses the entire flow of activities, evolving from the definition of a given design configuration to the finished product delivered to the consumer, and CAM applications include those activities that can be automated. Whereas the activities associated with process planning have been known and practiced for a long time, the use of computer technology in accomplishing these activities has been a relatively recent innovation.[12]

2. *Numerical control* (NC): within the production process, there may be many instances in which machine tools are required for milling, drilling, cutting, routing, welding, bending, or a combination of these operations. The application of computerized technology for the control of machine tools, with prerecorded coded information, to make a part has been practiced for many years. However, these activities have often been accomplished in isolation from other activities in the production process. NC instructions have been prepared by programmers taking information from engineering drawings, programs have been tested, revised, retested, and so on. As this can be quite expensive, the goal in the future is to be able to generate NC input instructions directly from the design database, developed through the CAD applications discussed in Section 4.3.

3. *Robotics:* at various stages in the production process, there may be applications where robots can be effectively employed for the purposes of materials handling (i.e., the carrying of parts from one location to the next), or for the positioning of tools and workpieces in preparation for NC applications. In some instances, robots are actually being used to operate drills, welders, and other tools. Computer technology is, of course, employed in the programming of robots for these and related production operations.

4. *Production management:* throughout the production process, there is an ongoing management activity where computer applications can be effectively utilized in support of production forecasting, scheduling, cost reporting, MRP activities, the generation of management reports, and so on. There is a requirement to develop a management information system (MIS) that enables the review and control of production functions.

The application of CAM methods to the production process will be different for a "flow shop" as compared to a "job shop" operation, or for the production of multiple quantities versus the construction of a "one-of-a-kind." Independent of the function, the entire process must be addressed as a system, possible CAM applications (i.e., combinations of NC, robotics, and data processing methods) must be identified, and a well-integrated and flexible manufacturing capability must be developed. The objective is to design a production capability with the proper mix of people and degrees of automation.

In the initial design, and subsequent monitoring and control, of a production

[12] In evaluating the entire production process, there may be some activities that are similar in nature and where standard procedures can be utilized. This is particularly true in the manufacture of components where, by appropriate grouping, a common and standardized process can be applied using CAM methods. This approach is often defined as "group technology."

capability, the principles covered under concurrent engineering and quality engineering must prevail. The manufacturing tolerances, allowable through the application of NC tools and robotics, must be consistent with the initial requirements for system design. Care must be exercised to ensure that unexpected variances, causing possible product degradation through the production process, do not occur. This is of particular concern relative to meeting the objectives of system engineering.

4.5 COMPUTER-AIDED LOGISTIC SUPPORT (CALS)[13]

Computer-aided logistic support (CALS) refers to the application of computerized technology to the entire spectrum of logistics and serviceability, as it is defined in Section 3.3.8. For years, the consideration for logistic support requirements has been relegated to an activity downstream in the system life cycle. Further, in the initial definition of system support requirements, the process that has been implemented includes the generation of an extensive amount of documentation, much of which is distributed through a network involving many different locations. As systems are produced and delivered for operational use, the overall logistics activity often assumes an additional degree of complexity, involving a great deal of data/documentation, the processing and distribution of many different components, and the requirements for a significant amount data communications. The logistic support requirements, particularly for large-scale systems, have not always been responsive to the system need and have been costly.

In addressing logistics in the context of the system life cycle, there are design-related activities, analysis activities, technical publications activities, provisioning and procurement activities, fabrication and assembly activities, inventory and warehousing activities, transportation and distribution activities, maintenance and product support activities, and management activities across the spectrum. In Figure 4.3, the applications are broad, and there is a certain degree of overlap with CAD and CAM requirements.

To provide an indication as to the variety of computer technology applications in the logistics field, the following examples are noted:[14]

1. *Logistics engineering:* The utilization of reliability, maintainability, and supportability models in the accomplishment of design trade-offs is a major requirement throughout system development (e.g., level-of-repair analysis, spares requirements, maintenance loading, transportation analysis, life-cycle cost models). This activity should be integrated into the CAD effort described in Section 4.3, and the

[13] CALS represents a concept that is being promoted by the Department of Defense to improve the quality, timeliness, and responsiveness in the acquisition of future system support requirements. A good reference is MIL-HDBK-59, Military Handbook, *Computer-Aided Acquisition and Logistics Support (CALS) Program Implementation Guide,* Department of Defense, Washington, DC.

[14] The organization of files of digital data into completed documents and reports is addressed in the defense industry by MIL-STD-1840A, Military Standard, "Automated Interchange of Technical Information," Department of Defense, Washington, DC.

use of graphics technology, analytical methods, and database management capabilities is required.

2. *Logistic support analysis* (LSA): Through the evaluation of a given design configuration, LSA data are developed with the objective of identifying the specific requirements for system support; for example, spare parts, test and support equipment, personnel quantities and skill levels. These data must be generated, processed, stored, retrieved, and fed back in a timely manner if the design for supportability is to be realized. Data processing and database management capabilities are required.

3. *Technical documentation:* This category covers the requirements for spares and repair parts provisioning data, support equipment provisioning data, design drawings and change notices, technical procedures, training manuals, and various reports. The development of technical manuals (i.e., system operating procedures and maintenance instructions) through automated processes is included. Computerized technology applications require the use of spreadsheets, word processing, graphics, and database management capabilities.

4. *Distribution, transportation, and warehousing:* The ongoing maintenance and support of the system throughout its planned life cycle requires the distribution, transportation, handling, and warehousing activities pertaining to spares and repair parts. In addition to the data processing and database management requirements associated with inventory control and MRP activities, the application of automated materials handling equipment and robotics can be effectively employed in the performance of warehousing functions. The automatic "ordering" and "picking" of components from the warehouse shelves, in response to a defined need, provides an excellent example of the use of computerized technology in the logistics area. Further, the opportunities for the future incorporation of "expert systems," or "artificial intelligence," are apparent when considering the "IF-THEN" type of decisions that are necessary throughout materials handling process.

5. *Maintenance and support:* The customer service activities, in support of the system in the field throughout its planned life cycle, include the accomplishment of scheduled and unscheduled maintenance actions. This, in turn, leads to the consumption of maintenance personnel resources, the utilization of test equipment, the requirements for spare parts, the need for formal maintenance procedures, and so on. Although computerized methods have been utilized in the generation of spares/repair parts provisioning data and technical publications, there are additional applications relative to test equipment. In a few instances, handheld testers, with supporting software, have been used for maintenance diagnostics purposes. This area, supported with some selected applications of artificial intelligence, is a prime candidate for future growth.

As is the case relative to CAD, the CALS objective is being implemented at the detail level (i.e., a bottom-up approach). Initially, there has been a great deal of effort expended in the development of requirements and interface specifications. Additionally, there are many activities underway that are directed toward the development of technical manuals and procedures using automated processes. The development and processing of logistic support analysis records (LSARs) and provisioning data are further examples where much has been accomplished. Whereas these efforts currently represent "islands of automation," the overall and objective is to integrate the CALS requirements with CAD/CAM, as illustrated in Figure 4.4.

4.6 SUMMARY

The successful implementation of the system engineering process is highly influenced by technology and the tools available to the designer. The advent of computer technology and the proper use of graphic methods and displays, analytical models, spreadsheets and word processing, database management capabilities, and so on, have enabled the designer to accomplish much more, in a shorter time frame, earlier in the system life cycle, and with less overall risk. As such, the design process has undergone the first phase in the transition from a long series of manually performed tasks to a more efficient integrated and automated process.

Relative to the future, the challenge is to complete the next phase. This involves the integration of the many analytical methods/tools, currently being utilized on an individual stand-alone basis, into the overall design process. The objective is to (1) develop a workstation concept providing the appropriate communications between all responsible design engineering functions (illustrated in Figure 4.2), and (2) develop a capability allowing the smooth and automated flow of information from the CAD capability to CAM and CALS (illustrated in Figure 4.4). Not only must these capabilities be able to "talk to each other" within the context of a given project, but the design information produced must be compatible and transferable between other projects having similar objectives. Thus, care must be exercised relative to the selection of an appropriate language, data format, and data structure. Nevertheless, a great deal of progress and future growth in this area can be anticipated.

QUESTIONS AND PROBLEMS

1. Define, in your own words, and describe what is included in each of the following: CAD, CAM, and CALS.

2. Identify and describe some of the "technologies" that are being applied in the design process. Provide some examples of typical applications.

3. Describe some of the benefits associated with the application of computerized methods in the design process. How do these methods relate to the objectives of system engineering?

4. Identify some of the problems associated with the application of computerized methods in the design process. What cautions must be observed?

5. For the design disciplines identified in Chapter 3, Section 3.3, a listing of typical program tasks is provided. In each of the following disciplines, identify those

tasks that can be accomplished using computerized methods. Include some specific examples.

(a) Reliability engineering tasks?
(b) Maintainability engineering tasks?
(c) Human-factors engineering tasks?
(d) Logistics, serviceability, and supportability engineering tasks?
(e) Quality engineering tasks?
(f) Value/cost engineering tasks?

6. Describe what is meant by "artificial intelligence." How can it be applied?

7. How does the application of computer technology impact "concurrent engineering"?

8. Define and describe the objective of IGES and PDES.

9. Refer to Figure 4.2. Select a system of your choice (describing the tasks in design), and develop a flow chart showing the application of CAD as it is being implemented.

10. How does CAM relate to system engineering? Describe some possible impacts.

11. How does CALS relate to system engineering? Describe some possible impacts.

12. Draw a flow chart showing the interface relationships among CAD, CAM, and CALS.

5 Design Review and Evaluation

System design is an evolutionary process, progressing from an abstract notion to something that has form and function, is fixed, and can be reproduced in specified quantities to satisfy a designated consumer need. Initially, a requirement (or need) is identified. From this point, design evolves through a series of phases; that is, conceptual design, preliminary system design, and detail design and development, illustrated in Figure 1.12.

As the design progresses, there are natural degrees of system definition. Requirements are defined leading to a "functional" baseline. This includes the definition of operational requirements and the maintenance concept, trade-off study reports and the results of the feasibility analysis, the identification of technical performance measures (TPMs), and the system specification (Type "A"). Functional analysis and requirements allocation are accomplished, the results of which are defined through an "allocated" baseline. This baseline may be defined through a combination of development, process, product, and/or material specifications (Type "B," "C," "D," and "E") as applicable. This configuration is progressively expanded, through numerous iterations, until a "product" baseline is defined, and so on. These natural phases of system definition are reflected by the activities and milestones identified in Figure 1.21.

In viewing the overall design process, the necessary "checks and balances" must be incorporated to ensure that the system configuration being developed will indeed fulfill the initially specified requirements. These checks and balances, accomplished through the conductance of design reviews, are provided early in the system life cycle when changes can be accomplished relatively easy and usually without great cost. A design review and evaluation function must be integral with the design process. Within the design review function, there must be feedback provisions for corrective action and the incorporation of design changes as necessary. The basic philosophy of design evolution, with the necessary review and feedback provisions, is shown in Figure 1.22. The purpose of this chapter is to explain this concept by describing evaluation methods, informal and formal design reviews, and the associated feedback and corrective-action loop.

5.1 DESIGN REVIEW AND EVALUATION REQUIREMENTS

One of the objectives in establishing a formal mechanism for design review and evaluation is to ensure, on a progressive and continuing basis, that the results of design reflect a configuration that will ultimately meet the stated consumer need.

Design evolves from the initial definition of requirements for a given system, through a series of iterations following a top-down approach, to a firm system configuration ready for production and/or construction. As one progresses through this series of steps, it is important that one initiate the requirements verification process from the beginning, because the earlier that potential problems are detected, the easier it will be to incorporate changes if needed. Thus, an ongoing design review and evaluation effort is required.

In evaluating the various stages of design, illustrated in Figures 1.12 and 3.1, the overall review process can be effectively accomplished through a combination of several approaches. First, there is an informal day-to-day review and evaluation activity that occurs as design decisions are made and data are developed (refer to Sections 2.9 and 2.10). This activity may involve many different design disciplines, making decisions on a relatively independent basis, and generating design data based on the results. Second, formal design reviews are conducted at designated stages in the evolution of design, and these serve as a vehicle for communications and the formal approval of design data. These two main areas of activity are reflected in Figure 5.1 and are discussed further in Sections 5.2 and 5.3, respectively.

In response to the "WHYs" of design review, the objective is to ensure that system requirements are being met. These requirements, which are included in the system specification (refer to Section 3.1), are stated in both quantitative and qualitative terms. The purpose of the design review process is to evaluate the system configuration at different stages in terms of these requirements.

When addressing the aspect of "requirements," there are program-level requirements, system-level technical requirements, detailed design requirements at the component level, and so on. These requirements are not only viewed in a hierarchical sense, but the level of emphasis placed on these requirements will shift as we progress from conceptual design to the detail design and development phase. For example, it may be appropriate to establish a hierarchical relationship of system parameters such as that shown in Figure 2.24. Many of these parameters can be expressed in terms of a specific quantitative measure of system performance; that is, the identification of a technical performance measure (refer to Section 2.6). Some of these measures are applicable at the system level, some are more appropriately applied at the subsystem level, and some are directly related to the assembly or component level. In any event, the system specification (and its supporting specifications) should establish the "order" of evaluation parameters on the basis of priority and importance.

From the desired hierarchical relationship(s) of evaluation parameters, it is now possible to establish some specific criteria against which the results of design are compared. This, of course, leads to the identification of design review requirements for conceptual design, for preliminary system design, and for the detail design and development phase. In conceptual design, the design review process must address top system-level performance measures, functional-level relationships, and so on (as included in the System Type "A" Specification). In the detail design and development phase, although the system-level requirements are still important, the emphasis may

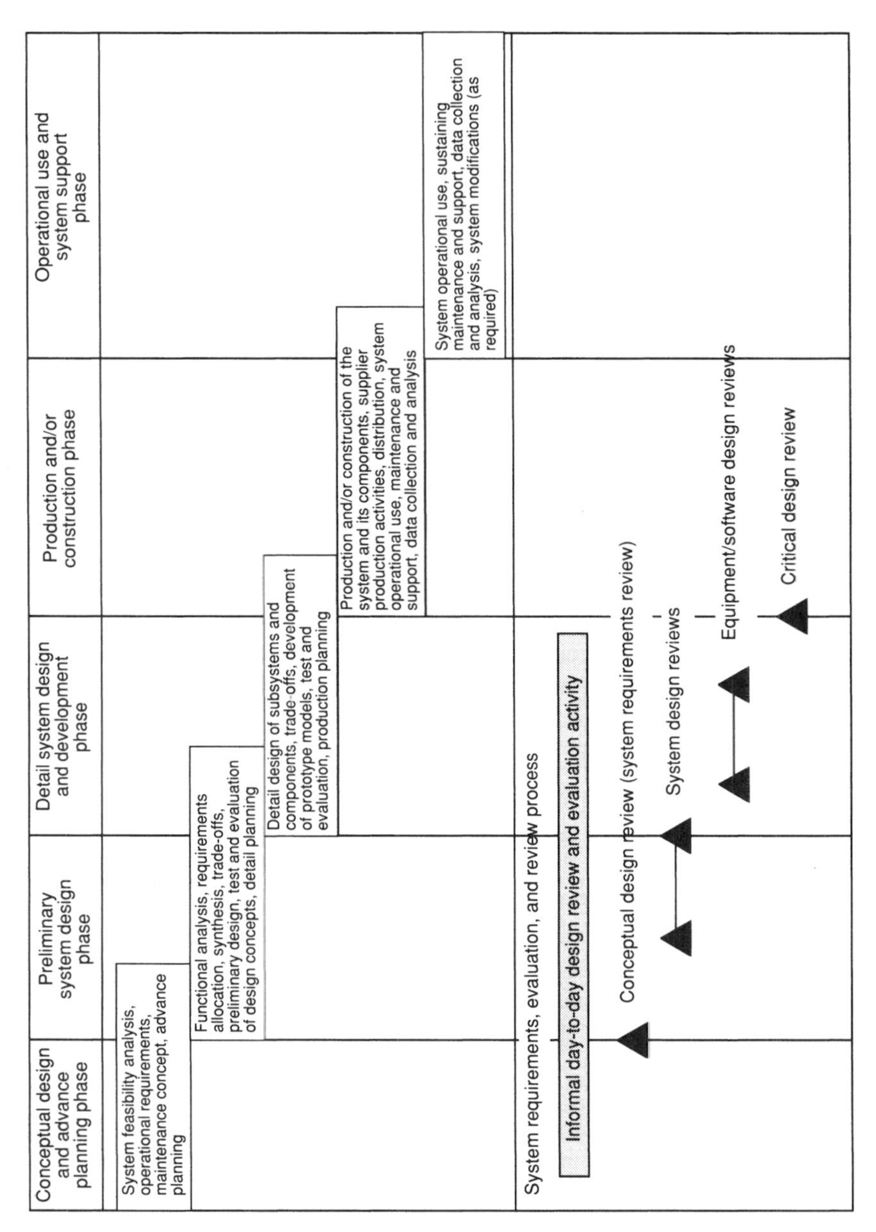

Figure 5.1 Design review and evaluation.

be on the selection and standardization of parts, the mounting of components in a module design, the accessibility of an item requiring frequent maintenance, and the labeling of panel displays and controls. These factors must be integrated into the overall design review and evaluation process presented in Figure 5.1.

Given the criteria against which the design is to be evaluated, it is important to identify the disciplines that have the greatest impact on the design relative to compliance with a specific requirement. For instance, in meeting an equipment diagnostics requirement in the design of an electronics system, electrical engineering, mechanical engineering, and maintainability engineering, in accomplishing their respective design tasks, may have the greatest impact on the corrective maintenance downtime ($\bar{M}ct$) figure of merit for the system. When assessing the level of participation in the design review process, it is necessary that these design disciplines be adequately represented. In other words, along with identification of the criteria for evaluation, the design "responsibility" must be identified.

Design responsibility (and the participation in design reviews) will be covered further, from the organizational perspective, in subsequent sections of this text. However, at this point, it is worthwhile adding a comment on the requirements for design review participation. By referring to Figure 2.24, a hierarchy of system evaluation parameters should be established and tailored for each major system being developed. Those parameters considered as being important can be identified, as shown in Figure 5.2. At the same time, a "degree-of-interest" relationship can be established between the various technical performance measures (TPMs) and the applicable disciplines participating in the design process. The level of interest indicated (i.e., high, medium, and low) pertains to the actual, or perceived, impact that the activity of the discipline has on a designated TPM for the system. This, in turn, should lead to establishing the organizational requirements for design review and evaluation as one progresses from conceptual design to the detail design and development phase. Sections 5.2 and 5.3 will cover this area in a more comprehensive manner.

5.2 INFORMAL DAY-TO-DAY REVIEW AND EVALUATION

From Figure 5.1, the design review and evaluation process includes two basic categories of activity: (1) an informal activity in which the results of design are reviewed and discussed on a day-by-day basis, and (2) a structured series of formal design reviews conducted at specific times in the overall system development process. The output from the day-to-day informal activity leads into the formal design reviews, and this relationship is shown in Figure 5.3.

Design is generally initiated by the electrical engineer, the mechanical engineer, the structural engineer, the process engineer, and/or others who are directly responsible for the design of various components of the system. The results, usually produced independently from these different sources, are described through a combination of drawings, parts lists, reports, computerized databases, and supporting design documentation. As this definition process evolves, there are several major objectives:

1. The results of design must be properly communicated in a clear, effective, and timely manner to all members of the design team. Everyone involved in the design process must work from the *same* database.

Technical Performance Measures (TPMs) \ Engineering Design Functions	Aeronautical Engineering	Components Engineering	Cost Engineering	Electrical Engineering	Human Factors Engineering	Logistics Engineering	Maintainability Engineering	Manufacturing Engineering	Materials Engineering	Mechanical Engineering	Reliability Engineering	Structural Engineering	Systems Engineering
Availability (90%)	H	L	L	M	M	H	M	L	M	M	M	M	H
Diagnostics (95%)	L	M	L	H	L	M	H	M	M	H	M	L	M
Interchangeability (99%)	M	H	M	H	M	H	H	H	M	H	H	M	M
Life Cycle Cost ($350K / unit)	M	M	H	M	M	H	H	L	M	M	H	M	H
$\bar{M}ct$ (30 min.)	L	L	L	M	M	H	H	M	M	M	M	M	M
MDT (24 hrs.)	L	M	M	L	L	H	M	M	L	L	M	L	H
MMH/ OH (15)	L	L	M	L	M	M	H	L	L	L	M	L	H
MTBF (300 hrs.)	L	H	L	M	L	L	M	H	H	M	H	M	M
MTBM (250 hrs.)	L	L	L	L	L	M	H	L	L	L	M	L	H
Personnel Skill Levels	M	L	M	M	H	M	H	L	L	L	L	L	H
Size (150 ft. by 75 ft.)	H	H	M	M	M	M	M	H	H	H	M	H	M
Speed (450 mph.)	H	L	L	L	L	L	L	L	L	L	L	M	H
System Effectiveness (80%)	M	L	L	M	L	M	M	L	L	M	M	M	H
Weight (150K pounds)	H	H	M	M	M	M	M	H	H	H	L	H	M

H= high interest; M= medium interest; L= low interest

Figure 5.2 The relationship between TPMs and responsible design disciplines.

2. The results of design must be compatible with the initially defined requirements for the system. Although each responsible designer should be familiar with the total spectrum of system requirements (e.g., electrical and reliability requirements), the physical separation of design disciplines and the lack of appreciation for the interfaces often result in discrepancies of one type or another (i.e., conflicts, omissions, incompatibilities between system components). These discrepancies must, of course, be corrected as soon as possible.

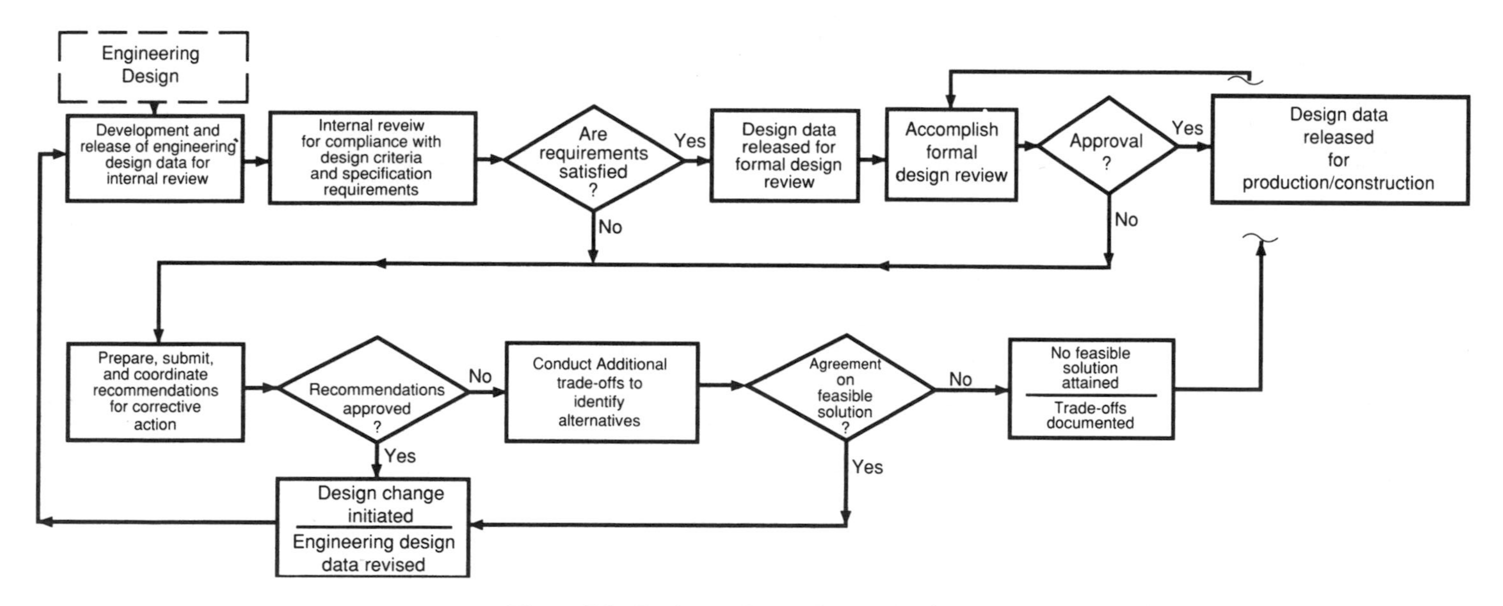

Figure 5.3 Design review and evaluation procedure.

The design review activity is intended to satisfy both objectives. This can be accomplished through a series of steps involving the distribution of drawings, parts lists, and data to all affected areas of design, the review and sign-off of data for approval, the generation of recommendations for change in the event of noncompliance with a given requirement or for the purposes of product improvement, review of the change recommendations by the responsible designer, and so on. This is a day-to-day process with design data evolving from many different sources, and the amount of data can be rather extensive depending on the nature of the system being developed and the size of the project.

In the past, particularly with regard to large projects in which members of the design team are remotely located from one another, the process of data distribution and approval has often been somewhat lengthy in terms of the time required to proceed through the cycle of events. Because of this, combined with the need for the designer to "get on with the design," many organizations have chosen to skip these steps of data distribution, review, and approval in the interest of saving time. In other words, the individual designer makes a decision (often independently), design documentation is prepared and released, component parts are procured and/or fabricated, and so on. Although it is hoped that all design interfaces have been recognized and that system requirements have been met, this has not always been the case! In the rush to complete the design, there have been omissions, conflicts, and/or problems associated with the incompatibility of system components. These problems have become evident later on during a formal design review (when formal design reviews have been conducted), or during system test and evaluation. Further, the implementation of design changes has been more costly than had these changes been incorporated earlier in the design process.[1]

Relative to the future, the implementation of the informal design review and evaluation process shown in Figure 5.3 is, of course, highly desirable. Yet, this procedure has to be accomplished both efficiently and in a timely manner. Although the series of data review steps accomplished in the past may have been somewhat time-consuming, the advent of the computerized methods described in Chapter 4 should result in a definite improvement. The utilization of computer-aided design (CAD) technology and the establishment of a communications network, as illustrated in Figure 4.2, will help to ensure the necessary information flow in an efficient manner. Design data can be distributed to many different locations expeditiously and on a concurrent basis, data review and approval sign-off can be accomplish through the electronics media, and data revisions can be implemented in a relatively short time frame. With these capabilities available, it is hoped that the process illustrated in Figure 5.3 can be implemented in an effective manner.

Concerning the review and evaluation itself, the depth of the review is a function of the complexity of design, whether the item being designed is new (i.e., promoting the state of the art) or is made up of existing "off-the-shelf" components, and whether the item is being developed by an outside supplier or designed "in-house."

[1] These problems can be partially solved through the implementation of an integrated database, as illustrated in Figure 2.30.

Items that are complex or include the application of new technology will be investigated to a greater extent than standard components that are available and have been used in other systems.

In evaluating a given design configuration for compliance with a specified set of requirements, the reviewer may wish to develop a series of "checklists" based on applicable criteria. For example, through the review of selected design standards, component parts data, human-factors anthropometric data, maintainability accessibility factors, safety standards, and so on, the various design review activities can develop criteria that are directly applicable to the system in question.[2] These criteria, summarized in the form of a checklist, are referenced as the evaluation of a given item is being conducted. The checklist serves as an aid in facilitating the review process. Figure 5.4 shows a sample checklist identifying typical topic areas for system-level reviews. Figure 5.5 presents an example of some specific questions that amplify the topics in Figure 5.4. In preparing for the various informal day-to-day design reviews, checklists of this nature can be very helpful.[3]

The results from the day-to-day informal review process, in the context of approved ("signed-off") design documentation, are identified as items to be addressed in the formal design review. This includes not only design drawings and parts lists, but trade-off study reports that support critical design decisions.

5.3 FORMAL DESIGN REVIEWS

The *formal design review* constitutes a coordinated activity (i.e., a structured meeting, or a series of meetings) directed toward the final review and approval of a given design configuration, whether it be the overall system configuration, a subsystem, or an element of the system. Although the informal day-to-day review process discussed in Section 5.2 covers specific aspects of the design, this coverage usually involves a series of independent fragmented efforts representing a variety of engineering disciplines. The purpose of the formal review is to provide a mechanism whereby *all* interested and responsible members of the design team can meet in a coordinated manner, communicate with each other, and agree on a recommended approach. The formal design review process usually includes the following steps:

1. A newly designed item, designated as being complete by the responsible design engineer, is selected for formal review and evaluation. The item may be the overall system configuration as an entity or a major element of the system, depending on the program phase and the category of review conducted.

[2] Three examples in which *general* design criteria/guidelines are included are (1) MIL-HDBK-338, Military Handbook, *Electronic Reliability Design Handbook,* Department of Defense, Washington, DC; (2) B. S. Blanchard, D. Verona, and E. L. Peterson, *Maintainability: A Key to Effective Serviceability and Maintenance Management,* John Wiley, New York, 1995; (3) W. E. Woodson, B. Tillman, and P. Tillman, *Human Factors Design,* 2nd Ed., McGraw-Hill, New York, 1991.

[3] A sample design review checklist is found in Appendix C of this text.

System Design Review Checklist			
General Requirements:			
Have the technical and program requirements for the system been adequately defined through			
1.	Feasibility Analysis	5.	Functional Analysis and Allocation
2.	Operational Requirements	6.	System Specification
3.	Maintenance Concept	7.	Supplier Requirements
4.	Effectiveness Factors	8.	System Engineering Management Plan (SEMP)
Design Features:			
Does the design reflect adequate consideration toward			
1.	Accessibility	19.	Panel Displays and Controls
2.	Adjustments and Alignments	20.	Personnel and Training
3.	Cables and Connectors	21.	Producibility
4.	Calibration	22.	Reconfigurability
5.	Data Requirements	23.	Reliability
6.	Disposability	24.	Safety
7.	Ecological Requirements	25.	Selection of Parts/Materials
8.	Economic Feasibility	26.	Servicing and Lubrication
9.	Environmental Requirements	27.	Societal Requirements
10.	Facility Requirements	28.	Software
11.	Fasteners	29.	Standardization
12.	Handling	30.	Storage
13.	Human Factors	31.	Supportability
14.	Interchangeability	32.	Support Equipment Requirements
15.	Maintainability	33.	Survivability
16.	Mobility	34.	Testability
17.	Operability	35.	Transportability
18.	Packaging and Mounting	36.	Quality
When reviewing the design (layouts, drawings, parts lists, reports), this checklist may be beneficial in covering major program requirements and design features applicable to the system. The items listed are supported with more detailed questions and criteria included in Appendix C. The response to each item listed should be YES.			

Figure 5.4 Sample design review checklist.

2. A location, date, and time for the formal design review meeting are specified.

3. An agenda for the review is prepared, defining the scope and anticipated objectives of the review.

4. A design review board (DRB) representing the organizational elements and the disciplines *affected* by the review is established. Representation from electrical engineering, mechanical engineering, structural engineering, reliability engineering, logistics engineering, manufacturing or production, component suppliers, management, and other appropriate organizations is included as applicable. This representation, of course, will vary from one review to the next. A well-qualified and unbiased chairman is selected to conduct the review.

Detailed Design Review Checklist

18. Packaging and Mounting

   a. Is the packaging design attractive from the standpoint of consumer appeal (e.g., color, shape, size)?
   b. Is functional packaging incorporated to the maximum extent possible? Interaction effects between packages should be minimized, and it should be possible to limit maintenance to the removal of one module (the one containing the failed part) when a failure occurs and not require the removal of two, three, or four modules in order to solve the problem.
   c. Are equipment modules and/or components that perform similar operations electrically, functionally, and physically interchangeable?
   d. Is the packaging design compatible with the level-of-repair analysis decisions? Repairable items are designed to include maintenance provisions such as test points, accessibility, and plug-in components. Items classified as "discard-at-failure" should be encapsulated and maintenance provisions are not required.
   e. Are disposable modules incorporated to the maximum extent practical? It is highly desirable to reduce overall product support through a "no-maintenance" design concept as long as the items involved are high in reliability and relatively low in cost.
   f. Are plug-in modules and components utilized to the maximum extent possible (unless the use of plug-in components significantly degrades the equipment reliability)?
   g. Are the accesses between modules adequate to allow for hand grasping (refer to Design Handbook "X" for recommended accessibility provisions)?
   h. Are modules and components mounted such that the removal of any single item for maintenance will not require the removal of other items? Component stacking should be avoided where possible.
   i. In areas where component stacking is necessary because of limited space, are the modules mounted in such a way that access priority has been assigned in accordance with the predicted removal and replacement frequency? Items that require frequent maintenance should be more accessible.
   j. Are modules and components, not of the plug-in variety, mounted with four fasteners or less? Modules should be securely mounted, but the number of fasteners should be held to a minimum.
   k. Are shock-mounting provisions incorporated where shock and vibration requirements are excessive?
   l. Are provisions incorporated to preclude the installation of the wrong module?
   m. Are plug-in modules and components removable without the use of tools? If tools are required, they should be of the standard variety.
   n. Are guides (slides or pins) provided to facilitate module installation?
   o. Are modules and components properly labeled?

Figure 5.5 Partial listing of design review questions.

5. The applicable specifications, drawings, parts lists, predictions and analysis results, trade-off study reports, and other data supporting the item being evaluated must be identified prior to the formal design review meeting, and made available during the meeting for reference purposes as required. It is hoped that each of the selected design review board members will be familiar with the data prior to the meeting.

6. Selected items of equipment (breadboards, service test models, prototypes), mockups, and/or software may be utilized to facilitate the review process. These items, of course, must be identified early.

7. Reporting requirements and the procedures for accomplishing the necessary follow-up action(s) stemming from design review recommendations must be defined. Responsibilities and action-item time limitations must be established.

8. Funding sources for the necessary preparations, for conducting the formal design review meetings, and for the subsequent processing of outstanding recommendations must be identified.

The formal design review meeting generally includes a presentation (or a series of presentations) on the item being evaluated, by the responsible design engineer, to the selected design review board members. This presentation should cover the proposed design configuration, along with the results of trade-off studies and analyses that support the design approach. The objective is to summarize what had been established earlier through the informal day-to-day design activity. If the design review board members are adequately prepared, this process can be accomplished in an efficient manner.

The formal design review must be well organized and firmly controlled by the design review board chairman. Design review meetings should be brief and to the point, objective in terms of allowing for *positive* contributions, and must not be allowed to drift away from the topics on the agenda. Attendance should be limited to those who have a direct interest in and can contribute to the subject matter being presented. Design specialists who participate should be authorized to speak and make decisions concerning their area of specialty. Finally, the design review activity must make provisions for the identification, recording, scheduling, and monitoring of corrective actions. Specific responsibility for follow-up action must be designated by the design review board chairman.

With the conductance of formal design review meetings, a number of purposes are served:

1. The formal design review meeting provides a forum for communications across the board! The necessary coordination and integration are not adequately accomplished through the informal day-to-day review process, even with the availability of computerized technology. The "person-to-person" contact is required.

2. It provides for the definition of a common configuration baseline for all project personnel; that is, everyone involved in the design process must work from the *same* baseline. The responsible design engineer is given the opportunity to explain the proposed design configuration, and representatives from the various supporting disciplines are provided the opportunity to learn of the designer's problems. This, in turn, creates a better understanding between design and support personnel.

3. It provides a means for solving outstanding interface problems, and it promotes the assurance that all elements of the system are compatible. Those conflicts

that were not resolved through the informal day-to-day review are addressed. Also, those disciplines not properly represented through earlier activity are provided the opportunity to be heard.

4. It provides a formalized check (i.e., audit) of the proposed system/product design configuration with respect to specification and contractual requirements. Areas of noncompliance are noted, and corrective action is initiated as appropriate.

5. It provides a formal report of major design decisions that have been made and the reasons for making them. Design documentation, analyses, predictions, and trade-off study reports that support these decisions are properly recorded.

The conductance of formal design review meetings tends to increase the probability of mature design, as well as the incorporation of the latest design techniques where appropriate. Group reviews may lead to the identification of new ideas, the application of simplier processes, and the realization of cost savings. A good "productive" formal design review activity can be very beneficial. Not only can it cause a reduction in the producer's risk relative to meeting specification and contractual requirements, but the results often lead to an improvement in the producer's methods of operation.

As stated earlier, formal design review meetings are generally scheduled prior to each major evolutionary step in the design process; for example, after the definition of a functional baseline, but prior to the establishment of an allocated baseline. Although the quantity and type of design reviews scheduled may vary from program to program, four basic types are easily identifiable and common to most programs. They are the conceptual design review, the system design review, the equipment or software design review, and the critical design review. The relative time phasing of these reviews is illustrated in Figure 5.1.

5.3.1 Conceptual Design Review

The *conceptual design review* (or system requirements review) is usually scheduled toward the end of the conceptual design and prior to entering into the preliminary system design phase of the program (preferably not longer than 1 to 2 months after program start). The objective is to review and evaluate the functional baseline for the system, and the material to be covered through this review should include the following:[4]

1. Feasibility analysis (the results of technology assessments and early trade-off studies justifying the system design approach being proposed).
2. System operational requirements.

[4]It is recognized that some of these requirements may not be adequately defined during the conceptual design phase, and that the review of such may have to be accomplished later. However, in promoting the desired generic approach described herein (and particularly with regard to system engineering), maximum effort should be made to complete these requirements early, even though changes may be necessary as system design progresses. The object is to encourage (or "force") early system definition, even if the "baseline" changes later.

3. System maintenance concept.
4. Functional analysis (top-level block diagrams).
5. Significant design criteria for the system (e.g., reliability factors, maintainability factors, and logistics factors).
6. Applicable effectiveness figures of merit (FOMs) and technical performance measures (TPMs).
7. System Specification (Type "A"; refer to Section 3.1 and Figure 3.2).
8. System Engineering Management Plan (SEMP).
9. Test and Evaluation Master Plan (TEMP).
10. System design documentation (layout drawings, sketches, parts lists, selected supplier components data).

The conceptual design review deals primarily with top *system-level requirements,* and the results constitute the basis for follow-on preliminary system design and development activity. Participation in this formal review should include selected representation from both the consumer and producer organizations. Consumer representation should involve not only those personnel who are responsible for the acquisition of the system (i.e., contracting and procurement), but those who will ultimately be responsible for the operation and support of the system in the field. Individuals with experience in operations and maintenance should participate in the system requirements review. On the producer side of the spectrum, those lead engineers responsible for *system* design should participate, along with representation from various design disciplines and production (as necessary). It is important to ensure that the disciplines identified in Chapter 3 are adequately represented in the formal design review process from the beginning.

In summary, the conceptual design review is extremely important for all concerned, as it represents the first opportunity for formal communication relative to system requirements from the top down! It can provide an excellent baseline for all subsequent design effort. Unfortunately, for many projects in the past, the conductance of conceptual design reviews has not been readily evident. Further, if such a review were conducted, the results were not always made available to responsible design engineering personnel assigned to the project. This, in turn, has resulted in a series of efforts conducted in somewhat of a vacuum and not well coordinated or integrated. Thus, with the objectives of system engineering in mind, it is essential that a good functional baseline for the system be defined and properly evaluated through the conductance of an effective conceptual design review.

5.3.2 System Design Reviews

System design reviews are generally scheduled during the preliminary design phase when functional requirements and allocations are defined, preliminary design layouts and detailed specifications are prepared, system-level trade-offs studies are conducted, and so on (refer to Figure 5.1). These reviews are oriented to the overall system configuration, in lieu of individual equipment items, software, and other components of the system. As the design evolves, it is important to ensure that the requirements described in the system specification are maintained. There may be one or more formal reviews scheduled depending on the size of the system and the complexity of design. System design reviews cover a variety of topics, a few of which are noted:

1. Functional analysis and the allocation of requirements (beyond what is covered through the conceptual design review).
2. Development, process, product, and material specifications as applicable (Types "B," "C," "D," and "E").
3. Design data defining the overall system (layouts, drawings, parts/material lists, supplier data).
4. Analyses, reports, predictions, trade-off studies, and related design documentation. This includes material that has been prepared in support of the proposed design configuration, and analyses/predictions that provide an assessment of what is being proposed. Reliability and maintainability predictions, logistic support analysis data, and so on are included.
5. Assessment of the proposed system design configuration in terms of applicable technical performance measures (TPMs).
6. Individual program/design plans (e.g., reliability and maintainability program plans, human factors program plan, and ILS Plan).

Participation in the system design reviews should include representation from both the consumer and producer organizations, as well as from major suppliers involved in the early phases of the system life cycle.

5.3.3 Equipment/Software Design Reviews

Formal design reviews covering equipment, software, and other components of the system are scheduled during the detail design and development phase of the life cycle. These reviews, usually oriented to a particular item, include coverage of the following:

1. Process, product, and material specifications (Types "C," "D," and "E"—beyond what is covered through the system design reviews).
2. Design data defining major subsystems, equipment, software, and other elements of the system as applicable (assembly drawings, specification control drawings, construction drawings, installation drawings, logic diagrams, schematic diagrams, material and detailed parts lists, and so on).
3. Analyses, reports, predictions, trade-off studies, and other related design documentation as required in support of the proposed design configuration and/or for assessment purposes. Reliability and maintainability predictions, human-factors task analysis, logistic support analysis data, and so on, are included.

4. Assessment of the proposed system design configuration in terms of the applicable technical performance measures (TPMs). An ongoing review and evaluation are required to ensure that these system-level requirements are maintained throughout the various stages of detail design and development.
5. Engineering breadboards, laboratory models, service test models, mockups, and prototype models used to support the specific design configuration being evaluated.
6. Supplier data covering specific components of the system as applicable (drawings, material and parts lists, analysis and prediction reports, and so on).

Participation in these formal reviews should include representation from the consumer (i.e., customer), producer (i.e., contractor), and applicable supplier organizations.

5.3.4 Critical Design Review

The *critical design review* is generally scheduled after the completion of detail design, but prior to the release of firm design data for production or construction. Design is essentially "frozen" at this point, and the proposed configuration is evaluated in terms of adequacy and producibility. The critical design review may address topics such as the following:

1. A complete set of final design documentation covering the system and its components (manufacturing drawings, material and parts lists, supplier component parts data, drawing change notices, and so on).
2. Analyses, predictions, trade-off studies, test and evaluation results, and related design documentation (final reliability and maintainability predictions, human-factors and safety analyses, logistic support analysis records, test reports, and so on).
3. Assessment of the final system design configuration (i.e., the product baseline) in terms of applicable technical performance measures (TPMs).
4. A detailed production/construction plan (description of proposed manufacturing methods, fabrication processes, quality control provisions, supplier requirements, material flow and distribution requirements, schedules, and so on).
5. A final Integrated Logistic Support Plan (ILSP) covering the proposed life-cycle maintenance and support of the system throughout the consumer utilization phase.

The results of the critical design review describe the final system/product configuration baseline prior to entering into production and/or construction. This review constitutes the last in a series of progressive evaluation efforts, reflecting design and development from a historical perspective and showing growth and maturity in design as the engineering project evolved. It is important to view the design review process in *total,* and to provide an overall evaluation of certain designated system attributes as the project progresses, particularly because close continuity is required between the various reviews. An example of designated system attributes that should be assessed on a continuing basis is presented in Figure 5.6.

5.4 THE DESIGN CHANGE AND SYSTEM MODIFICATION PROCESS

The objective thus far has been to develop a system on a progressive basis and to establish a firm configuration baseline through the formal review and evaluation process. In essence, the results from the conceptual design review lead to the definition of system-level requirements, the results from the system design reviews constitute a more in-depth description of the system packaging concepts, and so on. As we progress through the series of design reviews described in Section 5.3, the system definition becomes more refined, and the configuration baseline (updated from one review to the next) is established. This baseline, which constitutes a single point of reference for all individuals who are involved in the design process, is critical from the standpoint of meeting the system engineering objectives described earlier.

Once a configuration baseline has been established, it is equally important that any variations, or changes, with respect to that baseline be tightly controlled. It is certainly not anticipated that a given baseline will remain as such forever, particularly during the early stages of system development. However, in evolving from one design configuration to the next, it is important that all changes be carefully recorded and documented in terms of their possible impact on the initially specified system requirements. The process of configuration identification, the control of changes, and maintaining the integrity and continuity of design are accomplished through Configuration Management (CM).[5]

In the defense sector, configuration management is often related to the concept of "baseline management." From Figure 1.12, *functional, allocated,* and *product* baselines are established as the system development process evolves. These baselines are described through a family of specifications (Types "A," "B," "C," "D," and/or "E"), drawings and parts lists, reports, and related documentation. The formal design review process provides the necessary authentication of these baseline configurations, and the Configuration Identification (CI) function is accomplished. CI relates to a particular baseline, and the Configuration Status Accounting (CSA) function is a "management information system that provides traceability of configuration baselines and changes thereto, and facilitates the effective implementation of changes." CSA includes the documentation in evolving from one configuration baseline to the next.

Proposed design changes, or proposed changes to a given baseline (i.e., a CI

[5]Configuration Management (CM) constitutes the process that identifies the functional and physical characteristics of an item during its life cycle, controls changes to those characteristics, and records and reports change processing and implementation status.

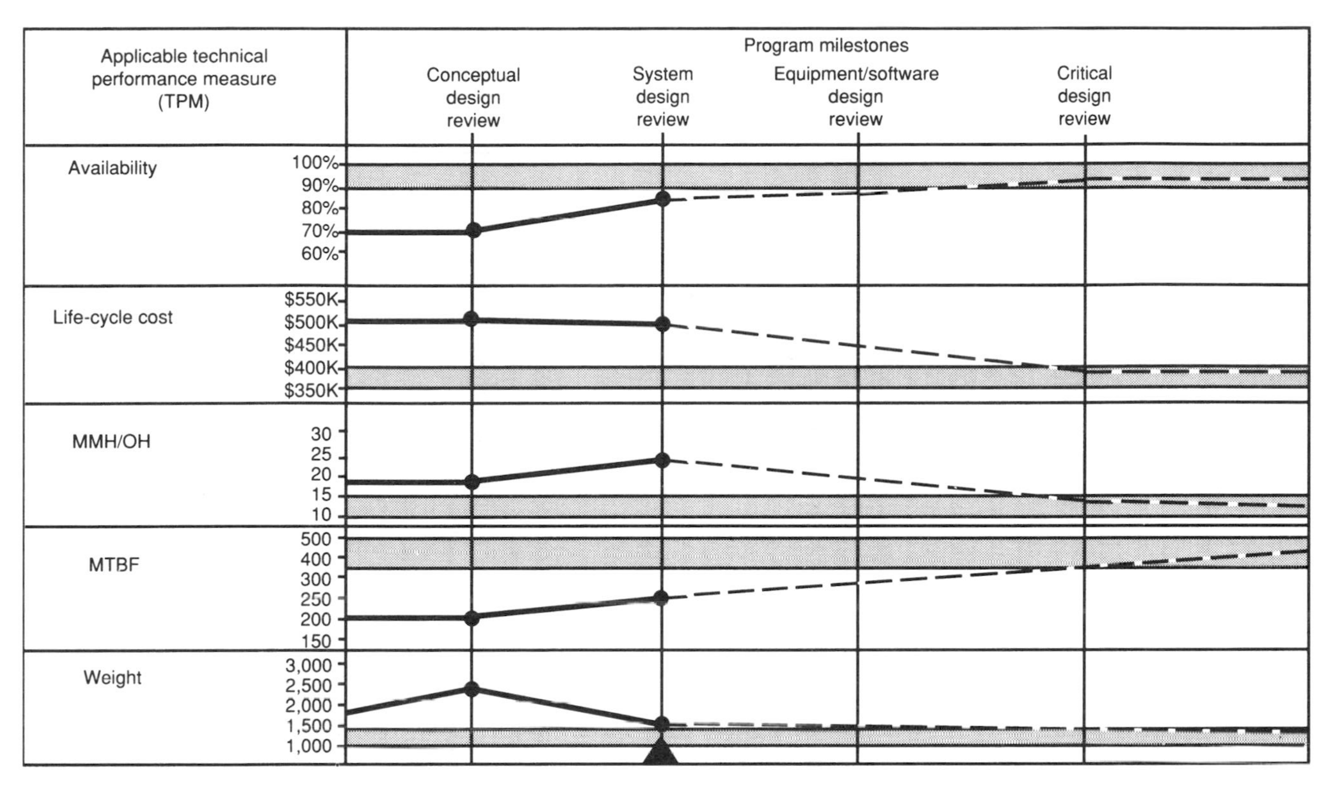

Figure 5.6 System parameter measurement and evaluation at design review (sample).

design), may be initiated from any one of a number of sources during any phase in the overall system life cycle. Such changes, prepared in the form of an Engineering Change Proposal (ECP), may be classified as follows:

1. *Class 1 changes:* design changes that will affect form, fit, and/or function (e.g., changes that will impact system performance, reliability, safety, supportability, life-cycle cost, and/or any other system specification requirement).
2. *Class 2 changes:* design changes that are relatively minor in nature and that will not affect system specification requirements (e.g., changes covering material substitutions, documentation clarifications, drawing nomenclature, producer deficiencies).

Changes may be categorized as "emergency," "urgent," or "routine," depending on priority and on the criticality of the change.

A simplified version of the system control procedure is illustrated in Figure 5.7. Proposed changes to a given baseline may be initiated during any phase of system development, production, and/or operational use. Each proposed change is presented in the form of an Engineering Change Proposal (ECP) submitted for review, evaluation, and approval. In general, each ECP should cover the following issues:[6]

1. A statement of the problem and a description of the proposed change.
2. A brief description of alternatives that have been considered in responding to the need.
3. An analysis showing how the change will solve the problem.
4. An analysis showing how the change will impact system performance, effectiveness factors, packaging concepts, safety, elements of logistic support, life-cycle cost, and so on. What are the impacts (if any) on system specification requirements? What is the effect on life-cycle cost?
5. An analysis to assure that the proposed solution will not cause the introduction of new problems.
6. A preliminary plan for incorporating the change; that is, proposed date of incorporation, serial numbers affected, retrofit requirements, and verification test approach (as applicable).
7. A description of the resources required to implement the change.
8. An estimate of the costs associated with implementing the change.
9. A statement covering the impact on the system if the proposed change is *not* implemented; that is, an identification of the possible risks associated with a "do-nothing" decision.

[6] In many organizations, the procedures related to configuration management and change control are a little more complex than what is presented here. The procedure may involve engineering change requests (ECRs), design revision notices (DRNs), interface control documents (ICDs), and so on. The objective here is to present a *simplified* approach, providing a basic understanding of the importance of change control as part of the system engineering process

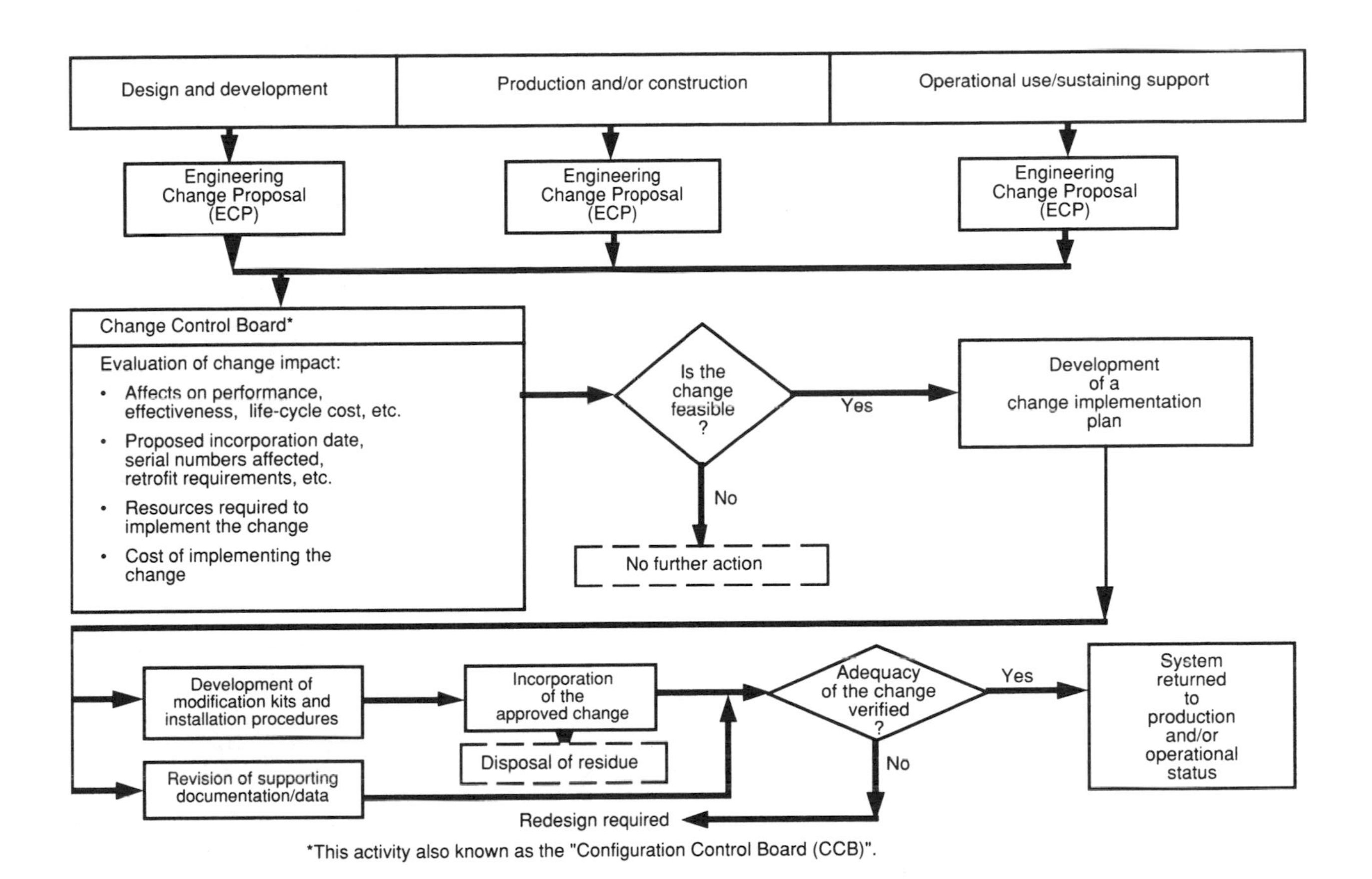

Figure 5.7 System change control procedure.

From Figure 5.7, engineering change proposals (ECPs) are processed through the Change Control Board (sometimes known as the "Configuration Control Board," or the CCB) for review and evaluation. The CCB should function in a manner similar to the Design Review Board (DRB) discussed in Section 5.3. Board representation should cover those design disciplines impacted by the change, to include customer and supplier representation as necessary. Not only is it necessary to review and evaluate the original design, but it is important to ensure that all proposed design changes are handled in a similar manner. On occasion when project schedules are "tight," the designer will generate data just to have something available for the record, and the *real* design configuration will be reflected through the "change process." Although this is not a preferred practice, it does occur in a number of instances when the objective is to save time. In any event, the review of design changes must be treated with the same degree of importance as is specified for the formal design review.

On completion of the formal design change review by the CCB, approved ECPs will be supported with the development of a plan for incorporating the change(s) in the system. This plan should include coverage of not only the modifications required for the prime equipment, but the modifications associated with test and support equipment, spares and repair parts, facilities, software, and technical documentation. All elements of the system must be addressed on an integrated basis.

The actual incorporation of changes to the system is accomplished using a variety of approaches depending on when the change is to be implemented. The time of implementation is a function of priority and/or criticality. Emergency or urgent changes may require immediate action, whereas routine changes may be grouped and incorporated at some convenient later point in time. Approved changes initiated during system design and development, prior to the availability of any hardware, software, or other physical components, may be incorporated through the preparation of design change notices (DCNs), or equivalent, attached to the applicable drawings/documentation covering those areas of design affected by the change. As the project progresses, these "paper" (or database) changes will be reflected in the new design configuration.

In the event that changes are initiated during the production/construction phase when multiple quantities of identical items are being produced, a designated serial-numbered item needs to be identified to indicate effectivity; that is, the change will be incorporated on the production line in Serial Number "X" and on later models. This should ensure that all applicable items scheduled to be produced in the future will automatically reflect the updated configuration.

For those system components that are already in use, changes may be incorporated through the installation of a modification kit in the field at the consumer's operational site. Such kits are installed and the system is tested to verify the adequacy of the change. At the same time, the system support capability (e.g., test equipment, spares, and technical data) needs to be upgraded for compatibility with the prime mission-oriented segments of the system. Optimally, the installation process should take place at a time when the system is not in demand or being utilized in the performance of a mission.

This overall process is illustrated in Figure 5.7. With the incorporation of validated changes, the system configuration is updated and a new baseline is established. In situations where the adequacy of the change is not verified, some additional redesign may be required.

5.5 SUMMARY

This chapter primarily addresses the basic review, evaluation, and feedback process illustrated in Figure 1.22. This process, which is critical with regard to the objectives of system engineering, must be tailored to the specific system development effort and must be properly controlled! An ongoing measurement and evaluation activity is essential, and must be initiated from the beginning. Performing a one-time review and evaluation after the system has been produced and is in operational use may be costly in terms of possible modifications for corrective action. Also, the incorporation of design changes on a continuing basis without the proper controls may be costly from the standpoint of system support. In essence, there must be a well-planned program approach, with the proper controls, in order to ensure a total integrated system configuration in the end.

QUESTIONS AND PROBLEMS

1. Describe the "checks" and "balances" in the design process (as you see them).
2. How is design review and evaluation accomplished? Why is it important relative to meeting system engineering objectives?
3. What is included in the establishment of a "functional" baseline? "Allocated" baseline? "Product" baseline? Why is baseline management important?
4. Select a system of your choice, and construct a sequential flow diagram of the overall system development process. Identify the major tasks in system development, and develop a plan/schedule of formal design reviews. Briefly describe what is covered in each.
5. Identify some of the benefits derived through formal design review. Describe some of the concerns.
6. When developing an agenda in preparing for a formal design review, what considerations must be addressed in the selection of items to be covered in the review process? How are review and evaluation criteria identified? Describe the steps and resources required in preparing for the design review.
7. How are technical performance measures (TPMs) considered in the design review process?
8. In the event that a deficiency is identified during design review, what steps are required for corrective action?

9. How are design changes initiated? How are priorities established?

10. How are design changes implemented? Identify the steps involved in system modification.

11. Describe the functions of the CCB.

12. What is configuration management" (CM)? Define "configuration identification" (CI) and "configuration status accounting" (CSA).

13. How does configuration management (CM) relate to system engineering? Why is it important? What is likely to occur if configuration management practices were not followed?

6 System Engineering Program Planning

The first five chapters of this text have dealt with the system engineering process, the major steps in the process, and some of the technologies and tools available that can be applied. Given this definition of the process as a baseline, the remaining challenge lies with its *implementation.* In Figure 1.20 (Chapter 1), the appropriate combination of both *technology* and *management skills* is required to meet the objectives described herein.

The key to the successful implementation of any program is *early planning.* Planning for system engineering activities commences at program inception. As the need for a system is identified and feasibility studies are conducted in selecting a technical design approach, requirements are being established to define a program structure that can be implemented to bring the system into being.

Planning is initiated with the definition of program requirements. System engineering functions and tasks are identified, work packages and a work breakdown structure (WBS) are developed, program schedules are prepared, an organizational structure is defined, key policies and procedures are described, and a detailed *System Engineering Management Plan* (SEMP) is prepared and implemented. The SEMP, which is generally developed during the conceptual design and advanced planning phase, covers all system engineering management activities throughout the system life cycle, including those pertaining to system operations, support, retirement, and any modifications to the system configuration as they may occur throughout these phases.[1]

This chapter covers system engineering program planning, the first step in system management, illustrated in Figure 6.1. The material leads into the organization for system engineering (Chapter 7), and the various aspects of supplier selection, contracting, management and control (Chapter 8). Implementing the requirements described in the first five chapters is highly dependent on the thoroughness of planning from the beginning and in the follow-on organization, management, and control later.

[1] In the preparation and implementation of the SEMP, it should be noted that system engineering activities may be implemented by the customer (consumer), by the prime producer (contractor), and/or by a major supplier. In some instances (particularly for large programs), an initial SEMP for the overall *system* may be prepared by the customer, with a lower-level SEMP prepared by the producer. Obviously, the second must evolve from and support the first. In any event, the intent herein is to describe what material might be included in the SEMP overall, and not attempt to differentiate between who does what! It is assumed that the proper "tailoring" will be accomplished depending on the individual program requirements.

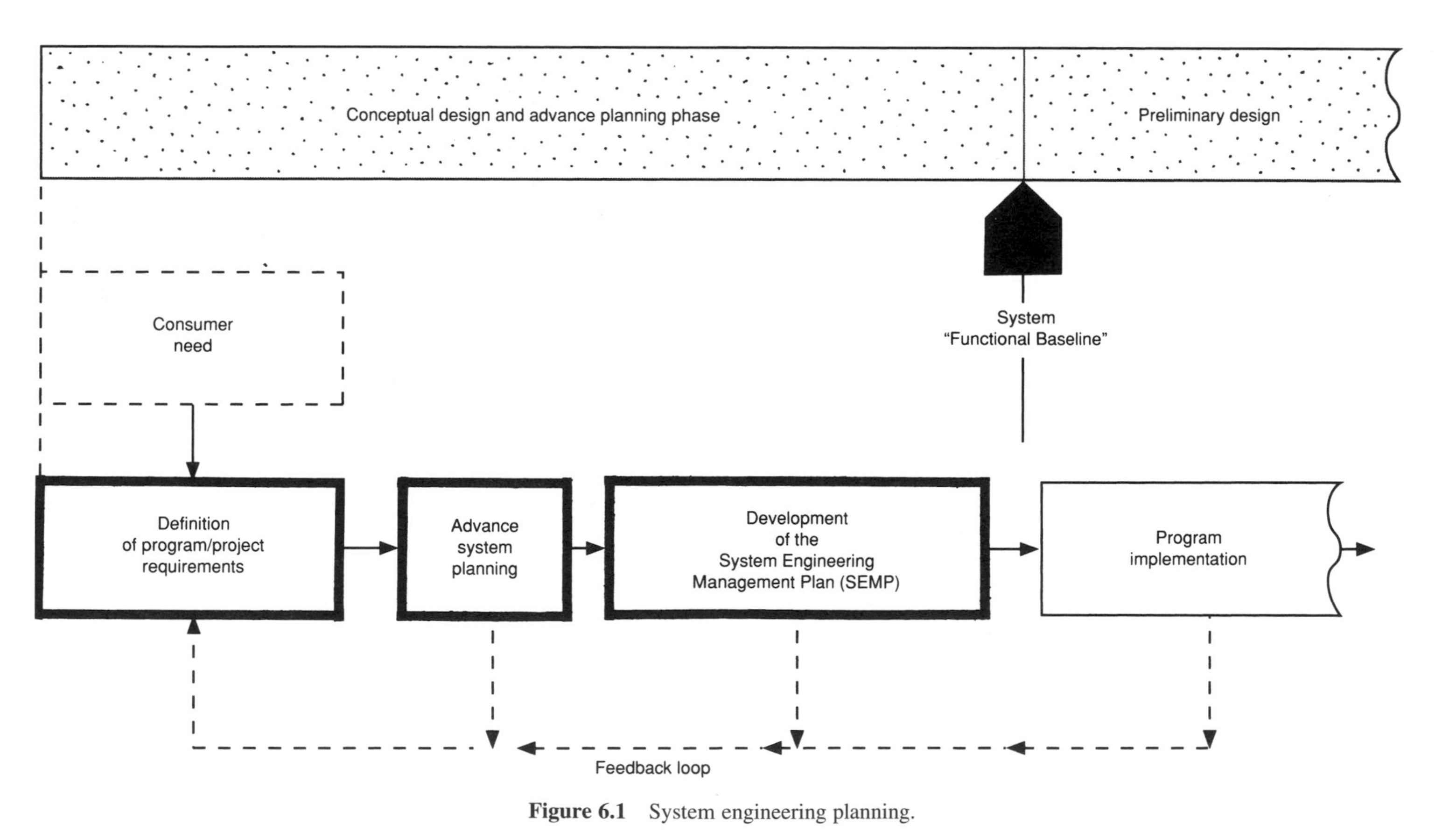

Figure 6.1 System engineering planning.

6.1 SYSTEM ENGINEERING PROGRAM REQUIREMENTS

From Figure 6.1, the first step in the planning process involves the definition of program (or project) requirements. Although this may appear to be rather basic in nature, every program is different and it is essential that system engineering requirements be tailored accordingly. The concepts and methods described throughout this text, however, are applicable to all programs. Only the nature and depth of application may vary from one program to the next.

6.1.1 The Need for Early System Planning

The successful implementation of the concepts and methods of system engineering is highly dependent on (1) employing a top-down approach in the development of a system, (2) integrating early the design and related supporting activities, (3) viewing requirements in terms of the entire system life cycle, and (4) preparing complete *requirements documentation* from the beginning (i.e., applicable specifications and plans). As early system concepts are generated and feasibility studies are conducted to determine possible alternative technical solutions in response to a given design problem, the appropriate level of planning must be initiated to ensure that the ideas generated through analysis are properly consumated and integrated into a final product configuration in a cost-effective manner.

System engineering planning commences at program inception during the needs analysis (described in Section 2.2). Liaison activities with the customer (consumer/user) are required to ensure that the defined need is interpreted correctly, described accurately, and that the translation from the stated need to the definition of system requirements is responsive! This is a critical step in the early definition of system requirements, as so often there is a communications gap and the *real* need is not thoroughly understood. This, of course, may result in the development of planning information for a design configuration that will not fulfill the functions intended.

Given the identification and description of the need, the next step involves the accomplishment of feasibility studies (refer to Section 2.3). Future technological opportunities are identified and alternative approaches are investigated for possible design application. The feasibility of each of the alternatives being considered is a function not only of meeting the necessary performance requirements, but being responsive to the following questions:

1. Have the resource requirements associated with each alternative been defined (i.e., human, material, equipment, software, and data needs)? Have the sources of supply been identified? Will the necessary resources be available when required?
2. Does the alternative being considered for selection reflect a cost-effective approach based on a life-cycle analysis?
3. Has an *impact analysis* been accomplished to determine whether there are possible secondary and/or tertiary effects as a result of selecting a given alter-

native? It is hoped that the selection of a specific alternative will not have a detrimental impact relative to the environment; that is, social, political, or ecological concerns as presented in Figure 3.44 (Section 3.3.10). Additionally, the interactions with other systems should be minimized.

It is at this stage in the life cycle when early system engineering planning is important. The analysis effort is directed toward the *system* level, potential suppliers are identified, the overall system integration process commences, the interaction effects (both internal and external) are assessed, and potential areas of risk are identified. As system definition continues through the development of operational requirements, the maintenance concept, and in the prioritization of technical performance measures (TPMs), the planning process evolves through another series of iterations. The requirements for system integration are greater in terms of both the *technical* integration of the various elements of the system and the integration of the many and varied organizational entities participating in the system development effort.

System planning is continuous, commencing with the definition of a need and extending through the development of the System Engineering Management Plan (SEMP). As system-level requirements are defined, the planning process leads to the identification of those activities that must be accomplished in order to provide a system configuration that will fulfill these requirements. Design and management decisions at this stage in the system life cycle have a great impact on program activities later on. Thus, it is imperative that a complete and well-integrated planning effort be implemented from the beginning.

6.1.2 Determination of Program Requirements

Although the concepts, methods, and processes describing system engineering are generally applicable to all categories of systems, they must be "tailored" for each individual application. Further, the applications are numerous and varied and include:

1. Large-scale systems with many different components such as a space system, an urban transportation system, a hydroelectric power generating system, and so on.
2. Small systems with relatively few components such as a local communications system, a computer system, a hydraulic system, a mechanical braking system, and so on.
3. Systems in which there is a great deal of new design and development effort required; that is, the application of new technologies.
4. Systems in which the design is based primarily on the utilization of existing standard "off-the-shelf" components.
5. Systems that are highly equipment-intensive, software-intensive, facilities-intensive, or human-intensive; for example, a production system versus a

ground command and control system, versus a data distribution system, versus a maintenance capability.

6. Systems in which there are a large number of suppliers involved in the design and development process, both at the national and international levels.
7. Systems in which there are a number of different organizations involved in the design and development process.
8. Systems being designed and developed for utilization in the government sector, the private sector, and so on.

Although this text basically addresses only a few of the major categories of systems described in Chapter 1 (i.e., the man-made, open-loop, dynamic system), there still are a wide variety of applications, as illustrated in Figure 6.2.

In each individual situation, the system engineering process described in Chapter 2 is applicable. Although the extent and depth of effort will vary, the steps required for bringing a system into being are basically the same. A needs analysis and feasibility analysis are accomplished, operational requirements and the system maintenance concept are defined, and functional analysis and requirements allocations are completed. Even though one may be dealing with a relatively simple case such as a small system made up of standard off-the-shelf components, there is still a need to accomplish a top-down requirements analysis, accomplish a functional analysis and allocation, and so on. In other words, there is a *system* design requirement, even though new design may not be required at the subsystem or component level.

Following the general steps reflected in Figure 1.12 (Chapter 1) represents a good overall approach to the design and development of any new system. As one progresses from the needs analysis through the accomplishment of feasibility studies and

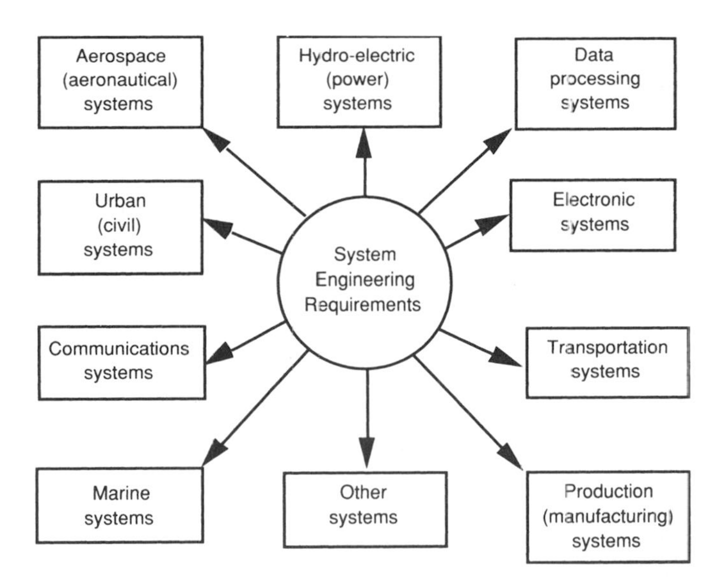

Figure 6.2 The application of system engineering requirements.

the definition of system-level requirements, the process evolves from the identification of the "WHATs" to the "HOWs;" that is, *what* is required in terms of a system functional capability, and *how* is this to be accomplished from the standpoint of a technical design approach. Evaluation of the early responses to the proposed technical design approach leads to the identification of specific *program* (or *project*) requirements.

Program requirements, in this context, refer to the management approach and the steps to be followed in the procurement and/or acquisition of the system in response to a stated need, along with the identification of the resources required to fulfill this objective. A program structure should be established that will enable the design and development, production and/or construction, and delivery of the system to the consumer in a cost-effective manner. This includes the identification of program functions and detailed tasks, the development of an organizational structure, the development of a work breakdown structure (WBS), the preparation of program schedules and cost projections, the implementation of a program evaluation and control capability, and so on. This information, presented in the form of a program plan, provides the necessary day-to-day management guidance required in the realization of any technical objective.

In fulfilling the system engineering objectives described in the earlier chapters of this text, a System Engineering Management Plan (SEMP) is developed as part of the early planning requirements for each program (refer to Figure 6.1). Although the detailed content may vary from one instance to the next, some of the major features of the SEMP are noted in Section 6.2.

6.2 SYSTEM ENGINEERING MANAGEMENT PLAN (SEMP)[2]

From Figure 3.2, the SEMP is developed from the basic Program Management Plan (PMP) and covers all management functions associated with the performance of system engineering activities for a given program.[3] The SEMP constitute's the Chief Engineer's plan for identifying and integrating all major engineering activities; that is, the top *technical* plan that causes the integration of the many more subordinate plans such as the mechanical engineering design plan, the reliability and maintainability program plans, the human factors and safety program plans, and so on.

Preparation of the SEMP is the responsibility of the "System Manager," and may be accomplished by the customer (consumer/user) or by a major contractor (pro-

[2] Three sources that include coverage of the SEMP are (1) IEEE-STD 1220, "Standard for the Application and Management of the Systems Engineering Process," Trial-Use Issue, September 1994; (2) IS-632, "Systems Engineering," Electronic Industries Association (EIA), Final Draft, December 1994; and (3) Defense Systems Management College, *Systems Engineering Management Guide,* DSMC, Fort Belvoir, VA (latest issue).

[3] For the purposes of discussion, it is assumed that the PMP represents the *top management* document for the program. In terms of the hierarchy of documentation, the SEMP should represent the *top engineering planning document,* supplementing the PMP.

ducer) depending on the program. The relationships among the consumer, prime contractors or major producers, subcontractors, suppliers, and so on, particularly for large-scale systems, may take the form illustrated in Figure 6.3. In such instances, the consumer/user is the "System Manager" and is responsible for the SEMP, but may delegate the overall system integration and management responsibility to a prime contractor (i.e., Contractor A or Contractor B).

In the event that the consumer prepares the SEMP, then Contractor A and Contractor B must each prepare a SEMP covering their respective system engineering activities, each being in response to the higher-level SEMP. On the other hand, if the system integration and management responsibility is delegated to Contractor A (for example), then the responsibility for preparation of the SEMP, and for implementing the activities defined therein, will be at this level.[4]

This discussion may initially appear to be rather trivial! However, if the SEMP is to be meaningful and accomplish its objectives, it must be developed directly from the top-level Program Management Plan (PMP). Further, the responsibility for the SEMP, and for the accomplishment of the functions and tasks described within, must be clearly defined and supported by the Program Manager (or Program Director). When system management responsibility is delegated to Contractor A in Figure 6.3, then Contractor A must be given both the *responsibility* and the *authority* to

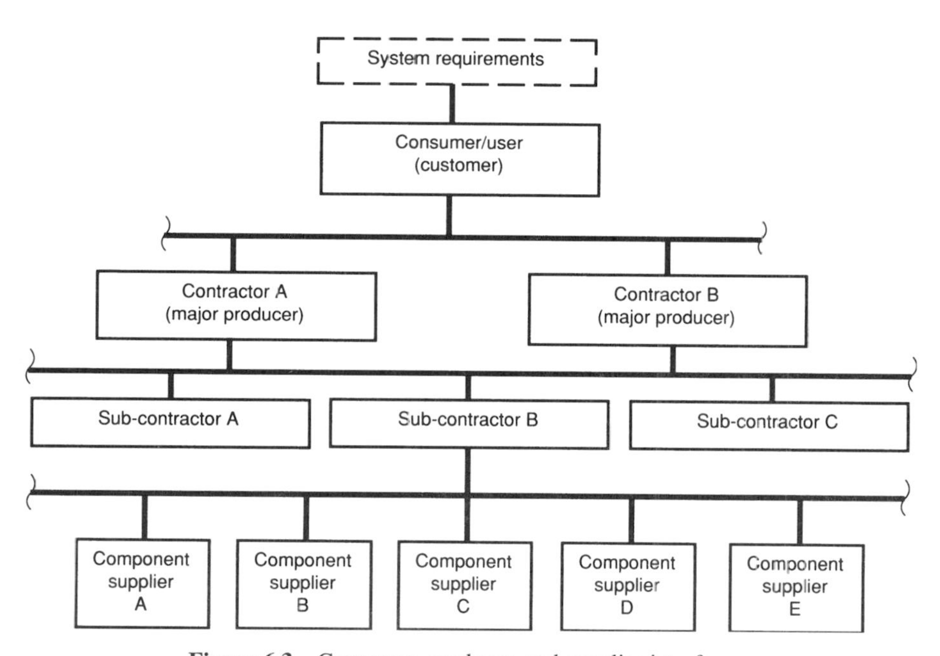

Figure 6.3 Consumer, producer, and supplier interfaces.

[4] In such instances, the SEMP is usually prepared and included as part of the contractor's proposal to the consumer (i.e., customer).

perform *all* system-level functions (described in the SEMP) on behalf of Contractor B, as well as for all subcontractors and applicable suppliers.[5] Finally, the SEMP must be appropriately identified as the key top-level design engineering plan in the overall program documentation tree.

In terms of material content, the SEMP should be *tailored* to the system requirements, the program size and complexity, and the nature of the procurement or acquisition process. Initially, the SEMP is prepared during the conceptual and advance planning phase (refer to Figure 6.1), and includes all system engineering activity planned from thereon. As the system development process progresses, the SEMP is updated to reflect new program requirements and changes in activity emphasis.

In order to convey an indication as to the nature of the information that may be included in the SEMP, two different outlines have been presented herein. Figure 6.4 conveys an abbreviated outline of the SEMP as presented in the Electronic Industries Association (EIA) IS-632 (refer to footnote 2), and Figure 6.5 illustrates a modification of the SEMP outline initially presented in MIL-STD-499A and in the 1990 DSMC *Systems Engineering Management Guide.*[6] Although both outlines are similar in content, your author has chosen to reference Figure 6.5 in the presentation of the material from hereon. Further, it is not the intent to discuss each topic in the detailed SEMP outline, but a few selected areas where it is felt that additional discussion is necessary.

6.2.1 Statement of Work

The Statement of Work (SOW) is a narrative description of the work required for a given project. With regard to the SEMP, it must be developed from the overall project SOW described in the PMP, and it should include the following:

1. A summary statement of the tasks to be accomplished. An identification of the major system engineering tasks is presented in Section 6.2.2. These, in turn, must be supported by elements of work included in the work breakdown structure (WBS) discussed in Section 6.2.3.
2. An identification of the input requirements from other tasks. These may include the results from other tasks accomplished within the project, tasks completed by the customer, and/or tasks accomplished by a supplier.
3. References to applicable specifications (to include the System "A" Specification), standards, procedures, and related documentation as necessary for

[5] Quite often, contractors are held responsible for the development and implementation of a systems engineering program (and the SEMP), but have not been delegated the *authority* to successfully complete the task. In such cases, it is essential that the proper level of authority be delegated along with the responsibility.

[6] The basis for the three-part outline is developed from MIL-STD-499A, Military Standard, "Engineering Management," U.S. Air Force/Air Force Systems Command, Andrews AFB, Maryland; and Defense Systems Management College, *Systems Engineering Management Guide,* DSMC, Fort Belvoir, VA. The detailed topics presented in Figure 6.5 represent an expansion developed by the author.

SYSTEMS ENGINEERING MANAGEMENT PLAN (SEMP)
(Outline,EIA Engineering Standard 632, "Systems Engineering")

1. Title Page, Table Of Contents, Applicable Documents
2. Systems Engineering Process
   2.1 Systems Engineering Process Planning -- major products and results from process, process inputs, technical objectives, work breakdown structure (WBS), training, standards and procedures, resource allocation, constraints, work authorization, verification planning, subcontractor/supplier technical effort.
   2.2 Requirements Analysis -- reliability and maintainability, survivability, electromagnetic compatibility, human engineering, safety, security, producibility, product support (supportability), test and evaluation, integrated diagnostics, transportability, infrastructure support, other areas of functionality.
   2.3 Functional Analysis And Allocation -- approach, methods, procedures, tools.
   2.4 Synthesis -- factor-dependent approaches and methods, use of leverage options (COTS/NDI, open systems architecture, reuse).
   2.5 Systems Analysis And Control -- approach, methods, procedures, and tools (trade studies, system cost-effectiveness analysis, risk management, configuration management, interface management, data management, SE master schedule, technical performance measurement, subcontractor/supplier control, requirements traceability).
3. Transitioning Critical Technologies -- activities, risks, and criteria for selecting technologies and for the transitioning of these technologies.
4. Integration Of Systems Engineering Effort -- organization and integration of design disciplines and related activities (concurrent engineering).
5. Implementation Tasks -- technology verifications, process proofing, manufacturing engineering test articles, development test and evaluation, generation and reuse of software, sustaining engineering and problem solution support, other systems engineering implementation tasks.
6. Additional Systems Engineering Activities -- long-lead items, engineering tools, design-to-cost, value engineering, system integration, other methods and controls.
7. Notes And Appendicies

Figure 6.4 Systems Engineering Management Plan (SEMP) outline.

the completion of the defined scope of work. These references should be identified as key requirements in the documentation tree described in Section 6.2.4.

4. A description of the specific results to be achieved. This may include deliverable equipment, software, design data, reports, and/or related documentation, along with the proposed schedule of delivery as presented in Section 6.2.6.

In preparing the SOW, the following general guidelines are considered to be appropriate.

1.0 Overview

2.0 Purpose and Scope

3.0 Application(s)

4.0 Technical Program Planning, Implementation, and Control (Part I)

- 4.1 Introduction
- 4.2 Program Requirements/Statement of Work
- 4.3 Organization (Project Organization, Functional Organization, Organizational Responsibilities and Authority, Organizational Interrelationships and Communications, Supplier Relationships and Control)
- 4.4 Program Planning (Specification Tree, Work Breakdown Structure, Program/Project Schedules and/or Milestone Charts)
- 4.5 Technical Interface Management (Relationships Among Internal Organizational Entities, Suppler and/or Subcontractor Management)
- 4.6 Technical Performance Measurement (Identification of Technical Performance Measures, Tracking)
- 4.7 Program Cost (Projections/Reporting)
- 4.8 Design Review and Technical Audit (Internal Reviews, Formal Reviews, Supplier/Subcontractor Reviews)
- 4.9 Technical Communications (Program Reports and Documentation, Monitoring and Control)
- 4.10 Configuration Management
- 4.11 Risk Management (Identification of Risk, Risk Assessment, and Abatement)

5.0 System Engineering Process (Part II)

- 5.1 System Needs Analysis and Feasibility Study
- 5.2 System Operational Requirements
- 5.3 Maintenance Concept
- 5.4 Functional Analysis
- 5.5 Requirements Allocation
- 5.6 System Synthesis, Analysis, and Trade-Offs
- 5.7 Design Integration and Support
- 5.8 System Test and Evaluation
- 5.9 Production/Construction Support
- 5.10 Retirement/Material Disposal and Recycling
- 5.11 System Modifications ("Change" Implementation and Control)

6.0 Engineering Specialty Integration (Part III)

- 6.1 Engineering Specialties (Identification of Key Engineering Specialties, How They Fit Into the System Engineering Process, and Their Relationships With Each Other)
- 6.2 Description of Individual Engineering Specialties (Objective, Organization, Task Descriptions, Schedules, etc., for Each Significant Specialty Area)
  - 6.2.1 Functional Engineering
  - 6.2.2 Reliability Engineering
  - 6.2.3 Maintainability Engineering
  - 6.2.4 Human Factors and Safety Engineering
  - 6.2.5 Logistics Engineering
  - 6.2.6 Software Engineering
  - 6.2.7 Quality Engineering
  - 6.2.8 Value/Cost Engineering
  - 6.2.9 Producibility
  - 6.2.10 Disposability
  - 6.2.11 Other Engineering Disciplines (as appropriate)

7.0 References (Specifications, Standards, Plans, and Other Pertinent References/Documentation)

Figure 6.5 Modified Systems Engineering Management Plan (SEMP) outline.

1. The SOW should be relatively short and to the point (not to exceed two to three pages), and must be written in a clear and precise manner.
2. Every effort must be made to avoid ambiguity and the possibility of misinterpretation by the reader.
3. Describe the requirements in sufficient detail to ensure clarity, considering both practical applications and possible legal interpretations. Do not *underspecify* or *overspecify!*
4. Avoid unnecessary repetition and the incorporation of extraneous material and requirements. This could result in unnecessary cost.
5. Do not repeat detailed specifications and requirements that are already covered in the applicable referenced documentation.

The SOW will be read by many different individuals with a variety of backgrounds (e.g., engineers, accountants, contract managers, schedulers, lawyers), and there must be no unanswered questions as to the scope of work desired. It forms a basis for the definition and costing of detailed tasks, for the establishment of subcontractor and supplier requirements, and so on.

6.2.2 Definition of System Engineering Functions and Tasks

System engineering, as defined throughout this text, covers a broad spectrum of activity. It may even appear that the "systems engineer," or the system engineering organization, does everything! Although this is not practical, the fulfillment of system engineering objectives does require some involvement, either directly or indirectly, in almost every facet of program activity. The challenge is to identify those functions (or tasks) that deal with the overall *system* as an entity and, when successfully completed, will have a positive impact on the many related and subordinate tasks that must be accomplished.

With the intent of identifying a select number of *key* tasks for system engineering, the process described in Chapter 2 can be considered as being a framework for further discussion. As a start, a review of some of the overall basic goals is appropriate. The objectives of system engineering are the following:

1. Ensure that the requirements for system design and development, test and evaluation, production, operation, and support are developed in a timely manner through a top-down, iterative requirements analysis.
2. Ensure that system design alternatives are properly evaluated against meaningful, quantifiable criteria that relate to all of the desired characteristics; for example, performance factors, effectiveness factors, reliability and maintainability characteristics, human factors and safety factors, supportability characteristics, and life-cycle cost.
3. Ensure that all applicable design disciplines and related specialty areas are appropriately integrated into the *total engineering effort* in a timely and effective manner.

4. Ensure that the overall system development effort progresses in a logical manner with established configuration baselines, formal design reviews, the proper documentation supporting design decisions, and the necessary provisions for corrective action as required.
5. Ensure that the various elements (or components) of the system are compatible with each other, and are combined to provide an entity that will perform its required functions in an effective and efficient manner.

Review of these general goals leads to the question: What detailed program/project tasks should be performed in order to successfully meet the objectives of system engineering? Although each individual program is different and activities must be tailored accordingly, the tasks presented in Figure 6.6 and discussed in what follows are considered to be applicable in most instances.

1. Perform a needs analysis and conduct feasibility studies (refer to Sections 2.1, 2.2, and 2.3). These activities should be the responsibility of the system engineering organization because they deal with the system as an entity and are fundamental in the initial interpretation and subsequent definition of system requirements.

2. Define system operational requirements, the system maintenance concept, and the technical performance measures (TPMs) (refer to Sections 2.4, 2.5, and 2.6). The results of these activities are included in the overall definition of system-level requirements and are the basis for top-down system design.

3. Prepare the System Type "A" Specification (refer to Section 3.1). This represents the top technical document for system design, and fulfilling the objectives of system engineering is dependent on the completeness and comprehensiveness of this specification. The "B," "C," "D," and "E" Specifications are based on the requirements of the "A" Specification.

4. Prepare the Test and Evaluation Master Plan (refer to Section 2.11). This document reflects the approach, methods, and procedures that are to be followed in the overall evaluation of the system in terms of compliance with the initially specified requirements. Although there are many different and relatively small facets of testing, it is the compilation of these that provides an overall evaluation of the system as an entity.

5. Prepare the System Engineering Management Plan (refer to Sections 1.3 and 6.2). This, of course, represents the top management document for all system engineering program activities.

6. Accomplish functional analysis and the allocation of requirements (refer to Sections 2.7 and 2.8). Functional analysis, which represents the process of translating system-level requirements into detailed design criteria, provides the foundation for the development of many different individual design disciplinary tasks (refer to Section 2.7.4). The allocation process defines the specific design requirements for different components of the system, whether developed through supplier activities or procured off the shelf. In any event, the responsibility for this effort is appropriate because it facilitates the necessary design integration effort by providing a common baseline definition of the system in functional terms.

System Engineering Tasks	Task Input Requirements	Task Output Requirements
1. Perform needs analysis and conduct feasibility studies.	Consumer/customer requirements documentation; technical information reports covering technology applications; selected research reports; trade-off study reports supporting design approach.	Feasibility study reports; trade-off study reports justifying system-level design decisions.
2. Define operational requirements and the system maintenance concept.	Consumer/customer requirements documentation; customer specifications and standards; feasibility study reports; trade-off study reports supporting design approach.	System requirements documentation (operational requirements and maintenance concept); trade-off study reports justifying system-level design decisions; list of prioritized TPMs; functional analysis (system-level).
3. Prepare the system Type "A" specification.	Technical information reports covering technology applications; feasibility study reports; system requirements documentation (operational requirements and maintenance concept); trade-off study reports justifying system -level design decisions; list of prioritized TPMs; functional analysis (system-level).	System Type "A" specification.
4. Prepare the Test and Evaluation Master Plan (TEMP).	System Type "A" specification; customer test specification and standard; test requirements sheets (individual discipline test requirements).	Test and Evaluation Master Plan (TEMP).
5. Prepare the System Engineering Management Plan (SEMP).	Consumer/customer requirements documentation; customer program specifications and standards; system requirements documentation (operational requirements and maintenance concept); system Type "A" specification; Test and Evaluation Master Plan (TEMP); advance system planning information ; Program Management Plan (PMP).	System Engineering Management Plan (SEMP).
6. Accomplish functional analysis and the allocation of requirements.	System requirements documentation (operational requirements and maintenance concept); system specification; trade-off study reports justifying system-level design decisions.	Functional analysis reports–functional flow diagrams (operational and maintenance functions), timeline analysis sheets, requirements allocation sheets (RASs), trade-off study reports, test requirements sheets, design criteria sheets.
7. Accomplish system analysis, synthesis, and system integration.	Consumer/customer requirements documentation; customer specifications and standards; functional analysis reports; system Type "A" specification; System Engineering Management Plan (SEMP); Test and Evaluation Master Plan (TEMP); individual design discipline program planning requirements.	Selected design data; system integration reports; supplier data and reports; trade-off study reports justifying design decisions; selected design discipline reports (predictions and analyses).
8. Plan, coordinate, and conduct formal design review meetings.	Program Management Plan (PMP); System Engineering Management Plan (SEMP); applicable design data (drawings, parts and material lists, reports, databases); trade-off study reports justifying design decisions;individual design discipline reports (predictions, analyses, etc.).	Design review meeting minutes; action-item lists with designated responsibilities; approved/released design data and supporting documentation.
9. Monitor and review system test and evaluation activities.	Test and Evaluation Master Plan (TEMP); System Engineering Management Plan (SEMP); individual test data and reports.	System test and evaluation report(s).
10. Plan, coordinate, implement, and control design changes.	Configuration management data and reports (description of design baseline); proposed engineering change proposals; change control requirements and actions.	Change implementation plans, change verification data/reports.
11. Initiate and maintain production/construction liaison; supplier liaison; and customer service activities.	System design data; production/construction requirements; approved design changes; system operating and maintenance procedures; consumer/customer operations and system utilization requirements; field data and failure reports.	Field data and failure reports; customer service reports on field operations.

Figure 6.6 System engineering tasks.

7. Accomplish system analysis, synthesis, and system integration functions on a continuing basis throughout the overall design and development process (refer to Sections 2.9 and 2.10, and Chapter 3). System integration is iterative by nature and includes both the technical considerations dealing with the physical and functional interfaces of equipment, software, personnel, facilities, and so on, and the management considerations pertaining to organizational interfaces. From the management perspective, the system engineering organization is responsible for ensuring that (a) all program design-related functions/tasks are initially defined, (b) appropriate responsibilities and working relationships are established, (c) organization and communication channels are identified, and (d) program requirements are completed in a satisfactory manner. From Chapter 3, the system engineering organization is responsible for ensuring that the proper level of communications, coordination, and integration exists between the various design disciplines described therein. Of particular interest are the task requirements for reliability engineering (Figure 3.13), maintainability engineering (Figure 3.19), human-factors engineering (Figure 3.24), safety engineering (Figure 3.26), logistic support (Figure 3.35), software engineering (Section 3.3.5), producibility engineering (Section 3.3.6), quality engineering (Section 3.3.7), value/cost engineering (Section 3.3.9), and environmental engineering (Section 3.3.10).

8. Plan, coordinate, and conduct formal design review meetings; for example, conceptual design review, system design reviews, equipment/software design reviews, and critical design review (refer to Chapter 5). The system engineering organization is responsible for ensuring that an ongoing design evaluation effort is performed. This is partially accomplished through the scheduling of periodic design review meetings. The conductance of these meetings must be accomplished by a unbiased individual, and the overall results must be supportive of *system-level* design objectives.

9. Monitor and review system test and evaluation activities (refer to Section 2.11). It is essential that the system engineering organization be involved from the standpoint of interpreting and integrating individual test results into the evaluation of the system as a whole.

10. Plan, coordinate, implement, and control design changes as they evolve from engineering change proposals (ECPs) initiated from either the informal day-to-day review activity or as a result of formal design reviews (refer to Section 5.4). The system engineering organization is responsible for establishing and maintaining system "baselines" through the design and development process; for example, "functional" baseline, "allocated" baseline, and the "product" baseline in Figure 1.12. System engineering is essentially responsible for configuration management as the system evolves through its planned life cycle.

11. Initiate and maintain production/construction liaison, supplier liaison, and customer service activities. As the system configuration progresses from the design and development phase into production and/or construction and subsequently into operational use, there is a requirement for a specified level of engineering support. The purpose is to provide some engineering assistance relative to training and the

understanding of system design, the incorporation of approved engineering changes into the system, and the acquisition of data from production activities and consumer operations in the field. The system engineering organization must be able to "track" the system throughout its planned life cycle.

The 11 basic program tasks just described constitute an example of what might be appropriate for a typical program, although the specific requirements may vary from one program to the next. The goal is to identify tasks that are oriented to the *system,* and that are *critical* relative to meeting the five major system engineering objectives stated earlier. More specifically, it is essential that an overall *systems* approach be followed from the initial establishment of requirements. As design progresses, it is essential that the system configuration being developed includes the desired characteristics. Finally, it is essential that the product output be verified in terms of meeting the initially established requirements.

In accomplishing this, there are requirements definition tasks, there are design review and approval tasks, there are configuration control tasks, and there are final test and evaluation tasks. These activities are undertaken through the combination of providing key documentation (specifications, plans, and reports), conducting carefully scheduled design reviews with the appropriate feedback provisions, and providing the necessary ongoing coordination and integration efforts. These activities must address all *system* functions accomplished throughout the various levels depicted in Figure 6.3.

In reviewing the proposed 11 tasks, it is realized that the system engineering organization per se may not, by itself, be able to successfully complete all of the work required. Also, it is not the intent to justify a large organization for working out the details. However, the system engineering organization, through its system-level technical expertise and its leadership abilities, must take the lead and ensure that these task requirements are completed in a timely manner. The organizational approach is discussed further in Chapter 7.

In order to provide a more in-depth understanding of the 11 tasks, Figure 6.6 presents a summary listing these tasks and showing typical input and output requirements. Although the majority of the input-output requirements are self-explanatory by reviewing the appropriate sections of this text, some additional discussion is necessary in support of the output requirements from Task 6 dealing with functional analysis and allocation.

Functional analysis encompasses the process of translating system-level requirements into detailed design criteria, and results in the complete definition of the system configuration in functional terms (refer to Section 2.7). The accomplishment of functional analysis is facilitated through the development of functional flow block diagrams, described in Section 2.7.1. From these diagrams, the system engineer may wish to evaluate the various functions further from the standpoint of series-parallel relationships, time durations, and ultimately the identification of major resource requirements. Additionally, specific functional requirements need to be communicated to program/project personnel through timeline analysis sheets, requirements allocation sheets (RASs), trade-off study reports, test requirements sheets, design criteria sheets, and the like. Timeline analysis sheets and requirements allocation sheets are briefly discussed in what follows.

1. *Timeline analysis sheets:* although the functional block diagrams convey general series-parallel relationships, these requirements may be developed further through the use of timeline analysis sheets. Timeline analysis adds considerable detail in defining the durations of various functions. Concurrency, overlap, and the sequential relationships of functions/tasks can be projected. Also, time-critical functions can be readily identified; that is, those functions that directly affect system availability, operating time, and maintenance downtime. An example of a timeline analysis sheet format is presented in Figure 6.7

2. *Requirements allocation sheets* (RASs): the requirements allocation sheet (RAS) is often used as the primary document for the identification of specific design requirements based on the functional analysis. A RAS is developed for each block in the functional flow block diagram. Performance requirements are described and include (a) the purpose of the function, (b) the detailed performance characteristics that the function must accomplish, (c) the criticality of the function, and (d) applicable design constraints. Performance requirements must address design characteristics such as size, weight, volume, output, throughput, reliability maintainability, human factors, safety, supportability, economic factors, and so on. Both qualitative and quantitative performance requirements resulting from an analysis of the function are identified by the RAS. These requirements are expanded in sufficient detail to allow for the synthesizing and evaluation of alternative concepts for satisfying each functional need, employing a combination of resources in terms of equipment, personnel, software, and facilities. An initial definition of these resource requirements is included in the RAS. Figure 6.8, which is an extension of Figure 2.20, presents an example of a requirements allocation sheet (RAS) format.

The specific output from Task 6 (Figure 6.6) will vary in structure and format depending on the type of system and the stage of design and development. For large-scale systems involving many different interfaces, the relationships illustrated in Figure 6.9 may exist. On the other hand, for smaller systems where the design is relatively simple in terms of complexity, the utilization of all of these data outputs may not be feasible.

6.2.3 Development of a Work Breakdown Structure (WBS)

One of the initial steps in the program planning process after the generation of the Statement of Work (SOW) is the development of the Work Breakdown Structure (WBS).[7] The WBS is a product-oriented family tree that leads to the identification of the activities, functions, tasks, subtasks, work packages, and so on, that must be

[7] The subjects of WBS and work packaging are covered in most texts dealing with project management. A good reference is H. Kerzner, *Project Management: A Systems Approach to Planning, Scheduling, and Controlling,* 5th Ed., Van Nostrand Reinhold, New York, 1995.

System:		Subsystem:	Description of requirement:											
Source (functional flow block diagram)	Function number	Location												
Task number	Task description	Personnel	Elapsed time (hours)											Total time
			0.5	1.0	1.5	2.0	2.5	3.0	3.5	4.0	4.5	5.0	5.5	6.0

Figure 6.7 Timeline analysis sheet.

<table>
<tr><td>System:</td><td colspan="4">Subsystem:</td><td colspan="7" rowspan="2">Description of requirement:</td></tr>
<tr><td>Source (functional flow block diagram)</td><td>Function number</td><td colspan="3">Location</td></tr>
<tr><td colspan="2" rowspan="2">Functional performance and design requirements</td><td colspan="4">Personnel requirements</td><td colspan="2">Equipment requirements</td><td colspan="2">Software/data requirements</td><td colspan="2">Facility requirements</td></tr>
<tr><td>Tasks</td><td>Task time</td><td>Perfromance requirements</td><td>Training</td><td>Nomenclature</td><td>Specification</td><td>Nomenclature</td><td>Specification</td><td>Nomenclature</td><td>Specification</td></tr>
<tr><td colspan="2"></td><td></td><td></td><td></td><td></td><td></td><td></td><td></td><td></td><td></td><td></td></tr>
</table>

Figure 6.8 Requirements allocation sheet.

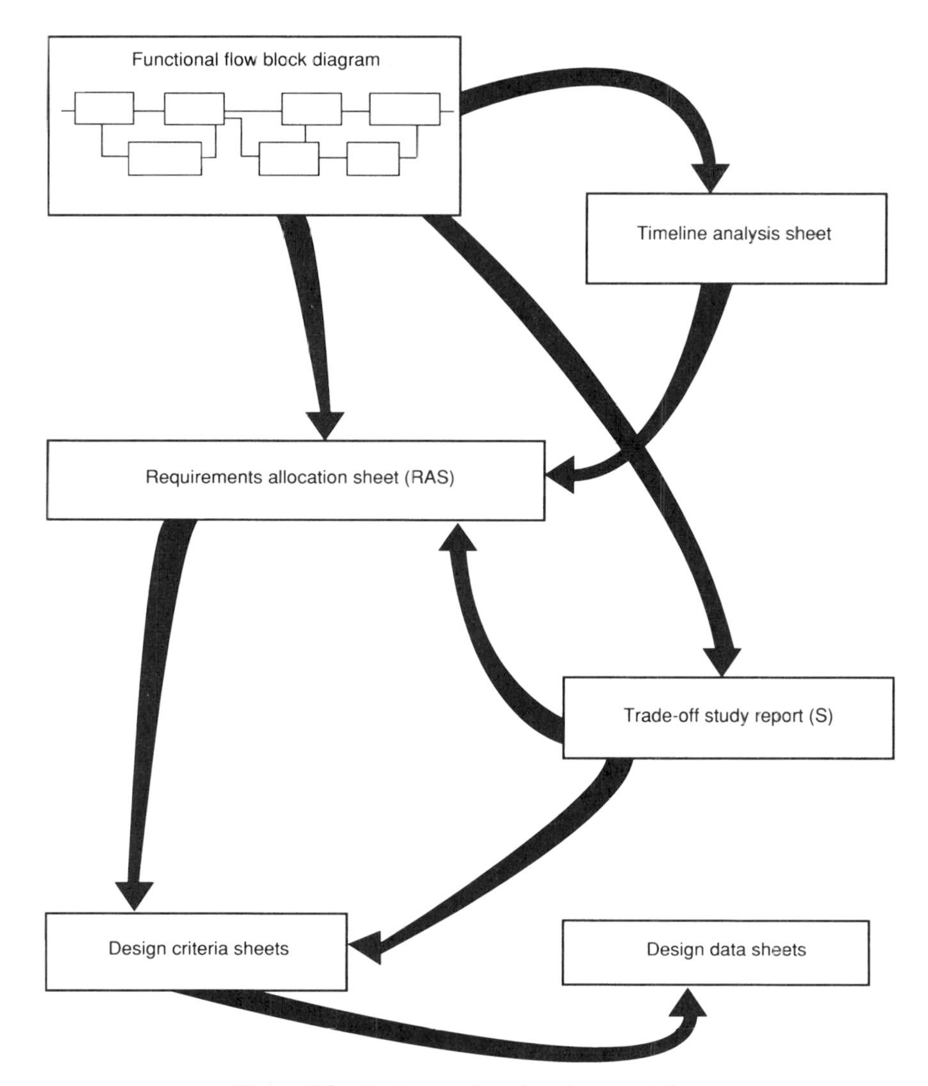

Figure 6.9 System engineering documentation.

performed for the completion of a given program. It displays and defines the system (or product) to be developed, and portrays all of the elements of work to be accomplished. The WBS is *not* an organizational chart in terms of project personnel assignments and responsibilities, but does represent an organization of work packages prepared for the purposes of program planning, budgeting, contracting, and reporting.

Figure 6.10 illustrates an approach to the development of the WBS. During the early stages of system planning, a Summary Work Breakdown Structure (SWBS) is usually prepared by the customer and included in a Request for Proposal (RFP) or an Invitation for Bid (IFB). This structure, developed from the top-down primarily

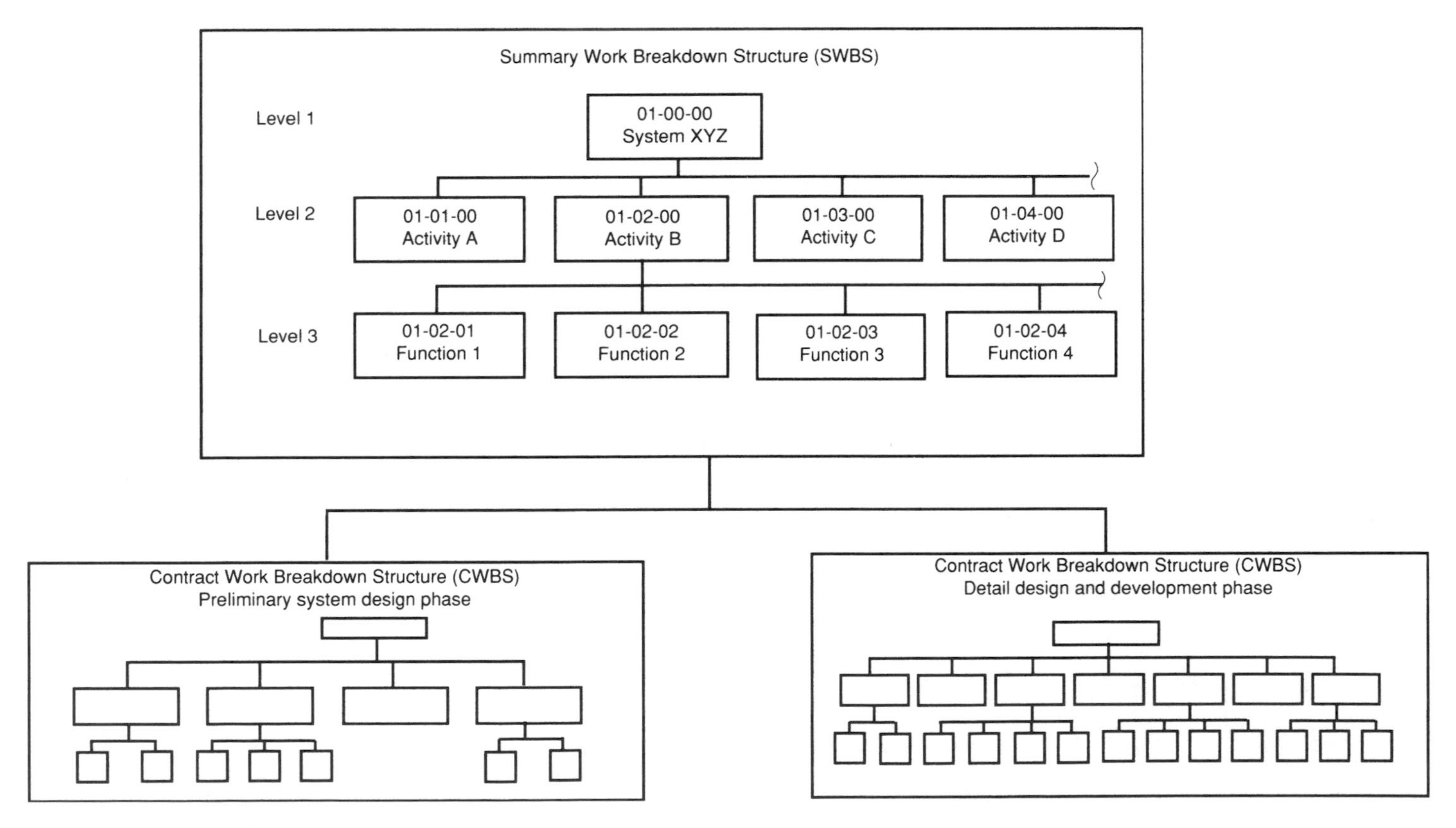

Figure 6.10 Partial work breakdown structure development.

for budgetary and reporting purposes, covers *all* programs functions and generally includes three levels of activity:

1. *Level 1:* identifies the total program scope of work, or the system to be developed, produced, and delivered to the customer. Level 1 is the basis for the authorization and "go-ahead" (or release) for all program work.
2. *Level 2:* identifies the various projects, or categories of activity, that must be completed in response to program requirements. It also may include major elements of the system and/or significant project activities; for example, subsystems, equipment, software, elements of support, program management, and system test and evaluation. Program budgets are usually prepared at this level.
3. *Level 3:* identifies the activities, functions, major tasks, and/or components of the system that are directly subordinate to the Level 2 items. Program schedules are generally prepared at this level.

As program planning progresses and individual contract negotiations are consumated, the SWBS is developed further and adapted to a particular contract or procurement action, resulting in a Contract Work Breakdown structure (CWBS). From Figure 6.3, for example, the customer may develop the SWBS with the objective of initiating program work activity. This structure will usually reflect the integrated efforts of all organizational entities assigned to the project and should not be related to any single department, group, or section. The SWBS, included in the customer's RFP, is the basis for the definition of all internal and contracted work to be performed on a given program. Through the subsequent preparation of proposals, contract negotiations, and related processes, Contractor A is selected to accomplish all work associated with the preliminary system design phase, and Contractor B is selected to complete all work associated with the detail design and development phase. From the definition of individual statements of work, a CWBS is developed to identify the elements of work for each program phase. The CWBS is tailored to a specific contract (or procurement action) and may be applicable to prime contractors, subcontractors, and/or suppliers, as shown in Figure 6.3.

The WBS constitutes a top-down hierarchical breakout of project activities that can be further divided into functions, functions into tasks, tasks into subtasks, subtasks into levels of effort, and so on. Conversely, detailed tasks (with defined starting and ending dates) can be combined into work packages, and work packages can be integrated into functions and activities, with the accumulation of all work being reflected at the top program or system level.

In developing a WBS, care must be exercised to ensure that (1) a continuous flow of work-related information is provided from the top down, (2) all applicable work is represented, (3) enough levels are provided to cause the identification of well-defined work packages for cost/schedule control purposes, and (4) the duplication of work effort is eliminated. If the WBS does not contain enough levels, then the management visibility and the integration of work packages may prove to be difficult. On the other hand, if too many levels exist, too much time may be wasted in performing program review and control actions.

Figure 6.11 presents an example of a Summary Work Breakdown Structure

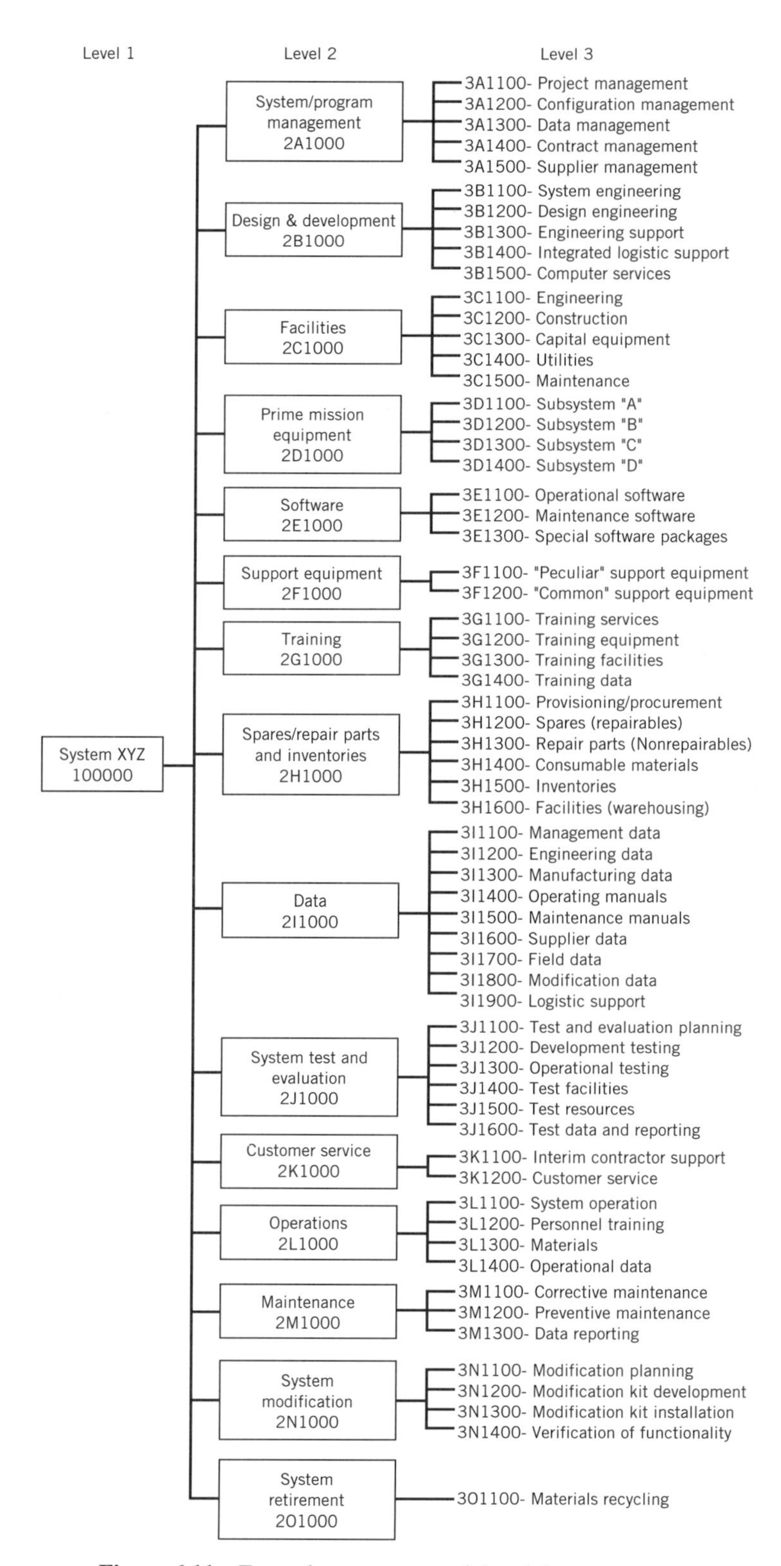
Level 1
Level 2
Level 3
System XYZ 100000
System/program management 2A1000
3A1100- Project management
3A1200- Configuration management
3A1300- Data management
3A1400- Contract management
3A1500- Supplier management
Design & development 2B1000
3B1100- System engineering
3B1200- Design engineering
3B1300- Engineering support
3B1400- Integrated logistic support
3B1500- Computer services
Facilities 2C1000
3C1100- Engineering
3C1200- Construction
3C1300- Capital equipment
3C1400- Utilities
3C1500- Maintenance
Prime mission equipment 2D1000
3D1100- Subsystem "A"
3D1200- Subsystem "B"
3D1300- Subsystem "C"
3D1400- Subsystem "D"
Software 2E1000
3E1100- Operational software
3E1200- Maintenance software
3E1300- Special software packages
Support equipment 2F1000
3F1100- "Peculiar" support equipment
3F1200- "Common" support equipment
Training 2G1000
3G1100- Training services
3G1200- Training equipment
3G1300- Training facilities
3G1400- Training data
Spares/repair parts and inventories 2H1000
3H1100- Provisioning/procurement
3H1200- Spares (repairables)
3H1300- Repair parts (Nonrepairables)
3H1400- Consumable materials
3H1500- Inventories
3H1600- Facilities (warehousing)
Data 2I1000
3I1100- Management data
3I1200- Engineering data
3I1300- Manufacturing data
3I1400- Operating manuals
3I1500- Maintenance manuals
3I1600- Supplier data
3I1700- Field data
3I1800- Modification data
3I1900- Logistic support
System test and evaluation 2J1000
3J1100- Test and evaluation planning
3J1200- Development testing
3J1300- Operational testing
3J1400- Test facilities
3J1500- Test resources
3J1600- Test data and reporting
Customer service 2K1000
3K1100- Interim contractor support
3K1200- Customer service
Operations 2L1000
3L1100- System operation
3L1200- Personnel training
3L1300- Materials
3L1400- Operational data
Maintenance 2M1000
3M1100- Corrective maintenance
3M1200- Preventive maintenance
3M1300- Data reporting
System modification 2N1000
3N1100- Modification planning
3N1200- Modification kit development
3N1300- Modification kit installation
3N1400- Verification of functionality
System retirement 2O1000
3O1100- Materials recycling

Figure 6.11 Example summary work breakdown structure.

(SWBS) covering the development of a large system. As program requirements become defined through a contractual (or procurement) arrangement, the SWBS can be readily converted into a CWBS to reflect the actual work required under the contract. The CWBS, as it appears in a contractual document, is also presented at three levels in order to provide a good baseline for planning purposes while allowing for some flexibility within the contractor's organization. An expansion of the CWBS can be accomplished as necessary to provide for internal cost/schedule controls.

Figure 6.12 shows an expansion of the system engineering activities to the fifth level; that is, those work packages under 3B1100 in Figure 6.11. The purposes are to recognize the major system engineering tasks presented in Figure 6.6 and to provide a breakout of these tasks in a CWBS format to the extent necessary for proper cost/schedule visibility. Please note that Figure 6.12 includes two different CWBSs, one covering the work to be performed during the conceptual design and advance planning phase and the other directed toward the work required during the preliminary system design phase. Each individual CWBS is derived from the SWBS and the overall program CWBS. Further, there must be a close tie between the two, as the CWBS for preliminary design must reflect the activities that evolve directly from the earlier phase.

The elements of the WBS may represent an identifiable item of equipment or software, a deliverable data package, an element of logistic support, a human service, or a combination thereof. WBS elements should be selected to permit the initial structuring of budgets and the subsequent tracking of technical performance measures (TPMs) against cost. Thus, in expanding the WBS to successfully lower levels, the requirements for day-to-day task management must be balanced against the overall reporting requirements for the program. In essence, program activities are broken down to the lowest level that can be associated with both an organization and a cost account, as illustrated in Figure 6.13. From this, schedules are developed, cost estimates are generated, accounts are established, and program activities are monitored for the purposes of schedule/cost control.

In developing the WBS, it is essential that a good comprehensive "WBS Dictionary" be prepared. This constitutes a document containing the terminology and definition of each element of the WBS. Traceability must be maintained from the top down, and all applicable work must be included. This is facilitated by assigning a number to each work package in the WBS. In Figure 6.10, the total program is represented by 01-00-00, and the numbers are broken down for activities, functions, tasks, subtasks, and so on. In Figure 6.11, a slightly different numbering system is used. Although the numbering systems will vary for different programs (and with different contractors), it is important to ensure that both activities and budgets/costs can be traced, both upward and downward. In the initial generation of a CWBS by a contractor during the preparation of a proposal, budgets may be allocated downward to specific tasks. After contract award as tasks are being accomplished, costs are being incurred and charged to the appropriate cost account. These costs are then collected upward for reporting purposes. The WBS provides the vehicle for measuring work package progress in terms of schedule and cost.

In summary, the work breakdown structure (WBS) provides many benefits:

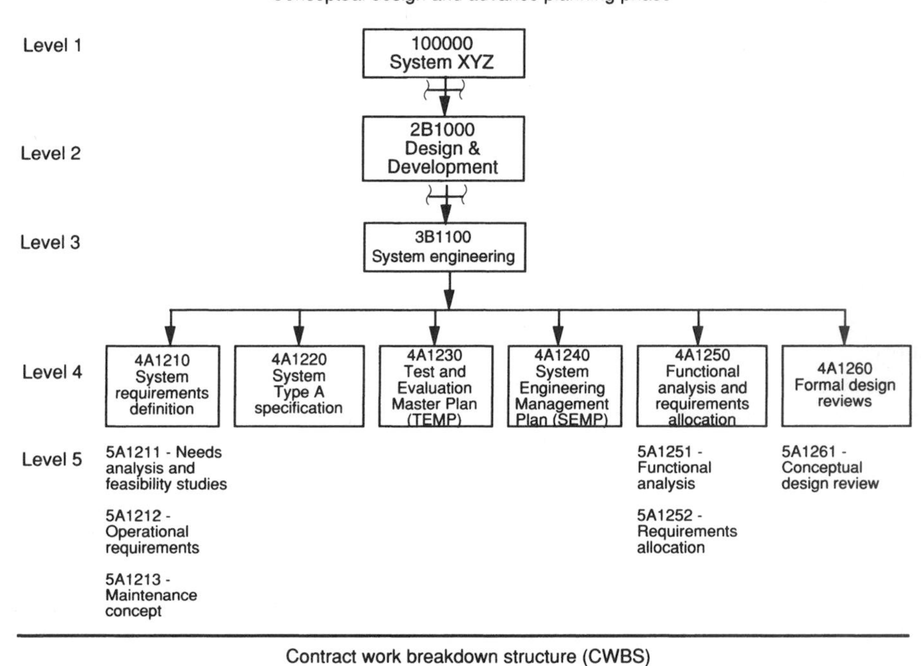

Contract work breakdown structure (CWBS)
Conceptual design and advance planning phase
Level 1
100000 System XYZ
Level 2
2B1000 Design & Development
Level 3
3B1100 System engineering
Level 4
4A1210 System requirements definition
4A1220 System Type A specification
4A1230 Test and Evaluation Master Plan (TEMP)
4A1240 System Engineering Management Plan (SEMP)
4A1250 Functional analysis and requirements allocation
4A1260 Formal design reviews
Level 5
5A1211 - Needs analysis and feasibility studies
5A1212 - Operational requirements
5A1213 - Maintenance concept
5A1251 - Functional analysis
5A1252 - Requirements allocation
5A1261 - Conceptual design review

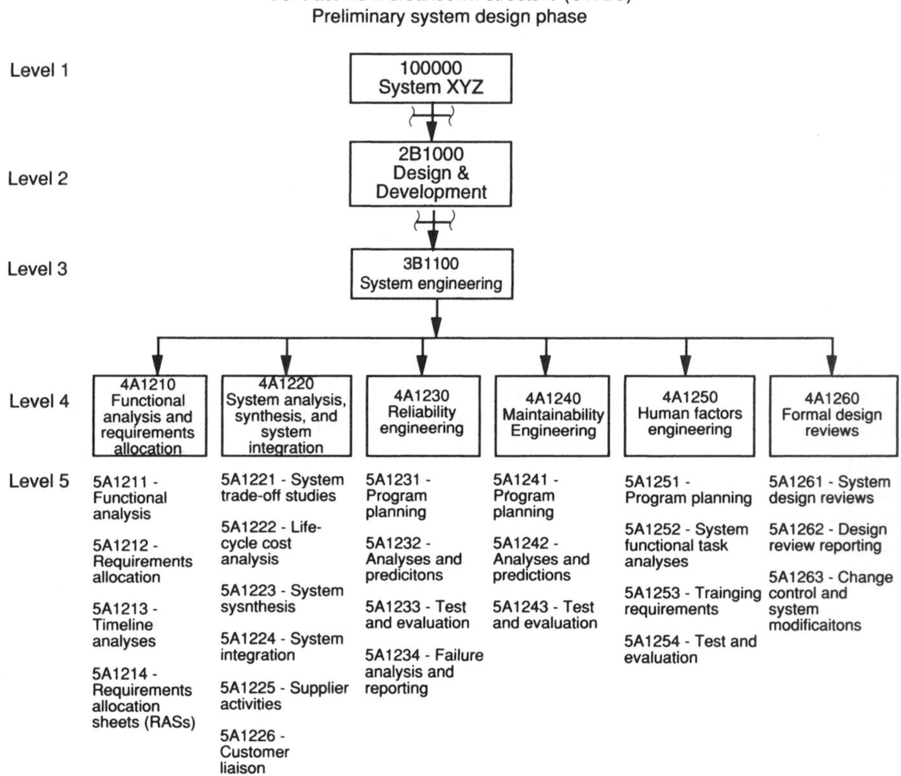

Contract work breakdown structure (CWBS)
Preliminary system design phase
Level 1
100000 System XYZ
Level 2
2B1000 Design & Development
Level 3
3B1100 System engineering
Level 4
4A1210 Functional analysis and requirements allocation
4A1220 System analysis, synthesis, and system integration
4A1230 Reliability engineering
4A1240 Maintainability Engineering
4A1250 Human factors engineering
4A1260 Formal design reviews
Level 5
5A1211 - Functional analysis
5A1212 - Requirements allocation
5A1213 - Timeline analyses
5A1214 - Requirements allocation sheets (RASs)
5A1221 - System trade-off studies
5A1222 - Life-cycle cost analysis
5A1223 - System sysnthesis
5A1224 - System integration
5A1225 - Supplier activities
5A1226 - Customer liaison
5A1231 - Program planning
5A1232 - Analyses and predicitons
5A1233 - Test and evaluation
5A1234 - Failure analysis and reporting
5A1241 - Program planning
5A1242 - Analyses and predictions
5A1243 - Test and evaluation
5A1251 - Program planning
5A1252 - System functional task analyses
5A1253 - Trainging requirements
5A1254 - Test and evaluation
5A1261 - System design reviews
5A1262 - Design review reporting
5A1263 - Change control and system modificaitons

Figure 6.12 CWBS expansion showing system engineering activities.

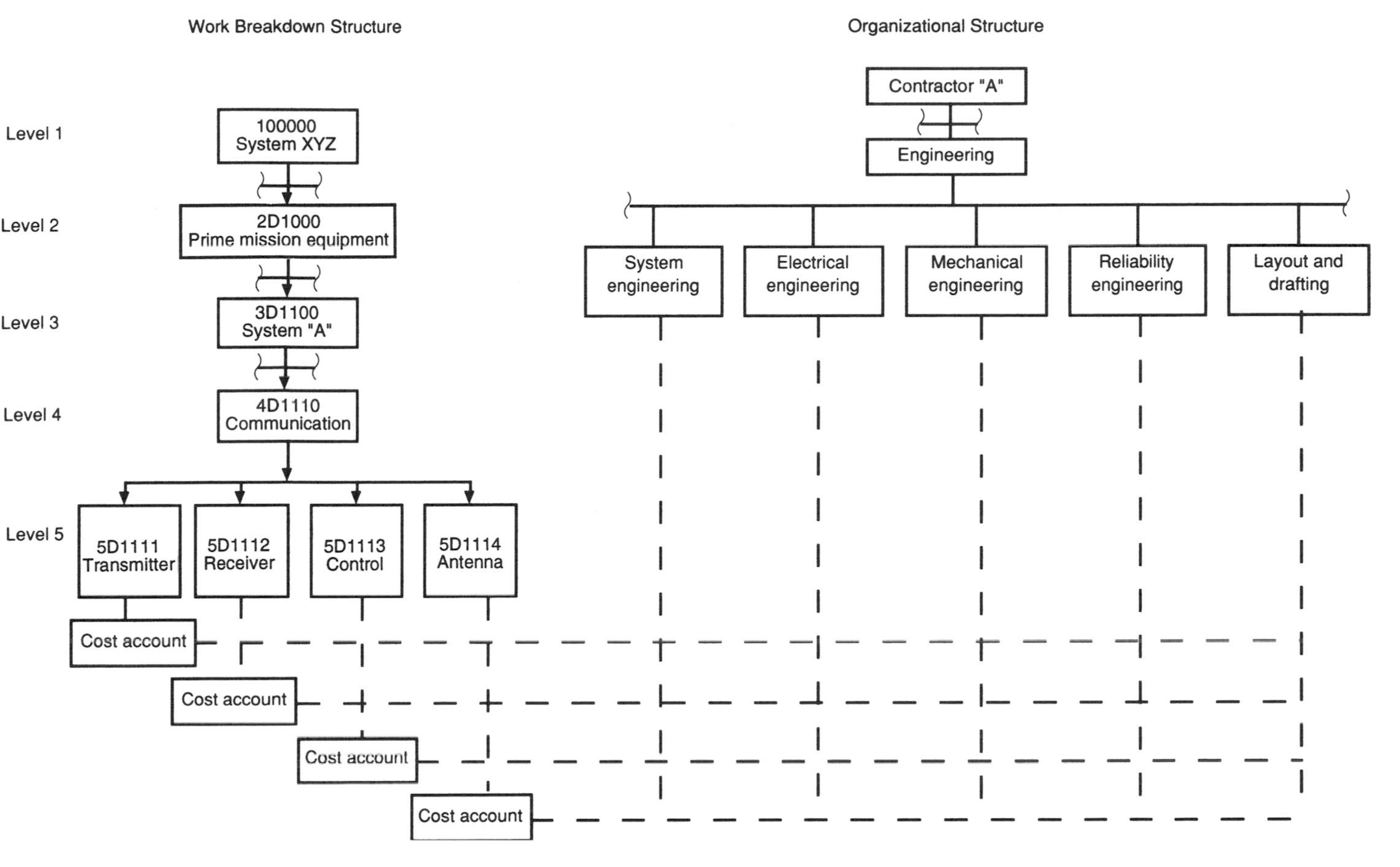

Figure 6.13 Organizational integration with CWBS.

1. The total program, or system, can be easily described through the logical breakout of its elements into nicely definable work packages.
2. The discipline associated with the development of the WBS provides a greater probability that every program activity will be accounted for.
3. The WBS is an excellent vehicle for linking program objectives and activities with available resources.
4. The WBS facilitates the initial allocation of budgets and the subsequent collection and reporting of costs.
5. The WBS provides an excellent matrix for the assignment of tasks and work packages to various organizational departments, groups, and/or sections. Responsibility assignments can be readily identified.
6. The WBS is an excellent vehicle for the reporting of system technical performance measures (TPMs) against schedule and cost.

Finally, the WBS is an excellent tool for the promotion of program communications at various levels. As such, it must be updated to reflect program/system changes, consistent with configuration management actions. Maintaining currency is essential in meeting system engineering objectives.

6.2.4 Specification/Documentation Tree

In Section 3.1 (Figures 3.2 and 3.3), *specifications* are utilized primarily for acquiring items and/or for some element of work; that is, contracting for the design and development of a new item, the procurement of a commercial off-the-shelf (COTS) item, the testing and verification of a product, the construction of a new facility, the production of "x" quantity of items, and so on. Specifications may be applied on contracts and imposed on major contractors, subcontractors, and the suppliers of goods and services. They are *requirements*-oriented, are *performance*-oriented, and should state the "WHATs" (i.e., *what* to do versus *how* to do it), and must be written in a clear and concise manner. Vague, redundant, nebulous, and ambiguous language should be eliminated. Requirements should be quantifiable and verifiable, and the need to use judgment for interpretation should be avoided; that is, the use of phrases such as "best design practices" or "good workmanship." Specifications establish requirements relative to both design and performance characteristics. Management information, statements of work, procedural data, schedules and cost projections, and so on, should not be included.

Relative to applications, there are (1) general specifications, (2) program-related specifications, (3) military specifications and standards, (4) industrial standards, (5) specific company standards, and (6) international specifications and standards.[8] The different categories of specifications described in Section 3.1 refer primarily to pro-

[8] Industrial standards can vary significantly and are developed by organizations such as the American National Standards Institute (ANSI), International Standards Organization (ISO), American Society for Testing and Materials (ASTM), Electronic Industries Association (EIA), Institute of Electrical and Electronics Engineers (IEEE), and the National Standards Association (NSA).

gram unique specifications, or those that are directed toward a particular program requirement and/or a specific system component. Additionally, there are specifications and standards that cover components, materials, and processes across the board independent of application. In any event, there may be a wide variety of specifications and standards applied on a given program.

In applying specifications, extreme care must be exercised to ensure that they are prepared to the proper depth of detail and applied at the appropriate level in the system hierarchy. Such documents must be detailed to the extent required to impact design in terms of component selection, the utilization of materials, and the identification of processes. On the other hand, applying specifications with too much detail and at too low a level in the system hierarchy can be extremely detrimental! This may not only tend to inhibit innovation and creativity by not allowing for possible trade-offs, but "overspecification" can be quite costly. Applying a detailed specification to a small commercially available off-the-shelf component may result in an overdesign situation, which, in turn, can significantly increase the cost of that component.[9]

Another concern pertaining to the application of specifications involves possible areas of conflict. Experience has indicated that conflicts (i.e., contradictions in direction) are sometimes introduced with the application of general specifications and standards across the board. These documents are prepared by different individuals at different times with different applications in mind, and are not necessarily consistent in terms of detailed requirements. Often in the development of program requirements, there is a tendency to follow the most expeditious and easiest approach by attaching a large list of these specifications and standards to a SOW. In other words, "the contractor must comply with the attached list of specifications and standards in fulfilling program requirements." This blind application can result in conflicts pertaining to component part selection, manufacturing process variations, test and evaluation parameters, and so on. In such instances, there is a question as to which specification takes precedence. What are the priorities in importance?

With the objective of promoting clarification and eliminating the areas of possible conflict, the preparation of a "specification tree" (or documentation tree) is recommended. This is a family tree of specifications and documents that supports the system hierarchy, establishes order of precedence in the event of conflicts, and relates to the elements of work in the work breakdown structure (WBS). Figure 6.14 illustrates a simplified specification tree.

In the figure, the "tree" is developed from the top down, commencing with the preparation of the system specification (refer to Section 3.1). Subsequently, additional specifications are applied, following the system hierarchy illustrated in Figure 2.21. As one progresses, the application of specifications must be consistent with

[9] The reference pertains to the number of instances in which, in the defense sector, military specifications and standards have been imposed on the procurement of small components, hand tools, and so on. The "blind" imposition of specifications on commercially available off-the-shelf items can turn out to be quite costly. Not only will an overdesign situation result, but the number of available suppliers will be reduced.

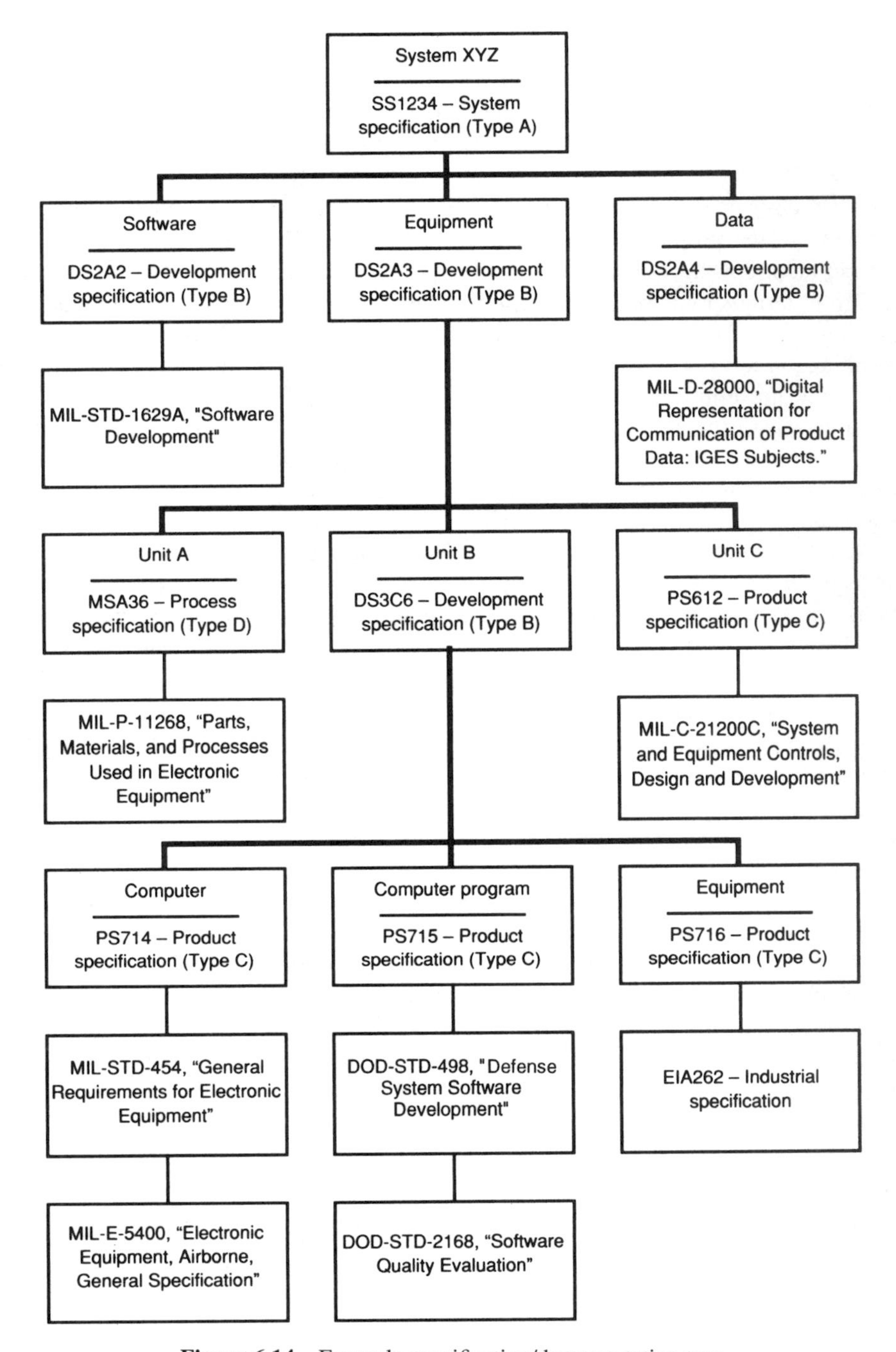

Figure 6.14 Example specification/documentation tree.

the work requirements described by the WBS in Figure 6.10. Further, this application must be adapted to the contracting structure between major contractors, subcontractors, and suppliers (refer to Figure 6.3).

The critical task here is the tailoring of specifications to the particular system application. Even though the design requirements may dictate the use of an available off-the-shelf item, the application of that item in this system may be quite different from comparable applications in other systems. Thus, the major components of the system should be described through a series of program-related specifications, as shown in Figure 6.14; for example, development specifications, product specifications, and process specifications. Below this level, it may be appropriate to apply general specifications as long as they support the overall requirements in system design. When there are a number of different specifications and/or standards applied to the same system component, they must be complementary and mutually supportive. In the event of conflicts in direction or concerns relative to priorities of importance, the specification tree must provide an indication as to which document takes precedence.

The development of design requirements from the top down is critical in meeting the system engineering objectives stated herein. Thus, extreme care must be exercised in the initial identification and application of specifications and standards. Although this function is sometimes viewed as being relatively minor, the results can be rather costly if the proper level of attention is not directed to this area from the beginning! Conflicts, changes in specification requirements resulting in contractual modifications, and so on, can be extremely detrimental to a program. The inclusion of a complete specification tree in the SEMP may assist in avoiding potential problems later.

6.2.5 Technical Performance Measurement

In Section 2.6 (Chapter 2), technical performance measures (TPMs) for the system are identified through the development of operational requirements and the maintenance concept, and are prioritized using QFD (or equivalent) methods. Figure 2.10 provides an example of the results. As indicated in the figure, *velocity, availability,* and *size* are the top three in priority. Employing the QFD approach will help in developing the criteria and characteristics in design that must be built-in in order to ensure that the velocity, availability, and size requirements are ultimately met.

As the system development effort progresses, periodic design reviews will be conducted as described in Chapter 5. The known design configuration at that time will be evaluated with the high-priority TPMs in mind. Checklists may be utilized to aid in the evaluation process, identifying those characteristics that have been incorporated and that relate to and directly support the TPM objectives (refer to Figure 5.4). Design parameters and the applicable TPMs will be measured and "tracked," as shown in Figure 5.6 (Chapter 5). This must be accomplished on a continuing basis, and the results should be included as an inherent part of the regular program management review process. Those high-priority TPMs, assumed as being critical, should receive the most attention in the review and evaluation process.

From Figure 5.6, if there is a deviation from the specified TPM value (either upward or downward), the "causes" for such must be identified and the appropriate corrective action needs to be initiated accordingly. Inherent within the program management information system structure is the requirement to plot future trends, predict potential deviations, and identify the possible consequences and associated risks in the event that no corrective action is initiated. In essence, technical performance measurement must be built into the regular program management and control process, and the prioritization of TPMs is a necessary input for the risk management plan (refer to Section 6.6).

6.2.6 Development of Program Schedules

In line with the Statement of Work (SOW) and the work breakdown structure (WBS), individual program tasks are presented in terms of a timeline; that is, a beginning time and an ending time. Schedules are developed to reflect the work requirements throughout all phases of a program.

Schedule planning commences with the identification of major program milestones at the top level and proceeds downward through successively lower levels of detail. A "System Engineering Master Schedule (SEMS)" is initially prepared laying out the major program activities on the basis of elapsed time. This serves as the frame of reference for a family of subordinate schedules, developed to cover subdivisions of work as represented by the WBS. Progress against a given schedule is measured at the bottom level, and task status information is related to the appropriate cost account identified by the WBS element and the responsible organization (refer to Figure 6.13).

Program task scheduling may be accomplished using one or a combination of techniques. Some of the more common methods are briefly described in what follows.

1. *Bar chart:* The simple bar chart presents program activities in terms of sequences and the time span of efforts. Specific milestones and the assignment of resources are not covered. Figure 6.15 illustrates a partial bar chart.

2. *Milestone chart:* A presentation of specific program events (i.e., identifiable outputs) and required start and completion times by calendar date is included. Deliverable items required under contract are noted. Figure 6.16 shows a sample milestone chart.

3. *Combined milestone/bar chart:* The combining of activities and milestones into an overall project schedule represents a common approach for many programs. Figure 6.17 presents the primary system engineering tasks, included in Figure 6.6, in a program timeline format. This, of course, serves as the basis for the assignment of resources and the development of cost projections.

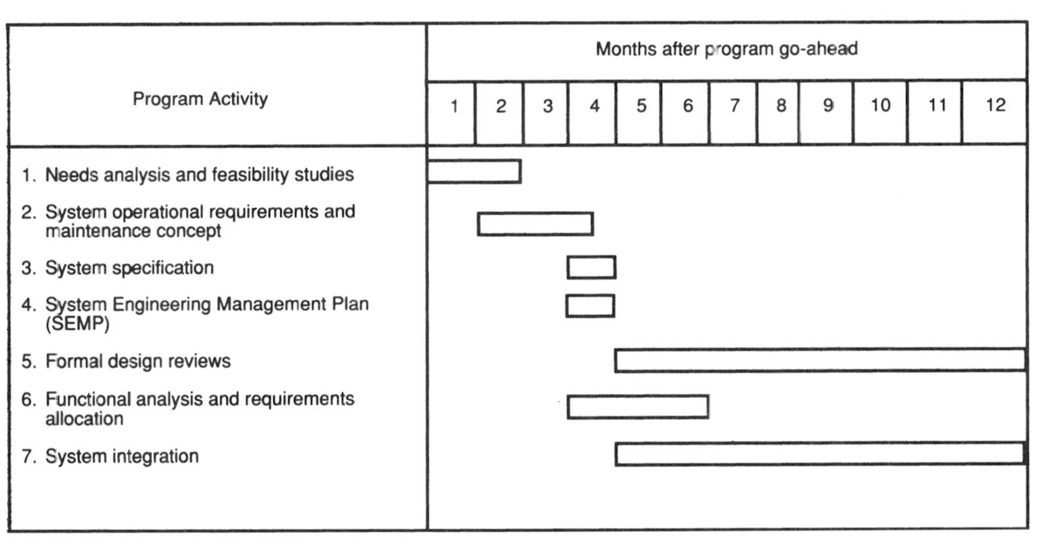

Figure 6.15 Partial bar chart.

4. *Program networks:* Network schedule methods include the Program Evaluation and Review Technique (PERT), the Critical Path Method (CPM), and various combinations of these. PERT and CPM are ideally suited for early planning where precise task time data are not readily available, and the aspects of probability are introduced to help define risk leading to improved decision making. These tech-

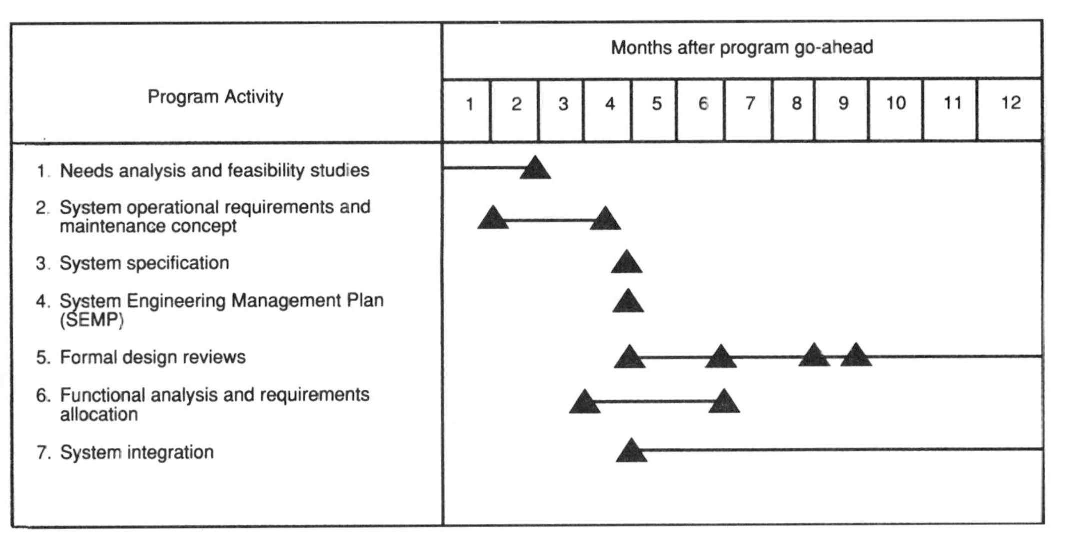

Figure 6.16 Sample milestone chart.

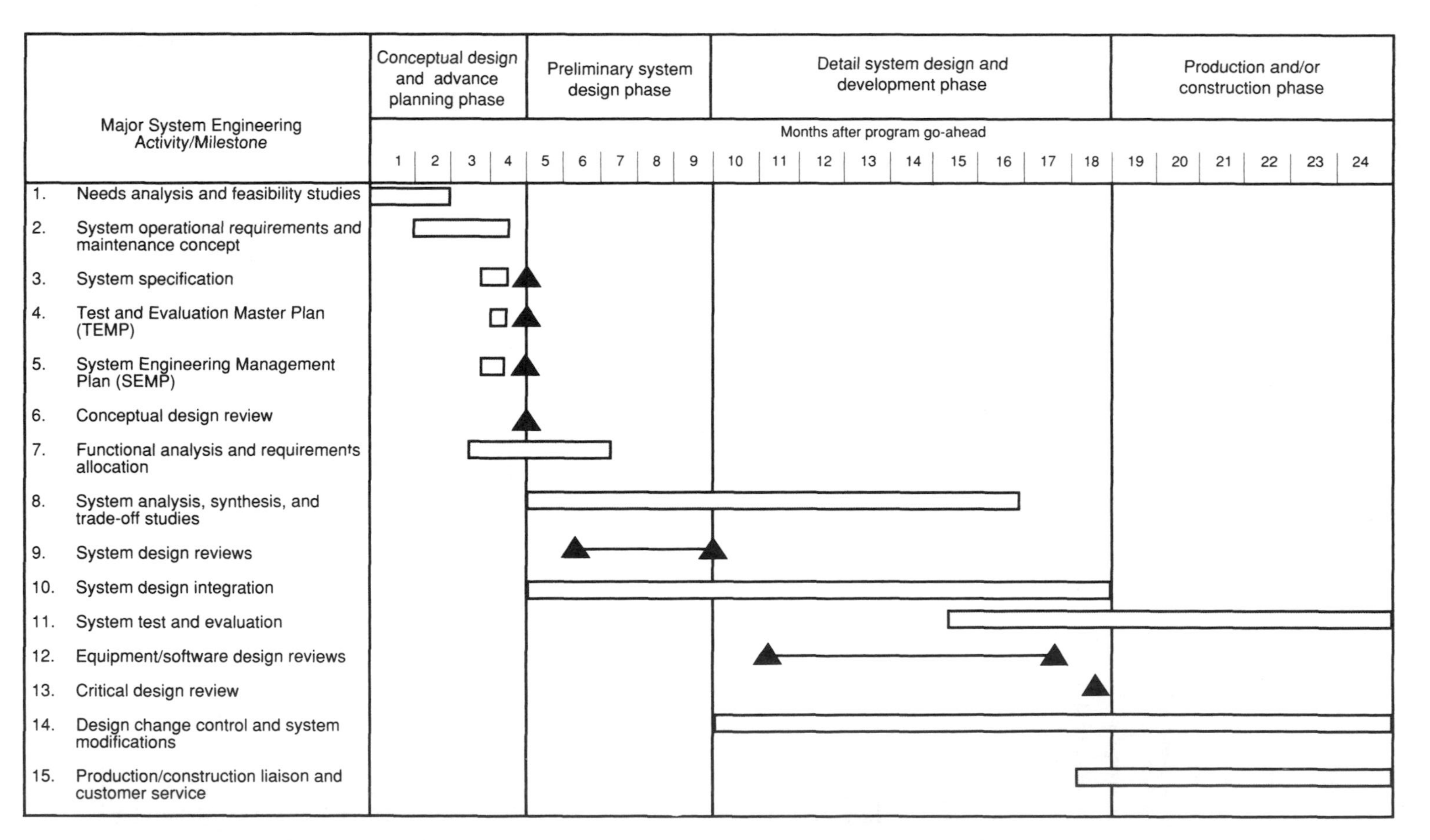

Major System Engineering Activity/Milestone
Conceptual design and advance planning phase
Preliminary system design phase
Detail system design and development phase
Production and/or construction phase
Months after program go-ahead
1 2 3 4 5 6 7 8 9 10 11 12 13 14 15 16 17 18 19 20 21 22 23 24
1. Needs analysis and feasibility studies
2. System operational requirements and maintenance concept
3. System specification
4. Test and Evaluation Master Plan (TEMP)
5. System Engineering Management Plan (SEMP)
6. Conceptual design review
7. Functional analysis and requirements allocation
8. System analysis, synthesis, and trade-off studies
9. System design reviews
10. System design integration
11. System test and evaluation
12. Equipment/software design reviews
13. Critical design review
14. Design change control and system modifications
15. Production/construction liaison and customer service

Figure 6.17 Major system engineering activities and milestones.

niques provide visibility and enable management to control one-of-a-kind projects as opposed to repetitive functions. Further, the network approach is effective in showing the interrelationships of combined activities.[10] Figure 6.18 reflects an example of a network diagram consisting of 17 "events" and 29 major "activities." Events are usually designated by circles and are considered as checkpoints showing specific milestones; that is, dates for starting a task, completing a task, and delivering an item under contract. Activities are represented by the lines between the circles, indicating the work that needs to be accomplished to complete an event. Work can start on the next activity only after the preceding event has been completed. The numbers on the activity lines indicate the time required in days, weeks, or months. The first number reflects an optimistic time estimate, the second number indicates the expected time, and the third number constitutes a pessimistic time estimate.[11]

In applying PERT/CPM to a project, one must identify all interdependent events and activities for each phase of the project. Events are related to program milestones dates that are based on management objectives. Figure 6.19 describes the major activities that are reflected by the lines in Figure 6.18. Managers and programmers work with engineering organizations to define these objectives and identify tasks and subtasks. When this is accomplished to the necessary level of detail, networks are developed, starting with a summary network and working down to detailed networks covering specific segments of a program. The development of networks is a team approach.

When actually constructing networks, one starts with an end objective (i.e., Event 17 in Figure 6.18) and works backward in developing the network until Event 1 is identified. Each event is labeled, coded, and checked in terms of program timeframe. Activities are then identified and checked to ensure that they are properly sequenced. Some activities can be performed on a concurrent basis, and others must be accomplished in series. For each completed network, there is "one beginning event" and "one ending event," and all activities must lead to the ending event.

The next step in developing a network is to estimate activity times and to relate these times in terms of probability of occurrence. An example of the calculations that support a typical PERT/CPM network is presented in Figure 6.20 and described in what follows.

[10] Three good references covering scheduling methods are (1) H. Kerzner, *Project Management: A Systems Approach to Planning, Scheduling, and Controlling,* 5th Ed., Van Nostrand Reinhold, New York, 1995; (2) J. E. Ullmann (Ed.), *Handbook of Engineering Management,* John Wiley, New York, 1986; (3) and D. I. Cleland, and W. R. King, *Project Management Handbook,* 2nd Ed., Van Nostrand Reinhold, New York, 1989.

[11] The level of detail and depth of network development (i.e., the number of activities and events included) are based on the criticality of tasks and the extent to which program evaluation and control are desired. Milestones that are critical in meeting the objectives of the program should be included along with activities that require extensive interaction for successful completion. The author has had experience dealing with PERT/CPM networks including from 10 to 700 events. The number of events/activities, of course, will vary with the project.

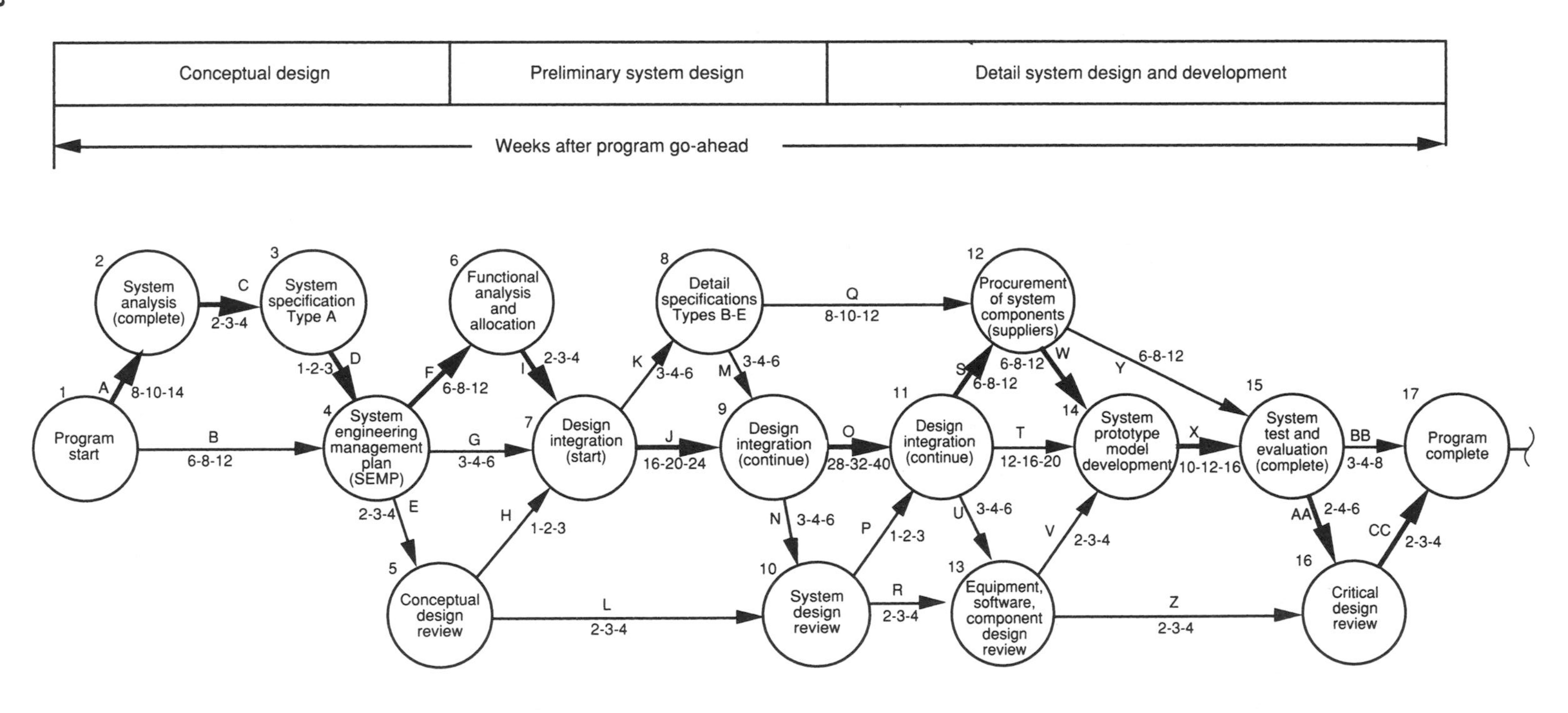

Figure 6.18 Partial program summary network.

Activity	Description of Program Activity
A	Perform needs analysis, conduct feasibility studies, and and accomplish systems analysis (operational requirements, maintenance concept, and functional definition of the system).
B	Conduct advance planning, perform initial management functions, and complete the System Engineering Management Plan (SEMP).
C	Prepare the System Specification (Type A).
D	Develop system-level technical requirements for inclusion in the System Engineering Management Plan (SEMP).
E	Prepare system-level design data and supporting materials for the Conceptual Design Review.
F	Accomplish functional analysis and the allocation of overall system requirements to the sub-system level and below (as required).
G	Develop the necessary organizational and related infra-structure in preparation for the accomplishment of the required program design integration tasks.
H	Translate the results from the Conceptual Design Review to the appropriate design activities (i.e., approved design data, recommendations for improvement/corrective action).
I	Translate the results from the functional analysis and allocation activity into specific design criteria required as an input for the design integration process.
J	Accomplish preliminary design and related design integration activities.
K	Translate the results from system-level design into specific requirements at the sub-system level and below. Prepare Development, Process, Product, and/or Material Specifications as required.
L	Conduct the necessary planning and prepare for the System Design Review.
M	Translate the requirements contained within the various applicable specifications into specific design criteria required as an input for the design integration process.
N	Prepare design data and supporting materials (as a result of preliminary design) for the System Design Review:
O	Accomplish detail design and related design integration activities.
P	Translate the results from the System Design Review to the appropriate design activities (i.e., approved design data, recommendations for improvement/corrective action).
Q	Identify the appropriate system component suppliers, impose the necessary specification requirements through contracts, and monitor supplier activities.
R	Conduct the necessary planning and prepare for the Equipment/Software/Component Design Reviews (there may be a series of individual design reviews covering different system components).
S	Provide detail design data (as necessary) to support supplier operations.
T	Develop a prototype model, with associated support, in preparation for system test and evaluation.
U	Prepare design data and supporting materials (as a result of detail design) for the Equipment/Software/Component Design Reviews.
V	Translate the results from the Equipment/Software/Component Design Reviews for incorporation into the prototype model(s) as applicable. The prototype model that is to be utilized in test and evaluation must reflect the latest design configuration.
W	Provide supplier components, with supporting data, for the development of the system prototype to be utilized in test and evaluation activities.
X	Prepare for and conduct System Test and Evaluation (implement the requirements of the Test and Evaluation Master Plan).
Y	Provide test data and logistic support, from the various suppliers, throughout the system test and evaluation phase. Test data are required to cover individual tests conducted at supplier facilities, and logistic support (i.e., spare/repair parts, test equipment, etc.) is necessary to support system testing activities.
Z	Conduct the necessary planning and prepare for the Critical Design Review.
AA	Test results, in the form of either design verification or recommendations for improvement/corrective action, are provided as an input the Critical Design Review.
BB	Prepare system test and evaluation report.
CC	Translate the results from the Critical Design Review for incorporation into the final system configuration prior to entering the Production and/or Construction Phase of the Program.

Figure 6.19 List of activities in the program network.

1	2	3	4	5	6	7	8	9	10	11	12
Event number	Previous number	t_a	t_b	t_c	t_e	s^2	TE	TL	TS	TC	Probability (%)
17	16	2	3	4	3.0	0.111	115.2	115.2	0	110	6.4
	15	3	4	8	4.5	0.694	112.1	115.2	3.1	115	47.9
16	15	2	4	6	4.0	0.444	112.1	112.2	0	120	91.9
	13	2	3	4	3.0	0.111	86.5	112.2	25.7		
15	14	10	12	16	12.3	1.000	108.2	108.2	0		
	12	6	8	12	8.3	1.000	95.9	108.2	12.3		
14	13	2	3	4	3.0	0.111	86.5	95.9	9.4		
	12	6	8	12	8.3	1.000	95.9	95.9	0		
	11	12	16	20	16.0	1.778	95.3	95.9	0.6		
13	11	3	4	6	4.2	0.250	83.5		13.6		
	10	2	3	4	3.0	0.111	53.8		42.1		
12	11	6	8	12	8.3	1.000	87.6	87.6	0		
	8	8	10	12	10.0	0.444	60.8	87.6	26.8		
11	10	1	2	3	2.0	0.111	52.8	79.3	26.5		
	9	28	32	40	32.7	4.000	79.3	79.3	0		
10	9	3	4	6	4.2	0.250	50.8		30.7		
	5	2	3	4	3.0	0.111	21.3		59.0		
9	8	3	4	6	4.2	0.250	35.0	46.6	11.6		
	7	16	20	24	20.0	1.778	46.6	46.6	0		
8	7	3	4	6	4.2	0.250	30.8		15.8		
7	6	2	3	4	3.0	0.111	26.6	26.6	0		
	5	1	2	3	2.0	0.111	20.3	26.6	6.3		
	4	3	4	6	4.2	0.250	19.5	26.6	7.1		
6	4	6	8	12	8.3	1.000	23.6	23.6	0		
5	4	2	3	4	3.0	0.111	18.3		9.3		
4	3	1	2	3	2.0	0.111	15.3	15.3	0		
	1	6	8	12	8.3	1.000	8.3	15.3	7.0		
3	2	2	3	4	3.0	0.111	13.3	13.0	0		
2	1	8	10	14	10.3	1.000	10.3	10.3	0		

Figure 6.20 Example of program network calculations.

a. Column 1
List each event, starting from the last event and working backward to the beginning (i.e., from Event 17 to Event 1 in Figure 6.18).

b. Column 2
List all previous events that lead into, or are shown as being prior to, the event listed in Column 1 (e.g., Events 15 and 16 lead into Event 17).

c. Columns 3 to 5
Determine the optimistic time (t_a), the most likely time (t_b), and the pessimistic time (t_c) in weeks or months for each activity. Optimistic time means that there is very little chance that the activity can be completed before this time, while pessimistic time means that there is little likelihood that the activity will take longer. The most likely time (t_b) is located at the highest probability point or the peak of the distribution curve. These times may be predicted by someone who is experienced in estimating. The time estimates may follow different distribution curves, where P represents the probability factor (see Figure 6.21). The three-time estimates are also included in Figure 6.18 for each activity (A, B, C, etc.).

d. Column 6
Calculate the expected or mean time, t_e, from

$$t_e = \frac{t_a + 4t_b + t_c}{6} \tag{6.1}$$

e. Column 7
In any statistical distribution, one may wish to determine the various probability factors for different activity times. Thus, it is necessary to compute the variance (σ^2) associated with each mean value. The square root of the variance, or the standard deviation, is a measure of the dispersion of values within a distribution, and is useful in determining the percentage of the total population sample that falls within a specified band of values. The variance is calculated from Equation (6.2):

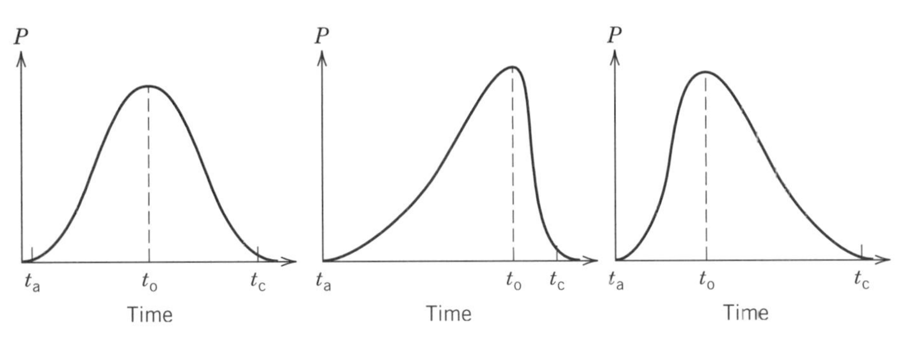

Figure 6.21 Sample distribution curves.

$$\sigma^2 = \left(\frac{t_c - t_a}{6}\right)^2 \tag{6.2}$$

f. Column 8

The earliest expected time for the project, *TE,* is the sum of all times, t_e, for each activity, along a given network path, or the cumulative total of the expected times through the preceding event remaining on the same path throughout the network. When several activities lead to an event, the highest time value (t_e) will be used. For instance, in Figure 6.18, Path 1–4–7–9–11–14–15–17 totals 98; Path 1–2–3–4–7–9–11–14–15–17 totals 105; and Path 1–2–3–4–6–7–9–11–12–14–15–16–17 totals 115.2. The highest value for *TE* (if one were to check all network paths) is 115.2 weeks, and this is the value selected for Event 17. The *TE* values for Events 16, 15, and so on, are calculated in a similar manner working backward to Event 1.

g. Column 9

The latest allowable time for an event, *TL,* is the latest time for completion of the activities that immediately precede the event. *TL* is calculated by starting with the latest time for the last event (i.e., where *TE* equals 115.2 in Figure 6.20) and working backward subtracting the expected time (t_e) for each activity, remaining on the same path. The *TL* values for Events 16, 15, and so on, are calculated in a similar manner.

h. Column 10

The slack time, *TS,* is the difference between the latest allowable time (*TL*) and the earliest expected time (*TE*):

$$TS = TL - TE \tag{6.3}$$

i. Columns 11 and 12

TC refers to the required scheduled time for the network based on the actual need. Assume that management specifies that the project reflected in Figure 6.18 must be completed in 110 weeks. It is now necessary to determine the likelihood, or probability (*P*), that this will occur. This probability factor is determined as follows:

$$Z = \frac{TC - TE}{\sqrt{\Sigma \text{ path variances}}} \tag{6.4}$$

where *Z* is related to the area under the normal distribution curve, which equates to the probability factor. The "path variance" is the sum of the individual variances along the longest path, or the critical path, in Figure 6.18 (i.e., Path 1–2–3–4–6–7–9–11–12–14–15–16–17).

$$Z = \frac{110 - 115.2}{\sqrt{11.666}} = -1.522$$

From the normal distribution tables, the calculated value of -1.522 represents an area of approximately 0.064, that is, the probability of meeting the scheduled time of 110 weeks is 6.4%. If the management requirement is 115 weeks, then the probability of success would be approximately 47.9%; or if 120 weeks were specified, the probability of success would be around 91.9%.

When evaluating the resultant probability value (Column 12 of Figure 6.20), management must decide on the range of factors allowable in terms of risk. If the probability factor is too low, additional resources may be applied to the project in order to reduce the activity schedule times and improve the probability of success. On the other hand, if the probability factor is too high (i.e., there is practically no risk involved), this may indicate that excess resources are being applied, some of which could be diverted elsewhere. Management must assess the situation and establish a goal.

From Figure 6.18, the critical path, which is reflected by the heavy arrows (i.e., Path 1–2–3–4–6–7–9–11–12–14–15–16–17), includes the series of activities requiring the greatest amount of time for completion. These are *critical* activities where slack times are zero, and a slippage of schedule in any one of these activities will cause a schedule delay in the overall program. Thus, these activities must be closely monitored and controlled throughout the program.

The network paths representing other program activities shown in Figure 6.18 include slack time (*TS*), which constitutes a measure of program scheduling flexibility. The slack time is the interval of time where an activity could actually be delayed beyond its earliest scheduled start without necessarily delaying the overall program completion time. The availability of slack time will allow for a possible reallocation of resources. Program scheduling improvements may be possible by shifting resources from activities with slack time to activities along the critical path.

As an additional point relative to program schedules, a hierarchy of individual networks may be developed following a pattern similar to the WBS development approach illustrated in Figure 6.10. To provide the proper monitoring and control actions, scheduling may be accomplished at different levels. Figure 6.22 shows a breakdown of the program network (illustrated in Figure 6.18) into a lower-level network covering reliability program requirements. A similar network may be developed for maintainability, another network for electrical design, and additional detailed networks as appropriate. These lower-level networks must, of course, directly support the overall program network.

The utilization of the PERT/CPM scheduling technique offers a number of advantages:

a. It is readily adaptable to advanced planning and essentially forces the detailed definition of tasks, task sequences, and task interrelationships. All levels of management and engineering are required to think through and evaluate the entire project carefully.
b. With the identification of task interrelationships, it tends to force the initial definition and subsequent management and control of the interfaces between

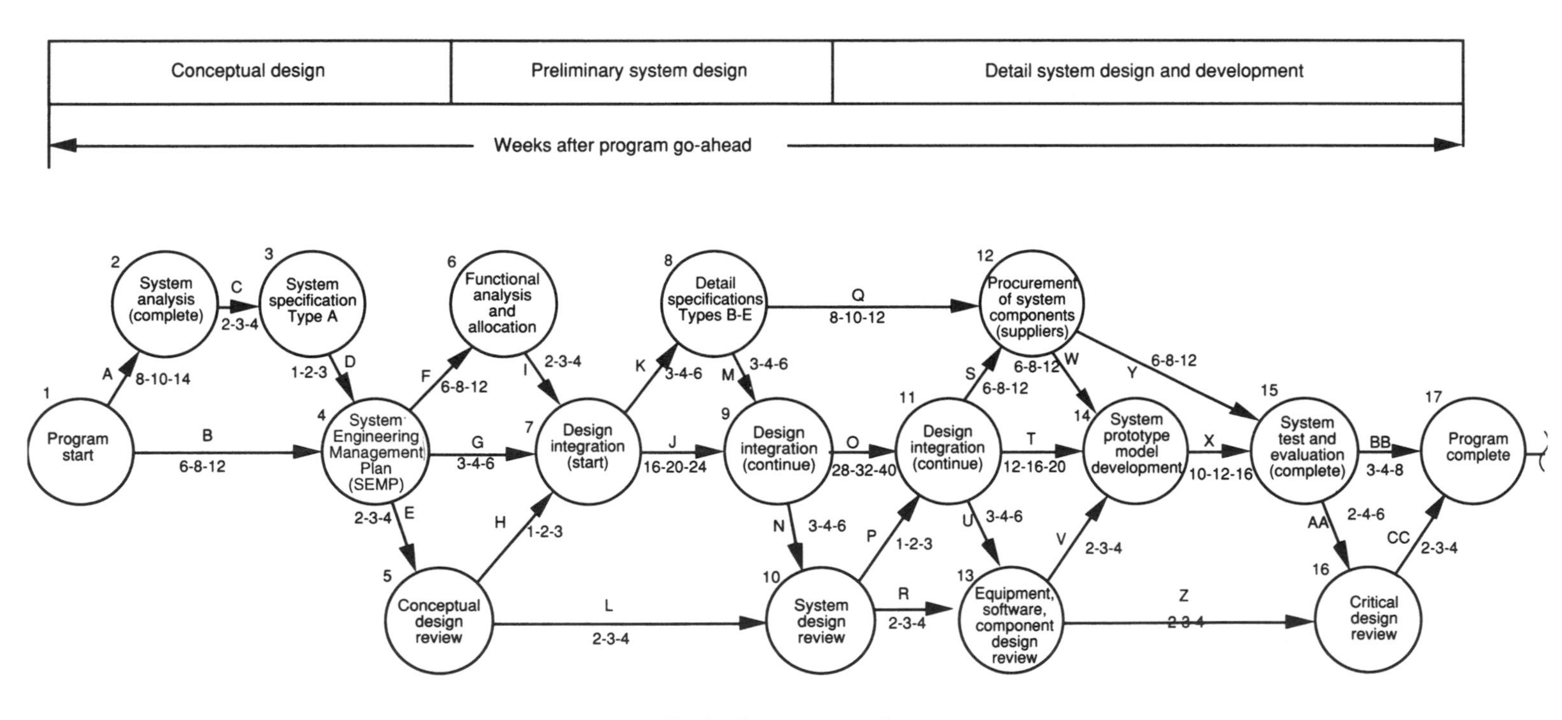

Top-level program network

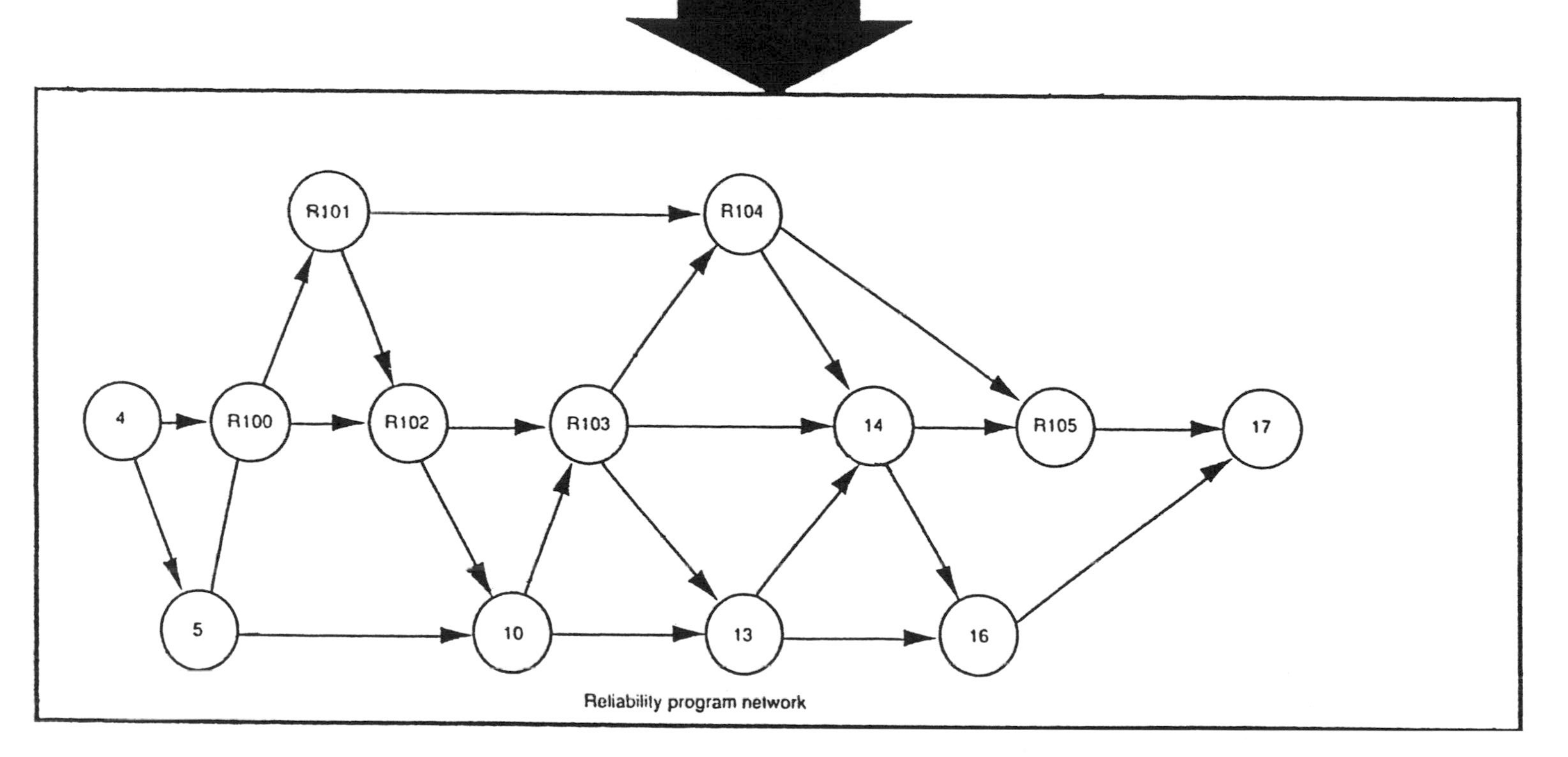

Figure 6.22 Top-level network breakdown by program element.

customer and contractor, organizations within the contractor's structure, and between the contractor and various suppliers. Management and engineering gain a greater appreciation of the project in terms of total resource requirements.

c. It enables management and engineering to predict with some degree of certainty the probable time that it will take to achieve an objective. Areas of program risk/uncertainty can be readily identified.

d. It enables the rapid assessment of progress and allows for the early detection of possible delays and problems.

The implementation of PERT/CPM in a comprehensive and timely manner is possible because the technique is particularly adaptable to computer methods. In fact, there are a number of computer models and associated software that are available for network scheduling.

5. *Network/cost:* PERT/CPM networks may be extended to include cost by superimposing a cost structure on the time schedule. When implementing this technique, there is always the time–cost option, which enables management to evaluate alternatives relative to the allocation of resources for activity accomplishment. In many instances, time can be saved by applying more resources. Conversely, cost may be reduced by extending the time to complete an activity.

The time–cost option can be attained through the following general steps:

a. For each activity in the network, determine possible alternative time and cost estimates (and cost slope) and select the lowest cost alternative.

b. Calculate the critical path for the network. Select the lowest cost option for each network activity, and check to ensure that the total of the incremental activity times does not exceed the allowable overall program completion time. If the calculated value exceeds the program time permitted, review the activities along the critical path and select the alternative with the lowest cost slope. Reduce the time value to be compatible with the program requirement.

c. After the critical path has been established in terms of the lowest cost option, review all network paths with slack time, and shift activities to extend the times and reduce costs wherever possible. Activities with the steepest time–cost slopes should be addressed first.

PERT/CPM–COST has proven to be a very useful technique in the planning of program events and activities, and it allows for the necessary program schedule–cost status monitoring and control requirements accomplished throughout system development.

6. *Gantt chart:* This technique is used primarily in production and/or construction planning to show activity or job requirements, facility loading, and work status on a day-to-day basis. It was designed for and is most successfully utilized to support highly repetitive operations. An example of one basic form of a Gantt chart is shown in Figure 6.23. Gantt charts, used for both long-range planning and short-range scheduling on a day-to-day basis, may take the form of machine-loading control charts, labor-loading control charts, and/or job progress control charts.

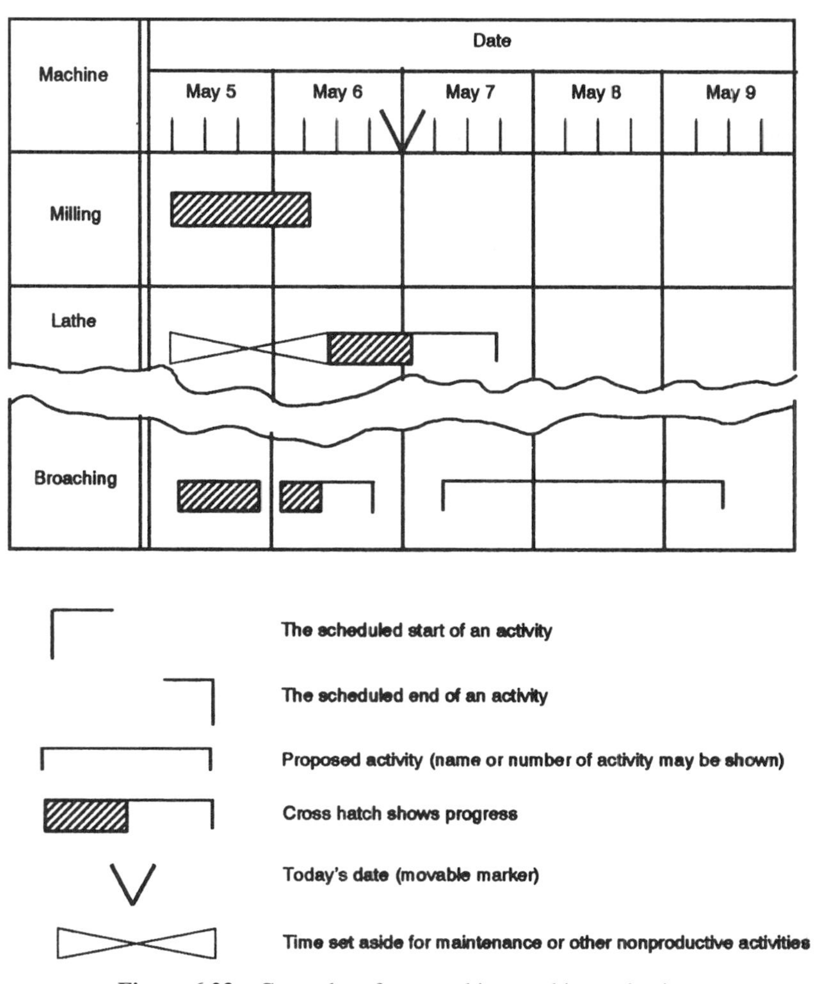

Figure 6.23 Gantt chart for a machine used in production.

7. *Line of balance (LOB):* This technique is similar to the Gantt chart relative to determining production/construction status. Although the Gantt technique primarily relates information on the effective and efficient utilization of resources expended (e.g., labor loading, machine loading), LOB is more product-oriented. LOB is not directly concerned with the resources expended, but is utilized in determining production progress in terms of percentage of task completion. Major "bottlenecks" in the production process are emphasized.

Application of the scheduling methods described herein will vary from project to project and from one organization to the next. Additionally, the technique used may be different for each phase of the system life cycle. For instance, the use of PERT/CPM may be readily adaptable to a research-and-development program, whereas Gantt charts are more appropriate for a production program.

When considering the objectives of system engineering, the use of PERT/CPM (or an equivalent network approach), as compared to bar charts or milestone charts, seems appropriate. There are many different one-of-a-kind tasks accomplished relatively early in the system life cycle, and the organizational task interfaces are numerous! There is a need for a high degree of visibility across the project, and it is important that potential problems be detected as early as possible. The use of the network scheduling technique should help in maintaining the necessary communications and in providing the appropriate monitoring and control functions.

6.2.7 Preparation of Program Cost Projections[12]

Good cost control is important to all organizations regardless of size. This is particularly true in our current environment where resources are limited and competition is high.

Cost control starts with the initial development of cost estimates for a given program and continues with the functions of cost monitoring and the collection of data, the analysis of such data, and the initiation of corrective action before it is too late. Cost control implies good overall cost management, which includes cost estimating, cost accounting, cost monitoring, cost analysis, reporting, and the necessary control functions. More specifically, the following activities are applicable.[13]

1. *Define elements of work:* Develop a Statement of Work (SOW) in accordance with the requirements described in Section 6.2.1. Detailed project tasks are defined and task schedules are developed as discussed in Section 6.26. System engineering tasks are identified in Section 6.2.2 (refer to Figure 6.6).

2. *Integrate tasks into the work breakdown structure* (WBS): Combine project tasks into work packages, and integrate these elements of work into the work breakdown structure (WBS). Work packages are identified against each block in the WBS. These packages and WBS blocks are then related to organizational groups, branches, departments, suppliers, and so on. The WBS is structured and coded in such a manner that project costs can be initially allocated (or targeted) and then collected against each block. Costs may be accumulated both vertically and horizontally to provide summary figures for various categories of work. WBS objectives and requirements are described in Section 6.2.3 (refer to Figures 6.11 and 6.12 for system engineering functions in the WBS).

3. *Develop cost estimates for each project task:* Prepare a cost projection for each project task, develop the appropriate cost accounts, and relate the results to elements of the WBS.

[12]Two references covering the subject of cost estimating are (1) P. F. Ostwald, *Engineering Cost Estimating,* 3rd Ed., Prentice Hall, Upper Saddle River, NJ, 1992; (2) and R. D. Stewart, *Cost Estimating,* 2nd Ed., John Wiley, New York, 1990. The emphasis here is primarily oriented to the costing of internal project tasks identified in the WBS versus the development of cost-estimating relationships (CERs) for life-cycle cost analysis purposes. LCC coverage is included in Appendix B.

[13]Cost control is covered in most texts on program/project management. One good reference is H. Kerzner, *Project Management: A Systems Approach to Planning, Scheduling, and Controlling,* 5th Ed., Van Nostrand Reinhold, New York, 1995.

4. *Develop a cost data collection and reporting capability:* Develop a method for cost accounting (i.e., the collection and presentation of project costs), data analysis, and the reporting of cost data for management information purposes. Major areas of concern are highlighted; that is, current or potential cost overruns and high-cost "drivers."

5. *Develop a procedure for evaluation and corrective action:* Inherent within the overall requirement for cost control is the provision for feedback and corrective action. As deficiencies are noted, or potential areas of risk are identified, project management must initiate the necessary corrective action in an expeditious manner. This requirement is discussed further in Section 6.2.9.

An initial step in developing a good cost control capability is that of cost estimating and the preparation of cost projections. Each task is broken down into subtasks and other detailed elements of work, and personnel projections are developed on a month-to-month basis. Figure 6.24 identifies selected activities for a project involving the design and development of a relatively large-scale system. In this instance, a 12-month design period is assumed, and projections are made in terms of the number of individuals by job classification required to complete the task, scheduled on a month-to-month basis. For instance, under system engineering there is a need for the assignment of four individuals with the grade of "Senior Engineer" during Month 3 of the project. Although not completely shown in the figure, *all* major program activities should be covered through an appropriate breakout of job classification requirements; that is, principal engineer, senior engineer, engineer, junior engineer, engineering technician, analyst, draftsman, data specialist, and shop mechanic. These resource requirements are projected for each project task and are related to the WBS (e.g., 3B1 100 in Figures 6.11 and 6.12).

Given a projection, presented in terms of labor requirements by grade, the next step is to convert these into cost factors on a month-to-month basis. Most organizations have established job classifications with computed salary pay scales. These factors are used in estimating the direct labor costs for a designated activity extended into the future. Additionally, material costs are determined for each month, and the appropriate inflationary factors are added to both labor and material. The net results include a projection of *direct labor costs* and *direct material costs,* inflated as necessary to cover future economic contingencies. These projections must, of course, support all program tasks identified in Sections 6.2.1 and 6.2.2, and should be compatible with the related task schedules described in Section 6.2.6.

As individual project activities are being further defined through the preparation of cost estimates, these activities must not only be tied to a particular block in the WBS, but the results must be assigned to a specific cost account (refer to Figure 6.13). A partial breakdown structure for a project is presented in Figure 6.25. The objective is to show the various project cost accounts in a hierarchical manner, indicating the structure that will be used for subsequent cost accounting and reporting purposes.

Relative to application, cost estimating may be accomplished at any time or during any phase of the system life cycle. Sometimes during the early phases of concep-

Program	WBS no.	Cost account	Projection (months)												Total
			1	2	3	4	5	6	7	8	9	10	11	12	
Project management	2A1000	2000	1	1	2	3	4	4	4	4	4	4	3	3	37
System engineering	3B1100	3000	-	-	-	-	-	-	-	-	-	-	-	-	
Principal engineer			1	1	1	1	1	1	1	1	1	1	1	1	12
Senior engineer			2	3	4	4	4	3	2	2	2	2	2	2	32
Design engineering	3B1200	7000	-	-	-	-	-	-	-	-	-	-	-	-	-
Principal engineer			2	3	3	3	3	2	2	2	2	2	2	2	28
Senior engineer			3	6	8	8	8	7	6	5	5	4	4	3	67
Engineer			5	7	10	15	17	20	20	20	20	17	16	9	176
Junior engineer			3	4	5	6	10	10	10	10	15	15	20	20	128
Engineering technician			1	1	1	5	7	7	8	9	10	10	12	15	86
Design data	2I1200	4000	2	2	3	5	5	10	15	15	20	25	25	25	152
System software	2E1000	8000	1	1	2	2	3	3	5	7	7	10	12	15	68
Design support	3B1300	5000	2	2	5	5	5	10	10	25	30	30	30	25	179
Intergrated logistic support	3B1400	6000	2	3	3	3	5	6	6	10	10	15	15	15	93
System test & evaluation	2J1000	9000	1	1	1	1	1	1	5	5	10	15	15	20	76
Total			26	35	48	61	73	84	94	115	136	150	157	155	1134

Figure 6.24 Project labor projection (man-months).

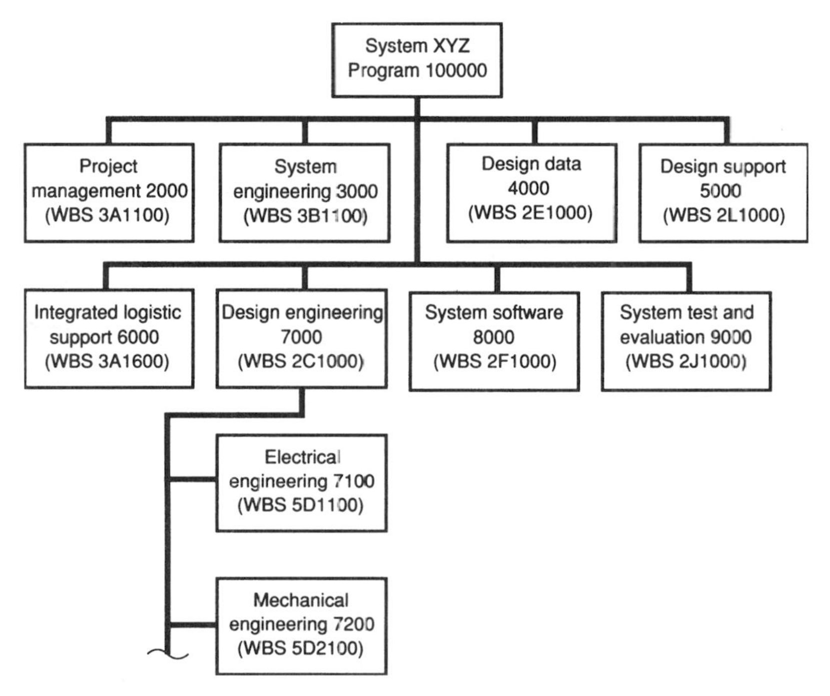

Figure 6.25 Partial cost account code breakdown structure.

tual and/or preliminary system design, when the availability of engineering data is limited, estimates may take the form of "rough orders of magnitude;" that is, approximations within plus or minus 30% of reality. The use of regression analysis, linear and nonlinear estimating relationships, learning curves, parametric analysis, or a combination of these, aids in the development of cost figures of merit (FOMs). Later, as engineering experience is acquired, estimating methods are more precise. Plans, specifications, design data, supplier cost proposals, updated project "cost-to-complete" reports, and so on are available. Cost estimates, using actual engineering data and/or the development of data through analogous methods, are prepared with an expected accuracy in the order of plus or minus 5%.

On completion of the cost projections for individual tasks, one can then combine these into an overall cost projection for the project as a whole, as shown in Figure 6.26. Initially, an estimate for all direct labor is developed, with an organizational overhead factor applied on top. Direct material costs are then determined, and a second burden rate (i.e., a general and administration factor) is applied to cover some additional indirect costs associated with both labor- and material-related activities. The net result is an overall cost projection for the project, to include both direct and indirect costs.[14]

[14] In the cost projection shown in Figure 6.26, direct labor costs were determined from the personnel labor figures in Figure 6.24. An average rate of $4,000 per man-month was used to calculate the monthly cost figures. A 200% overhead rate and a 20% general-and-administration rate were used for illustrative

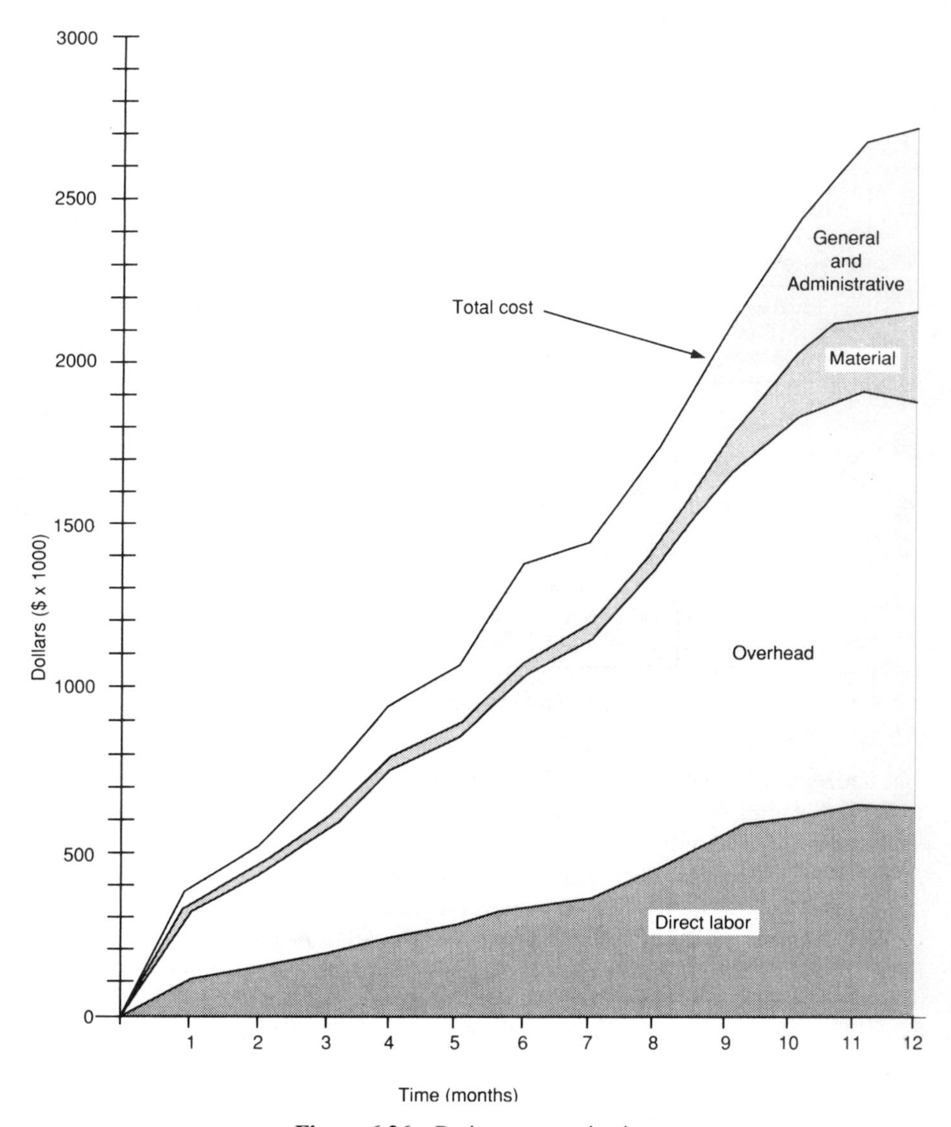

Figure 6.26 Project cost projection.

6.2.8 Technical Reviews

Technical reviews are an integral part of the system engineering process. These reviews can vary from the very formal design reviews described in Chapter 5 to the very informal reviews concerned with specific project activities or task elements of the work breakdown structure (WBS). All such reviews share the common objective

purposes. In reality, each company, government organization, or equivalent, will have different rates based on individual audited criteria.

of determining the technical adequacy of the existing system design configuration and whether or not its meets the initially specified requirements. Further, as the design and development effort evolves, the reviews become more detailed and definitive.

The type, number, and basic objectives of the formal design reviews conducted for a given program will vary with the nature and complexity of the system being developed, the organizational structure and type of contracting mechanism in place, and so on. In Chapter 5, formal design reviews included four basic categories of reviews; that is, *conceptual, system, equipment/software,* and *critical.* These were considered to be *basic* and representative for most programs. On the other hand, for many large-scale defense programs, there may be many more reviews to include system requirements review (SRR), system functional review (SFR), system design reviews (SDRs), preliminary design reviews (PDRs), software specification reviews (SPRs), system verification review (SVR), critical design review (CDR), test readiness review (TRR), production readiness review (PRR), and so on.[15] Although the scheduling of design reviews has many benefits, as conveyed in Chapter 5, it is essential that care be taken so as not to schedule so many that they become meaningless! The conductance of such reviews may be quite costly when considering the personnel time and resources required.

In addition to the formal design reviews conducted on many projects, there may be a number of formal *program management reviews* scheduled as well. Sometimes, the design reviews are perceived as being oriented to only *engineering* and involving responsible engineers representing the appropriate *engineering specialties.* Key levels of program management are not involved, even though many of the design decisions discussed may have significant implications from an overall program management perspective. On the other hand, during the periodic management-oriented reviews, where the emphasis is often directed to current status in terms of *performance, cost,* and *schedule,* there are decisions made that can have a direct impact on design. Under certain conditions, the two categories of reviews can be counterproductive unless care is taken to ensure that both the technical and management reviews are mutually supportive. A system engineering goal is to facilitate the communications process and the scheduling of both categories of reviews such that the results are complementary in meeting the overall program/project objectives.

6.2.9 Program Reporting Requirements

Inherent within the planning process is the establishment of both *technical* and *management* requirements at program inception. Additionally, one needs to review progress against these requirements on a periodic basis as system design and development evolves. In the event of problems, a procedure needs to be established for the initiation of corrective action as necessary.

In response, a management information system (MIS) should be developed to

[15] Defense Systems Management College, DSMC, *Systems Engineering Management Guide,* Fort Belvoir, VA (latest edition).

provide ongoing visibility and the reporting of progress against designated cost, schedule, and performance measures. Schedule and cost information is derived in accordance with the procedures described in Sections 6.2.6 and 6.2.7. Periodic reports are necessary for the purposes of assessing *current* status against *planned* status. The frequency of reporting is a function of the overall project schedule and the risks associated with various design activities. The comparison process should address such questions as: Is the project on schedule? Are the program costs within the established budget limitations? Assuming that the current personnel loading continues as is, what tasks are likely to be in a "cost overrun" position 6 months from now? These and other questions of this nature will have to be answered on many occasions throughout the program.

Figure 6.27 presents an extract from a report covering schedule and cost data. The schedule (or time status) information reflects the output from a typical PERT/CPM network. Relative to performance, the technical performance measures (TPMs) identified in the system specification, and selected as being critical from a periodic review and control standpoint, must be included within the program reporting structure. These TPMs may include factors such as range, accuracy, weight, size, reliability (MTBF/MTBM), maintainability ($\bar{M}ct$, MLH/OH), downtime (MDT), availability, cost, power output, process time, and other parameters that relate directly to the mission of the system being developed. Figure 5.6 (Chapter 5) illustrates the TPM evaluation process as it is tied in with formal design reviews. The measurement, evaluation, and control of these parameters also must be covered through periodic program reporting.

The management information system (MIS) should readily point out existing problems, as well as potential areas in which problems are likely to occur if program operations continue as originally planned. To deal with such contingencies, planning should be initiated to establish a corrective-action procedure that will include the following steps:

1. Identify problems (or potential problem areas) and rank these in order of importance. Ranking should consider the criticality of the system function.

2. Evaluate each problem on the basis of ranking, addressing the most critical problems first. Alternative possibilities for corrective action are considered in terms of (a) effects on program schedule and cost, (b) impact on performance and effectiveness of the system, and (c) the risks associated with the decision as to whether to take corrective action. The most feasible alternative is identified.

3. Given the decision to take corrective action, planning is accomplished to initiate the steps required to resolve the problem. This may be in the form of a system configuration change, a change in management policy, a contractual change, and/or an organizational change.

4. After corrective action has been implemented, some follow-up activity is required to (a) ensure that the incorporated change(s) actually has resolved the problem, and (b) assess other aspects of the program to ensure that additional problems have not been created as a result of the change.

Network/Cost Status Report

Project: System XYZ				Contract Number: 6BSB-1002					Report Date: 3/15/91			
Item/Identification				Time Status					Cost Status			
WBS. No.	Cost Account	Beginning Event	Ending Event	Exp. Elap. Time (te) (weeks)	Earliest Completion Date(D_E)	Latest Completion Date (D_L)	Slack D_L-D_E (Weeks)	Actual Date Completed	Cost Est. ($)	Actual Cost to Date ($)	Latest Revised Est. ($)	Overrun (Underrun) ($)
4A1210	3310	8	9	4.2	3/4/91	4/11/91	11.6	4/4/91	2500	2250	2250	(250)
4A1230	3762	R100	R102	3.0	5/15/91	4/28/91	-3.3		4500	4650	5000	500
5A1224	3521	7	9	20.0	6/20/91	8/3/91	0		6750	5150	6750	0

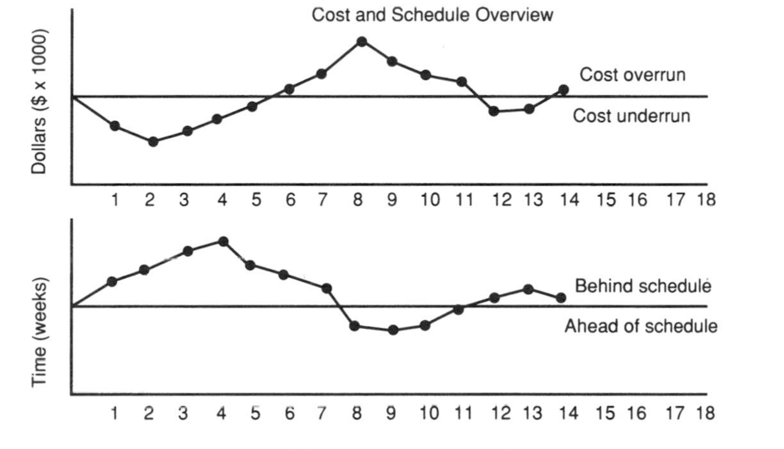

Figure 6.27 Program cost–schedule reporting.

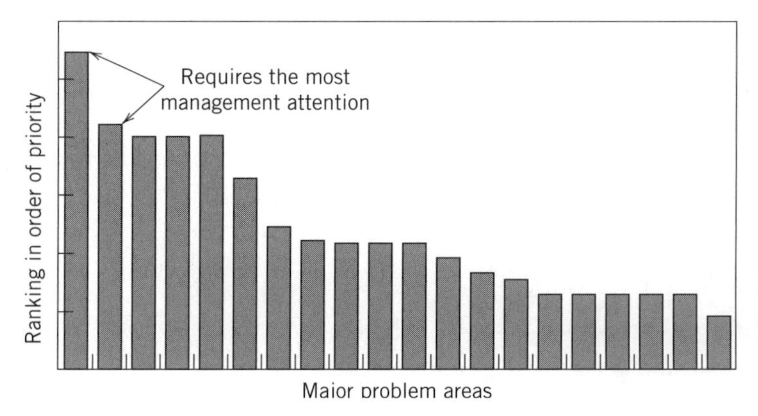

Figure 6.28 Pareto diagram identifying problem areas.

Relative to the ranking of problems (and their priorities) that need to be addressed, a Pareto analysis approach might be beneficial in creating visibility pertaining to degrees of importance. From Figure 6.28, the highest-ranked items need the most management attention. The implementation of any changes, of course, must be compatible with the procedures described in Section 5.4.

6.3 INTEGRATION OF DESIGN SPECIALTY PLANS

From the basic definition of system engineering in Chapter 1, a major objective is to ensure the proper integration of all applicable engineering disciplines into the total design effort (refer to Figure 2.28, Chapter 2). Although there are some variations with each program, those disciplines identified in Figure 6.5 are considered as being critical!

Along with the hierarchy of specifications illustrated in Figure 6.14, there is also a hierarchy of program plans, with the Program Management Plan (PMP) at the top, the System Engineering Management Plan (SEMP) next, and a number of subordinate plans supplementing the SEMP. Figure 6.29 illustrates this hierarchical relationship.[16]

In Chapter 3, there is a description of the requirements for each of these supporting disciplines. An evaluation of the requirements illustrates the importance of relia-

[16]It should be emphasized that within the broad spectrum of system engineering, there may be a wide variety of different design plans covering not only the prime functional design disciplines (e.g., civil engineering, electrical engineering, chemical engineering, mechanical engineering), but covering some of the basic engineering supporting disciplines as well. Relative to the latter, many of the supporting disciplines (e.g., reliability, maintainability, logistics) have been treated as separate entities, each requiring a stand-alone plan, and being implemented not as an integral part of the total engineering effort, but on an independent basis. The emphasis in Figure 6.29 is to cause the integration of these plans into the overall process.

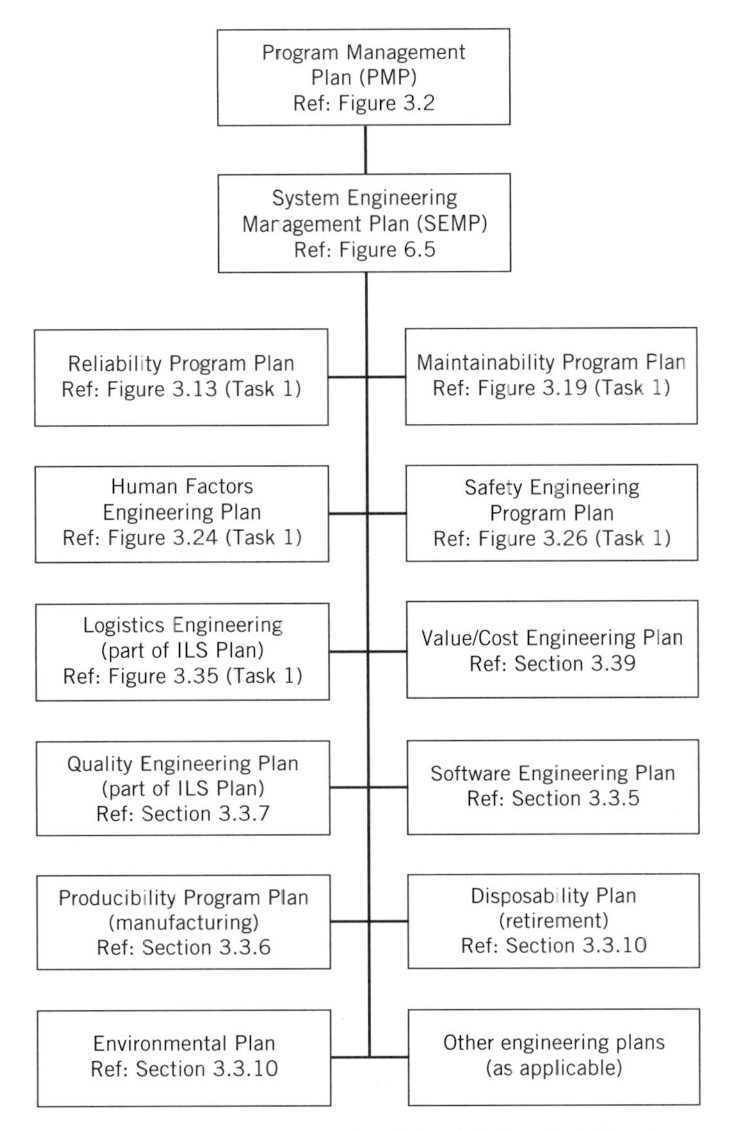

Figure 6.29 The integration of individual design discipline plans.

bility, maintainability, human factors, supportability/serviceability, producibility, quality, and so on, in design. Further, these requirements are closely interrelated (with each other), in addition to being "key" to the system engineering process. Thus, these factors must be properly addressed in the design process, in a timely manner, and commencing from the beginning during the conceptual design phase.

From a review of the individual sections in Chapter 3, one can see that there are many tasks that are similar across the board. The first task in each of the design disciplines represented is the *preparation of a program plan,* a document that specifies the tasks to be accomplished in order to fulfill program requirements. Within each area of activity, there are analysis tasks, prediction tasks, design review and evaluation tasks, and test and demonstration tasks. In the accomplishment of design trade-offs, as part of the ongoing analysis effort, the net results must reflect a *balanced* approach. This, in turn, forces the proper mix of reliability characteristics in design, maintainability characteristics in design, and so on. In other words, these factors (as they impact the design process) must be carefully integrated throughout!

In the past, a common practice has resulted in the preparation of these program plans on a separate and independent basis, and often at different times in the system development process. This has caused some differences in stated objectives, inconsistencies in schedules, redundancies in program task requirements, and conflicts in output. In view of the importance of these disciplines in meeting system engineering objectives, it is recommended that the respective plans be prepared and integrated into the System Engineering Management Plan (SEMP). This is illustrated in Figure 6.29.

From the figure, the intent is not to hamper or in any way curtail the efforts of the individual disciplines in fulfilling program requirements. The purpose is to ensure the proper relationships among the many tasks that must be accomplished, as well as eliminating possible redundancies. For instance, the results of reliability prediction must feed into the accomplishment of maintainability prediction and the logistic support analysis; the preparation of the FMECA constitutes an input to other reliability tasks, the maintainability analysis, the logistic support analysis, and the safety analysis; the fault-tree analysis (FTA) constitutes an input to the safety hazard analysis; the accomplishment of the human-factors operator task analysis (OTA) must be compatible and directly support the maintenance task analysis (MTA); and the reliability analysis (model), maintainability analysis, operator task analysis, and logistic support analysis must evolve from the system-level functional analysis (refer to Section 2.7). It is essential that the "system engineer" completely understand the many interrelationships that exist among these disciplines, and that such activities be properly integrated through the SEMP.

6.4 INTERFACES WITH OTHER PROGRAM PLANNING ACTIVITIES

Although it is important to provide the proper integration of the individual design discipline plans as conveyed in Figure 6.29, it is also necessary to ensure that the proper communications link exists between the SEMP and other related program plans. Of particular interest are those noted in Figure 6.30 and identified in what follows.

1. *Individual design plans:* For some programs, individual plans may be prepared by the traditional design disciplines such as civil engineering, electrical engineering, industrial engineering, mechanical engineering, and other such disciplines. It is essential that these plans be supportive of the material presented in the SEMP

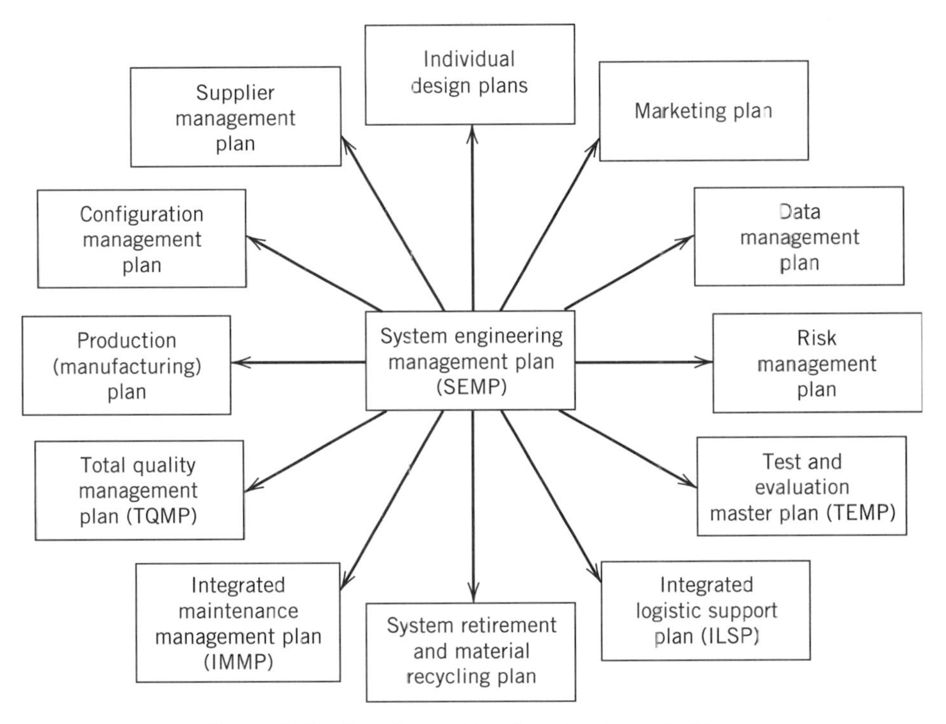

Figure 6.30 Interfaces with other planning activities.

and be referenced accordingly. An objective of system engineering is to cause the integration of *all* engineering disciplines, requiring that these plans "communicate" with each other (refer to Figure 6.5).

2. *Configuration management plan:* The importance of configuration management (CM), or "baseline" management, is critical to the fulfillment of system engineering objectives, and has been emphasized throughout this text. Maintaining the design baseline and controlling design changes are essential for system evaluation and cost control.

3. *Data management plan:* The proper integration of all design and supporting data is necessary to ensure that the various elements are compatible (i.e., "track" where applicable), are available at the right place and in a timely manner, that data redundancies are minimized (if not eliminated), and that data costs are minmized to the extent possible. Although the data environment is rapidly changing with the advent of new technologies (i.e., the conversion of data to a digital format and using a shared database approach), there is still a need for some degree of consistency in presentation, data format, control of data/documentation changes, and so on (refer to Section 2.10 and Figure 2.30 in Chapter 2).

4. *Test and Evaluation Master Plan* (TEMP): From Section 1.3 and Figure 2.31, there is a need for *integrated* test planning from the beginning. As system-level

requirements are initially defined and TPMs are established, one has to determine how the system will ultimately be evaluated to ensure that the initially specified requirements will be met. This constitutes the *validation* loop within the system engineering process; hence, the preparation of the TEMP has been identified as one of the critical tasks in implementing a system engineering program (refer to Figure 6.6, task 4).

5. *Marketing plan:* Individual company/agency marketing plans are often directed toward the "short term" and do not address the necessary long-term, *life-cycle* considerations that are essential in system/product acquisitions. Further, some marketing plans may be directed toward the establishment of certain *partnership* relations with organizations that may, or may not, be sympathetic to the concepts methods of system engineering. Thus, all marketing plans must prepared to convey the *system engineering approach.*

6. *Supplier management plan:* With the increasing trends toward "outsourcing" and the selection of suppliers from various sources around the world, it may be feasible to develop a plan (in support of the marketing plan) incorporating the criteria for initial supplier selection, and for the follow-on procedures leading to some form of contracting, subsequent monitoring and control activities, and so on. For some programs, over 50% of the components of a given system may be subcontracted, and it is imperative that the specifications being imposed on a contract are complete, well-prepared, and include performance-based requirements, and are supportive of the system specification (Type "A"). The objective is to ensure that all of the components of the system that are procured from an outside source will properly *fit* or can be *integrated* into the system as an entity without unexpected problems occurring. Thus, it is essental that system engineering criteria be included in the supplier selection process, in the preparation of specifications for the purposes of subcontracting, and for the follow-on supplier monitoring and control activity. Supplier activities are discussed further in Chapter 8.

7. *Production/manufacturing plan:* What may initially appear to be a well-designed and configured set of prime mission-oriented elements of a system may turn out not to be so after being subjected to the production process. The production process is highly *dynamic,* with variances introduced throughout, and may have a significant impact on the products being manufactured. This would certainly be the case if *producibility* considerations were not initially incorporated in the design of the applicable product(s). The concurrency approach conveyed in Figure 3.31, where the product and the manufacturing life cycles must be properly integrated, is critical in the implementation of system engineering concepts and methods.

8. *Integrated logistic support plan* (ILSP): From Section 3.3.8, the ILSP covers all of the activities associated with the design of the prime elements of the system for *supportability,* design of the support infrastructure, the procurement and acquisition of the elements of support (maintenance personnel, spares and repair parts, test equipment, facilities, transportation and handling provisions, computer resources, and technical data), and for the sustaining maintenance and support of that system throughout its planned life cycle. Within the ILSP, there is a *logistics engineering*

section that should either be incorporated within the SEMP or constitute a major reference required for the successful completion of system engineering requirements (refer to Figure 6.29). Further, as one progresses through the development process, there is an ongoing need for the evaluation of the overall support infrastructure, and for the data collection and feedback process as the system is being utilized by the consumer in the field. From Figure 3.31, the *logistics, service, and support* life cycle must be appropriately integrated with the product development and manufacturing life cycles on a concurrent basis.

9. *Integrated maintenance management plan* (IMMP): Whereas the ILSP deals with the major issues pertaining to the system support infrastructure in the defense sector, the IMMP may serve in a comparable role in the commercial sector. Whether one is dealing with a transportation system, a communications system, a health-care or hospital complex, or manufacturing plant, there is still a need to *plan for maintenance.* This includes the design for reliability and maintainability, the design of the maintenance and support infrastructure, the procurement of the elements of support, and the sustaining maintenance and support of the system throughout its planned life cycle.[17]

10. *Total quality management plan* (TQMP): Within the context of total quality management (TQM), there are activities associated with *quality engineering* dealing with the design of products, the design of the manufacturing process, and the design of the maintenance and support infrastructure.[18] These areas should either be incorporated within the SEMP or constitute a major reference required for the successful completion of system engineering requirements (refer to Figure 6.29). Further, as one progresses through the development, manufacturing, and support processes, there is an ongoing requirement to maintain the *quality* that has been built into the design.[19]

11. *System retirement and material recycling plan:* Although the fourth life cycle in Figure 3.31 (retirement and material disposal) has not been properly addressed on many programs, the environmental concerns described in Section 3.3.10 are assuming an increasing degree of importance. The engineering aspects pertaining to the *design for the environment* should be addressed within the context of the SEMP, and the life-cycle activities associated with system development, construction and/or production, operations and support, and retirement should be monitored in terms of their impact on the environment. Further, as the system evolves through the con-

[17] The concepts of *total productive maintenance* (TPM) and *total asset management* (TAM) are included within this area. The objectives of TPM and TAM are directly comparable to those specified within the context of ILS, except that the terminology is somewhat different in each case. If one can break down the perceived "semantics barrier," it is believed that much can be gained through the transfer of technology and management practices between the defense and the commercial sectors.

[18] Two good references for the design of production systems with quality in mind are (1) G. E. A. Taguchi, E. A. Elsayed, and T. C. Hsiang, *Quality Engineering in Production Systems,* McGraw-Hill, New York, 1989; and (2) P. J. Ross, *Taguchi Techniques for Quality Engineering,* McGraw-Hill, New York, 1988. Refer to Appendix F for additional references in the area of *quality.*

[19] The concepts of *total quality control* (TQC) and *total quality leadership* (TQL) are comparable to the objectives of TQM except for some of the terminology.

sumer utilization and support phase, there needs to be an ongoing assessment relative to the possible impact of external environmental factors on the system; that is, the impact of ecological, technological, political, economic, and related factors on system operations. Is the system still performing as initially intended, or has there been some degradation due to external factors?

12. *Risk management plan:* Inherent within any system development effort is the aspect *risk;* that is, risk due to technical decisions, risk due to management decisions, and so on. The objective is, of course, to minimize risk throughout, and a major goal in system engineering is to implement a *risk management plan* that will allow for the early identification of potential areas of risk, the assessment of risk, and for risk abatement. From Figure 6.5 (item 4.11), the SEMP should address the area of risk management. This subject is discussed further in Section 6.6.

Although the successful implementation of a system engineering program requires close coordination with all of the design-related activities, a special emphasis is required to ensure close working relationships with those organizations responsible for the activities covered in these plans.[20]

6.5 MANAGEMENT METHODS/TOOLS

A major system engineering "challenge" is to be able to evolve through the process illustrated in Figure 1.12 (Chapter 1) following an organized, logical, and methodical approach and utilizing whatever techniques/tools that are available at the time. Inherent within this process is the requirement for accomplishing analysis, synthesis, and evaluation efforts as early in the life cycle as practicable! The goal is to gain early visibility, be able to make better design and management decisions as system development evolves, reduce the time that it takes in the system acquisition process (i.e., from the identification of need to the delivery of the system for consumer use), and minimize the risks associated with the decision-making process. On the other hand, too much activity in the front-end may turn out to be meaningless and costly. The challenge is to be able to quickly assess what is required, when, and to what extent? At the same time, it is essential that one be familiar with the techniques/tools that are available to assist in accomplishing whatever is desired!

Although it is impossible (and certainly not feasible) to mention all of the analytical techniques/tools that may be utilized to assist the system engineer in successfully fulfilling the previous goal, it is recommended than he/she become familiar with the following:[21]

[20] One of the roles of the system engineering organization is to provide the necessary day-to-day liaison and coordination with these activities. A few of the areas, considered by the author as being significant, have been identified although they may be addressed in different ways from one program to the next. Further, it is hoped that the material presented provides some general guidance should the subjects not be adequately covered elsewhere.

[21] Refer to the bibliography in Appendix F for selected references in these areas.

1. Statistical process control (SPC) and the use of statistical methods in design. This includes the techniques used for the *design of experiments.*
2. The use of operations research methods in design: linear and dynamic programming, optimization techniques, queuing analysis, and control theory.
3. The application of economic analysis methods in design (refer to Section 3.3.9 and Appendix B).
4. The use of simulation methods as described in Section 4.2.1.
5. Rapid prototyping and its applications as described in Section 4.2.2.
6. The use of the analytical methods as described in Section 4.3.
7. The use of mockups and scaled models as described in Section 4.2.3.

The challenge is to know what techniques/tools to use in response to attaining certain design objectives and which methods can be integrated into a total integrated design workstation configuration. These "design-enhancement" capabilities should be identified, integrated (in terms of their application), and described in the System Engineering Management Plan (SEMP).

6.6 RISK MANAGEMENT PLAN

Risk is the potential that something will go wrong as a result of one or a series of events. It is measured as the combined effect of the probability of occurrence and the assessed consequence given that occurrence. The potential for risk becomes increasingly higher as complexities and new technologies are introduced in the design of systems. Risk, as used in the context described herein, refers to the potential of not meeting a specified technical and/or program requirement; for example, not meeting a requirement specified by a TPM, a schedule, or a cost projection.

Risk management is an organized method for identifying and measuring risk, and for selecting and developing options for handling risk. Risk management is not a separate program thrust by itself, but should be an inherent part of any sound management activity. Risk management includes the following basic activities:

1. *Risk assessment:* This involves the ongoing review of technical design and/or program management decisions, and the identification of potential areas of risk.

2. *Risk analysis:* This includes conducting an analysis to determine the probability of events and the consequences associated with their occurrence. The purpose of risk analysis is to identify the cause(s), the effects, and the magnitude of the risk perceived, and to identify alternative approaches for risk avoidance. There are many tools that are available and can be used as an aid in conducting risk analyses; for example, scheduling network analysis, life-cycle cost analysis, FMECA, the Ishikawa cause-and-effect or "fishbone" diagram, hazard analysis, and trade-off studies in varying forms.

3. *Risk abatement:* This involves the techniques and methods developed to reduce (if not eliminate) or control the risk. A plan must be implemented for the handling of risk.

One of the first steps in risk management is the identification of the potential areas of risk. Although there is some degree of risk associated with any program area of activity where decisions are being made, one needs to identify those in which the potential consequences of failure could be significant! Program areas of risk may include funding, schedule, contract relationships, political, and technical. Technical risks relate primarily to the potential of not meeting a design requirement, not being able to produce an item in multiple quantities, and/or not being able to support a product in the field. Design engineering risks can be tied directly to the technical performance measures (TPMs) identified in Section 2.6 and in Figure 5.2. These TPMs, which reflect critical factors in design, can be prioritized to reflect relative degrees of importance.

Given the identification of performance characteristics to which the system is to be designed (i.e., those parameters that require monitoring on a regular basis), the next step is to evaluate these by indicating possible causes for failure. In the event of a failure to meet a specific design requirement, what are the possible causes and what are the probabilities of occurrence? Although the output measure being monitored may be a high-priority TPM, the cause for a possible failure may be the result of a misapplication of a new technology in design, a schedule delay on the part of a major supplier, a cost overrun, or a combination of these.

The causes are evaluated independently to determine the degree to which they can impact the TPM(s) being monitored. Sensitivity analyses are conducted, using various analytical models as appropriate, to determine the magnitude of the potential risk. This, in turn, will lead to the classification of factors in terms of "high," "medium," or "low" risk. These classifications of risk are then addressed within the program management review and reporting structure. High-risk items are monitored to a greater extent, with a higher priority relative to initiating a risk abatement plan, than low-risk items.

To facilitate the risk management implementation process, it is often feasible to develop a model of some type. One approach is to address risk in terms of two major variables: the probability of failure (P_f), and the effect or consequence of that failure (C_f). Consequences may be measured on the basis of technical performance, cost, or schedule. Mathematically, this model can be expressed as[22]

$$\text{Risk Factor } (RF) = P_f + C_f - (P_f)(C_f) \qquad (6.5)$$

where P_f is the probability of failure, and C_f is the consequence of failure. The quantitative relationships of these parameters are described in Figure 6.31.

[22] This model was adapted from the procedure in Defense Systems Management College, *Systems Engineering Management Guide,* DSMC, Fort Belvoir, VA, 1986. Although the later issues of this document do cover the various aspects of risk management, this particular model has been deleted for reason unknown to the author.

(1) Risk Factor = $P_f + C_f - P_f \bullet C_f$

(2) $P_f = (a)(P_{Mhw}) + (b)(P_{Msw}) + (c)(P_{Chw}) + (d)(P_{Csw}) + (e)(P_D)$

where

P_{Mhw} = Probability of failure due to degree of hardware maturity

P_{Msw} = Probability of failure due to degree of software maturity

P_{Chw} = Probability of failure due to degree of hardware complexity

P_{Csw} = Probability of failure due to degree of software complexity

P_D = Probability of failure due to dependency on other items

and where: a, b, c, d, and e are weighting factors whose sum equals one.

(3) $C_f = (f)(C_t) + (g)(C_c) + (h)(C_s)$

where

C_t = Consequence of failure due to technical factors

C_c = Consequence of failure due to changes in cost

C_s = Consequence of failure due to changes in schedule

and where f, g, and h are weighting factors whose sum equals one.

Magnitude	Maturity Factor (P_M)		Complexity Factor (P_C)		
	Hardware PMhw	Software PMsw	Hardware PChw	Software PC_{SW}	Dependency Factor (P0)
0.1	Existing	Existing	Simple design	Simple design	Independent of existing system, facility, or associate contractor
0.3	Minor redesign	Minor redesign	Minor increases in complexity	Minor increases in complexity	Schedule dependent on existing system, facility, or associate contractor
0.5	Major change feasible	Major change feasible	Moderate increase	Moderate increase	Performance dependent on existing system performance, facility, or associate contractor
0.7	Technology available, complex design	New software similar to existing	Significant increase	Significant Increase/major increase in # of modules	Schedule dependent on new system schedule, facility, or associate contractor
0.9	State of art some research complete	State of art never done before	Extremely complex	Extremely complex	Performance dependent on new system schedule, facility, or associate contractor

Magnitude	Technical Factor (C_f)	Cost Factor (C_c)	Schedule Factor (C_s)
0.1 (low)	Minimal or no consequences, unimportant	Budget estimates not exceeded, some transfer of money	Negligible impact on program, slight development schedule change compensated by available schedule slack
0.3 (minor)	Small reduction in technical performance	Cost estimates exceed budget by 1 to 5 percent	Minor slip in schedule (less than 1 month), some adjustment in milestones required
0.5 (moderate)	Some reduction in technical performance	Cost estimates increased by 5 to 20 percent	Small slip in schedule
0.7 (significant)	Significant degradation in technical performance	Cost estimates increased by 20 to 50 percent	Development schedule slip in excess of 3 months
0.9 (high)	Technical goals cannot be achieved	Cost estimates increased in excess of 50 percent	Large schedule slip that affects segment milestones or has possible effect on system milestones

Figure 6.31 A mathematical model for risk assessment. (*Source:* Defense Systems Management College, *Systems Engineering Management Guide,* DSMC, Fort Belvoir, VA, 1986.)

To illustrate the model application, with Figure 6.31 as the prime source of information, consider the following system design characteristics:[23]

1. System design uses off-the-shelf hardware with minor modifications to the software.
2. The design is relatively simple involving the use of standard hardware.
3. The design requires software of somewhat greater complexity.
4. The design requires a new data base to be developed by a supplier (subcontractor).

The characteristics of the system suggest that there is potential risk associated with the software development task. Using the criteria in Figure 6.31 (and applying the weighting factors as indicated), the probability of failure (P_f) is calculated as follows:

$$P_{M_{hw}} = 0.1, \text{ or (a) } (P_{M_{hw}}) = (0.2)(0.1) = 0.02$$
$$P_{M_{sw}} = 0.3, \text{ or (b) } (P_{M_{sw}}) = (0.1)(0.3) = 0.03$$
$$P_{C_{hw}} = 0.1, \text{ or (c) } (P_{C_{hw}}) = (0.4)(0.1) = 0.04$$
$$P_{C_{sw}} = 0.3, \text{ or (d) } (P_{C_{sw}}) = (0.1)(0.3) = 0.03$$
$$P_D = 0.9, \text{ or (e) } (P_D) = (0.2)(0.9) = \underline{0.18}$$
$$0.30$$

Given the preceding criteria, P_f of this item is 0.30.

If the consequence of the item's failure due to technical factors causes problems of a corrective nature, but the correction results in an 8% cost increase and a 2-month schedule slippage, the C_f is calculated as follows:

$$C_t = 0.3, \text{ or (F) } (C_t) = (0.4)(0.3) = 0.12$$
$$C_c = 0.5, \text{ or (g) } (C_c) = (0.5)(0.5) = 0.25$$
$$C_s = 0.5, \text{ or (h) } (C_s) = (0.1)(0.5) = \underline{0.05}$$
$$0.42$$

Based on the preceding (using the weighting factors indicated), the C_f factor is 0.42 and, from Equation (6.5), the calculated Risk Factor (*RF*) is 0.594. This can be classified within the category of medium risk, as noted in Figure 6.32. In this in-

[23] In applying this model to your own application, it is recommended that Figure 6.31 be replaced with some form of a checklist tailored, with the appropriate weighting factors, to your specific system and in support of the TPMs specified for that system.

stance, the risk is primarily associated with the system software and the reliance on a supplier.

A similar approach can be applied in performing a risk analysis on all other applicable parameters. The net result is the development of a list of critical items, presented in order of priority, that require special management attention. Risk reports are prepared at different times (i.e., frequency of distribution) depending on the nature of the risk. High-risk items require frequent reporting and special management attention, whereas low-risk items can be handled through the normal program review, evaluation, and reporting process.

For items classified under "high" and "medium" risk, a risk abatement plan should be implemented. This constitutes a formal approach for eliminating (if possible), reducing, and/or controlling risk. The accomplishment of such may involve one or a combination of the following:

1. Provide increased management review of the problem area(s) and initiate the necessary corrective action through an internal allocation or shift in resources.

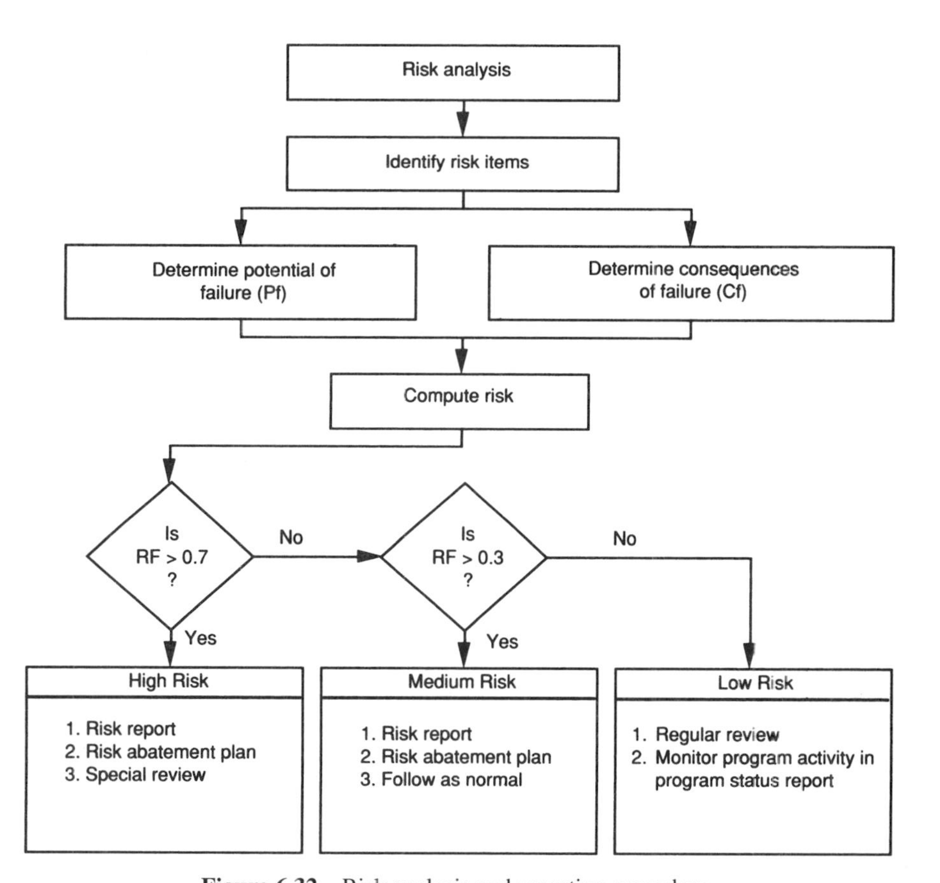

Figure 6.32 Risk analysis and reporting procedure.

2. Hire outside consultants or specialists to help resolve existing design problems.
3. Implement an extensive testing program with the objective of better isolating the problem and eliminating possible causes.
4. Initiate special research-and-development activities, conducted in parallel, in order to provide a "fall-back" position.

The purpose of the Risk Abatement Plan is to *highlight* those areas where special management attention is required. The identification of technical risks is of particular interest with regard to system engineering, because the fulfillment of design objectives is highly dependent on the proper and expeditious handling of these risks. In this respect, risk management should be an inherent aspect of system engineering management.

6.7 PROGRAM EVALUATION FACTORS

As part of any planning activity, there should be a mechanism whereby (1) an assessment can be made as to the current status of an organization, a product, and/or a process; (2) a comparison of the results with other comparable entities can be accomplished; and (3) where specific goals and objectives can be developed for the future. *Where are we today? How do we compare with the competition (relative to both product and organization)? Where would we like to be in the future?*

In response, one should address both the technical characteristics of an existing system/product configuration and the issues pertaining to the organization and processes that are being implemented to develop and produce the system/product. This can be facilitated through the establishment of a *benchmarking* capability and the implementation of a model for the measurement and evaluation of the organization and its operations.

6.7.1 Benchmarking

The term *benchmarking* may be defined in different ways depending on one's individual background and experience. Webster's dictionary defines it as "a point of reference from which measurements may be made; something that serves as a standard by which others may be measured." Although this definition primarily refers to a surveyor's mark or point of reference, the term has also been used in the context of setting and measuring standards related to product characteristics and organizational performance. In the early 1970s, the Xerox Corporation (and others) promoted the concept of benchmarking as a "business practice." According to Camp, benchmarking can be defined as "the continuous process of measuring products, services, and practices against the toughest competitors or those companies recognized as industry leaders."[24] Balm provided a more comprehensive definition where

[24] Robert C. Camp, *Benchmarking—The Search for Industry Best Practices That Lead to Superior Performance,* ASQC Quality Press, Milwaukee, WI, 1989.

benchmarking is "the ongoing activity of comparing one's own process, product, or service against the best known similar activity, so that by challenging attainable goals can be set and a realistic course of action implemented to efficiently become and remain the best of the best in a reasonable time."[25] This definition includes the element of *time,* which is critical if improvement is to be made in a competitive manner.

With regard to system engineering, there have been a number of benchmarking studies, and a few companies who practice the concepts and principles described throughout this text have implemented an active benchmarking effort internally.[26] The emphasis in most of these instances has been oriented directly to organizations and the processes that they use in accomplishing their day-to-day functions. Although this is appropriate, care must be taken to first define the company's objectives in terms of *product output* and then address the organizational characteristics that are considered to be essential in order to successfully meet the overall product goals. It is often tempting to launch into an evaluation of organizational effectiveness, employing some measures that may or may not be relevant to the ultimate objectives, and then initiating changes that may turn out to provide negative results because of not having defined the proper goals in the beginning.

From Figure 6.33, the general approach to benchmarking commences with the development of a plan for implementation (see block 2). This is based on a definition of the organization's objectives as they pertain to product goals. Product goals may be specified in terms of the TPMs for a given system, or some equivalent set of measures for one or more products. For example, Figure 2.10 identifies the TPMs for a system/product that resulted from a QFD analysis. Assuming that the specified quantitative requirements in the second column represents current status and that the immediate objectives include progressing to the requirements specified in the third column, then a plan needs to be developed covering the steps that must be accomplished in progressing from the current status to the level of performance ultimately desired. These steps relate to the organizational structure and the processes that are currently being implemented to support the product-oriented goals. For the purposes of this text, these include the system engineering functions and tasks described in Section 6.2.2.

In block 3 of Figure 6.33, one of the first steps is to define what is meant by *system engineering,* what is included, and what tasks must be accomplished in order to properly implement the concepts and principles described herein as they pertain to the product goals? This may lead to the development of a questionnaire, or series

[25] Gerald J. Balm, *Benchmarking: A Practitioner's Guide for Becoming and Staying the Best of the Best,* QPMA Press, Schaumburg, IL, 1992.

[26] Kenneth Jones, *Benchmarking Systems Engineering in United States Industry,* Systems Engineering Design Laboratory, Virginia Poltechnic Institute and State University, Blacksburg, VA, 1994. Forty individuals from 21 different companies who had previously indicated that they were implementing the concepts and principles of systems engineering participated in this project study. For additional references relative to the application of benchmarking in system engineering, it is recommended that you research the literature contained within the *Proceedings* from the annual conferences sponsored by the International Council on Systems Engineering (INCOSE), Seattle, Washington.

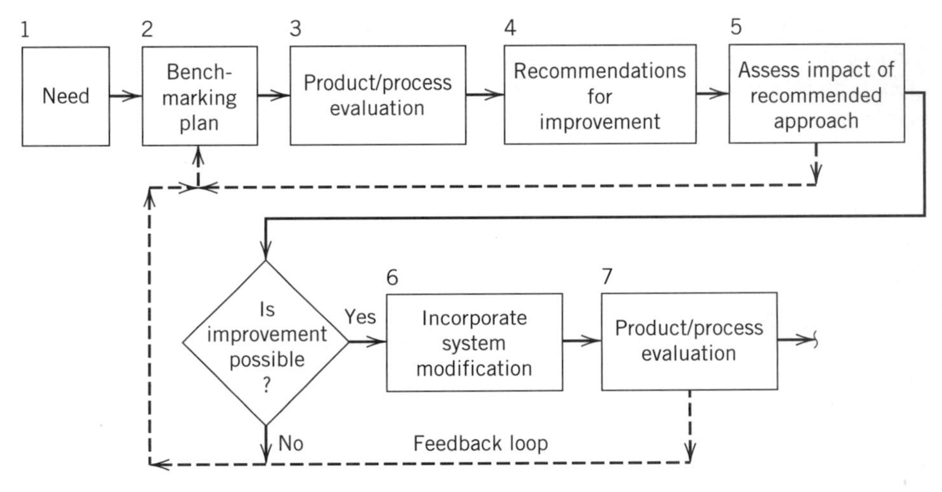

Figure 6.33 Benchmarking process.

of checklists, used to facilitate the evaluation process. An assessment of the current processes is accomplished, possible problem areas are noted, recommendations for process/product improvement are developed (block 4), the potential impact of these proposed changes is assessed (block 5), and, if feasible, modifications are incorporated as appropriate. This may be a continuous process until the desired level of performance is attained.

Figure 6.34 illustrates a benchmarking plan showing the current status in terms of some level of performance, the status of the major competition, and the desired objective. It can be assumed that the competitor is also involved in a benchmarking

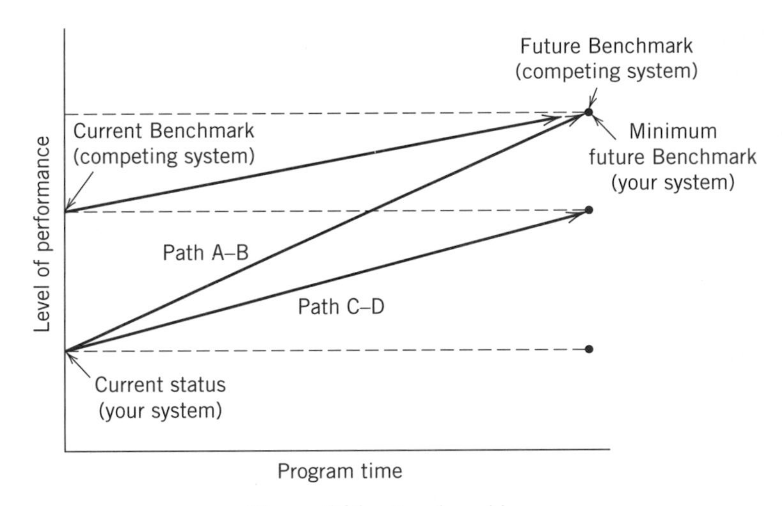

Figure 6.34 Benchmarking.

effort and has established some higher-level goals. Thus, for the system/product in question, a plan needs to be developed that will enable one to follow Path A–B in lieu of Path C–D.

6.7.2 System Engineering Capability Maturity Model (SE-CMM)

Having established certain company/agency goals, the next step is to discover the extent to which your organization has progressed toward meeting those goals; that is, the measure of your organization's capability to meet the desired level of performance. Given the objectives of system engineering and the recommended tasks that need to be performed, to what extent is your organization completing these tasks effectively and efficiently? Does your management understand the concepts and principles of system engineering? Is there a commitment from the top down toward the implementation of the system engineering process? If so, what tasks are currently being implemented? Have standards, measurable goals, and the appropriate processes been established for the successful accomplishment of system engineering objectives? Although many questions of this nature can be asked, the goal is to determine your organization's *level of maturity,* where it may "fit" in the hierarchical structure when comparing it with other organizations in a similar area, and where there are weaknesses that need to be addressed. In other words, although the benchmarking process aids in establishing specific goals, there is a need to develop a model to assist in the evaluation of an organization's current capability.

In response, there has been a concerted effort during the past few years to develop a model that will address the organizational assessment issue. Although there are numerous models being used to varying degrees, two specific projects come to mind. Under the auspices of Carnegie Mellon University, and through the efforts of many from industry, government, and academia, the "Systems Engineering Capability Maturity Model (SE-CMM)" was developed and released for use in 1994.[27] A second effort involved the development of the "Systems Engineering Capability Assessment Model (SECAM)," through the efforts of INCOSE's Capability Assessment Working Group, with the first version being released early in 1994.[28] Although both models are considered to be effective in their application, the author has chosen to briefly describe the SE-CMM in order to provide an overview as to the objectives of organizational assessment.

The SE-CMM describes the essential *elements* of an organization's system engineering process that must exist to ensure that the principles and practices of *good system engineering* are being implemented. The model does not specify a particular process or sequence of activities, but provides a reference for comparing actual

[27] Software Engineering Institute (SEI), "A Systems Engineering Capability Maturity Model (SE-CMM)," Version 1.1, SECMM-95-01, Carnegie Mellon University, Pittsburgh, 1995.

[28] A good reference that provides a historical basis for the SECAM and its applications is B. A. Andrews and E. R. Widmann, "A Synopsis of Metrics and Observations from Systems Engineering Process Assessments Conducted Using the INCOSE SECAM," in *Proceedings of the Sixth Annual International Symposium of the INCOSE,* Vol. 1, 1996, p. 1071. Additional references are included in the *Proceedings* from earlier INCOSE symposia.

system engineering practices against these essential elements. These essential elements are broken down into 18 *process areas* (PAs), against which organizations are evaluated. Is the organization:

1. Analyzing candidate solutions in response to a need?
2. Deriving and allocating requirements?
3. Evolving system architecture?
4. Integrating design disciplines?
5. Integrating system elements?
6. Understanding customer/consumer needs and expectations?
7. Verifying and validating the system?
8. Ensuring good quality?
9. Managing configurations?
10. Managing risk?
11. Monitoring and controlling technical effort?
12. Planned technical effort?
13. Defining its system engineering process and activities?
14. Improving its system engineering processes?
15. Managing product line evolution?
16. Managing system engineering support environment?
17. Providing ongoing skills and knowledge?
18. Providing coordination with suppliers?

In the evaluation of an organization, an assessment is made to determine the extent to which it is addressing each of the preceding 18 process areas. A questionnaire is developed, with specific questions supporting each area. A numerical scale is used to rate the degree of compliance, numerical values are combined and compiled, and an overall rating is calculated.[29]

From the results of the combined assessment relative to how well an organization is complying with the objectives in the 18 process areas, a *capability index* is determined and the organization is placed in any one of six *levels of maturity,* as shown in Figure 6.35. As indicated, the step function illustrated in the figure evolves from "Level 0," where there is no evidence of system engineering activity being performed, to "Level 5," where all of the appropriate system engineering activities are being effectively implemented and where there is a *continuous process improvement* mechanism in place. Figure 6.36 is included to show how a rating is applied to each of the process areas. Areas of deficiency can be easily detected, leading to the need for process improvement.[30]

[29] A good reference covering the questionnaire development and application is Software Productivity Consortium, "Systems Engineering Maturity And Benchmarking," Version 01.00.06, SPC-95075-CMC, SPC, Herndon, VA, 1996.

[30] The material presented herein is extremely cursory in nature. It is highly recommended that you review the references for the SECAM and SE-CMM in depth, and then determine whether either of these models

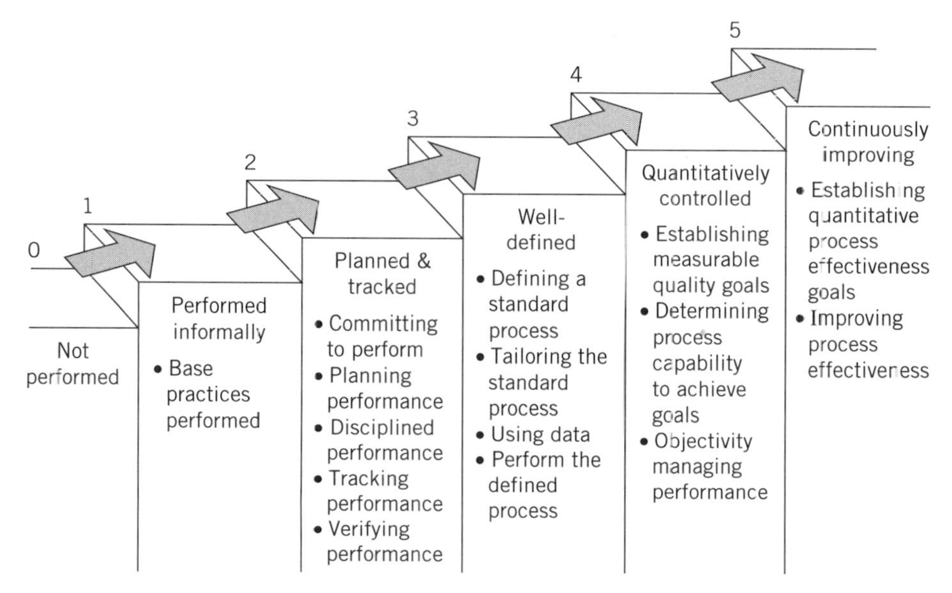

Figure 6.35 Improvement path for process capability. (*Source:* Software Engineering Institute, "A Systems Engineering Capability Maturity Model (SE-CMM)," Version 1.1, SECMM-95-01, SEI, Carnegie Mellon University, Pittsburgh, 1995.)

6.8 SUMMARY

With the thrust of this chapter primarily oriented to the subject of *planning,* the major emphasis has been applied to the development of the System Engineering Management Plan (SEMP). This document is the vehicle through which system engineering functions/tasks are initially defined and later implemented. As this document is the key to the objectives described throughout this text, it is appropriate to summarize through a review of the checklist that follows. The checklist questions pertain mainly to the content of the SEMP, and the response to each question should be *yes.* For additional related questions, refer to Appendix C.

1. Does the SEMP include
   (a) A Statement of Work (SOW)?
   (b) The definition of system engineering tasks?
   (c) A description of work packages and a work breakdown structure (WBS)?
   (d) A description of the program/project organization, the system engineering organization, the critical organizational interfaces (i.e., customer interfaces, producer/contractor interfaces, supplier interfaces), and applicable operating procedures?

can be effectively utilized in determining your organization's level of maturity in the application of system engineering concepts and principles.

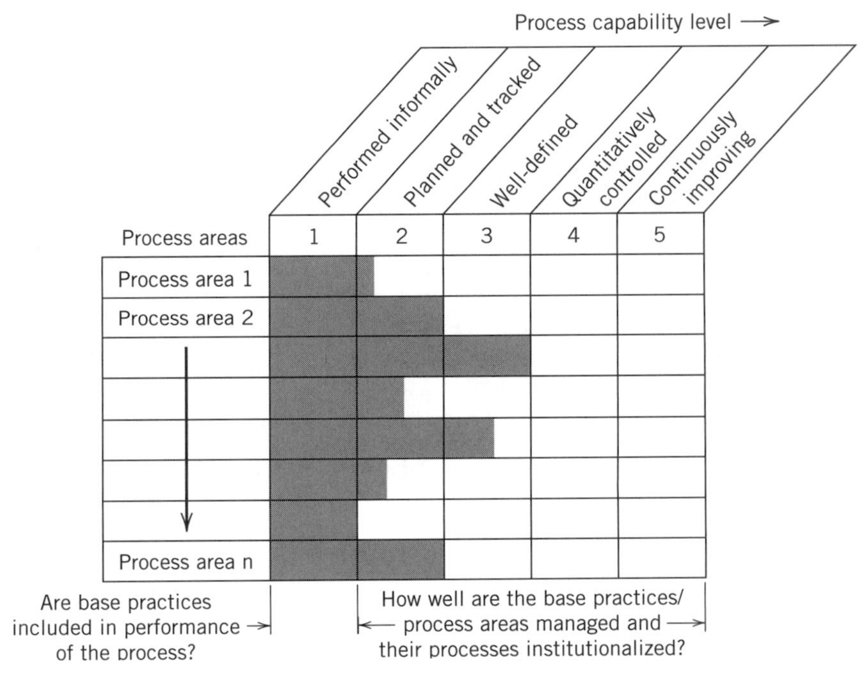

Figure 6.36 Determining process capability. (*Source:* Software Engineering Institute, "A Systems Engineering Capability Maturity Model (SE-CMM)," Version 1.1, SECMM-95-01, SEI, Carnegie Mellon University, Pittsburgh, 1995.)

(e) A specification/documentation tree?
(f) A detailed program schedule?
(g) Program/task cost projections?
(h) A procedure for cost/schedule/technical performance measurement, evaluation, and control?
(i) A description of program reporting requirements?
(j) A risk management plan?

2. Does the SEMP adequately describe the system engineering process to include coverage of
   (a) The needs analysis?
   (b) Feasibility analysis?
   (c) Operational requirements?
   (d) System maintenance concept?
   (e) Functional analysis and allocation?
   (f) System synthesis, analysis, and trade-offs?
   (g) Design integration and support?
   (h) Design reviews?
   (i) System test and evaluation?
   (j) Production and/or construction?

(k) System modification(s)?
(l) System retirement and materials disposal?

3. Does the SEMP include the requirements for the integration of applicable engineering specialties into the total design process through
   (a) Reliability engineering?
   (b) Maintainability engineering?
   (c) Human-factors and safety engineering?
   (d) Logistics, serviceability, and supportability engineering?
   (e) Software engineering?
   (f) Quality engineering?
   (g) Value/cost engineering?
   (h) Producibility?
   (i) Disposability?
   (j) Environmental engineering?
4. Does the SEMP describe the necessary communications link with other program plans such as
   (a) Program Management Plan (PMP)?
   (b) Individual functional design plans?
   (c) Marketing and supplier management plans?
   (d) Configuration management plan?
   (e) Integrated Logistic Support (ILS) and/or maintenance management plans?
   (f) Test and Evaluation Master Plan (TEMP)?
   (g) Data management plan?
   (h) Production/manufacturing plan?
   (i) Total quality management plan (TQMP)?
   (j) System retirement and material disposal plan?
5. Does the SEMP support the requirements of the System Specification (Type "A")?
6. Does the SEMP cover a method for benchmarking and organizational assessment?
7. Does the SEMP adequately support the objectives of system engineering?

QUESTIONS AND PROBLEMS

1. System engineering planning commences early at program inception with the definition of overall program requirements. Why is it essential that this planning activity start as soon as possible? What is likely to happen if system engineering planning is initiated later?

2. How do the System Specification (Type "A") and the System Engineering Management Plan (SEMP) relate to each other?

3. Who is responsible for preparing the SEMP—consumer, producer, contractor, subcontractor, or supplier? Describe some of the conditions and interfaces as applicable.

4. Select a system of your choice, describe the acquisition process, and develop a detailed outline of a SEMP for the program in question.

5. Select a program of your choice, and describe the system engineering tasks for that program (justify the tasks selected). Identify some of the key interfaces that exist.

6. For the tasks identified in Question 5, develop (a) a detailed schedule in the form of a program network, and (b) a cost estimate for the proposed scheduled activity.

7. The following data are available:

Event	Previous Event	t_a	t_b	t_c
8	7	20	30	40
	6	15	20	35
	5	8	12	15
7	4	30	35	50
	3	3	7	12
6	3	40	45	65
	2	25	35	50
5	2	55	70	95
4	1	10	20	35
3	1	5	15	25
2	1	10	15	30

Figure 6.37 Problem 7 data.

(a) Construct a PERT/CPM chart from the data.
(b) Determine the values for standard deviation, *TE, TL, TS, TC,* and *P.*
(c) What is the critical path? What does this value mean?

8. When employing PERT/COST, the cost–time option applies. What is meant by the cost–time option? How can it affect the critical path?

9. Describe the functional relationships between system engineering and each of the following: reliability engineering, maintainability engineering, human-factors engineering, safety engineering, logistics engineering, software engineering, value/cost engineering, quality engineering, producibility, and disposability.

10. What is the purpose of a WBS? What is the difference between a WBS, a SWBS, and a CWBS? How do work packages relate to the WBS? Construct a WBS for a program of your choice.

11. What is the relationship between a WBS and a cost account breakdown structure?

12. What is the purpose of timeline analysis sheets? Requirements allocation sheets? Design criteria sheets? How do these relate to the system functional analysis?

13. Why is a good cost/schedule/performance measurement, evaluation, and control capability important? What attributes should be incorporated?

14. Assume that you are a manager responsible for a system engineering organization. What program/project reporting requirements would you impose in order to gain the necessary visibility to ensure that system engineering objectives are being fulfilled?

15. Why is it important that a good communications link be established between the SEMP and the TEMP? Configuration management plan? ILSP? Production/manufacturing plan? Total quality management plan (TQMP)?

16. Why is it important to develop a risk management plan? What is included?

17. How does technical performance measurement fit into the system engineering planning process? What is the significance of the TPMs?

18. Describe some of the methods/tools that can be used to facilitate implementation of the system engineering process.

19. Assume that you have just been assigned to manage a system engineering organization. Describe the process that you would implement to assess your organization as to its current status (i.e., level of maturity). What steps would you take for the purposes of improvement (i.e., to attain the level of maturity desired)?

7 Organization for System Engineering

The initial planning for system engineering commences during the early phases of conceptual design and evolves through the development of the System Engineering Management Plan (SEMP) described in Chapter 6. To implement this plan successfully requires an organizational structure that will promote, support, and generally enhance the application of system engineering principles and concepts on in-house programs. The proper organizational *environment* must be created that will (1) allow for the accomplishment of system engineering requirements, and (2) will facilitate the implementation of these requirements in an effective and efficient manner. In Figure 7.1, there are two sides of the spectrum; that is, the *technology* issues that can be applied to enhance the implementation of the system engineering process and the *management* issues that are necessary to meet the objectives in this area. Inherent within this overall spectrum is the organizational element.

This chapter addresses the *organizational* issues pertaining to system engineering. Organization is the combining of resources in such a manner as to fulfill a need. Organizations constitute groups of individuals of varying levels of expertise combined into a social structure of some form to accomplish one or more functions. Organizational structures will vary with the functions to be performed, and the results will depend on the established goals and objectives, the resources available, the communications and working relationships among the individual participants, the motivation of the personnel, and many other factors. The ultimate objective is to achieve the most effective and efficient utilization of human, material, and monetary resources through the establishment of decision-making and communications processes designed to accomplish specific objectives.

The fulfillment of system engineering objectives is highly dependent on the proper mix of resources, the establishment of good communications, and on the development of good interpersonal skills by the participants. The uniqueness of tasks and the many different interfaces that exist requires not only good communication skills, but an understanding of the system as an entity and the many design disciplines that contribute to its development.

This chapter deals with the system engineering organization, its functions, organizational interfaces, and the personnel skills necessary in meeting the objectives described throughout this text. The material presented within is supportive of the planning process described in Chapter 6.[1]

[1] The level of discussion of organizational concepts in this chapter is very cursory in nature, and is intended to provide the reader with an overview of some of the key points with respect to system

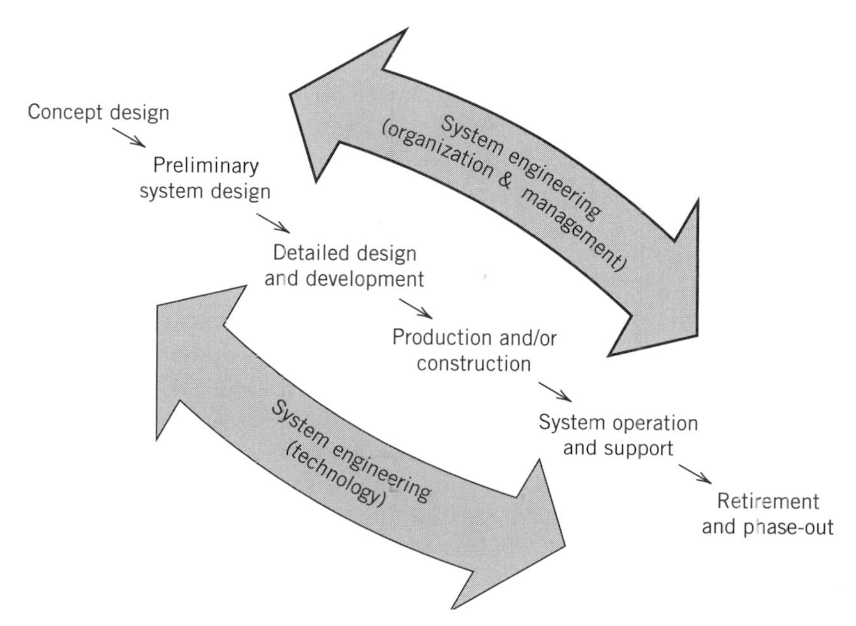

Figure 7.1 Management and technology applied to the system engineering process.

7.1 DEVELOPING THE ORGANIZATION STRUCTURE

An initial step in the development of any type of an organizational structure is the determination of goals and objectives for the overall company/agency involved, along with the functions and tasks that must be accomplished. Depending on the complexity and size of programs, the structure may assume a pure *functional* model, a *project* orientation, a *matrix* approach, or a combination thereof! Further, the structure may change in context as the system development effort evolves from the conceptual design phase through detail design and development, production, and so on. The ultimate goal, of course, is to achieve the most effective utilization of human, material, and monetary resources in accomplishing the functions that are required at the time.

With regard to system engineering, a prime objective during the early stages of conceptual design is to ensure the proper development of system-level requirements; that is, the needs analysis, feasibility analysis, operational requirements, maintenance concept, identification of TPMs, and the preparation of the System Specification (Type "A"), as described in Sections 2.1 to 2.6, Chapter 2. These activities are highly *customer/user* focused, are directed toward the *system* as an entity, and the accomplishment of such does not require a large organization per se. On other hand, the selection of a few key personnel with the appropriate skills, background, and experience levels is essential.

engineering. A more in-depth coverage of organization, organizational dynamics, and management theory is provided in several of the references in Appendix F.

As the program evolves into the preliminary and detail design and development phases, the number of assigned personnel may increase as the design requirements at the subsystem level (and below) may dictate the necessity for including expertise from many different design disciplines; for example, reliability, maintainability, human factors, and supportability. In this context, the organizational structure may transition from a pure *project* configuration to a mixed *functional-project* or *matrix* approach. As the system and its components enter the production phase, the organizational structure may shift once again.

In addressing the organizational issue overall, the emphasis herein is intended to stress the accomplishment of the many and varied tasks described in Section 6.2.2, independent of which organizational element (department or group of personnel) accomplishes the work. Experience has indicated that there are organizational departments or groups located within industrial firms and/or government agencies that have been designated as "System Engineering" and assigned the appropriate responsibilities, but are not actually performing the tasks required. Conversely, there are organizational elements with different identities that are, in actuality, performing the desired functions. Further, for small projects, where a single individual must assume many different roles, the system engineering responsibilities may be accomplished by an electrical engineer, a mechanical engineer, or someone equivalent. On one hand, the project manager may serve as the "system engineer," or there may be a designated group of people performing the required tasks.

Sections 7.2 through 7.5 are presented to provide a more in-depth discussion on the various types of organization structures, the advantages and disadvantages of each, the personnel staffing issues, and so on.

7.2 CONSUMER, PRODUCER, AND SUPPLIER RELATIONSHIPS

To properly address the subject of "organization for system engineering," one needs to understand the environment in which system engineering functions are performed. Although this may vary somewhat depending on the size of the project and the stage of design and development, this discussion is primarily directed to a large project operation, characteristic in the acquisition of many large-scale systems. By addressing large projects, it is hoped that a better understanding of the role of system engineering in a somewhat complex environment will be provided. The reader, of course, must adapt and structure an approach for his/her program requirements.

For a relatively large project, the system engineering function may appear at several levels, as shown in Figure 7.2. The requirements for system engineering and the responsibility for implementing the tasks described in Chapter 6 lie with the customer (or consumer). The customer may establish a system engineering organization to accomplish the required tasks, or these tasks may be relegated (in part or in total) to the producer through some form of contractual arrangement. In any event, the responsibility, along with the authority, for accomplishing system engineering functions must be clearly defined from the beginning.

In some instances, the customer may assume full responsibility for the overall

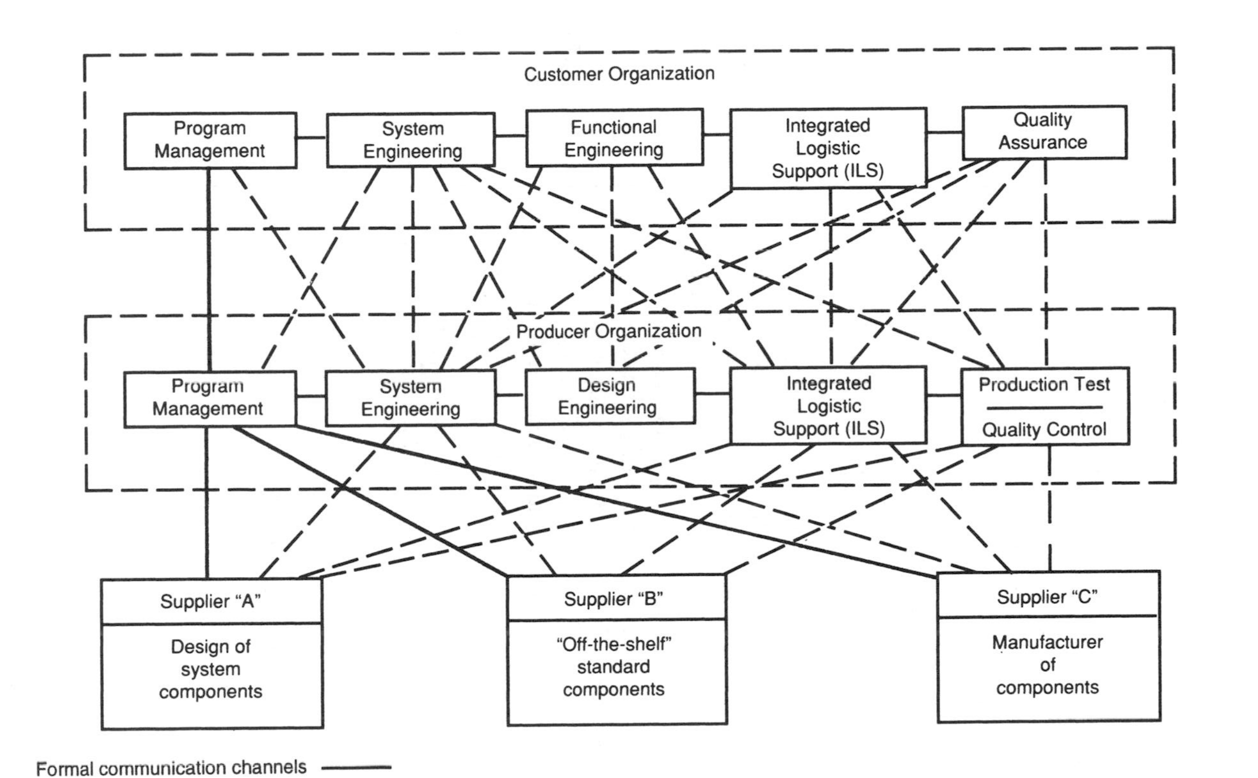

Figure 7.2 Consumer/producer/supplier interfaces.

design and development, production, and installation of the system for operational use. The needs analysis, feasibility studies, definition of operational requirements and the maintenance concept, identification and prioritization of TPMs, preparation of the system specification (Type "A"), and preparation of the system engineering management plan (SEMP) are accomplished by the customer. Top-level functions are defined and specific program requirements are allocated to individual producers, subcontractors, and component suppliers.

In other cases, whereas the customer provides the overall guidance in terms of issuing a general Statement of Work (SOW) or a contractual document of an equivalent nature, the producer (or *prime contractor*) is held responsible for entire system design and development effort and for completing the tasks described in Chapter 6. In other words, while both the customer and the producer have established system engineering organizations, the basic responsibility for fulfilling the objectives described throughout this text lies with the producer's organization, with supporting tasks being accomplished by individual suppliers as required. To accomplish such, the customer must not only delegate the appropriate level of *responsibility* for completing the functions specified, but the necessary *authority* as well. Further, the customer must make accessible all of the necessary input data in order for the producer to successfully complete the conceptual design tasks noted earlier.[2]

From Figure 7.2, it should be noted that there is not only an extensive amount of communications required within each of the customer and producer organizations, but among the various customer, producer, and supplier organizations as well. Although the solid lines pertain primarily to the more formal program management direction and that of contractual nature, there are many informal channels of communication that must exist to ensure that the proper dialogue is established between the numerous and varied entities involved in the system development effort. The successful implementation of a *teaming* or *partnership* approach, along with the fostering of concurrent engineering principles, is heavily dependent on good communications (both downward and upward) from the beginning.

7.3 CONSUMER ORGANIZATION AND FUNCTIONS (THE "CUSTOMER")

The customer/consumer organization may vary from one or a small group of individuals to an industrial firm, a commercial business, an academic institution, a government laboratory, the Department of Defense, or a military service. The customer may be the ultimate "user" of the system or may be the procuring agency for a user. An example of the latter is found within the defense sector, where the Air Force, Army, and Navy each have acquisition agencies that are responsible for the con-

[2] It is not uncommon for the customer to perform a requirements analysis, prepare a report describing the requirements for a new system, place it in a file somewhere, and then fail to pass on the necessary information later to the responsible producer. Thus, the producer has to generate a new set of requirements that may, or may not, be consistent with those initially developed by the customer.

tracting and procurement of systems, while the "user" is the operating command in the field/fleet responsible for the utilization and sustaining maintenance and support of the system throughout its planned life cycle.

In Figure 7.2, the acquisition agency may be represented by the top block, with a chain of industrials firms, small businesses, and component suppliers providing the materials and services necessary for the development of the system and its elements. In such instances, it is incumbent on the procuring agency to ensure that the early contracting and acquisition process will result in satisfying the needs of the ultimate "user," and not just respond to the short-term desires of the procuring agency. In this case, the procuring agency must be responsive to the "user" organization (as the customer), the producer or industrial firm in Figure 7.2 must be responsive to the acquisition agency (as the customer), and the suppliers must be responsive to the producer (as the customer). The question is: Who is the *ultimate customer,* who is *your customer,* and do the requirements associated with the latter support the objectives specified for the first? It is essential that this overall "chain" of organizational entities be addressed in the planning and development of systems.

There are a variety of approaches and associated organizational relationships involved in the design and development of new systems. The objective is to identify the overall "Program Manager," and to pinpoint the responsibility for *system engineering management.* In the past, there have been numerous instances where the procuring agency (e.g., the "customer" in Figure 7.2) has initiated a contract with an industrial firm (e.g., the "producer") for the design and development, and/or reengineering, of a large system, but has not delegated the complete responsibility (or corresponding *authority*) for system engineering management. The industrial firm has been held responsible for the design, development, production, and delivery of a system in response to certain specified requirements. However, the customer has not always provided the producer with the necessary data and/or controls to allow the development effort to proceed in accordance with good system engineering practices. At the same time, the customer has not performed the necessary functions of system engineering management. The net result has led to the development of systems without the consideration of many of the characteristics discussed throughout Chapter 3; that is, a system that is unreliable, not maintainable, not supportable, not cost-effective, and not responsive to the needs of the ultimate user.

The fulfillment of system engineering objectives is highly dependent on a *commitment* from the top down! These objectives must be recognized from the beginning by the customer and an organizational entity needs to be established to ensure that these objectives are met. The Program Manager must first "understand" and "believe in" the concepts and principles of system engineering, and then must create the appropriate environment and take the lead by initiating either one of the following courses of action:

1. Accomplish the system engineering functions within the customer's organizational structure (see Figure 7.2). This may include completing the basic activities reflected in Figure 1.12 and described in Figure 6.6; that is, the needs analysis and feasibility studies, development and operational requirements

and the maintenance concept, the identification and prioritization of TPMs, functional analysis and allocation, synthesis, design optimization, and so on. In other words, the customer (or procuring agency) will prepare the system specification (Type "A") and will perform all of the tasks required at the *system level,* and will delegate requirements for the subsystem level and below.

2. Accomplish the system engineering functions within the industrial firm or the producer's organizational structure shown in Figure 7.2. They may include the completion of the system engineering tasks reflected in Figure 1.12 and described in Figure 6.6; that is, development of operational requirements and the maintenance concept, functional analysis and allocation, synthesis, design optimization, and so on. Although the customer will define the program requirements in the form of a Statement of Work (SOW), all of the system engineering tasks and associated management functions will be delegated to and be accomplished by the producer.

Although these two options represent the extremes, there may be any combination of models where the responsibilities for accomplishing system engineering management functions have been split. In such cases, it is essential that the responsibility for system engineering be established from the beginning. The customer must clarify system objectives and program functions, and the requirements for system engineering must be well-defined. It is critical that the process described in Chapter 2 be implemented properly, independent of organizational splits, the sharing of responsibilities, or whatever!

In the event that the system engineering responsibility is delegated to the producer (i.e., the preceding second option), then the customer must completely support this decision by providing the necessary top-down guidance and managerial backing. Responsibilities must be properly delineated, system-level data generated through earlier customer activities and studies must be made available to the producer (e.g., the results of feasibility analyses, the documentation of operational requirements), and the producer must be given the necessary leeway relative to making decisions at the system level. The challenge for the customer is to prepare a good comprehensive, well-written, and clear statement of work to be implemented by the producer. The emphasis should be on the issues of producer *performance,* specifying *what* needs to be accomplished and *when,* versus telling the producer *how* to perform the job. Additionally, the various lines of communication between the customer and producer shown in Figure 7.2 must support a unified and consistent approach throughout!

7.4 PRODUCER ORGANIZATION AND FUNCTIONS (THE "CONTRACTOR")

For the purposes of discussion, it is assumed that the producer (or contractor) in Figure 7.2 will undertake the bulk of the system engineering activities associated with the design and development of large-scale systems. The customer will specify the necessary system-level and program requirements through the preparation of a *Request for Proposal* (RFP) or an *Invitation for Bid* (IFB), and various industrial firms will respond by submitting a formal proposal. The response may represent the results of a *teaming* arrangement involving a designated number of industrial firms and component suppliers. As there may be a number of responding proposals, a formal competition is initiated, individual proposals are reviewed and evaluated, contractual negotiations are consummated, and a selection is made. The successful contractor (i.e., producer) will then proceed with the proposed level of effort.[3]

In addressing program requirements, it is essential that the successful contractor have access to all information and data leading to the requirements specified in the technical portions of the RFP/IFB. In some instances, the RFP will include a system specification covering the technical aspects of system development, along with a Statement of Work (SOW) directed toward project tasks and the management aspects of a program. The preparation of the specification will result from the completion of those activities described in Sections 2.1 through 2.7.1. In other words, by the time that the contractor gets involved in this case, the customer will have completed the first three tasks in Figure 6.6. The main objective here is to assure continuity in the transition from the activities accomplished by the customer to those to be performed by the contractor.

This transition process is one of the most critical points in a program. First, the process described in Chapter 2 must be maintained, and a thorough understanding of this process by both customer and contractor personnel is essential! Second, the specification and Statement of Work prepared by the customer must be complete and easily understandable, they must "talk to each other," and they must jointly *promote* the system engineering process. Often, in attempting to meet a schedule, specifications and Statements of Work will be hurriedly put together without the benefit of a complete review and the proper level of integration. The results are usually diasterous, and the follow-on activities reflect inconsistencies and the lack of the proper integration of those activities described in Chapter 3. Finally, given a good specification and Statement of Work, the key system engineering activities must not be *negotiated out* in the development of a contractual agreement between the customer and the contractor (i.e., the development of a contract work breakdown structure; refer to Section 6.2.3). Sometimes, there is a tendency to eliminate system engineering tasks to save money, which reflects a lack of understanding of the process and its objectives. This must not be allowed to happen.

Given that system-level requirements have been properly defined and that a prime contractor has been selected to accomplish the design and development effort, the next step is to address the subject of system engineering in the context of the contractor's organizational structure. Organizational structures will vary from the pure *functional,* to the *project,* combined *project-functional,* the *matrix,* and so on. These organizational patterns are discussed in the sections to follow, keeping the objectives of system engineering in mind.

[3] This process is described further in Chapter 8.

7.4.1 Functional Organization Structure

The primary building block for most organizational patterns is the *functional structure* reflected in Figure 7.3. This approach, sometimes referred to as the "classical" or "traditional" approach, involves the grouping of specialties or disciplines into separately identifiable entities. The intent is to perform similar activities within one organizational component. For example, all engineering work would be the responsibility of one executive, all production or manufacturing work would be the responsibility of another executive, and so on. Figure 7.4 shows a further breakout of engineering activities for illustrative purposes.

From the figures, the depth of the individual elements of the organization will vary with the type of project and level of emphasis required. For projects involving the conceptual and/or preliminary design of new systems, there will be a great deal of emphasis in marketing and engineering. Within engineering, the system engineering organization should be highly influential in the design decision-making process as compared to some of the individual design disciplines. Later, as the development process phases into detail design, the individual design disciplines will assume a greater degree of importance, and the interest in production and manufacturing increases. Figure 7.5 conveys the degree of design influence presented on a relative basis. This relationship is also conveyed through the projected manpower loading in Figure 6.24 (Chapter 6).

As for any organizational structure, there are advantages and disadvantages. Figure 7.6 identifies some of the pros and cons associated with the pure functional approach illustrated in Figure 7.3. As shown, the president (or general manager) controls all the functional entities necessary to design and develop, produce, deliver, and support a system. Each department maintains a strong concentration of technical expertise and, as such, a project can benefit from the most advanced technology in the field. Additionally, levels of authority and responsibility are clearly defined, communication channels are well structured, and the necessary controls over budgets and costs can be easily established. In general, this organizational structure is well-suited for a single project operation, large or small.

On the other hand, the pure functional organization may not be as appropriate for large multiproduct firms or agencies. Where there are a large number of different projects, each competing for special attention and the appropriate resources, there are some disadvantages. The main problem is that there is no strong central authority or individual responsible for the total project. As a result, the integration of activities, which cross functional lines, becomes difficult. Conflicts occur as each functional activity struggles for power and resources, and decisions are often made on the basis of what is best for the functional group rather than what is best for the project. Further, the decision-making processes are sometimes slow and tedious because all communications must be channeled through upper-level management. Basically, projects may fall behind and suffer in the classical functional organization structure.

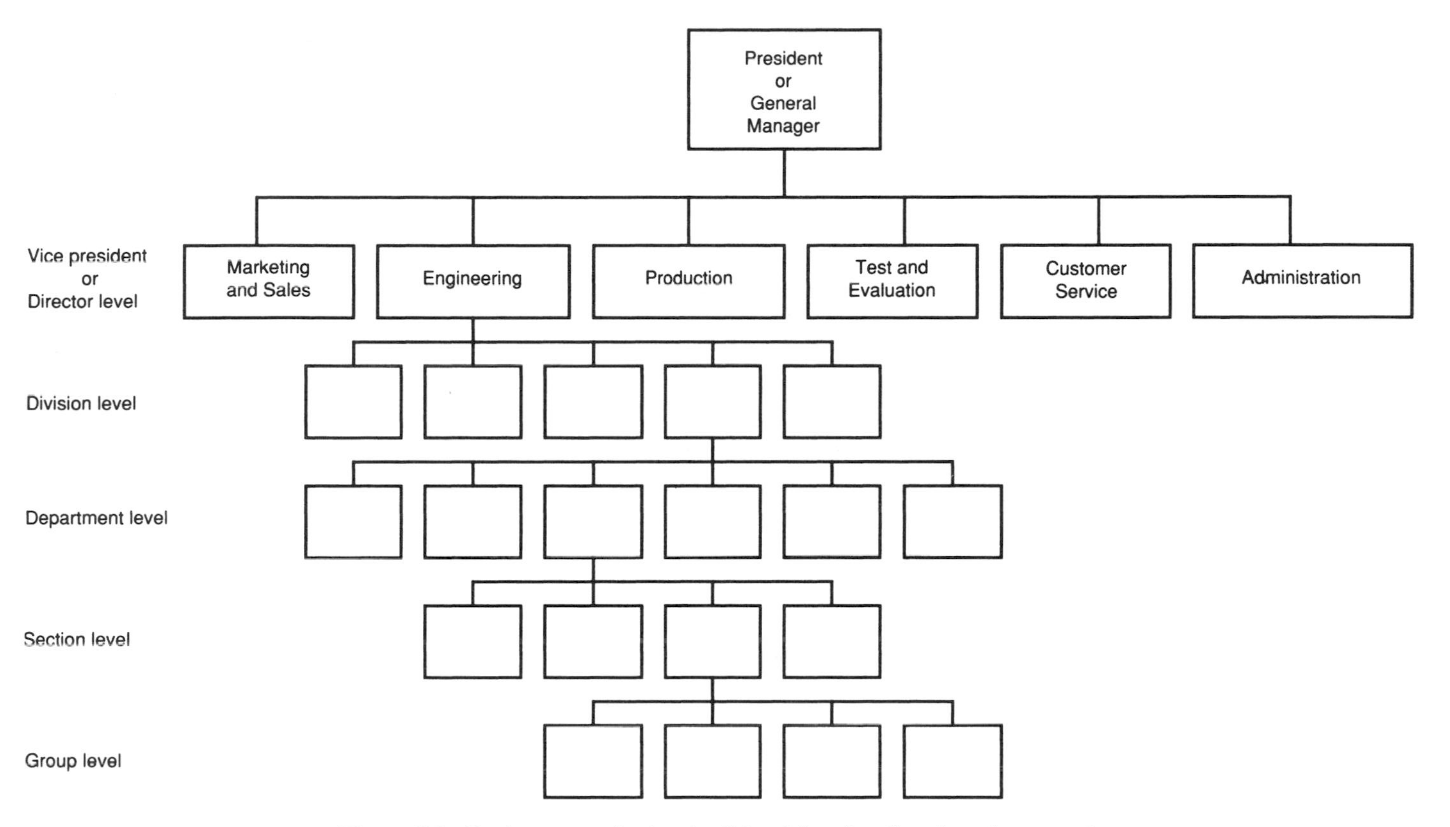

Figure 7.3 Producer organization (traditional functionally oriented structure).

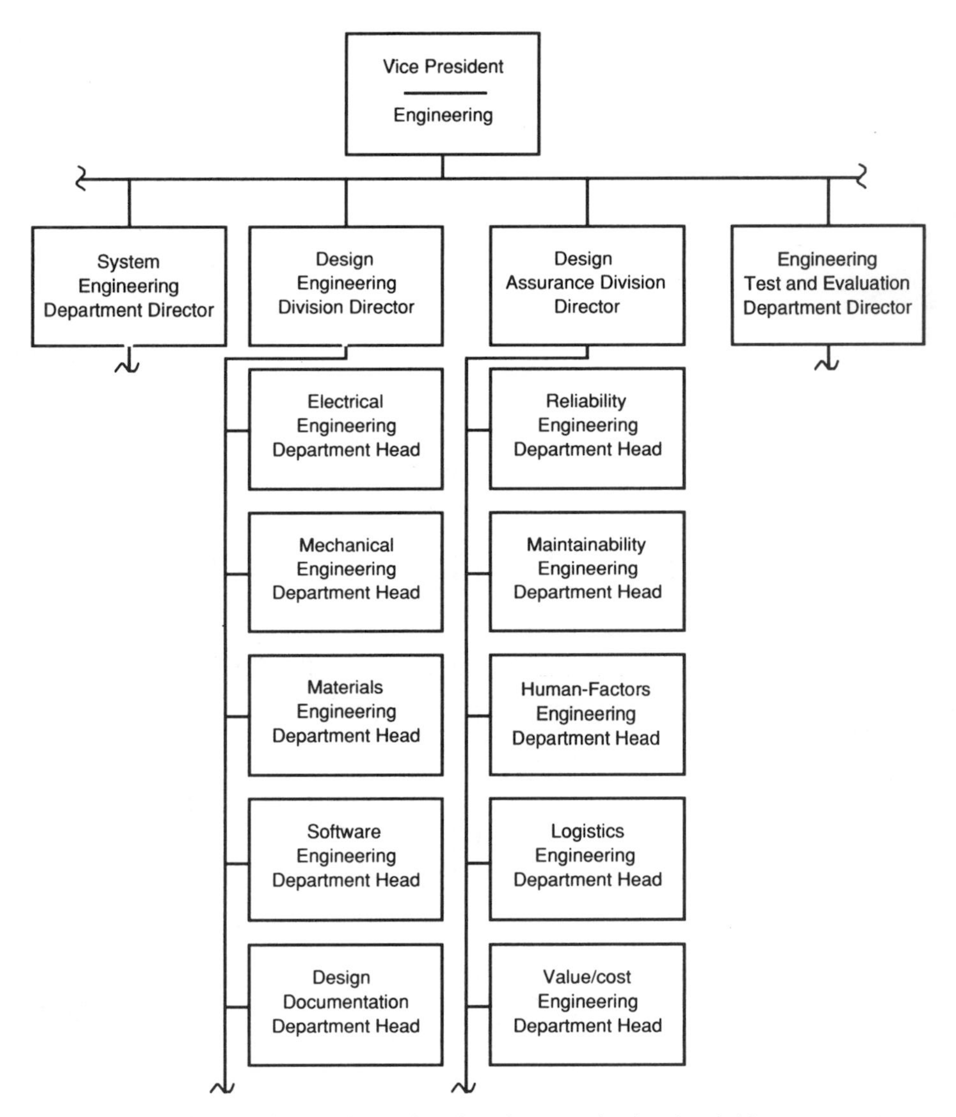

Figure 7.4 Breakout of engineering organizational activities.

7.4.2 Product-Line/Project Organization Structure

As industrial firms grow and there are more products being developed, it is often convenient to classify these products into common groups and to develop a product-line organization structure, as shown in Figure 7.7. A company may become involved in the development of communication systems, transportation systems, and electronic test and support equipment. Where there is functional commonality, it may be appropriate to organize the company into three divisions, one for each product line. In such instances, each division will be self-sufficient relative to system

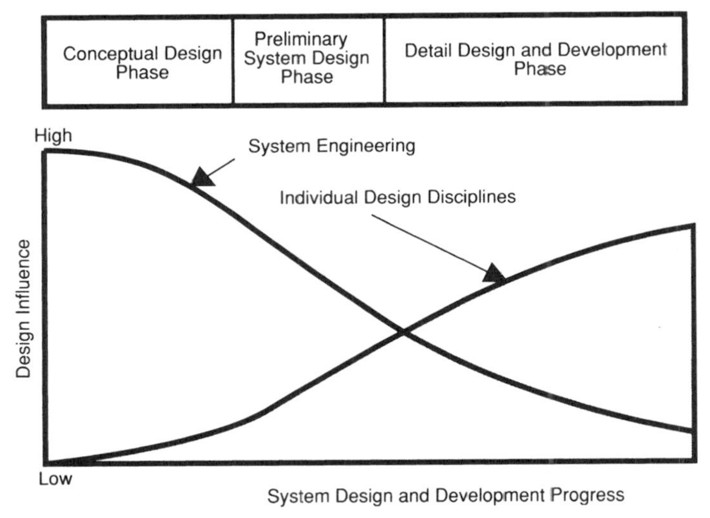

Figure 7.5 System engineering influence on design.

design and support. Further, these divisions may be geographically separated, and each may serve as a functional entity with operations similar to those described in Section 7.4.1.

In divisions where there are large systems being developed, the product-line responsibilities may be subdivided into projects, as illustrated in Figure 7.8. In such cases, the project will be the lowest independent entity.

A project organization is one that is solely responsive to the planning, design and development, production, and support of a single system or a large product. It is time-limited, directly oriented to the life cycle of the system, and the commitment of personnel and material is purely for the purposes of accomplishing tasks peculiar to that system. Each project will contain its own management structure, its own engineering function, its own production capability, its own support function, and so on. The project manager has the authority and the responsibility for all aspects of the project, whether it is a success or a failure.

In the case for both the product-line and the project structures, the activities are organized as presented in Figure 7.7. The lines of authority and responsibility for a given project are clearly defined, and there is no question as to priorities. On the other hand, there is potential for the duplication of activities within a firm that can be quite costly. Emphasis is on individual projects as compared to the overall functional approach illustrated in Figure 7.3. Some of the advantages and disadvantages of project/product-line structures are presented in Figure 7.9.

7.4.3 Matrix Organization Structure

The matrix organizational structure is an attempt to combine the advantages of the pure functional organization and the pure project organization. In the functional

Advantages

1. Enables the development of a better technical capability for the organization. Specialists can be grouped to share knowledge. Experiences from one project can be transferred to other projects through personnel exchange. Cross-training is relatively easy.

2. The organization can respond quicker to a specific requirement through the careful assignment (or reassignment) of personnel. There are a larger number of personnel in the organization with the required skills in a given area. The manager has a greater degree of flexibility in the use of personnel and a broader manpower base with which to work. Greater technical control can be maintained.

3. Budgeting and cost control are easier due to the centralization of areas of expertise. Common tasks for different projects are integrated, and it is easier not only to estimate costs but to monitor and control costs.

4. The channels of communication are well established. The reporting structure is vertical, and there is no question as to who is the "boss."

Disadvantages

1. It is difficult to maintain an identity with a specific project. No single individual is responsible for the total project or the integration of its activities. It is hard to pinpoint specific project responsibilities.

2. Concepts and techniques tend to be functionally oriented with little regard toward project requirements. The "tailoring" of technical requirements to a particular project is discouraged.

3. There is little customer orientation or focal point. Response to specific customer needs is slow. Decisions are made on the basis of the strongest functional area of activity.

4. Because of the group orientation relative to specific areas of expertise, there is less personal motivation to excel and innovation concerning the generation of new ideas is lacking.

Figure 7.6 A functional organization—advantages and disadvantages.

organization, technology is emphasized and project-oriented tasks, schedules, and time constraints are sacrificed. For the pure project, technology tends to suffer because there is no single group for the planning and development of such! Matrix management is an attempt to acquire the greatest amount of technology, consistent with project schedules, time and cost constraints, and related customer requirements. Figure 7.10 presents a typical matrix organization structure.

Each project manager reports to a vice president, and has the overall responsibility and is accountable for project success. At the same time, the functional departments are responsible for maintaining technical excellence and for ensuring that all available technical information is exchanged between projects. The functional managers, who also report to a vice president, are responsible to ensure that their personnel are knowledgeable of the latest accomplishments in their respective fields.

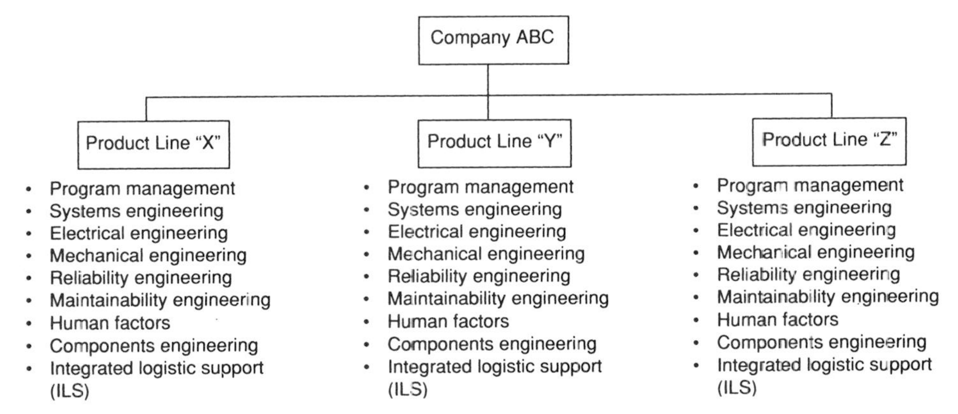

Figure 7.7 Traditional project/product-line organization.

The matrix organization, in its simplest form, can be considered as being a two-dimensional entity with the projects representing potential profit centers and the functional departments identified as cost centers. For small industrial firms, the two-dimensional structure may reflect the preferred organizational approach because of the flexibility allowed. The sharing of personnel and the shifting back and forth are often inherent characteristics. On the other hand, for large corporations with many product divisions, the matrix becomes a multidimensional structure.

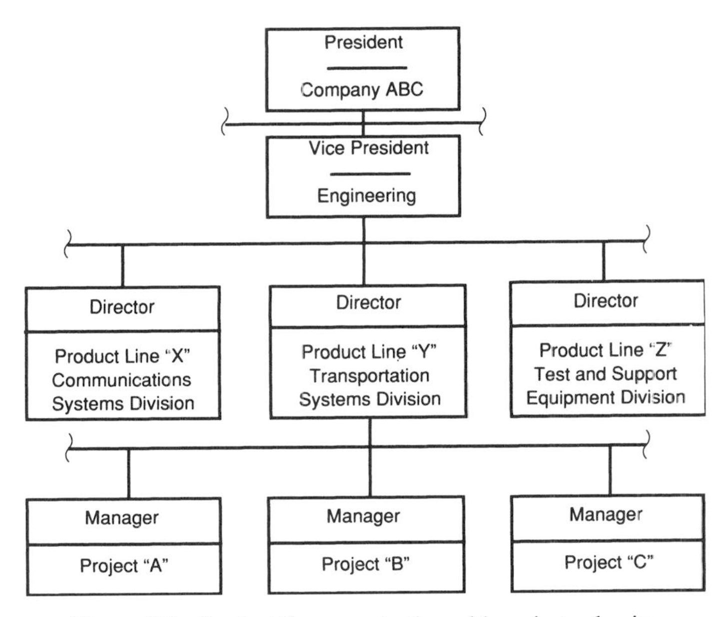

Figure 7.8 Product-line organization with project subunits.

Advantages

1. The lines of authority and responsibility for a given project are clearly defined. Project participants work directly for the project manager, communication channels within the project are strong, and there is no question as to priorities. A good project orientation is provided.

2. There is a strong customer orientation, a company focal point is readily identified, and the communication processes between the customer and the contractor are relatively easy to maintain. A rapid response to customer needs is realized.

3. Personnel assigned to the project generally exhibit a high degree of loyalty to the project, there is strong motivation, and personal morale is usually better with product identification and affiliation.

4. The required personnel expertise can be assigned and retained exclusively on the project without the time sharing that is often required under the functional approach.

5. There is greater visibility relative to all project activities. Cost, schedule, and performance progress can be easily monitored, and potential problem areas (with the appropriate follow-on corrective action) can be identified earlier.

Disadvantages

1. The application of new technologies tends to suffer without strong functional groups and the opportunities for technical interchange between projects. As projects go on and on, those technologies that are applicable at project inception continue to be applied on a repetitive basis. There is no perpetuation of technology, and the introduction of new methods and procedures is discouraged.

2. In contractor organizations where there are many different projects, there is usually a duplication of effort, personnel, and the use of facilities and equipment. The overall operation is inefficient and the results can be quite costly. There are times when a completely decentralized approach is not as efficient as centralization.

3. From a managerial perspective, it is difficult to effectively utilize personnel in the transfer from one project to another. Good qualified workers assigned to projects are retained by project managers for as long as possible (whether they are being effectively utilized or not), and the reassignment of such personnel usually requires approval from a higher level of authority which can be quite time-consuming. The shifting of personnel in response to short-term needs is essentially impossible.

4. The continuity of an individual's career, his or her growth potential, and the opportunities for promotion are often not as good when assigned to a project for an extended period of time. Project personnel are limited in terms of opportunities to be innovative relative to the application of new technologies. The repetitiousness of tasks sometimes results in stagnation.

Figure 7.9 A project/product-line organization—advantages and disadvantages.

As the number of projects and functional departments increases, the matrix structure can become quite complex. In order to ensure success in implementing matrix management, a highly cooperative and mutually supportive environment must be created within the company. Managers and workers alike must be committed to the objectives of matrix management. A few key points follow:

1. Good communication channels (vertical and horizontal) must be established to allow for a free and continuing flow of information between projects and the

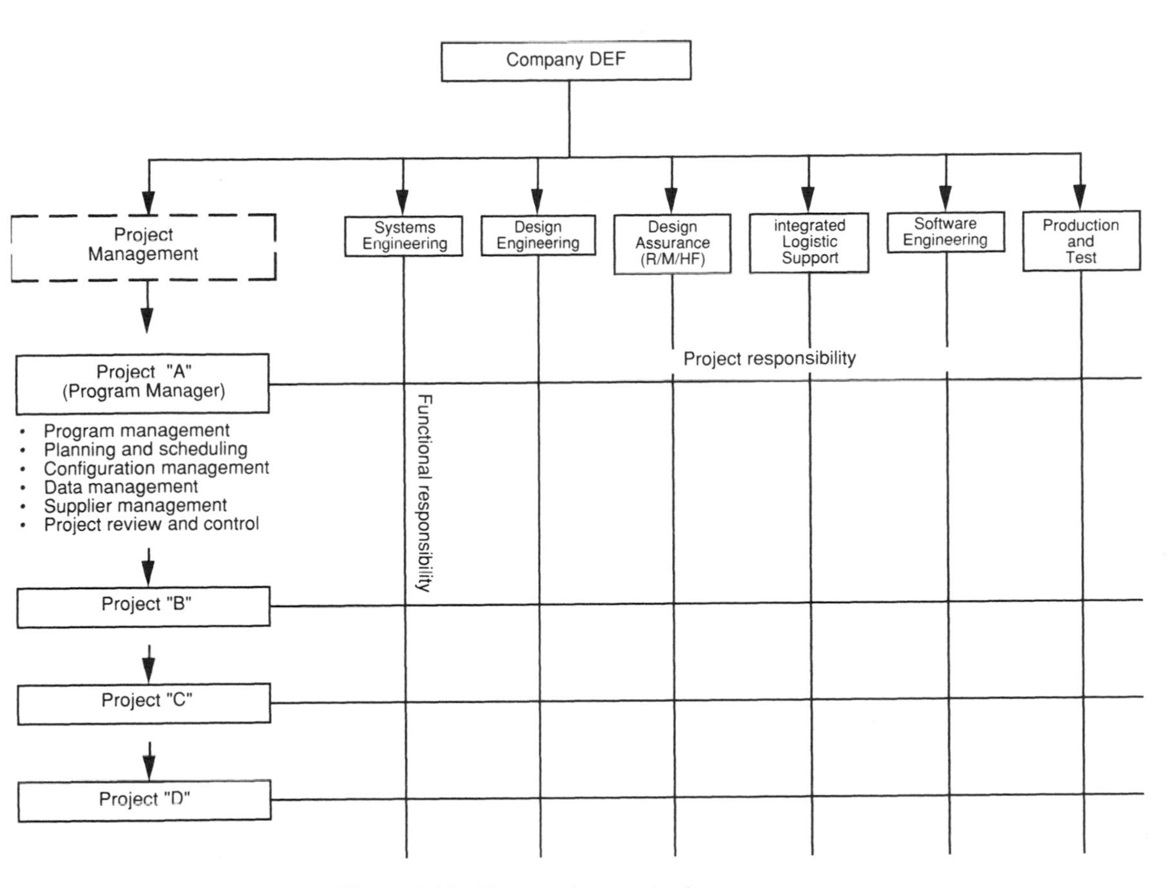

Figure 7.10 Pure matrix organization structure.

functional departments. Good communications must also be established from project to project.

2. Both project managers and functional department managers should participate in the initial establishment of companywide and program-oriented objectives. Further, each must have an input and become involved in the planning process. The purpose is to help ensure the necessary commitment on both sides. Additionally, both project and functional managers must be willing to negotiate for resources.

3. In the event of conflict, a quick and effective method for resolution must be established. A procedure must be developed with the participation and commitment from both project and functional managers.

4. For personnel representing the technical functions and assigned to a project, both the project manager and the functional department manager should agree to the duration of assignment, the tasks to be accomplished, and the basis upon which the individual(s) will be evaluated. The individual worker must know what is to be expected of him/her, the criteria for evaluation, and which manager will be conducting the performance review (or how will the performance review be conducted?). Otherwise, a "two-boss" situation (each with different objectives) may develop and the employee will be caught in the middle!

The matrix structure provides the best of several worlds; that is, a composite of the pure project approach and the traditional functional approach. The main advantage pertains to the capability of providing the proper mix of technology and project-related activities. At the same time, a major disadvantage relates to the conflicts that arise on a continuing basis as a result of a power struggle among project and functional managers, changes in priorities, and so on. A few advantages and disadvantages are noted in Figure 7.11.

7.4.4 Integrated Product and Process Development (IPPD)

With the objectives of concurrent engineering in mind, the Department of Defense (DOD) initiated the concept of *Integrated Product and Process Development* (IPPD) in the mid-1990s. IPPD can be defined as "a management technique that simultaneously integrates all essential acquisition activities through the use of multidiscipline teams to optimize the design, manufacturing, and supportability processes."[4] The concept promotes the communications and integration of the key functional areas, as they apply to the various phases of program activity from conceptual design through detail design and development. Although the specific nature of the activities involved and the degree of emphasis exerted will change somewhat as the system design and development effort evolves, the concept conveyed in Figure 7.12 is maintained throughout in order to foster the necessary communications across the more traditional functional lines of authority. In this regard, the concept of IPPD is

[4] Department of Defense, "Manditory Procedures for Major Defense Acquisition Programs (MDAPs) and Major Automated Information System (MAIS) Acquisition Programs," DOD Regulation 5000.2, DOD, Washington, DC, 1996.

Advantages

1. The project manager can provide the necessary strong controls for the project while having ready access to the resources from many different functionally-oriented departments.

2. The functional organizations exist primarily as support for the projects. A strong technical capability can be developed and made available in response to project requirements in an expeditious manner.

3. Technical expertise can be exchanged between projects with a minimum of conflict. Knowledge is available for all projects on an equal basis.

4. Authority and responsibility for project task accomplishment are shared between the project manager and the functional manager. There is mutual commitment in fulfilling project requirements.

5. Key personnel can be shared and assigned to work on a variety of problems. From the company top-management perspective, a more effective utilization of technical personnel can be realized and program costs can be minimized as a result.

Disadvantages

1. Each project organization operates independently. In an attempt to maintain an identity, separate operating procedures are developed, separate personnel requirements are identified, and so on. Extreme care must be taken to guard against possible duplication of efforts.

2. From a company viewpoint, the matrix structure may be more costly in terms of administrative requirements. Both the project and the functional areas of activity require similar administrative controls.

3. The balance of power between the project and the functional organizations must be clearly defined initially and closely monitored thereafter. Depending on the strengths (and weaknesses) of the individual managers, the power and influence can shift to the detriment of the overall company organization.

4. From the perspective of the individual worker, there is often a split in the chain of command for reporting purposes. The individual is sometimes "pulled" between the project boss and the functional boss.

Figure 7.11 A matrix organization—advantages and disadvantages.

directly in line with system engineering objectives; that is, to cause the integration of the various features of design and the organizations involved in the design process.

Inherent within the IPPD concept is the establishment of *Integrated Product Teams* (IPTs), with the objective of addressing certain designated and well-defined issues.[5] An IPT, constituting a selected team of individuals from the appropriate

[5] The term IPT is also used as a designator for "Integrated Process Team." Another term that is often used in a similar context is "Process Action Team (PAT)."

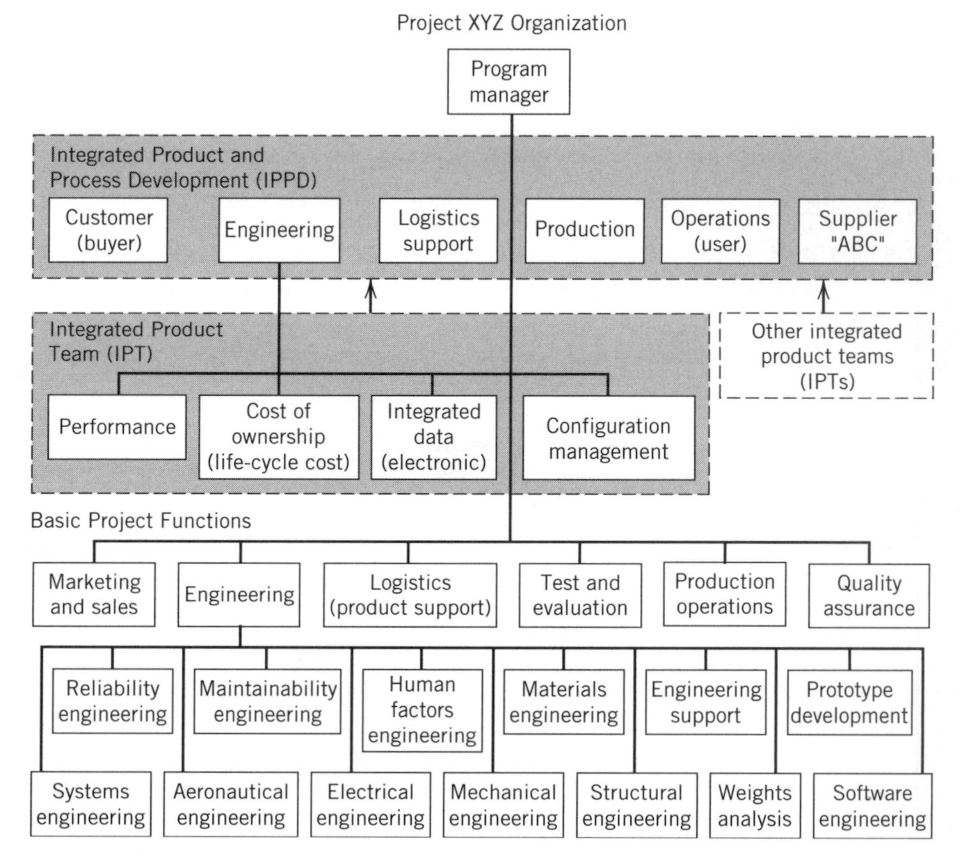

Figure 7.12 Functional organization structure showing IPPD/IPTs.

disciplines, may be established to investigate a specific segment of design, a solution for some outstanding problem, design activities that have a large impact on a high-priority TPM, and so on. The objective is to create a *team* of qualified individuals that can effectively work together to solve some problem in response to a given requirement. Further, there may be a number of different teams established to address issues at different levels in the overall system hierarchical structure; that is, issues at the system level, subsystem level, and/or component level. From Figure 7.12, an IPT may be established to concentrate on those activities that significantly impact selected *performance* factors, *cost of ownership,* and *configuration management.* There may be another IPT assigned to "track" the *integrated data* environment issue. The objective is to provide the necessary emphasis in critical areas, and to reap the benefits of a *team* approach in arriving the best solution possible.

IPTs are often established by the Program Manager, or by some designated high-level authority in the organization. The representative team members must be well-qualified in their respective areas of expertise, must be empowered to make on-the-spot decisions when necessary, must be proactive relative to team participation, success-oriented, and be resolved to addressing the problem assigned. The Program Manager must clearly define the objectives for the team, expectations in terms of results, and the team members must maintain a continuous "up-the-line" communications channel. The longevity of the IPT will depend on the nature of the problem and the effectiveness of the team in progressing toward meeting its objective. Care must be taken to avoid the establishment of too many teams, as the communication processes and interfaces become too complex when there are many teams in place. Additionally, there often are conflicts when it comes to issues of importance and that issue is "traded off" as a result. Further, as the team ceases to be effective in accomplishing its objectives, it should be disbanded accordingly. An established team that has "outlived its usefulness" can be counterproductive.

7.4.5 System Engineering Organization

Sections 7.4.1 through 7.4.4 provide an overview of the major characteristics of the functional, project, and matrix organization structures. Some of the advantages and disadvantages of each are identified. It is important to thoroughly understand and have these characteristics in mind when developing an organizational approach involving system engineering. More specifically, when considering system engineering objectives, the following points should be noted:

1. The function of system engineering must be oriented to the objective of "bringing a system into being in an effective and efficient manner." In this regard, there is a natural close association with the "project" type of organizational structure. System engineering is heavily involved in the initial establishment of requirements, and in the follow-on integration of design engineering and supporting activities throughout system development, production, and operational use. System engineering influences design to a significant degree (see Figure 7.5), and this is best accomplished through a project organizational structure.

2. The nature of the system engineering function, its objectives in terms of design integration, its many interfaces with other program activities, and so on, require the existence of good communication channels (both vertically and horizontally). The personnel within the system engineering organization must maintain effective communications with all other project organizational elements, with many different functional departments, with a variety of suppliers, and with the customer. These requirements are facilitated through the project organization approach.

3. The successful fulfillment of system engineering objectives requires the specification of technical requirements for the system, the conductance of trade-off studies, the selection of appropriate technologies, and so on. Personnel within the system engineering organization must be current (i.e., up to date) relative to the latest technology applications and/or must have access to technical expertise in the appro-

priate disciplines. A strong technical thrust is required, and good communications must be established with the functional departments (as applicable). Thus, the preferred organization structure should include selected functional elements, in addition to the project orientation.

Although the implementation of system engineering requirements can actually be fulfilled through any one of a number of organizational structures, the preferred approach should respond to these three major considerations. It appears that the best organizational structure constitutes a combination of project requirements and functional requirements. Although a major *project* orientation is required in response to customer needs, a *functional* orientation is necessary to ensure consideration of the latest technology applications. The combined project-functional organization approach may vary somewhat depending on the size of the industrial firm. For a large firm, the organization structure illustrated in Figure 7.13 may be appropriate. Project activities are relatively large in scope (and in personnel loading) while there are supporting functional activities covering selected areas of expertise where centralization is justified. For smaller firms, the functional departments are relatively large, and they provide support to individual projects on a demand basis. This support is assigned on a task-by-task basis. Figure 7.14 illustrates an organizational structure in which the emphasis is on the functional end of the spectrum. In essence, the degree of "project" emphasis and "functional" emphasis often shifts back and forth depending on both the size of the firm and the nature of activity; that is, whether conceptual design, preliminary system design, or detailed design and development activities are in progress.

From the figures, there may be a variation of approaches within the same firm. One or two large projects may exist along with numerous smaller projects. The large projects will tend to support an organizational structure similar to that presented in Figure 7.13, whereas the smaller projects will likely follow the format in Figure 7.14. Where the larger projects can afford to support significant numbers of personnel on a full-time basis, the smaller projects may be able to support a select number of individuals on only a part-time basis. The specific requirements are dictated through the generation of program tasks by the project organization; that is, a request for assistance is initiated by the project manager, with the task(s) being completed within the functional department.

Project size will vary not only with the type and nature of the system being developed, but with the specific stage of development. A large-scale system in the early stages of conceptual design may be represented by a small project organization as shown in Figure 7.14. As system development progresses into the phases of preliminary system design and detail design and development, the organization structure may shift somewhat, replicating the configuration in Figure 7.13. In other words, the characteristics and structure of organizations often are dynamic by nature. The organization structure must be adapted to the needs of the project at the time, and these needs may shift as system development evolves.

With regard to system engineering, the tasks identified in Figure 6.6 (Chapter 6) can be allocated by phase as follows:

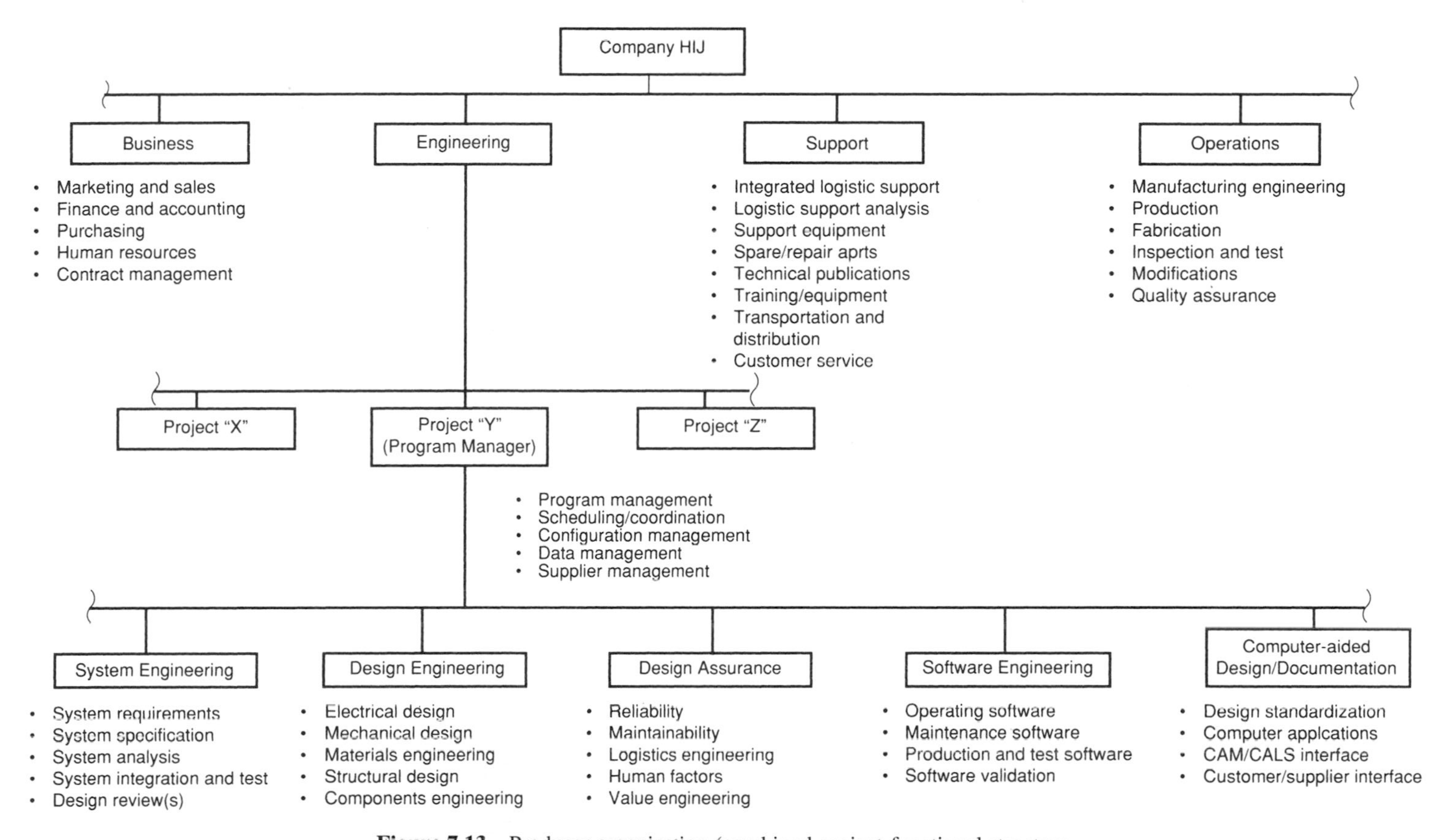

Company HIJ
Business
Engineering
Support
Operations
• Marketing and sales
• Finance and accounting
• Purchasing
• Human resources
• Contract management
• Integrated logistic support
• Logistic support analysis
• Support equipment
• Spare/repair aprts
• Technical publications
• Training/equipment
• Transportation and distribution
• Customer service
• Manufacturing engineering
• Production
• Fabrication
• Inspection and test
• Modifications
• Quality assurance
Project "X"
Project "Y" (Program Manager)
Project "Z"
• Program management
• Scheduling/coordination
• Configuration management
• Data management
• Supplier management
System Engineering
Design Engineering
Design Assurance
Software Engineering
Computer-aided Design/Documentation
• System requirements
• System specification
• System analysis
• System integration and test
• Design review(s)
• Electrical design
• Mechanical design
• Materials engineering
• Structural design
• Components engineering
• Reliability
• Maintainability
• Logistics engineering
• Human factors
• Value engineering
• Operating software
• Maintenance software
• Production and test software
• Software validation
• Design standardization
• Computer applcations
• CAM/CALS interface
• Customer/supplier interface

Figure 7.13 Producer organization (combined project-functional structure.

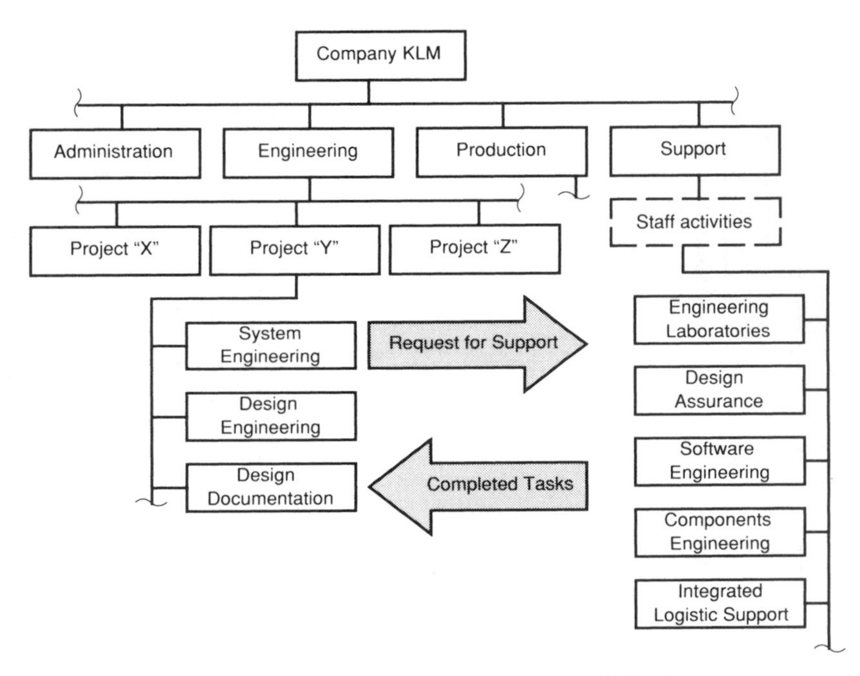

Figure 7.14 Producer organization (work flow).

1. *Conceptual design phase:*
   (a) Perform needs analysis and conduct feasibility studies.
   (b) Define operational requirements, the system maintenance concept, and accomplish system-level functional analysis.
   (c) Accomplish system integration.
   (d) Prepare the system specification (Type "A").
   (e) Prepare the Test and Evaluation Master Plan (TEMP).
   (f) Prepare the System Engineering Management Plan (SEMP).
   (g) Plan, coordinate, and conduct the conceptual design review.
2. *Preliminary system design phase:*
   (a) Accomplish functional analysis and the allocation of requirements.
   (b) Accomplish system analysis, synthesis, and trade-off studies.
   (c) Accomplish system integration; that is, the integration of design disciplines, supplier activities, and data.
   (d) Plan, coordinate, and conduct system design reviews.
3. *Detail design and development phase:*
   (a) Accomplish system analysis, synthesis, and trade-off studies.
   (b) Accomplish system integration; that is, the integration of design disciplines, supplier activities, and data.
   (c) Monitor and review system test and evaluation activities.
   (d) Plan, coordinate, implement, and control design changes.
   (e) Plan, coordinate, and conduct equipment/software design reviews, and the critical design review.

(f) Initiate and maintain production and/or construction liaison and customer service activities.

4. *Production and/or construction phase:*
   (a) Monitor and review system test and evaluation activities.
   (b) Plan, coordinate, implement, and control design changes.
   (c) Maintain production and/or construction liaison and conduct customer service activities (i.e., field service).
5. *System utilization and life-cycle support:*
   (a) Monitor and review system test and evaluation activities.
   (b) Plan, coordinate, implement, and control design changes.
   (c) Conduct customer service activities (i.e., field service).
   (d) Collect, analyze, and process field data. Prepare reports on system operations in the user environment.
6. *System retirement and material disposal:*
   (a) Prepare plan for system retirement.
   (b) Monitor material disposal and recycling activities.
   (c) Prepare reports on environmental impact(s).

As noted in Section 6.2.2, these tasks reflect what is envisioned as being critical to the successful implementation of the system engineering process. This does not mean to imply a tremendous level of effort! The tasks must be tailored to the particular application, and the accomplishment of many of these requires a significant input from other organizations. A major objective is to provide a mechanism for *integration,* and the interfaces and relationships between system engineering and other organizations are extensive. As an example, the conductance of a life-cycle cost analysis as part of the overall system analysis task involves a data exchange with all project-related engineering organizations, finance and accounting, logistic support, production, the customer, and so on. The goal of the system engineering organization is to provide the necessary technical management and guidance to ensure that these activities are completed in a timely manner.

To further illustrate the many interfaces that exist, the combined project-functional organization structure in Figure 7.13 has been extended to include a sampling of the required communication channels that must be established between the system engineering function and other organizational elements. These communication links are shown in Figure 7.15, and an abbreviated description covering the nature of the necessary communications is presented in Figure 7.16. The system engineering function, as an integrating agency, must not only provide the technical leadership throughout the system development activity, but must initially establish and subsequently maintain open and free-flowing communications across the board.[6]

The ultimate success in meeting these system engineering objectives is, of

[6] It is not intended to imply that the system engineering organization does everything! The emphasis here is on providing a *technical* thrust and on assuming a *technical* leadership role in the design and development of the system. The Project/Program Manager must, of course, provide the necessary leadership from the overall organizational standpoint.

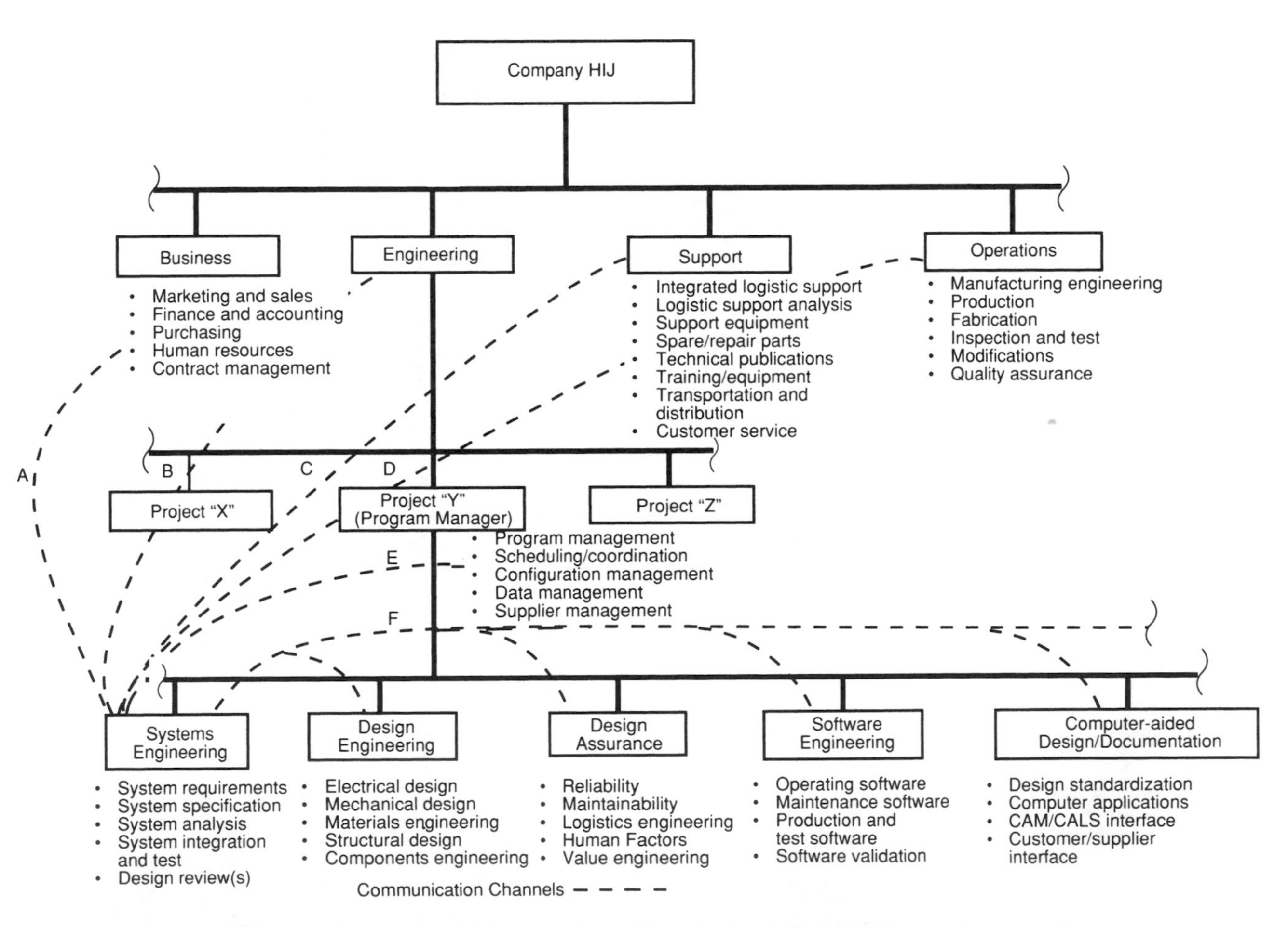

Figure 7.15 Major system engineering communication links (producer organization).

Communication Channel (Figure 7.13)	Supporting Organization (Interface Requirements)
A	1. *Marketing and sales* – to acquire and sustain the necessary communications with the customer. Supplemental information pertaining to customer requirements, system operational and maintenance support requirements, changes in requirements, outside competition, etc., is needed. This is above and beyond the formal "contractual" channel of communications. 2. *Accounting* – to acquire both budgetary and cost data in support of economic analysis efforts (e.g., life-cycle cost analysis). 3. *Purchasing* – to assist in the identification, evaluation, and selection of component suppliers with regard to technical, quality, and life-cycle cost implications. 4. *Human resources* – to solicit assistance in the initial recruiting and hiring of qualified project personnel for system engineering, and in the subsequent training and maintenance of personnel skills. To conduct training programs for all project personnel across the board relative to system engineering concepts, objectives, and the implementation of program requirements. 5. *Contract management* – to keep abreast of contract requirements (of a technical nature) between the customer and the contractor. To ensure that the appropriate relationships are established and maintained with suppliers as they pertain to meeting the *technical* needs for system design and development.
B	To establish and maintain ongoing liaison and close communications with other projects with the objective of transferring knowledge that can be applied for the benefit of Project "Y". To solicit assistance from other companywide functionally oriented engineering laboratories and departments relative to the application of new technologies in support of system design and development.
C	To provide an input relative to project requirements for system support, and to solicit assistance in terms of the functional aspects associated with the design, development, test and evaluation, production, and sustaining maintenance of a support capability through the planned system life cycle.
D	To provide an input relative to project requirements for production (i.e., manufacturing, fabrication, assembly, inspection and test, and quality assurance), and to solicit assistance relative to the design for producibility and the implementation of quality engineering requirements in support of system design and development.
E	To establish and maintain close relationships and the necessary on going communications with such project activities as scheduling (the monitoring of critical program activities through a network scheduling approach); configuration management (the definition of various configuration baselines and the monitoring and control of changes/modifications); data management (the monitoring, review, and evaluation of various data packages to ensure compatibility and the elimination of unnecessary redundancies); and supplier management (to monitor progress and ensure the appropriate integration of supplier activities).
F	To provide an input relative to *system-level* design requirements, and to monitor, review, evaluate, and ensure the appropriate integration of system design activities. This includes providing a *technical* lead in the definition of system requirements, the accomplishment of functional analysis, the conductance of system-level trade-off studies, and the other project tasks presented in Figure 6.6.

Figure 7.16 Description of major project interface requirements.

course, highly dependent on managerial support from the top down! The president (or general manager), the vice president of engineering, Project "Y" Manager, and other high-level managers must each *understand* and *believe in* the concepts and objectives of system engineering. If the System Engineering Manager is to be successful, these higher-level managers must be directly supportive all the time. There will be many occasions in which individual design engineers and/or middle managers will go off on their own, making decisions that will conflict with system engineering objectives. When this occurs, the System Engineering Manager must have the necessary support to ensure that actions are taken to "get things back on track!" This area is discussed further in Section 7.6.

7.5 SUPPLIER ORGANIZATION AND FUNCTIONS

The term "supplier," as defined herein, refers to a broad category of organizations that provide system components to the producer (i.e., the contractor). These components may range from a large element of the system (e.g., a facility, an intermediate maintenance shop full of test equipment, Unit "B" of System "XYZ") to a small nonrepairable item (e.g., a resistor, fastener, bracket, cable). In some instances, the component may be newly developed and require detailed design activity. The supplier will design and produce the component in the quantities desired. In other cases, the supplier will serve as a manufacturing source for the component. The supplier will produce the desired quantities from a given data set. In other words, the component design has been completed (whether by the same firm or some other), and the basic service being provided is "production." A third scenario may involve the supplier as being an inventory source for one or more common and standard commercial off-the-shelf (COTS) components. There are no design or production activities, but just distribution and materials handling functions. As one can see, individual supplier roles can vary significantly.

In the process of identifying and selecting suppliers to provide system components, there are a number of issues that must be addressed. In addition to having the right component and demonstrating a desired level of technical competency, there are economic considerations, political considerations, environmental considerations, and so on. Suppliers are often selected based on geographic location with economic need in mind. It may be desirable to establish a new manufacturing capability to help stimulate the economy in a depressed area. Suppliers may be selected due to their location within a politician's jurisdiction. The selection of a supplier may be based on an environmental issue, particularly if there is a production process that impacts the environment in a detrimental manner. More recently, there has been a significant trend toward "globalization," and suppliers have been selected based on nationality. This globalization and international exchange is likely to expand further in view of the growth in activities along the Pacific Rim, with the advent of the European Union, and when considering the technology advances in communications, data processing methods, and transportation systems.

The quantity of suppliers and the nature of their activities are a function of the type and complexity of the system being developed. For large, highly complex systems, there may be many different suppliers located throughout the world, as reflected in Figure 7.17. Some of these suppliers may be heavily involved in the design and development of major elements of the system, whereas other suppliers serve as sources for manufacturing and for selected inventories. For a system of this type, there is usually a large variety and mix of activities.

With regard to system engineering, the requirements specified by the customer and imposed on the system prime contractor must be passed on to the various suppliers as applicable. The contractor will prepare an appropriate specification (Type "B," "C," "D," or "E") for each supplier requirement, along with a supporting Statement of Work (SOW) with specific supplier program tasks identified. Supplier responses, in the context of formal proposals, are submitted to the contractor and evaluated, selections are made, contracts are negotiated, and programs are implemented as a result. This process and related areas of activity are discussed further in Chapter 8.

In determining system engineering task requirements for large programs where design and development are required, a review of both the basic tasks in Figure 6.6 and the proposed tasks as they are applied to different program phases in Section 7.4.5 is recommended. From this information, specific task requirements as they apply to the supplier organization can be developed. A sample listing of the task requirements for a large supplier is presented as follows:

1. Conduct feasibility studies and define design criteria for the system component(s) being developed. This information is based on system operational requirements, the maintenance concept, the functional analysis, and the allocation of re-

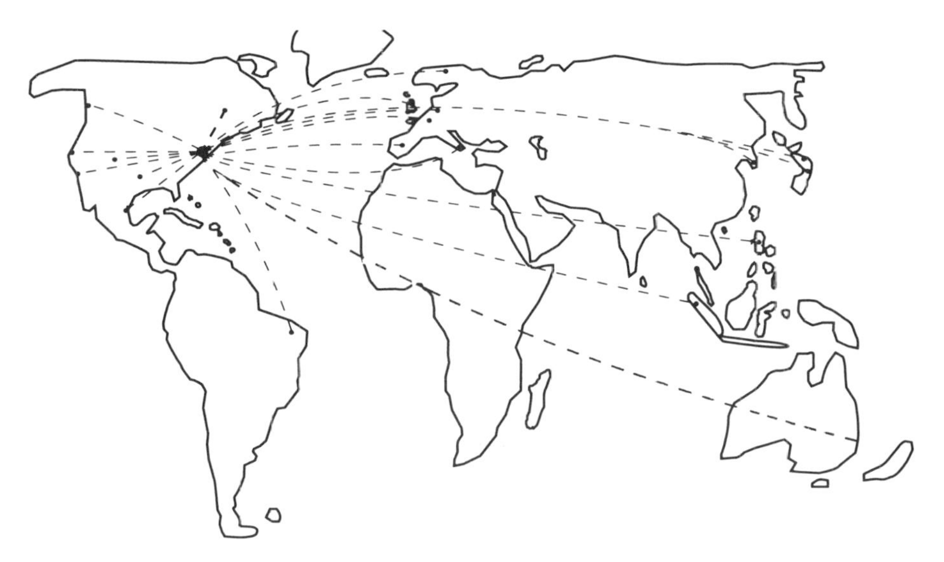

Figure 7.17 Potential suppliers for System "XYZ."

quirements as it applies to the supplier-produced item. Supplier requirements are also included in the development specification (Type "B") as applicable. Refer to Sections 2.3 through 2.8.

2. Prepare the supplier engineering plan, or equivalent. This plan is an expansion of the SEMP, and must reflect those requirements that support system engineering objectives. Refer to Section 6.2.

3. Accomplish analysis, synthesis, and trade-off studies in support of component design decisions, and as they impact higher-level system requirements. Refer to Section 2.9.

4. Accomplish design integration; that is, the ongoing integration of design disciplines, activities, and data. Refer to Section 2.10.

5. Prepare the test and evaluation plan covering the system component being developed. Integrate testing activities into system-level testing requirements where feasible. Monitor, review, and evaluate component testing activities conducted at the supplier's facility. Refer to Section 2.11.

6. Participate in equipment/software design reviews and the critical design review; that is, formal design reviews that cover the system component being developed and its interfaces with other elements of the system. Refer to Chapter 5.

7. Plan, coordinate, and monitor proposed design changes as they apply to the component and impact the overall system. Refer to Section 5.4.

8. Initiate and maintain liaison with production/manufacturing activities that support the system component.

9. Initiate and maintain liaison with the system contractor throughout all phases of the program when supplier activities are in progress.

Although large projects may require the supplier to complete all of these tasks, the level of activity will obviously be scaled down for smaller projects. For suppliers involved in production or manufacturing only, the system engineering focus is directed primarily to total quality management (TQM). It is important to ensure that the characteristics initially designed into system components are maintained through the subsequent production of multiple quantities of these items. As the production process is dynamic by nature and material substitutions often occur, it is necessary to guarantee that each of the components produced does indeed reflect the quality characteristics initially specified through design (refer to Section 3.3.6). Thus, close communications must be established with the supplier's quality control or quality assurance organization.

In dealing with the suppliers of standard off-the-shelf components, it is important to prepare a good specification for the initial procurement of these items. Input-output parameters, size and weight, shape, density, and so on, must be covered in detail along with allowable tolerances. Uncontrolled variances in component characteristics can have a significant impact on total system effectiveness and quality. Complete electrical, mechanical, physical, and functional interchangeability must be maintained where applicable. Although there are many different components that are currently in inventory and fall under the category of "commercial off-the-shelf (COTS)" components, extreme care must be exercised to ensure that like components (i.e., those with the same part number) are actually manufactured to the *same* standards. Also, the allowable variances around key parameters must be minimized.

In addressing the category of "supplier," one often finds a layering effect as illustrated in Figure 7.18. There is a wide variety of suppliers with varying objectives and organizational patterns. Although many are functionally oriented, the practices and procedures of each will be different. With regard to system engineering per se, it is unlikely that the supplier organization will include a department, group, or section identified as such! However, accomplishment of the functions described previously (as applicable) is necessary. Although the overall responsibilities for system engineering are assigned to the prime contractor, there must be some identifiable organizational element responsible for the *technical integration* of those tasks assigned to the supplier. It is important that system engineering concepts and objectives be established and understood from the beginning, and the contribution of each supplier is essential to the realization of these objectives.

From the perspective of the prime contractor, the organization and management of supplier activity present major challenges! Not only are there many different types of suppliers with varying levels of responsibility, but these suppliers may be located across the United States, Canada, Mexico, Africa, Asia, Australia, Europe, South America, and so on. This is particularly true for large projects where the supplier is responsible for major elements of the system.

In such instances, the system engineer must ensure that the appropriate level of design integration is maintained throughout the system development process. This commences with the initial definition of supplier requirements and the preparation of specifications. Subsequently, design integration is realized through good commu-

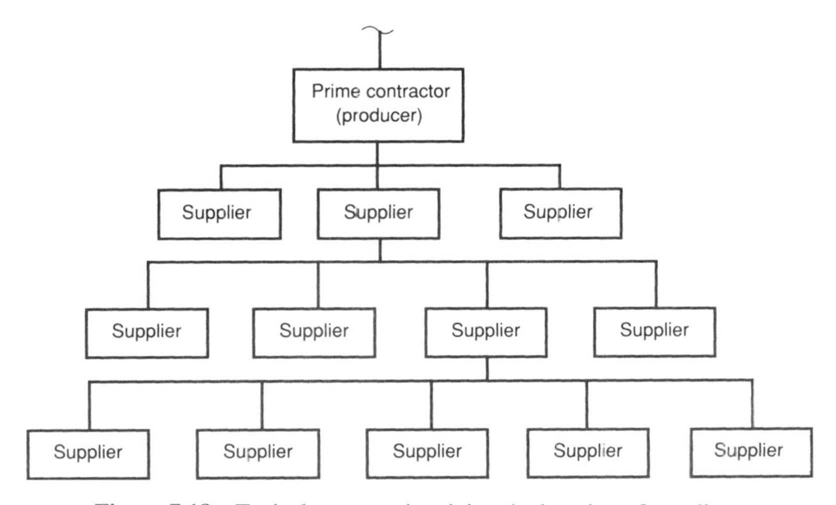

Figure 7.18 Typical structure involving the layering of suppliers.

nications and the periodic review of design documentation, the conductance of design reviews, and so on. The establishment of successful communications is usually dependent on the system engineer's complete understanding of the environment in which the supplier operates. For instance, in the international arena, the system engineer should have a basic understanding of the culture, customs and practices, export–import requirements, geographic layout, communication and transportation links, and available resources within each of the nationalities providing system components. Some knowledge of the language, and interpretations thereof, as it is applied in specifications and design data is important. In any event, this area of activity is increasing in importance as the emphasis on globalization becomes greater.

7.6 HUMAN RESOURCE REQUIREMENTS

To complete our discussion of the organizational elements of system engineering, it is necessary to provide some coverage of human resource requirements, the staffing of a system engineering organization, and personnel development. Although there will be variations in each situation, there are certain objectives that should be met from the employee standpoint.

7.6.1 Creating the Organizational Environment

The nature of system engineering activities requires consideration of the following characteristics when developing an organizational structure:

1. The personnel selected for the system engineering group must, in general, be highly professional senior-level individuals with varied backgrounds and a wide breadth of knowledge—for example, an understanding of research, design, manufacturing, and system support applications. The emphasis is on overall system-level design and technology applications, with knowledge of user operations and sustaining life-cycle support in mind.
2. The system engineering group must incorporate "vision" and be "creative" in the selection of technologies for design, manufacturing, and support applications. Group personnel are constantly searching for new opportunities, must be innovative, and applied research is often required in order to solve specific technical problems.
3. A teamwork approach must be initiated within the system engineering group. The personnel assigned must be committed to the objectives of the organization, there is a certain degree of interdependence required, and there must be mutual trust and respect.
4. A high degree of communications must prevail, both within the system engineering group and with the many other related functions associated with a given project (refer to Figure 7.15). Communications is a two-way process and may be accomplished via written, verbal, and/or nonverbal means. Good communications must first exist within the system engineering organization. With that established, it

is then necessary to develop two-way communications externally, using both vertical and horizontal channels as required.[7]

Given the objectives described throughout this text, and with these considerations in mind, the appropriate "environment" must be created to allow for the accomplishment of system engineering tasks in an effective manner. Environment in this instance refers both to (1) the working environment external to the system engineering function but within the contractor's organizational structure, and (2) the working environment within the system engineering group itself.

The creation of a favorable environment within the contractor's organization, or within any other organizational structure being addressed, must start from the *top!* The president, or general manager, must initially believe in and subsequently support the concepts and objectives of system engineering. On numerous occasions, power struggles will occur, conflicting organization goals and objectives will develop, and there will be a lack of communication between the key organizational entities. This, in turn, usually leads to redundancies, the hiring of unqualified personnel, and the expenditure of unnecessary resources, resulting in waste. A mechanism must be established for quick conflict resolution, and all project personnel must know that the system engineering philosophy *will* prevail, in spite of individual interests. Top management must create this understanding from the beginning!

Additionally, the appropriate level of responsibility and authority must be delegated from the top, through the vice president of engineering and the program manager, and to the system engineering department manager (refer to Figure 7.13). Responsibilities must be defined, and the commensurate level of authority to control and direct the means by which the activity is to be accomplished must be delegated accordingly. Quite often, a manager is willing to delegate the responsibility for a particular activity to a subordinate, but will retain the authority for controlling the resources necessary for task completion. In this situation, the individual assigned the responsibility is powerless when it comes to the full utilization of available resources and, when things go wrong, is unable to respond relative to initiating the appropriate corrective action. He (or she) becomes discouraged, loses motivation, and the level of productivity ultimately falls off. In essence, the system engineering manager must be given the *responsibility,* the *authority,* and the *resources* to do the job assigned.

As a third point, the proper relationships must be established between the vice president of engineering, the program manager, and the system engineering manager, as identified in Figure 7.13. These relationships are, of course, heavily dependent on the managerial styles of the individuals in these positions. Although there are many variations, the two managerial styles most often discussed are the "autocratic" approach and the "democratic" approach.

The autocratic concept is basically dictatorial in nature, and is restrictive in that

[7] Basic organizational communication processes are discussed in detail in many texts covering organizational theory or organizational design. One good reference is R. W. Griffin, and G. Moorhead, *Organizational Behavior,* Houghton Mifflin, Boston, 1986.

decisions are generally unilateral; that is, decisions are made from the top down without input from those who are required to carry out these decisions. Managers control, direct, coerce, and even threaten employees to force them to work toward specific organizational goals. On the other hand, the democratic concept is participative, nonthreatening, and organizational interests are group-centered. The general theme is that individuals in the group have some voice in matters that directly affect them.[8]

Although both styles of management are prevalent for certain situations, the democratic style has been accepted as being more effective from the motivational standpoint. In general, people work harder, are more cooperative, and are more willing to accept changes if they feel that they have some influence in the results. Democratic leadership implies an organizational environment in which employees have a chance to grow and develop their skills, where formal supervision is considerate and the application of dictates is not arbitrary, and where individual opinions are solicited and respected. As compared to the autocratic approach, management is committed to the recognition of employees as high-level professionals and not merely as factors in the production scheme.[9]

With regard to system engineering, an environment must be created that will allow for individual initiative, creativity, flexibility, personnel growth, and so on. A democratic and participative approach seems appropriate in meeting the objectives stated. Although the manager must maintain authority and provide the necessary direction and control to effectively accomplish the organization's goals and objectives, he or she can introduce some practices that directly support the democratic style. The selected style, it is hoped, will help to create a favorable environment for the accomplishment of system engineering tasks. This is influenced not only by the managerial style employed within the system engineering group itself, but the approach employed by higher-level management that has an impact on the group. Creating such an environment is critical to the objectives stated throughout this text. There are examples in which two (or more) similar organizations have the same basic objectives, the same structure, the same position titles, and so on—but one is productive and the other is not! High-level productivity is a function of the working environment.

7.6.2 Leadership Characteristics

The system engineering organization is composed of a group of individuals with varying abilities, different roles and expectations, diverse personal goals, and dis-

[8] These concepts are related to the managerial views described as "Theory X" and "Theory Y" by Douglas McGregor in his classic book, *The Human Side of Enterprise,* McGraw-Hill, New York, 1960. This is recommended reading for a more in-depth understanding of the human relations movement developed in the 1940s.

[9] A review of the literature on human motivation is recommended at this point. Three good references are (1) A. H. Maslow, "A Theory of Human Motivation," *Psychological Review* (July 1943); (2) F. Herzberg, B. Mausner, and B. Snyderman, *The Motivation to Work,* John Wiley, New York, 1959; (3) and M. S. Myers, "Who Are Your Motivated Workers," *Harvard Business Review* (1970).

tinct behavioral patterns. Although individuals within the organization are highly dependent on one another, they often push in different directions because of some of these factors. The challenge for the system engineering manager is to integrate these various characteristics into a cohesive force, leading to the accomplishment of organizational objectives. The manager must not only ensure that the job is completed in a satisfactory manner, but it is hoped that he (or she) will inspire and motivate his (her) subordinates to *excel* in the fulfillment of organizational objectives. It is apparent that the manager must create a climate that presents challenges (not threats) for the individuals involved.

In facilitating this objective, the manager should initiate certain practices that are responsive to both the organization and the individual needs of people within the organization. As a start, the democratic style of leadership, discussed in Section 7.6.1, tends to promote the necessary environment for the organization. Within this framework, the manager should encourage the participation of individuals in goal setting and in decision making, and should promote communications. By doing so, he (or she) will not only tend to improve the motivation from within, but will acquire a better understanding of the individuals in the organization. This understanding can be enhanced by taking the following steps:

1. Recognize the personal characteristics of each individual in the organization in order to better match the individual with the job requirements. A person may excell in one situation while doing only a mediocre job in another situation, even though the ability to do both and the overall organization climate remain relatively constant. In essence, the manager must assign employees to the type of work they do most effectively. A high-quality output is essential if the system engineering organization is to gain the respect and retain a leadership role on a project.
2. Inspire each individual to excell in his/her job by creating an atmosphere of personal interest. An employee will tend to perform better if he/she knows that the boss is personally interested. The personal interest is developed through involvement at the employee level.
3. Be sensitive to employee problems such that each can be addressed on personal terms. The solution to a problem should, if at all possible, consider the effects on the individual employee.
4. Evaluate employees on a personal basis and initiate rewards promptly when warranted. Promotions and merit raises should not be oriented to the organizational hierarchy alone, but should be directed to the best performers.

Good communications and a good rapport with employees must be established from the beginning. Obtaining the desired organization environment not only depends on the intent of the manager in initially establishing such practices, but is highly dependent on his/her personal leadership ability in directing the activities of the organization over time. The best planning in the world will have little benefit unless the actions that follow in the implementation phase are directly supportive.

Leadership Characteristics
1. Acceptance: Earns respect and has the confidence of others. 2. Accomplishment: Effectively uses time in meeting goals and objectives. 3. Acuteness: Mentally alert and readily comprehends instructions, explanations, and unusual circumstances. 4. Administration: Organizes his own work and that of his subordinates; delegates responsibility and authority; measures, evaluates, and controls position activities. 5. Analysis and judgment: Performs critical evaluation of potential and current problem areas; breaks problem into components; weighs alternatives; relates, and arrives at sound conclusions. 6. Attitude: Enthusiastic; optimistic; and loyal to firm/agency, superior, position, and associates. 7. Communication: Promotes communication within and between organization elements. 8. Creativeness: Has inquiring mind; develops original ideas; and initiates new approaches to problems. 9. Decisiveness: Makes prompt decisions when necessary. 10. Dependability: Meets schedules and deadlines in consistent manner and adheres to firm/agency policies and procedures. 11. Developing others: Develops competent successors and replacements. 12. Flexibility: Adaptable; quickly adjusts to changing conditions,; and copes with the unexpected. 13. Human relations: Is sensitive to and understands personnel interactions; has "feel" for individuals and recognizes their problems; considerate of others; ability to motivate and get people to work together. 14. Initiative: Self-starting; prompt to take hold; seeks and acts on new opportunities; exhibits high degree of energy in work; not easily discouraged, and possesses basic urge to get things done. 15. Knowledge: Possesses knowledge (breadth and depth) of functional skills needed to fulfill position requirements; uses information and concepts from other related fields of knowledge; and generally understands the "big picture." 16. Objectivity: Has an open mind and makes decisions without the influence of personal or emotional interests. 17. Planning: Looking ahead; developing new programs; preparing plans; and scheduling requirements. 18. Quality: Accuracy and thoroughness of work; and maintains high standards consistently. 19. Self-confidence: Self-assurance; inner security; self-reliant; and takes new developments in stride. 20. Self-control: Calm and poised, under pressure. 21. Self-motivation: Has well planned goals; willingly assumes greater responsibilities; realistically ambitious; and generally is eager for self-improvement. 22. Socialness: Makes friends easily; works well with others; and has sincere interest in people. 23. Verbal ability: Articulate; communicative; and is generally understood by persons at all organizational levels. 24. Vision: Possesses forsight; sees new trends and opportunities; anticipates future events; and is not bound by tradition or custom.

Figure 7.19 Checklist of leadership characteristics.

As a goal, the manager should strive to exhibit the characteristics listed in Figure 7.19.

7.6.3 The Needs of the Individual

With the discussion thus far primarily covering the organization environment and the desired characteristics of leadership, it is now important that some attention be given to the needs of the individual employee. If the manager is to inspire, motivate, and deal successfully with subordinates, then a good understanding of these needs is necessary.

As a starting point, the reader should first review A. H. Maslow's theory dealing with the hierarchy of needs. The theory was developed to identify relatively separate and distinct drives that motivate individuals in general. The following five basic needs are identified:[10]

1. The physiological needs such as thirst, hunger, sex, sleep, and activity. These constitute the needs of the body and, unless these needs are basically satisfied, they remain as the prime influencing factors in the behavior of an individual.
2. The safety and security needs, which include protection against danger, threat, and deprivation. Having satisfied the bodily needs, the safety and security needs become a dominant goal.
3. The need for love and esteem by others, or social needs. This includes the belonging to groups, giving and receiving friendship, and the like.
4. The need for self-esteem or self-respect, and for the respect of others (i.e., ego need). An individual wishes to consider himself/herself strong, able, competent, and basically worthy by his/her own standards.
5. The need for self-fulfillment or the achieving of one's full potential through self-development, creativity, and self-expression. This relates to man's desire to grow, develop to the point of full potential, and ultimately attain the highest level possible.

The needs are related to each other, and are arranged in an order that will stimulate consciousness and activity. As a need becomes satisfied, activity emphasis shifts to the next need category. In other words, a satisfied need is no longer a motivator, and the next need becomes a driving factor.

While Maslow addressed the overall hierarchy of needs from a general standpoint, Herzberg conducted research, resulting in his "motivation-hygiene" theory, which identifies factors commonly known as "satisfiers" and "dissatisfiers." The theory was developed from research pertaining to the job attitudes of 200 engineers and accountants, and was based on two questions: (1) "Can you describe, in detail, when you felt exceptionally *good* about your job?" and (2) "Can you describe, in detail, when you felt exceptionally *bad* about your job?" The results were classified in the two categories identified in what follows:[11]

1. *Satisfiers:*
   (a) Achievement—personal satisfaction in job completion and problem solving.
   (b) Recognition—acknowledgment of an accomplishment (e.g., a job well done).

[10] A. H. Maslow, "A Theory of Human Motivation," *Psychological Review* (July 1943).

[11] F. Herzberg, "Work and Motivation," in *Studies in Personnel Policy Number 316: Behavioral Science, Concepts and Management Application,* National Industrial Conference Board, New York, 1969.

   (c) Work itself—actual content of the job and its positive/negative effect on the employee.
   (d) Responsibility—both responsibility and authority in relation to the job.
   (e) Advancement—promotion on the job.
   (f) Growth—learning new skills offering greater possibility for advancement.
2. *Dissatisfiers:*
   (a) Company policy and administration—feelings about the adequacy or inadequacy of company organization and management, policies and procedures, and so on.
   (b) Supervision—competency or technical ability of supervision.
   (c) Working conditions—physical environment associated with the job.
   (d) Interpersonal relations—relations with supervisors, subordinates, and peers.
   (e) Salary—pay and fringe benefits.
   (f) Status—miscellaneous items such as size of office, having a secretary, and a private parking place.
   (g) Job security—tenure, company stability or instability.
   (h) Personal life—personal factors that affect the job (e.g., family problems, social problems).

Most of these factors have some bipolar affects. For instance, "advancement" certainly is a *satisfier* when it happens and may be somewhat of a *dissatisfier* when it does not occur. However, this category weighs much heavier as a satisfier. The item, "salary," is definitely a dissatisfier when pay scales and fringe benefits are poor, and is a mild satisfier when the compensation is good. In this case, the predominant classification for salary is that of a dissatisfier. In any event, all of the factors listed represent needs of the individual and should be considered by management.

Myers conducted research involving 282 subjects (including engineers, scientists, and technicians) interviewed at Texas Instruments commencing in 1961. The categories used were "motivators" and "dissatisfiers," and results as they pertain to engineers are noted in what follows. The items listed and identified with a "M" are clearly motivators, and those with a "D" are definitely dissatisfiers. For instance, the item of "pay" is clearly a dissatisfier if considered as being inadequate, and "advancement" is definitely a motivator when it occurs. Once again, there are bipolar affects; however, the categorization does indicate where the greatest impact occurs.[12]

1. Work itself (M)
2. Responsibility (D)
3. Company policy and administration (D)
4. Pay (D)
5. Advancement (M)

[12] M. S. Myers, "Who Are Your Motivated Workers," *Harvard Business Review* (1970). This study employs factors comparable to those used by Herzberg.

6. Recognition (D)
7. Achievement (M)
8. Competence of supervision (D)
9. Friendliness of supervision (D)

In summary, the needs of the individual employee will vary somewhat, depending on his (or her) situation. If one need is satisfied, then another need becomes predominant, and so on. In addition, these needs are often related to the business position of the company where he/she is employed (or the success of the organization overall). If the firm is in a growth posture, the individual's perceived needs may be somewhat different than if the firm is experiencing a business decline and the prospect of laying off employees is apparent. Finally, the manager's job is twofold. He/she must (1) be aware of the individual needs in the organization and create the necessary conditions for employee motivation, and (2) satisfy those needs on a continuing basis to the extent possible. Human motivation is a key to organizational success, and an understanding of the concepts in this section should help in meeting this objective.

7.6.4 Staffing the Organization

The requirements for staffing an organization initially stem from the results of the system engineering planning activity described in Chapter 6. Tasks are identified from both short- and long-range projections (refer to Figure 6.24), combined into work packages and the work breakdown structure (WBS), and the work packages are grouped and related to specific position requirements. The positions are, in turn, arranged within the organizational structure considered to be most appropriate for the need (refer to Figures 7.3 through 7.15).

With regard to specific position requirements for a system engineering organization, one should first have a good understanding of the basic functions of the organization. These are discussed throughout the earlier chapters of this text and, more specifically, in Chapter 6. Review of the assigned tasks, the nature and challenges of the organizational structure, and so on, indicate that, in general, an entry-level "System Engineer" should have the following:

1. A basic formal education at the undergraduate and graduate levels in some recognized field of engineering; that is, a masters degree in engineering or equivalent.[13]
2. A high level of general technical competence in the engineering fields being pursued by the organization, project, and so on.

[13] Recognized accredited programs in engineering are defined by the Accreditation Board for Engineering and Technology (ABET), United Engineering Center, 345 East 47th Street, New York, NY 10017. Refer to the latest *Annual Report.*

3. Relevant design experience in the appropriate areas of activity. For example, if the company is involved in the development of electrical/electronic systems, then it is desirable for the candidate to have had some prior design experience in electrical/electronic systems. A different type of experience would be required for aeronautical systems, civil systems, hydraulic systems, and so on.
4. A basic understanding of the design requirements in areas such as reliability engineering, maintainability engineering, human factors, safety engineering, logistics engineering, software engineering, quality engineering, and value/cost engineering.
5. An understanding of the system engineering process and the methodologies/tools that can be effectively employed in bringing a system into being; for example, the definition of system requirements and functional analysis and allocation.
6. An understanding of the relationships among functions to include marketing, contract management, purchasing, integrated logistic support, configuration management, production (manufacturing), quality control, customer and supplier operations, and so on.

As the specific definition of a "System Engineer" will often vary from one organization to the next, individual perceptions as to the requisites will differ! Based on experience, it is believed that a good solid technical engineering education is a necessary foundation, some design experience is essential, a thorough understanding of the system life cycle and its elements is required, and knowledge of the many design interfaces is appropriate. If an individual is to successfully implement the functions identified in Chapter 6 (Figure 6.6), then some prior experience in these areas is highly recommended.

Given the basic requisites, the system engineering department manager will prepare an individual position description for each open slot in the organization. A sample position description format is illustrated in Figure 7.20. The position title, responsible supervisor, areas of responsibility and job objectives, background requirements, and the date of need should be clearly identified. The system engineering position requirements are completed and forwarded to the human resources department (or equivalent) in order to proceed with the necessary steps for recruiting and employment.

In staffing the organization, possible sources include (1) qualified personnel from within the company and ready for promotion, and (2) personnel from outside and available through the open market. It is the responsibility of the system engineering department manager to work closely with the human resources department in establishing the initial requirements for personnel, in developing position descriptions and advertising material, in recruiting and the conducting of interviews, in the selection of qualified candidates, and in the final hiring of individuals for employment within the system engineering organization. In the process of conducting interviews and selecting system engineering personnel, the characteristics identified in Figure 7.20 should be kept in mind.[14]

7.6.5 Personnel Development and Training

Nearly every engineer wants to know how he (or she) is doing on a day-to-day basis and what are the opportunities for growth? Response to the first part of the question is derived through a combination of the "formal performance review," which is often conducted on a regularly scheduled basis (either semiannually or annually), and through the ongoing day-to-day "informal communications process" with the boss. The engineer is given responsibility and seeks recognition and approval from

Date of need:

Position title:

Senior system engineer

Supervisor:

System Engineering Department Head

Broad Function:

Responsible for the performance of system engineering functions in the design and development of communication products.

Functional Objectives:

1. Perform system feasibility studies and evaluate alternative technology applications.
2. Develop operational requirements and maintenance concepts for new communication systems/ equipments.
3. Interpret and translate system level requirements into functional design requirements.
4. Prepare system and subsystem specifications and plans.
5. Accomplish system integration activities (to include supplier functions).
6. Determine requirements and conduct formal design reviews for all system elements.
7. Prepare system test and evaluation requirements, monitor test functions, and evaluate test results to determine system performance and effectiveness. Make recommendations for corrective action and/or improvement as appropriate.
8. Provide assistance to marketing in product sales activities, and fulfill customer service requirements as necessary.

Requirements:

Degree in electrical engineering (masters degree or higher and some training in managment skills and practices), plus at least ten years of experience in communication systems design.

Figure 7.20 Sample position description.

[14]The human resources department in most companies is responsible for establishing job classifications and salary structures, for the recruiting and hiring of personnel, for initiating employee benefit coverage, for providing employee opportunities for education and training, and so on. It is incumbent on the system engineering department manager to ensure that his/her organizational requirements are initially understood and subsequently met through recruiting, employment, and training activities.

the supervisor. As discussed in Section 7.6.4, there needs to be close communications, and the boss needs to provide some reinforcement that he (or she) is doing a good job. Also, the employee needs to know as soon as possible when his (her) work is unsatisfactory and improvement is desired. Waiting until the formal performance review is conducted to learn that one's work is not satisfactory is a poor practice and demoralizing because, by virtue of not having heard any comments to the contrary, it has been assumed that all is well! In a system engineering organization, it is particularly important that the appropriate close level of communications is established from the beginning.

The second question pertaining to the opportunities for growth depends on (1) the climate provided within the organization and the actions of the manager that allow for individual development, and (2) the initiative on the part of the engineer to take advantage of the opportunities provided. Within a system engineering department, it is *essential* that individual personal growth take place if that department is to function effectively. The climate (or environment) must allow for individual development, and the individual system engineer must seek opportunities accordingly. The system engineering department manager should work with each employee in preparing a tailored *development plan* for that employee. The plan adapted to each person's specific needs should allow for (and promote) personal development by providing a combination of the following:

1. Formal internal training designed to familiarize the engineer with the policies and procedures applicable to the overall company as a whole, as well as the detailed operating procedures of his/her own organization. This type of training should enable the individual to function more successfully within the framework of the total organization through familiarization with the many interfaces that he/she will encounter on the job.

2. On-the-job training through selective project assignments. Although the extensive shifting of personnel from job-to-job (or project-to-project) can be detrimental, it is sometimes appropriate to reassign an individual to work where he/she is likely to be more highly motivated! Every employee needs to acquire new skills, and occasional transfers may be beneficial as long as the overall productivity of the organization does not suffer.

3. Formal technical education and training designed to upgrade the engineer relative to the application of new methods and techniques in his/her own field of expertise. This pertains to the necessity for the engineer to maintain currency (and avoid technical obsolescence) through a combination of (a) continuing education short courses, seminars, and workshops; (b) formal off-campus graduate engineering programs provided at the local level (leading to an advanced degree); and (c) long-term training involving opportunities for research and advanced education on some university or college campus. The opportunities for acquiring continuing education, while on the job, are greater now than ever before with the availability of satellite TV, two-way compressed video (VTEL/PICTEL), video tapes, and computer-based delivery capabilities. If an individual is motivated, he/she can acquire a great deal of support in this area.

4. A technical exchange of expertise with others in the field through participation in technical society activity, industry association activity, symposia and congresses, and the like.

The system engineering manager must recognize the need for the ongoing development of personnel in his/her organization, and should encourage each individual to seek a higher level of performance by offering not only challenging job assignments, but opportunities for growth through education and training. The long-term viability of such an organization is highly dependent on personnel development. This, in turn, should enhance individual motivation and result in the fulfillment of system engineering functions in a high-quality manner.

QUESTIONS AND PROBLEMS

1. Describe what is meant by "organization"? What are its characteristics, objectives, and so on?
2. There are various types of organizational structure to include the "pure functional," the "product line," the "project," and the "matrix." Briefly describe the structure and identify some of the advantages and disadvantages of each.
3. Refer to Question 2. Which type of organizational structure is preferred from a system engineering perspective? Why?
4. Refer to Figure 7.2. Where is system engineering accomplished? Who is responsible for the accomplishment of system engineering functions? Identify some of the major concerns associated with the organizational relationships shown in the figure.
5. Assume that you have been assigned the responsibility for the design and development of a new system. Develop a project organizational structure (using any combination of approaches desired), identify the major elements contained therein, and describe some of the key interfaces; describe how you plan to organize for system engineering. Construct an organizational chart.
6. Refer to Question 5. Describe the system engineering functions (or tasks) that should be accomplished.
7. From an organizational standpoint, identify and describe some of the conditions that must exist in order to accomplish system engineering objectives in an effective manner.
8. Suppose that you have just been assigned to develop a System Engineering Department for a new program:
   (a) What type of an organizational structure would you develop (given a choice)?
   (b) What policies and procedures would you implement?

(c) What type of people would you need in terms of quantity, skill levels, individual backgrounds, and so on?

(d) What type of management style would you impose? Describe some of the characteristics.

9. In terms of "organizational environment" for system engineering, what factors need to be considered? Briefly describe the "organizational environment" that is appropriate for the successful implementation of system engineering functions.

10. Describe some of the major task interfaces between the system engineering organization and contract management? Purchasing? Production (manufacturing)? Integrated logistic support (ILS)? Configuration management? Marketing and sales?

11. Refer to Section 7.5. Describe some of the major challenges associated with supplier management.

12. In terms of styles of management, what is meant by "Theory X" and "Theory Y"? Which is preferred relative to the system engineering application?

13. Assume that you, as the Vice President of Engineering, are looking for a new System Engineering Department Manager. What leadership characteristics would you identify as being critical (identify in order of importance)?

14. Refer to Figure 7.19. Select and list in order of importance the top 10 characteristics based on your experience.

15. Review the results of Herzberg's research in Section 7.6.3. From your own experience, list the "satisfiers" and the "dissatisfiers" in order of importance. Show additional factors as you see fit. A bar chart showing the bipolar relationships is a good way to present your thoughts.

16. Based on your own perspective, describe the characteristics of a "System Engineer" (background, personal characteristics, motivational factors, etc.).

17. As Manager of the System Engineering Department, what steps would take to ensure that your organization maintains a lead position relative to technical competency?

8 Supplier Evaluation, Selection, and Control

The term *supplier* refers to a broad class of external organizations that provide products, components, materials, and/or services to the producer (or prime contractor). This may range from the delivery of a major subsystem or configuration item down to a small component part. In Figure 7.2, suppliers may provide services to include the design, development, and manufacture of a major element of a system; the production and distribution of items already designed (providing a manufacturing source); the distribution of commercial and standard component parts from an established inventory (serving as a warehouse and providing parts from various sources of supply); and/or the implementation of a process in response to some functional requirement. For many systems, suppliers provide a large number of the elements that make up the system (i.e., over 50% of the components in some instances), as well as the spares and repair parts that are required to support maintenance activities.

From Figure 7.18, for large programs, there may be a *layering* of suppliers, with one or more component suppliers providing services to the supplier of a configuration item or major subsystem. Additionally, recent trends in the economy have favored the practices of increased "outsourcing," or the process of seeking external sources of supply as compared to the completion of work internally with a producer's organization. This, along with the current degree of emphasis in *globalization,* has resulted in suppliers being selected from various geographical centers around the world (refer to Figure 7.17).

With the involvement of many different suppliers in the design, development, manufacture, and support of systems, there is an ever increasing need for the implementation of system engineering practices and techniques. Major suppliers, as key participants in the design process, must be involved from the beginning. The System Engineering Management Plan (SEMP) must include the coverage of supplier functions and activities. The System Specification (Type "A") must provide a good functional baseline, from which the various lower-level specifications can be developed. In essence, major suppliers must be brought into the early design process, participate as members of the design team, and must be commited to the system engineering process.

Chapters 6 and and 7 provided an introduction to supplier activities within the context of the requirements for a program as an entity. This chapter, which is an extension of these two earlier chapters, presents additional material highlighting the evaluation and selection of suppliers, the contracting for supplier services, and the follow-on supplier monitoring and control activities.

8.1 PROGRAM REQUIREMENTS

Review of the system engineering process described in Chapter 2 illustrates a number of steps commencing with the identification of a consumer need and extending through the definition of system operational requirements, the maintenance concept, the identification of technical performance measures, functional analysis and allocation, and the preparation of the system specification (Type "A"). These steps are shown in Figure 8.1.

As shown in the figure, the system is described in *functional* terms identifying the "WHATs," and each functional entity is evaluated and trade-off studies are conducted with the objective of determining "HOW" the function(s) can best be accomplished (refer to Section 2.7, Chapter 2). The basic question in each instance is: Should the function be accomplished through the application of equipment, the use of software, the utilization of human resources, or through a combination of these? The results of these trade-off studies are presented in the form of specific resource requirements.

The next step is to identify the source of supply, or from "WHERE" can these required resources be obtained? Should the design or manufacture of an item of equipment, the development of a software package, or the completion of a process be accomplished *in-house* by the producer or prime contractor, or should an *external* source of supply be selected? In many industrial organizations, a "Make-or-Buy" Committee is established to evaluate the alternatives and select a preferred approach.

In such instances, the Make-or-Buy Committee, or some equivalent activity within the producer's organization, is established with representation from program management, engineering, logistics, manufacturing, purchasing, and quality assurance. Engineering participation should include the system engineering organization and the appropriate design disciplines. Decisions are made based on an evaluation of a combination of factors such as the criticality of need (when is the item required?), item complexity, the availability of internal technical capabilities and required resources versus the use of potential outside suppliers, related social and political factors, and cost.[1] From the system engineering perspective, items that are relatively complex in nature, involving the application of new technologies, and are critical to the overall system development effort should be handled in-house if at all possible. These items will, in all likelihood, require frequent monitor-

[1] On certain occasions, decisions may be based on social, economic, and/or political considerations such as the identification of a need to improve the local economy by selecting a supplier in a given geographical area; the desire to increase the amount of subcontracting; the need to establish a manufacturing and/or support capability in a designated foreign nation; the need to respond to an existing unemployment crisis; the desire to support a given political position; and so on.

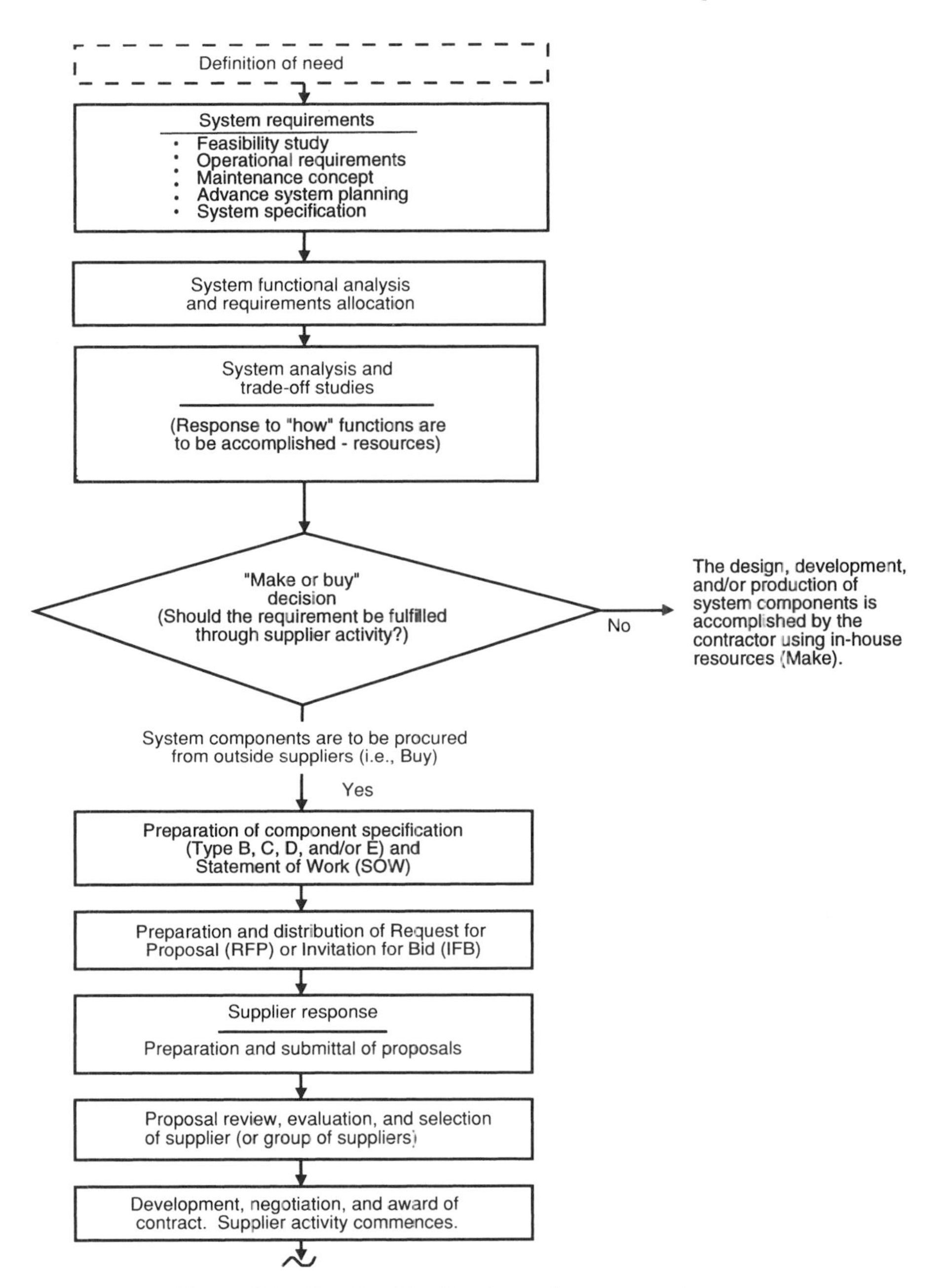

Figure 8.1 Supplier identification and procurement process.

ing and the application of tight controls (both management and technical), which may be difficult to accomplish should a remotely located supplier be selected for the task.

As shown in Figure 8.1, the results of the Make-or-Buy Committee will lead to a recommendation as to the source of supply: Should a product or service be developed within the producer's organization or should it be procured from an outside source? Ultimately, after a number of requirements are evaluated in a similar manner, the system development effort may include a wide variety of suppliers participating in varying capacities. Figure 8.2 identifies the candidates for "outsourcing" for a selected system. Note that suppliers are involved in the design and development of a major subsystem and lower-level components, the delivery of commercial off-the-shelf (COTS) items, the delivery of materials, and in providing a service (i.e., completion of a process).[2]

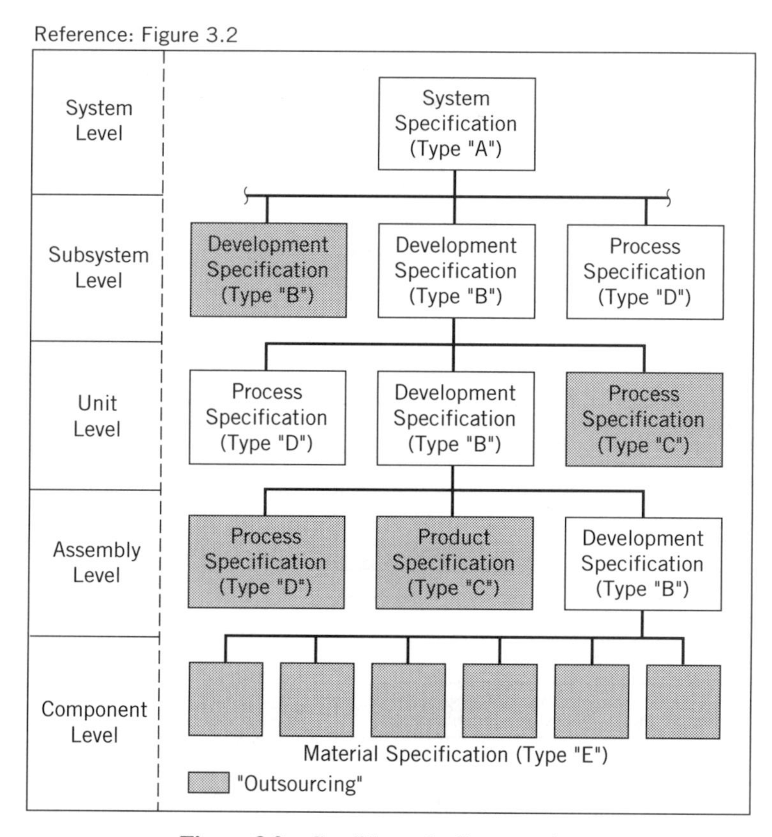

Figure 8.2 Candidates for "outsourcing."

[2]The specifications identified in Figure 8.2 should be presented in the form of a specification/documentation tree similar to the illustration in Figure 6.14.

8.2 PROPOSALS AND SUPPLIER SELECTION

From Figure 8.1, the recommendations from the Make-or-Buy Committee lead to the preparation of detailed specifications covering the desired efforts to be accomplished by suppliers through "outsourcing." The example presented in Figure 8.2 indicates the need to prepare the appropriate specifications for the design and development of system elements (Type "B"), the procurement of commercial off-the-shelf items (Type "C"), the completion of a service or a process (Type "D"), and for the procurement of materials (Type "E").

In the preparation of specifications, it is critical that they be presented in some form of hierarchy and that there is a *traceability* of requirements from the top system-level specification (Type "A") down to the more detailed material specifications (Type "E"). Through the establishment of the appropriate technical performance measures (TPMs) and the requirements allocation process described in Section 2.8, the application of specific quantitative and qualitative design criteria should be specified for each level in the overall system hierarchical structure; that is, subsystem level, configuration item level, unit level, assembly level. These criteria should be included in the applicable development, product, process, and material specifications accordingly.

The preparation of good and complete specifications is particularly important in not only defining the input-output, performance-related requirements for each of the system elements being acquired, but in defining the functional *interface requirements* as well. In many instances, the process utilized in the development and production of systems has shifted away from a bottom-up approach of designing individual components, combining these into a next-higher assembly, and then constructing the system through the final integration of the various subsystems, all within the producer's facility. With the advent of many different suppliers from various parts of the world, the activity within the producer's facility has shifted more toward the *integration* of subsystems, products, and/or the results of processes into the ultimate system configuration as an entity. Referring to the manufacturing system illustrated in Figure 2.19, the challenge is to design such that the proper integration efforts can be applied effectively and efficiently in the final assembly of the total system. Where there may be efficiencies within the lower-level processes, the total overall integration effort has been a problem in many instances in the past. In other words, the design and development of systems involves the proper selection of components and the final integration of these components into a functioning entity. Thus, a good and comprehensive set of specifications must be prepared, including a clear definition of the functional interfaces.

Given a good set of specifications, the next step is to prepare a supporting Statement of Work (SOW), develop a Request for Proposal (RFP), identify potential sources of supply, solicit responses from qualified suppliers, evaluate responding supplier proposals, and select the appropriate suppliers to accomplish the work desired.

8.2.1 Preparation of Request for Proposal (RFP)

Having evaluated the alternatives with an ultimate decision to "BUY," the contractor (in this instance) must develop the necessary materials for incorporation into a Request for Proposal (RFP). The objective is to develop a data package that can be distributed to potential suppliers for the purposes of soliciting a proposal.[3]

In general, the RFP is a formal mechanism by which the contractor specifies the requirements for a product, or for a service, in response to a designated need. The need for a system component has been identified, a decision has been made to procure the item from an outside source, and the contractor must translate the requirements for this item in a detailed and precise manner. These requirements are described in a data package, attached to a letter of invitation to bid and sent to prospective suppliers interested in responding to the RFP. More specifically, the content of the data package should include the following:

1. A technical specification describing the product, its performance and effectiveness characteristics, physical features, logistics and quality provisions, and so on. This document, tailored to the application, may constitute a Type "B," "C," "D," or "E" Specification, depending on the particular requirement (refer to Figure 6.14).
2. An abbreviated management plan describing overall program objectives, contractor organizational responsibilities and interfaces, the WBS, program tasks, task schedules, applicable policies and procedures, and so on. This information primarily relates to contractor activities; however, individual suppliers must understand their respective roles in the context of the overall program.
3. A Statement of Work (SOW) describing detailed tasks, task schedules, deliverable items, supporting data, and reports that are to be provided by the supplier. This information, derived from a combination of the specification and the management plan, constitutes a summary of the work to be performed and serves as the basis for the supplier's proposal.

Meeting the objectives of system engineering is highly dependent on initial supplier selection, applicable follow-on activities, and the ongoing evaluation and control efforts imposed by the contractor. As an input to this process, the technical specification (i.e., the Type "B," "C," "D," or "E" Specification, as applicable) must be *comprehensive* in covering *all* of the system-level requirements as they are allocated (or apportioned) down to the element of the system being procurred. A top-down approach is an important aspect of system engineering, and the technical specification must support system requirements to the extent applicable.

The degree of influence of the System Specification (Type "A") on the lower-tier specifications is, of course, dependent on the item being procured from the supplier. A large developmental effort will require a very comprehensive Type "B" Specifica-

[3]The solicitation package may be presented as a Request for Proposal (RFP), a Request for Quotation (RFQ), or an Invitation for Bid (IFB). In each case, the objective is to provide a data package on which to base a proposal.

tion, whereas a standard, commercial, off-the-shelf component may be covered by a relatively short and simple Type "C" Specification. It is important to ensure that the appropriate "traceability" is maintained as one progresses down through the applicable specification tree (refer to Figure 6.14).

Although the top-down technical requirements are maintained through the "specification track," the appropriate management-oriented requirements must be imposed on the supplier through the management plan and the Statement of Work (SOW). Organizational continuity must be ensured from the top down, tasks specified for the supplier must directly support those tasks being accomplished by the contractor, schedules must be compatible, the WBS must show the relationships between the supplier and contractor activities, and so on. In other words, a close continuity must be ensured in the transition of work from the contractor to the supplier.

The RFP data package, prepared by the contractor to cover planned supplier activity, is extremely important in maintaining the necessary continuity from the top system-level requirements down to the lowest-level component of the system. One of the prime tasks in system engineering is that of *system integration,* and it is an objective in developing the RFP that the appropriate level of system integration be recognized and addressed. So often a document such as this is compiled in a "hurry-up" manner, proposals are generated, contracts are negotiated, and the necessary system integration requirements are put off until the end. This, of course, can be a costly practice. The RFP data package must be considered as an extension of the System Type "A" Specification and the SEMP.

8.2.2 Development of Supplier Proposals[4]

After the RFP data package has been developed and distributed to interested and qualified suppliers, each recipient must make a "bid/no-bid" decision. Those suppliers deciding to respond will establish a proposal team and will proceed with the preparation of a proposal. The results, of course, must be responsive to the instructions included in the RFP.

The nature of the supplier's proposal activity will depend on the type and scope of the effort described in the RFP. When the acquisition process is directed toward large elements of the system involving some design and development (e.g., major subsystems), the supplier proposal activity could be rather extensive. A formal project-type organization may be established, specific project tasks are identified, and the level of effort may be somewhat similar in approach to the project configuration(s) described in Chapters 6 and 7.

[4]The subject of proposals (i.e., proposal requirements, bidders conferences, establishing the proposal team, proposal preparation activities, proposal processing and review) is rather extensive. Only those functions as they relate to the fulfillment of system engineering objectives are discussed here. Two good references covering this subject indepth include (1) R. D. Stewart and A. L. Stewart, *Proposal Preparation,* 2nd Ed., John Wiley, New York, 1992; and (2) W. C. Wall, *Proposal Preparation Guide,* John Wiley, New York, 1990.

In situations in which large proposal efforts exist, there is usually a requirement for some design and development activity. If the RFP (through a Type "B" Development Specification) dictates the need for the design of a major system element, the supplier will often attempt to design and construct a prototype model of the item as part of the proposal effort. A miniproject is organized, design and development tasks are completed on an expeditious basis, and a physical model is delivered to the contractor along with the written proposal. Design decisions are consummated early, with the objective of impressing the contractor (i.e., the customer in this instance) relative to both design approach and the capabilities of the supplier. Should the supplier be successful and be selected in this case, the constructed prototype may well be considered as the baseline configuration leading into follow-on detailed design.

In the preceding scenario, subsystem requirements were specified as part of the RFP, design and development activities were completed during the proposal phase, a formal design review occurred through the contractor's review and evaluation of the supplier's proposal, and the resultant configuration became somewhat fixed relative to the possibility of incorporating any design changes. This scenario can be related to the development process described in Chapter 2, except that the time element is compressed significantly. Because of this type of scenario, the preparation of the RFP assumes a great degree of importance from the system engineering viewpoint (as indicated in Section 8.2.1). Further, the ongoing design activity accomplished during the proposal phase must consider the necessary design characteristics supportive of system engineering objectives (e.g., reliability characteristics and maintainability characteristics). Finally, the formal evaluation of supplier proposals must serve as a final check for compliance with system engineering requirements as they apply to the item, or the service, being procured.

8.2.3 Evaluation and Selection of Suppliers

On receipt of all proposals (solicited and unsolicited) from perspective suppliers, the contractor proceeds with the review and evaluation process. When competitive bidding occurs, the contractor generally establishes an evaluation procedure directed toward selecting the best proposed approach. Initially, each supplier proposal is reviewed in terms of *compliance* with the requirements specified in the Request for Proposal (RFP). Noncompliance may result in automatic disqualification, or the contractor may approach the potential supplier and recommend a proposal revision and/or addition.

When two or more suppliers meet the basic RFP requirements, an evaluation of each proposal is then completed employing certain preestablished criteria. One may commence with the the preparation of a supplier checklist such as presented in Figure 8.3. The items identified cover some general criteria, design characteristics of the subsystem or product being considered for procurement, the supplier's proposed maintenance and support infrastructure for the subsystem/product, and the qualifications of the supplier. Each of these items in Figure 8.3 are supported by the questions

Supplier Evaluation Checklist

Refer to Appendix D for supporting questions.

D.1 General criteria

D.2 Product design characteristics
- D.2.1 Technical performance parameters
- D.2.2 Technical applications
- D.2.3 Physical characteristics
- D.2.4 Effectiveness factors
  1. Reliability
  2. Maintainability
  3. Human factors
  4. Safety factors
  5. Supportability/ serviceability
  6. Quality factors
- D.2.5 Producibility factors
- D.2.6 Disposability factors
- D.2.7 Environmental factors
- D.2.8 Economic factors

D.3 Product maintenance and support infrastructure
- D.3.1 Maintenance and support requirements
- D.3.2 Data/documentation
- D.3.3 Warranty/guarantee provisions
- D.3.4 Customer service
- D.3.5 Economic factors

D.4 Supplier qualifications
- D.4.1 Planning/procedures
- D.4.2 Organizational factors
- D.4.3 Available personal and resources
- D.4.4 Design approach
- D.4.5 Manufacturing capability
- D.4.6 Test and evaluation approach
- D.4.7 Management controls
- D.4 8 Experience factors
- D.4.9 Past performance
- D.4.10 Maturity
- D.4.11 Economic factors

Figure 8.3 Supplier evaluation checklist.

presented in Appendix D and are weighted relative to degrees of importance based on the requirements for the system overall.[5]

The contractor develops a list of topic areas considered to be relevant in the evaluation and assigns weighting factors as shown in Figure 8.4. Note that *supplier qualifications, product design characteristics, product maintenance and support infrastructure,* and *general criteria* have been identified in order of precedence.

[5] The questions in Appendix D are similar in nature to the design review questions in Appendix C, except that a "supplier orientation" has been provided.

Through a review of each supplier proposal, using the questions in Appendix D as a guide, the analyst can assess degree to which the supplier's proposal responds to the desired features conveyed through the questions. From Figure 8.4, the topics listed under *evaluation criteria,* taken from Figure 8.3, are weighted on the basis of level of importance, an assessment is made, and a *rating* is given in each area. A more detailed checklist for each topic may be developed to support the designated rating factor. Figure 8.5 shows an example covering item E in Figure 8.4.

In Figure 8.4, the assigned ratings are multiplied by the weighting factors to

Evaluation Criteria	Weighting Factor (%)	Proposal "A"		Proposal "B"		Proposal "C"	
		Rating	Score	Rating	Score	Rating	Score
A. General Criteria	10	7	70	5	50	6	60
B. Product Design Characteristics	30						
1. Performance factors	6	3	18	5	30	4	24
2. Technology applications	3	7	21	8	24	6	18
3. Physical characteristics	2	3	6	4	8	5	10
4. Effectiveness factors	7	7	49	7	49	8	56
5. Producibility factors	2	4	8	6	12	5	10
6. Disposability factors	3	5	15	4	12	8	24
7. Environmental factors	2	2	4	3	6	5	10
8. Economic factors	5	4	20	4	20	6	30
C. Product Maintenance and Support Infrastructure	20						
1. Maintenance and support requirements	7	6	42	8	56	7	49
2. Data/documentation	3	3	9	4	12	6	18
3. Warranties/guarantees	3	2	6	3	9	5	15
4. Customer service	5	5	25	8	40	30	30
5. Economic factors	2	4	8	3	6	6	6
D. Supplier Qualifications	34						
1. Planning/procedures	3	5	15	4	12	5	15
2. Organizational factors	2	6	12	5	10	5	10
3. Personnel and resources	2	4	8	3	6	2	4
4. Design approach	4	6	24	4	16	3	12
5. Manufacturing capability	3	7	21	5	15	4	12
6. Test and evaluation	2	6	12	5	10	4	8
7. Management controls	6	7	42	6	36	4	24
8. Experience factors	4	6	24	4	16	4	16
9. past performance	5	6	30	5	25	7	35
10. Maturity	3	7	21	7	21	6	18
E. Life-Cycle Cost	12	5	60	7	84	4	48
Grand Total	100		570		595		562

Figure 8.4 Proposal evaluation results.

Refer to Figure 8.4, Item E

Rating (Points)	Evaluation Criteria—Life-Cycle Cost*
10–12	The supplier has justified his design on the basis of life-cycle cost, and has included a complete life-cycle cost analysis in his proposal (i.e., cost breakdown structure, cost profile, etc.).
8–9	The supplier has justified his design on the basis of life-cycle cost, bud did not include a complete life-cycle cost analysis in his proposal.
6–7	The supplier's design has not been based on life-cycle cost; however, he plans to accomplish a complete life-cycle cost analysis and has described the approach, model, etc., that he proposes to use in the analysis process.
3–5	The supplier's design has not been based on life-cycle cost, but he intends to accomplish a life-cycle cost analysis in the future. No description of approach, model, etc., was included in his proposal.
0–2	The subject of life-cycle cost (and its application) was not addressed at all in the supplier's proposal.

* Refer to Figure 8.2 for individual criteria

Figure 8.5 Sample checklist of evaluation criteria for supplier proposals.

provide a score for each item. The individual scores are then added and the highest score reflects the supplier with the best overall approach. In this instance, Supplier B appears to be the preferred alternative.[6]

In the evaluation of supplier proposals from the system engineering perspective, the following general questions, as they apply to the subsystem or product being procured, are appropriate.

1. Is the supplier's proposal responsive to the contractor's needs as specified in the Request for Proposal (RFP)?
2. Is the supplier's proposal directly supportive of the system requirements specified in the System Type "A" Specification and the System Engineering Management Plan (SEMP)?
3. Have the performance characteristics been adequately specified for the item(s) proposed? Are they meaningful, measurable, and traceable from system-level requirements?
4. Have effectiveness factors been specified (e.g., reliability, maintainability, supportability, and availability)? Are they meaningful, measurable, and traceable from system-level requirements?
5. In the event that new design is required, has the design process within the supplier's organization been adequately defined? Does the process incorporate the utilization of CAD/CAM/CALS technologies where appropriate?

[6] Refer to Case Study A.6, Appendix A, for the results of a similar evaluation.

Have reliability, maintainability, human factors, supportability, life-cyle cost, and related characteristics been properly integrated into the design where appropriate? Have design change procedures been developed, and are changes properly controlled through good configuration management practices?

6. Is the design adequately defined through good documentation; that is, drawings, parts lists, reports, software, tapes, disks, and databases? Are the required data available? Have the data rights been specified?
7. Has the supplier addressed the requirement for the test and evaluation of the proposed system element a component? If testing has been accomplished in the past, are the test results documented and available? Have the plans for future testing been properly integrated into the system Test and Evaluation Master Plan (TEMP)?
8. Have the life-cycle support requirements been identified for the item being proposed; that is, maintenance resource requirements, spare/repair parts, test and support equipment, personnel quantities and skill levels, training, facilities, data, maintenance software, and so on? Have these requirements been minimized to the extent possible through good design?
9. Does the design configuration reflect good growth potential; reconfigurability?
10. Has the supplier developed a comprehensive production/construction plan? Are key manufacturing processes identified along with their characteristics?
11. Does the supplier have a good quality assurance program? Are statistical quality control methods utilized where appropriate? Does the supplier have a good rework plan to handle rejected items as necessary?
12. Does the supplier's proposal include a good comprehensive management plan? Does the plan cover program tasks, organization structure and responsibilities, a WBS, task schedules, program monitoring and control procedures, and so on? Has the responsibility for system engineering tasks (as applicable) been defined?
13. Does the supplier's proposal address all aspects of *total cost;* that is, acquisition cost, operation and support cost, and life-cycle cost?
14. Does the supplier have previous experience in the design, development, and production of system elements/components that are similar in nature to the item proposed? Was that experience favorable in terms of delivering high-quality products in a timely manner and within cost?

Although these questions may be helpful in the evaluation of a supplier's proposal, there are some additional factors that need to be considered before recommending a specific procurement approach:

1. Should a single supplier be selected (i.e., sole source), or should two or more suppliers be selected to fulfill the requirements as stated in the RFP? If the level of

effort specified covers a relatively large element of the system and involves some design and development activity, the selection of two (or more) suppliers to perform the same tasks may be rather costly. On the other hand, for smaller standard off-the-shelf components, it may be appropriate to establish several sources of supply. The objective is to ensure a source of supply that will meet the need for as long as required, with a minimum of risk associated with the possibility of the supplier "going out of business"!

2. Will the supplier be able to provide the necessary support for the proposed item, both during and after production, throughout the planned life cycle of that item? Of particular interest is the source for spare/repair parts to support sustaining maintenance requirements after the initial production has been completed and the capability no longer exists; that is, postproduction support. If such support will not be available, then the procurement policy may dictate that enough spare/repair parts be purchased initially to support maintenance operations for the entire life cycle.

3. Should a supplier be selected on the basis of political, social, and/or economic factors? In this era of international involvement (or globalization), there may be certain political pressures encouraging the procurement of components, or services, from a particular foreign source. On the other hand, it may be feasible to select a prospective supplier on the basis of geographic location and economic need. On occasion, it may be specified that at least "X"% of the total volume of system development effort must be subcontracted. In any event, supplier selection is sometimes influenced by political, social, and/or economic factors.

The evaluation of supplier proposals may be accomplished using the approach conveyed in Figure 8.4, modified to take into consideration these additional factors; that is, single versus multiple suppliers, postproduction support requirements, and the influence of political and economic factors on supplier selection. This evaluation activity usually includes not only a review of the written proposal itself, but one or more on-site, inspection-type visits to the supplier facility. A recommendation is made and contract negotiations between the contractor and the supplier are initiated.

As the results of the supplier evaluation and selection process have a significant impact on program success and meeting the objectives of system engineering, it is important that the system engineering organization be represented throughout this process. The proper coordination and the integration of supplier activity into the total engineering design and development effort are essential.

8.3 CONTRACT NEGOTIATIONS

Having identified prospective suppliers through the evaluation and selection process, it is now incumbent on the contractor to develop a formal contractual arrangement with the supplier. A Request for Proposal (RFP) was initiated, proposals from potential suppliers were generated and evaluated, and a contractual structure (in some form) needs to be established. The type of contractual agreement negotiated can have a significant impact on supplier performance, particularly in the procurement of large system components involving design and development activity.

The objective of contract negotiation is to achieve the most advantageous contractual agreement from the standpoint of technical requirements, deliverables, pricing, the type of contract imposed, and payment schedule. Obviously, the contractor and the prospective supplier each views this objective relative to his own individual position in terms of the risks associated with the numerous options that are available to him. At one extreme in contracting is the firm-fixed-price (FFP) contract where the program risks are primarily assumed by the supplier. At the other end, there is a cost-plus-fixed-fee (CPFF) structure where the contractor assumes most of the risk. Between these two extremes, there are a number of relatively flexible options.

The type of contract negotiated is important because the results may well impact supplier performance, which, in turn, may influence the contractor's ability to develop and produce a system that will meet the specified requirements and in a timely manner. Supplier performance, particularly in the acquisition of large subsystems, is critical to the successful accomplishment of system engineering objectives. Further, the risk factors associated with the type of contractual structure negotiated should be considered in the development of the risk management plan included as part of the System Engineering Management Plan (SEMP)—refer to Section 6.2.

With this in mind, it is important that the system engineer have some understanding of contracts because he (or she) is not only affected by the type of contract negotiated, but is often directly involved in the negotiation process itself. Fixed-price contracts are tightly controlled, with the supplier assuming most of the risk (in this instance). Engineering design should be fairly well-defined, as changes subsequent to contract negotiation may be quite costly. On the other hand, cost-reimbursement type of contracts (i.e., cost-plus-fixed-fee, cost-plus-incentive-fee) are more flexible in terms of making changes after initial contract negotiation, and the bulk of the risk is assumed by the contractor. In any event, the system engineer should have some feel relative to the extent of design definition required and what can and cannot be done by virtue of various contractual arrangements.

As an additional point, the system engineer often participates, from both a technical and a cost-estimating standpoint, in the initial preparation of the Request for Proposal (to include the preparation of the development specification, the management plan, and the Statement of Work) that leads to contract negotiations. When the negotiations actually take place, the system engineer often participates once again relative to the interpretation of specifications and the technical aspects of task accomplishment. Throughout the negotiation process, the intended scope of work may change, and these changes must be evaluated for their impact on other system design and development activities.

In order to provide some additional understanding, the major categories of contracts are briefly described in what follows:

1. *Firm-fixed-price* (FFP) *contract:* a legal agreement to pay a specified amount of money when the items called for by the contract have been delivered and are accepted. No price adjustments are allowed for the contracted work after award

regardless of the actual costs experienced by the supplier. At a specified price, the supplier assumes all financial risks for performance, and his profits depend on his ability to initially predict cost, negotiate, and to subsequently control costs. Concerning application of this type of contract, the component design should be fairly well established through appropriate specifications.

2. *Fixed-price-with-escalation contract:* similar to the FFP contract except that an escalation clause may be added to cover uncontrollable price increases or decreases. Escalation can be applied to both labor and material. Because there are many uncertainties relative to predicting the magnitude of escalation, an escalation ceiling is often established with the supplier and the contractor sharing the risks up to that point. Unexpected costs above the established ceiling are assumed by the supplier.

3. *Fixed-price-incentive contract:* applied in situations in which some cost uncertainties exist and there is an excellent possibility that cost reduction can be attained through good supplier management and by providing the supplier with some profit incentive. A target cost, a minimum cost, and a ceiling price are negotiated, along with a profit-adjustment formula. Profit adjustment, from the initial targeted profit, can be made based on total cost performance.

4. *Cost-plus-fixed-fee* (CPFF) *contract:* a cost-reimbursement contract where the supplier is reimbursed for all allowable costs associated with the project. A negotiated fixed fee (e.g., 10% of the estimated cost) is paid to the supplier on completion of work. Although this fee is fixed in terms of a percentage of the total cost, fee increases or decreases may occur as changes occur to the scope of work and the contract. This is particularly applicable when the contractor is willing to accept supplier-generated engineering change proposals to perform work beyond the scope of the initial contract.

5. *Cost-plus-incentive-fee* (CPIF) *contract:* intended to cover situations in which uncertainties in program performance exist. Allowable costs are paid to the supplier, together with additional incentive fee payments based on designated accomplishments. At the time of negotiation, individual factors such as schedule milestones and specific performance measures may be identified as items where incentives are to be specified in order to motivate suppliers to excel in these areas. Contract negotiation will result in a defined target cost, target fee, a minimum and maximum fee, and a fee-adjustment formula. On completion of the contract, the supplier's performance will serve as the basis for fee adjustment. An application of this type of contract could include the negotiation of incentives against each of the Technical Performance Measures (TPMs) specified for the system, as they apply to the item being procured.

6. *Cost-sharing contract:* is primarily designed for research-and-development work conducted with educational institutions and nonprofit organizations. Such work is jointly sponsored, and reimbursement to the supplier is in accordance with a predetermined sharing agreement. No fee is awarded; however, in lieu thereof, the supplier anticipates that the work accomplished will derive other benefits (e.g., a patentable item, acquisition of technical know-how, a good publication).

7. *Time and material contract:* allows for the payment for actual materials and services expended in the performance of designated tasks. This type of contract is employed when the extent and duration of work cannot be determined ahead of time and when costs cannot be estimated to any degree of accuracy. Appropriate applications include specific subcontracted research and development tasks, maintenance repair and overhaul services, and so on.

8. *Letter agreement:* often used as a preliminary contractual document initiated with the intent of authorizing the supplier to start work on a project immediately. These agreements serve as an interim means for providing a rapid response to an identified need that otherwise might be delayed pending the negotiation of a definitive contract. Letter agreements usually do not include total pricing information; however, an upper-limit dollar amount is usually specified to preclude excessive spending. Under this type of agreement, all costs incurred by the supplier for work accomplished are fully reimbursed by the contractor.

Associated with each major type of contract is the question concerning schedule of payments. When will the supplier be reimbursed for the successful completion of contracted tasks? What is the magnitude of expected payments? For incentive contracting, what type of incentive/penalty plan should be applied? These and comparable questions are significant, particularly for the larger contracts, because the contractor is generally tied to a specific budgeting cycle and the supplier must offset operating costs without going too far into debt. Thus, a payment schedule of some type should be developed.

Figure 8.6 presents an example of one type of plan, where progress payments are tied to the successful completion of formal design reviews; that is, the system design review, the last equipment/software design review, and the critical design review. These particular design reviews will include coverage of supplier activity and, by tying progress payments to these events, it should motivate the supplier to place emphasis in this area with the intent of ensuring success.

If incentive contracting is used, an incentive/penalty plan should be developed as a supplement to the schedule for progress payments. Such a plan should specify the application of incentive and penalty payments to significant project milestones and/or demonstrated system performance and effectives characteristics. From Figure 5.2 (Chapter 5), the TPMs applied at the system level should be allocated to the subsystem, or to the level applicable to the item being provided by the supplier. Performance measures that are realistic for the item being procured may be appropriate factors for consideration in the development of an incentive/penalty plan for the supplier.

In developing an incentive/penalty payment plan, it is necessary to identify the parameters to which incentives and penalties are to be applied. In many instances, there is more than one parameter, resulting in a multiple structure. The appropriate sum of money for each incentive is difficult to determine and will depend on the type of component (or service) and the importance of the item to which the incentive is to be applied. It is unlikely that all selected parameters will be equally important; therefore, it will be necessary to assign an "importance value" or "weighting" for

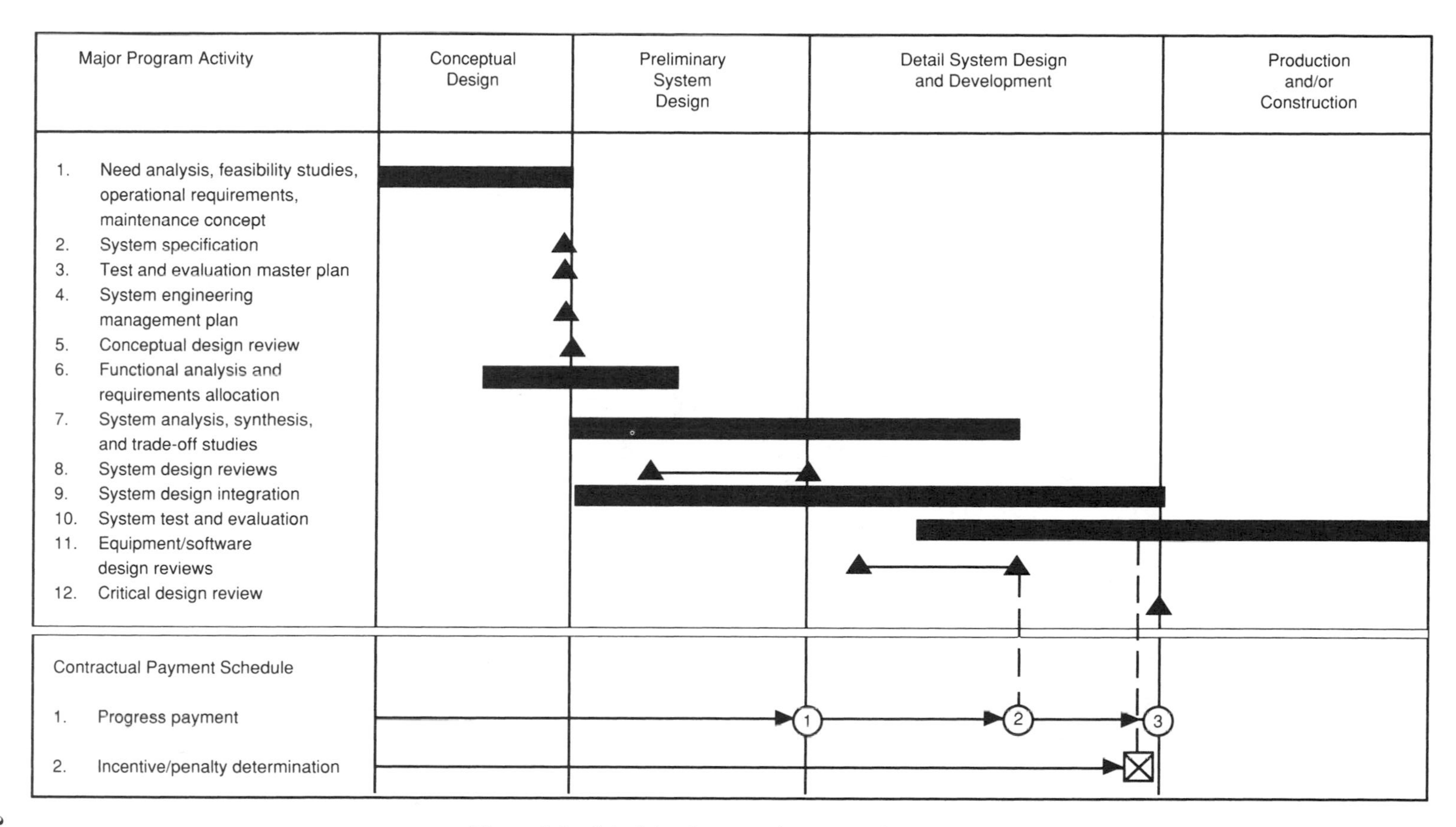
Major Program Activity
Conceptual Design
Preliminary System Design
Detail System Design and Development
Production and/or Construction
1. Need analysis, feasibility studies, operational requirements, maintenance concept
2. System specification
3. Test and evaluation master plan
4. System engineering management plan
5. Conceptual design review
6. Functional analysis and requirements allocation
7. System analysis, synthesis, and trade-off studies
8. System design reviews
9. System design integration
10. System test and evaluation
11. Equipment/software design reviews
12. Critical design review
Contractual Payment Schedule
1. Progress payment
2. Incentive/penalty determination
1
2
3

Figure 8.6 Schedule of proposed contractual payments.

each parameter and to estimate the magnitude of the incentive/penalty values accordingly. An example of the multiple approach, involving two component characteristics, is illustrated in Figure 8.7. A target value is established based on specification requirements, which also may be considered as a "contracted value." If, after test and evaluation, the actual measured value is an improvement over the target

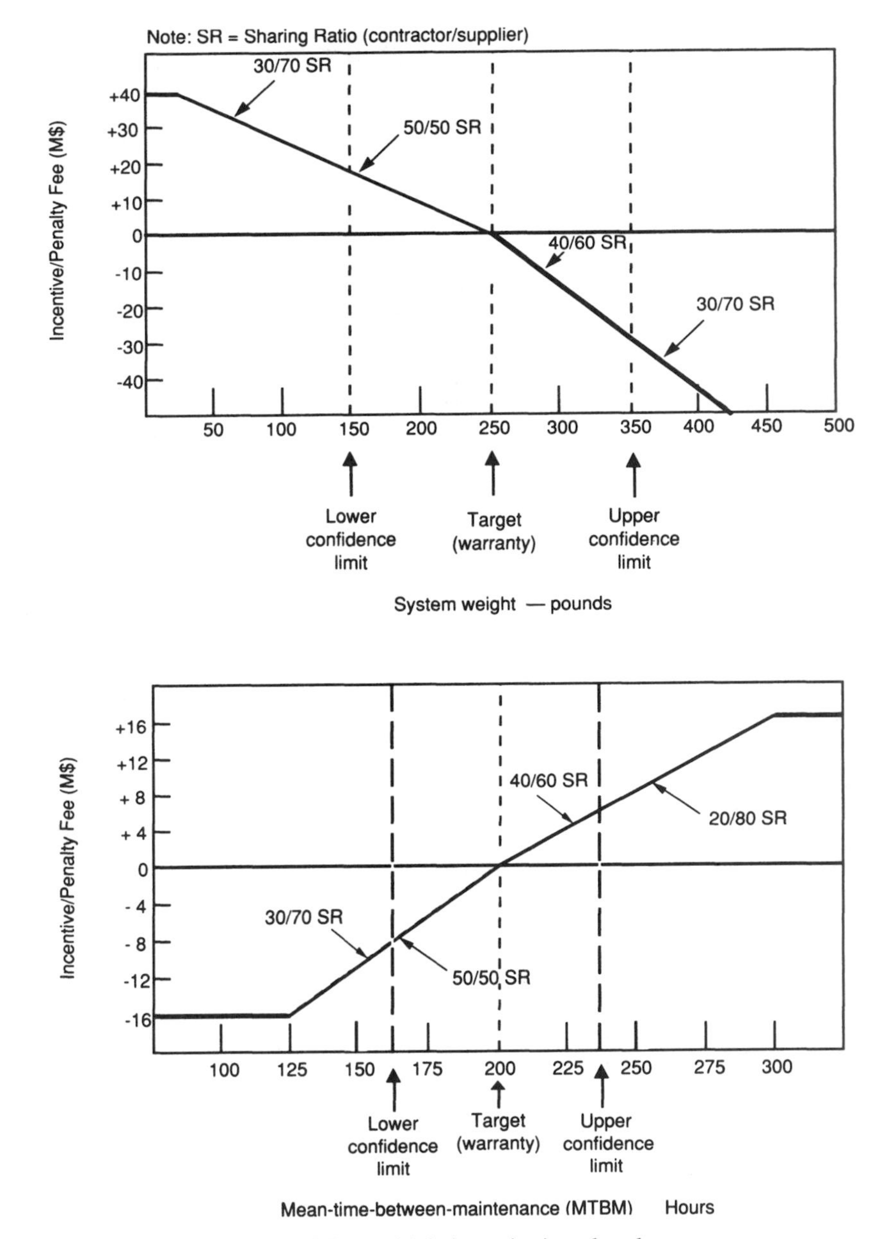

Figure 8.7 Multiple incentive/penalty plans.

value, an incentive fee is awarded to the supplier at the designated time, as indicated in the schedule in Figure 8.6. More specifically, if the measured MTBM exceeded the upper confidence limit of approximately 238 hours in Figure 8.7, the assigned fee would be split, with 20% going to the contractor and 80% going to the supplier. Conversely, if the measured MTBM fell below the target objective, a penalty of 50% of the indicated ordinate value would be paid by the supplier. Similar type applications involving other key parameters may be covered through incentive/penalty contracting.

Although there are a variety of possible contract types, care must be taken in adapting the appropriate contract structure to the particular procurement action. For example, if design and development activity is required on the part of the supplier, then it may be appropriate to negotiate either a cost-plus-fixed-fee (CPFF) or a cost-plus-incentive-fee (CPIF) type of contract. In such instances, when considering the flexibility, the contractor may have to implement tighter monitoring and control activities to ensure the timely completion of tasks by the supplier. At the same time, the contractor needs to be careful not to impose (or cause) any design changes that might have an impact on the supplier. If the contractor even suggests a possible improvement to the supplier's product, or a change in direction relative to activity, then the supplier is likely to claim a change in the scope of work and charge the contract accordingly. Additionally, a supplier knowledgeable of contracting may initially submit a proposal representing a "minimal effort" in order to keep the price low and win the competition! At the same time, the supplier is planning to initiate changes and/or additions at a later time to cover items that perhaps should have been included in the initial proposal. These changes are likely to be processed through a series of individual engineering change proposals (ECPs), and the ultimate costs will increase accordingly. In such situations, it is important that the system engineer not only be familiar with contracting methods in general, but should be thoroughly familiar with the item(s) being proposed, its technical makeup and how it fits into the system hierarchy, and the various interface and support requirements that are applicable.

At the other end of the spectrum, there will be undoubtedly many different system components that are well-defined and no additional design effort is required. In this instance, the implementation of a firm-fixed-price (FFP) contract may be preferred. For the performance of services, such as in the accomplishment of maintenance and repair actions, the basic time and materials type of contract may be the most appropriate.

The ultimate achievement of definitive contract terms and conditions is accomplished through formalized negotiations between the contractor and the supplier. Negotiations per se can assume a simplified approach involving several representatives from each side meeting on a given day to discuss requirements in general. On the other hand, for relatively large subsystems and/or major system components, the contract negotiation process can become quite complex. In a more formalized negotiation, the contractor, in response to the supplier's proposal, will interrogate the supplier relative to the validity of his proposed technical approach, management approach, and/or price. Questions along a technical line will attempt to ascertain whether or not the supplier has justified that his technical approach is the best (based on the results of design trade-off studies), and that he has the technical expertise and experience to follow through in developing and producing the proposed item. Concerning cost, the object is to verify that the supplier's price is fair and reasonable, and that it was developed through a logical cost analysis. From the supplier's standpoint, the negotiation initially takes the form of defending the proposal as submitted to the contractor. The supplier may be required to provide any amount of supporting material to help convince the contractor that he is thorough, honest, and offering the best deal possible.

Negotiation, in general, is an art and usually requires some strategy on both sides. Initially, a plan is developed that identifies the location where the negotiations are to be held and includes an agenda for each meeting that is scheduled. The contractor and the supplier each identifies the personnel who will participate in the negotiations process. Both technical and administrative personnel will be included, and a representative from the contractor's system engineering organization should be present for technical discussions covering system-oriented requirements. During the formal negotiations at the "bargaining table," both sides will assume a minimum-risk position, considering the contractual terms and conditions mentioned before. Interruptions will occur, short strategy meetings will be held to discuss events that have taken place, attempts will be made to gain some sympathy from the opposition, and it is hoped that an agreement will be made after some compromises on both sides. This process may evolve through a number of iterations, perhaps consuming more time than initially anticipated. However, the final objective is to realize a signed formal contract between the contractor and the supplier.

8.4 SUPPLIER MONITORING AND CONTROL

With the identification, approval, and the establishment of formal contractual relationships with suppliers, the contractor's main activity assumes the role of program coordination, evaluation, and control. This ongoing activity can be rather significant for the following reasons:

1. The magnitude of supplier activity and the number of individual product/component suppliers for a given system may be extensive. For some systems, as much as 50 to 75% of the planned development and production activity will be accomplished by suppliers.

2. In addition to the large number of suppliers involved in system acquisition, the geographic distribution of these suppliers may be worldwide! Many systems utilize components that are developed and manufactured in Pacific Rim countries, Europe, Africa, Canada, Mexico, South America, and so on. The requirements in system acquisition may dictate a truly international communications and distribution network.

3. In the acquisition of relatively large-scale systems, where there are many different component suppliers, the variety of tasks being accomplished at any given point in time can be rather extensive. Some suppliers may be undertaking a full-scale design and development effort, others may be performing manufacturing and

production functions, and there are many suppliers providing standard off-the-shelf components in response to routine purchase orders. There are some programs that are staggered and discontinuous, and there are other programs that are continuous over a long period of time. Figure 8.8 presents a sample plan of supplier project activities.

In this type of environment (i.e., many different suppliers, located worldwide, performing a wide variety of functions), the contractor is faced with a formidable and challenging task. As discussed earlier, specific supplier requirements must be carefully developed and clearly stated from the beginning, and an appropriate contracting structure must be established to ensure that the requirements will be met. The type of contract, of course, should be tailored to the supplier level of effort.

From Figure 8.8, Suppliers A, C, D, F, and G are each involved in a project that includes some design and development activity. As part of this effort, trade-off studies are conducted, reliability and maintainability prediction reports are prepared, design reviews are scheduled, test and evaluation functions are accomplished, and so on. The process described in Chapter 2 and many of the activities discussed throughout Chapters 6 and 7 are applicable, although the effort must be scaled down to be compatible with the particular needs of the supplier's program.

With regard to supplier program evaluation and control, the contractor must incorporate supplier activities as part of the overall design review process described in Chapter 5. For large design and development efforts, individual selected design reviews may be conducted at the supplier's facility, with the results of these reviews being included in the higher-level reviews conducted at the contractor's plant (refer to Figure 5.1). For smaller programs, the review process may not be as formal, with the results of the supplier's effort being integrated into the evaluation of a larger element of the system. When addressing projects involving the manufacture and production of components (i.e., each of the projects in Figure 8.8), the contractor's primary concern is that of incoming inspection and quality control. It is essential that the characteristics designed into the component, or as "advertised" in an off-the-shelf item, be maintained throughout!

In essence, supplier evaluation and control are merely extensions of the program review and control activities initiated by the customer and imposed on the contractor. The contractor, in turn, must impose certain requirements on the supplier. Large suppliers must impose the necessary controls on smaller suppliers in the event that a "layering of suppliers" exists. The objectives are to (1) ensure that system-level requirements are being properly allocated from the top down, and (2) that compliance with these requirements is being realized from the bottom up. The SEMP must describe the necessary procedures, technical reviews, and so on, as related to supplier activities (refer to Section 6.2).

8.5 SYSTEM INTEGRATION

Inherent within the concepts associated with system engineering is the function of "integration." Initially, as part of conceptual design, emphasis is placed on the defi-

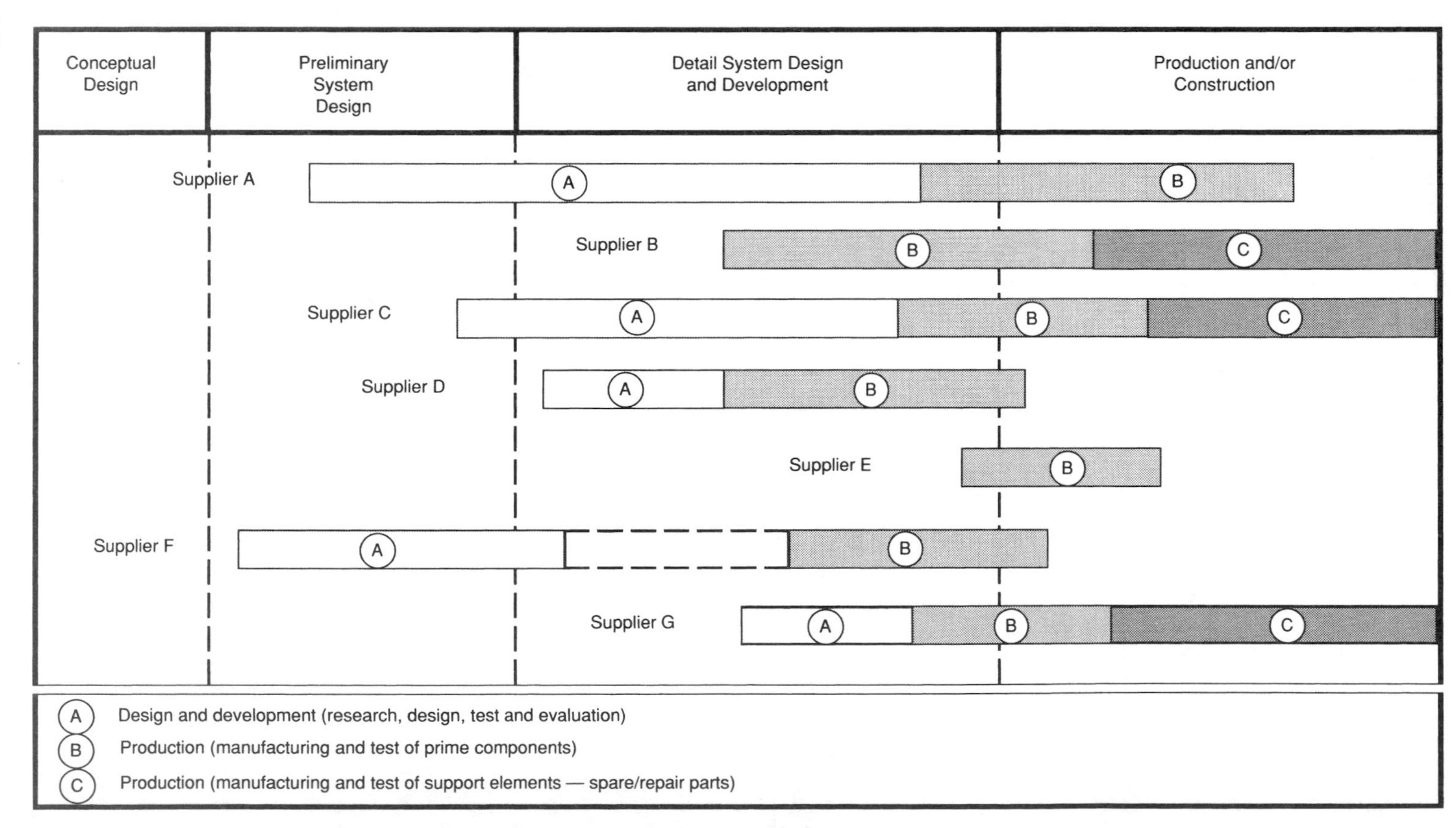

Figure 8.8 A sample of supplier project activities.

nition of system requirements and the proper integration of these requirements through the development of the System Type "A" Specification and the SEMP. Then, throughout preliminary and detail design, the necessity for integration continues. From the technical perspective, there is an ongoing integration effort associated with the proper interfacing of subsystems, units, assemblies, modules, software, data, facilities, test and support equipment, personnel, and other elements of the system. The software must be compatible with the hardware, the test and support equipment must be compatible with the prime equipment, the personnel must be compatible with the equipment and software, and so on. There is also a management thrust dealing with the proper integration of the various design disciplines and other related program activities. Finally, there is a critical integration function associated with combining the various components of the system into an operating entity that is responsive to consumer needs.

In the early phases of design, the emphasis is primarily on a top-down approach. System requirements are defined, functional analyses are completed, and top system-level requirements are allocated to the depth necessary to provide criteria for the purposes of design guidance. As design progresses, trade-off studies are conducted and system components are selected. Through synthesis, design concepts are verified and components are combined to form larger elements of the system. At this stage, there is a shift in emphasis to a bottom-up approach, as conveyed in Figure 8.9. Components are identified at the lowest level and integrated into subassemblies, subassemblies are integrated into assemblies, assemblies are integrated into units, and so on.

In Figure 8.9, the objective of the integration is to ensure that Components 1, 2, 3, . . . , and 8 are not only compatible with each other, but can be properly integrated into Subassembly G, considering all of the tolerances and interchangeability requirements. Although this goal should be obvious, it has not always been easily attained. Historically, systems have been developed mostly from the bottom up without benefit of some higher-order structure and, in many instances, the integration process has occurred through the application of "brute-force" methods, utilizing a "trial-and-error" approach! Component integration has not been very easy, component substitutions have been made on the spot, there has been much waste, and the resultant costs have been high!

Because of these past experiences with system integration and test, it is essential that a top-down system engineering approach be initiated from the beginning. Through the proper definition and allocation of requirements early in system design, many of the problems experienced in the past, it is hoped, can be avoided in the future. With the early identification and elimination of potential problem areas, the system integration process accomplished later toward the end of the detail design and development stage can be significantly enhanced. Instead of relying on the final system integration and test activity to resolve *all* of the problems, it is necessary that the bottom-up approach to system design follow the requirements established earlier through the top-down process. Without the proper consideration of system integration from the *beginning,* the problems associated with the final integration and test activity will continue as in the past.

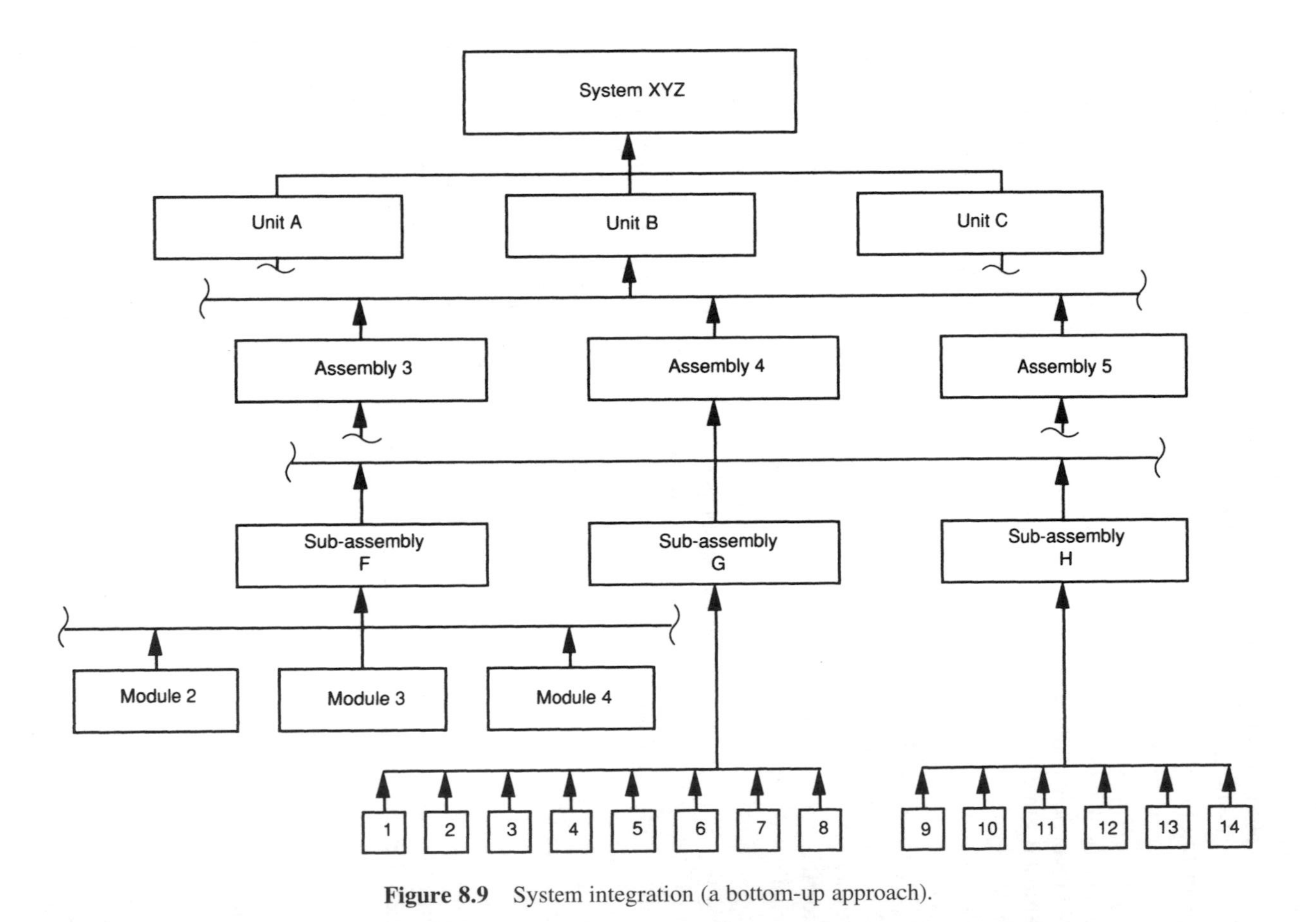

Figure 8.9 System integration (a bottom-up approach).

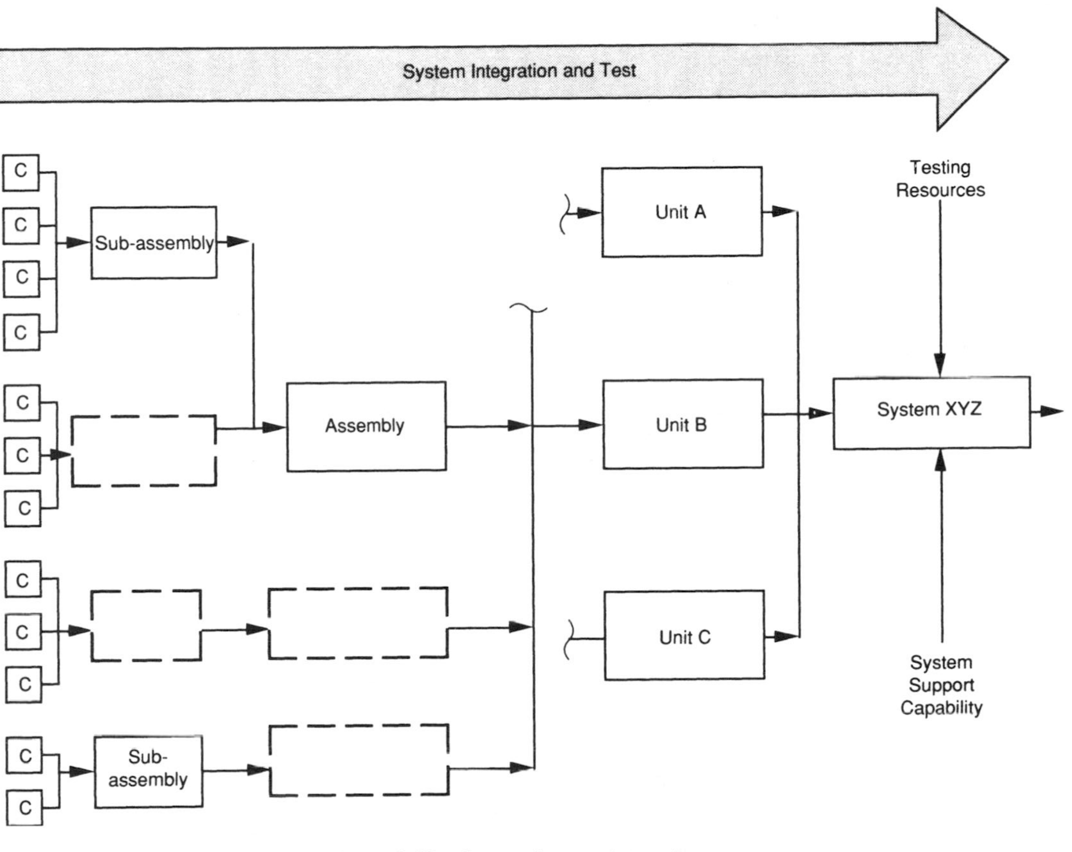

Figure 8.10 System integration and test.

With these considerations in mind, an overall system integration and test plan should be developed and included within the SEMP. The objective is to describe the flow process and the tasks associated with the integration of system components from the bottom up, as illustrated in Figure 8.9. This activity prevails throughout the design and development phase, commencing with the integration of components using simulation techniques in conjunction with computer-aided design methods. As components are identified and selected, they must be properly integrated into the system, and eventually they will be verified through final system test and evaluation (refer to Section 2.11).

This system integration flow process is illustrated in Figure 8.10. Components are introduced early in design (entering the figure from the left), combined and integrated into higher-level elements of the system, and a final system test and evaluation effort is accomplished in accordance with the TEMP. This integration effort is accomplished using a step-by-step approach, evolving from left to right, employing a combination of individual evaluations via analytical means and/or by conducting a series of components tests. The objective of this integration effort is to ensure that all components are compatible with each other, will fit properly into the next higher assembly, are interchangeable (as applicable), and will function as required.

In summary, the system engineering organization must not only ensure that a good and comprehensive system integration effort is initially planned, but that the various integration tasks are accomplished throughout the entire system design and development effort. Of particular significance are the challenges associated with the many supplier functions identified earlier in this chapter. These supplier activities, the delivery of components, the integration of the components into the system, and so on, must be covered in the contractor's system integration and test plan.

QUESTIONS AND PROBLEMS

1. Describe in your own words the steps that should be followed in determining the "supplier requirements" associated with the acquisition of a new system.
2. Identify and describe some of the factors that should be considered in "make-or-buy" or "outsourcing" decisions.
3. Why is the development of a "make-or-buy plan" important? What is included?
4. Should the system engineering organization be involved in "make-or-buy" decisions? If so, in what capacity? If not, why not?
5. Development of an RFP in preparation for supplier proposals, evaluation, and selection is extremely critical from a system engineering standpoint. Identify and explain the reasons for such, and briefly describe key features that should be included.
6. Refer to Figure 8.3 and Appendix D. Develop a supplier proposal evaluation checklist for application in your own organization (prepare the checklist in the

format shown in Figure 8.3, and provide a breakout of factors for each item in your checklist, as illustrated in Appendix D).

7. How can political, social/societal, and economic factors influence the supplier selection process? Provide a few examples.
8. Describe in your own words some of the trends that are occurring in the world today as they relate to customer, contractor, supplier activities.
9. What is meant by "postproduction support"? Provide some examples.
10. There are a number of challenges associated with supplier management and control. Identify and describe a few of these.
11. Assume that you are a newly appointed Manager of System Supplier Activities in a contractor's organization. Describe your approach to planning, organizing, identifying tasks, implementing controls, evaluating, and so on.
12. As part of a supplier evaluation effort, you are planning to visit the supplier's facility. What would you do in preparation, and what information would you solicit during the on-site visit?
13. There are various types of contract structures that can be imposed through the contract negotiation process to include FFP, FP, CPFF, CPIF, cost sharing, and time and material. Describe each and include some discussion as to applications.
14. When establishing multiple incentives under incentive contracting, what steps would you follow? How will you determine the specific factors or characteristics on which to establish incentives?
15. Under incentive contracting, what is meant by an incentive/penalty sharing ratio? How is it applied? How does the SR relate to supplier/contractor risks?
16. Should the system engineer participate in the contract negotiation process? If so, in what capacity?
17. Should the system engineering organization participate in supplier management and control activities? If so, in what capacity?
18. Refer to Chapter 5. How are suppliers (and supplier activity) covered through the formal design review process?
19. Describe what is meant by "system integration." How is it accomplished? What is included? When is it accomplished? Why is it important?
20. Discuss some of the concepts associated with the "top-down" and "bottom-up" approaches to system design and development. How are they related (if at all)? How is system engineering involved?

Appendixes

APPENDIX A
Case Studies

Throughout the early stages of the life cycle, as part of the system engineering process, there are a number of applications of different tools that can facilitate the conductance of trade-off studies. Of particular interest here are some of the tools that address the downstream aspects of system support but can be effectively utilized earlier. In Figure A.1, seven abbreviated examples have been selected to illustrate the utilization of analytical methods in the engineering decision-making process.[1]

A.1 FAILURE MODE, EFFECTS, AND CRITICALITY ANALYSIS (FMECA)

A.1.1 Definition of the Problem

Company ABC, a manufacturer of gaskets for automobiles, was experiencing problems related to declining productivity and increased product costs. At the same time, competition was increasing and the company was losing its share of the market. As a result, the company decided to implement a *continuous process improvement program* with the objective of identifying potential problem areas, and their impact and criticality on both internal company operations and on the product being delivered to the customer. To aid in facilitating this objective, the company's manufacturing operations were evaluated using the failure mode, effects, and criticality analysis (FMECA).

A.1.2 The Analysis Process

An initial step included the identification of the the major functions performed in the overall gasket manufacturing process by completing a functional flow diagram in accordance with the procedures described in Section 2.7 (Chapter 2). In this instance, there are 13 major functions that were subject to evaluation. For each function, required input factors and expected outputs were identified, along with the appropriate metrics. This led to the initial selection of 1 of the 13 functions, based

[1] Case studies A.1, A.2, and A.3 were taken in part from B. S. Blanchard, D. Verma, and E. L. Peterson, *Maintainability: A Key to Effective Serviceability and Maintenance Management*, John Wiley, New York, 1995.

Analysis Tools	Description of Application
A.1 Failure Mode, Effects, and Criticality Analysis (FMECA)	Identification of potential product and/or process failures, the expected modes of failure and "causes," failure effects and mechanisms, anticipated frequency, criticality, and the steps required for compensation (i.e., the requirement for redesign and/or the accomplishment of preventive maintenance). An Ishikawa "cause-and-effect" diagram may be used to facilitate the identification of "causes," and a Pareto analysis may help in identifying those areas requiring immediate attention.
A.2 Fault-Tree Analysis (FTA)	A deductive approach involving the graphical enumeration and analysis of different ways in which a particular system failure can occur, and the probability of its occurrence. A separate fault tree may be developed for every critical failure mode, or undesired top-level event. Attention is focused on this top-level event and the first-tier causes associated with it. Each of these causes is next investigated for its causes, and so on. The FTA is narrower in focus than the FMECA and does not require as much input data.
A.3 Reliability-Centered Maintenance (RCM)	Evaluation of the system/process, in terms of the life cycle, to determine the best overall program for preventive (scheduled) maintenance. Emphasis is on the establishment of a cost-effective preventive maintenance program based on reliability information derived from the FMECA; that is, failure modes, effects, frequency, criticality, and compensation through preventive maintenance.
A.4 Maintenance Task Analysis (MTA)	Evaluation of those *maintenance* functions that are to be allocated to the human. Identification of maintenance functions/tasks in terms of task times and sequences, personnel quantities and skill levels, and supporting resources requirements (i.e., spares/repair parts and associated inventories, tools and test equipment, facilities, transportation and han-

Figure A.1 Design analysis methods (case study applications).

Analysis Tools	Description of Application
	dling requirements, technical data, training, and computer software). Identification of high resource-consumption areas.
A.5 Level-of-Repair Analysis (LORA)	Evaluation of maintenance policies in terms of levels of repair; that is, should a component be repaired in the event of a failure or discarded and, given the "repair" option, should the repair be accomplished at the intermediate level of maintenance, at the supplier's factory, or at some other level? Decision factors include economic, technical, social, environmental, and political considerations. The emphasis here is based on life-cycle cost factors.
A.6 Design Evaluation of Alternatives	Evaluation of alternative design configurations using multiple criteria. Weighting factors are established to specify levels of importance.
A.7 Life-Cycle Cost Analysis (LCCA) Refer to Appendix B	Determination of the system/product/process life-cycle cost (design and development, production and/or construction, system utilization, maintenance and support, and retirement/disposal costs); high-cost contributors; cause-and-effect relationships; potential areas of risk; and identification of areas for improvement (i.e., cost reduction).

Figure A.1 (*Continued*)

on a perception by company personnel as to the area causing the most problems. Given this, the sequence of steps conveyed in Figure A.2 was followed in completing an FMECA on the selected function.

Figure A.3 represents the function, or portion of the overall manufacturing process, that was selected for evaluation. Note that, although the emphasis is on the manufacturing process and its impact on the gasket, one must also consider the impact of a faulty gasket on the automobile. Thus, the FMECA needs to address both the *process* and the *product.*

In Figure A.2, the approach selected for conducting the FMECA was in accordance with the practices followed in the automotive industry.[2] This included the following:

[2] Three references were used including (1) *Potential Failure Mode and Effects Analysis,* Instruction Manual, Ford Motor Company, 1988; (2) *Failure Mode and Effects Analysis,* Instruction Manual, Saturn

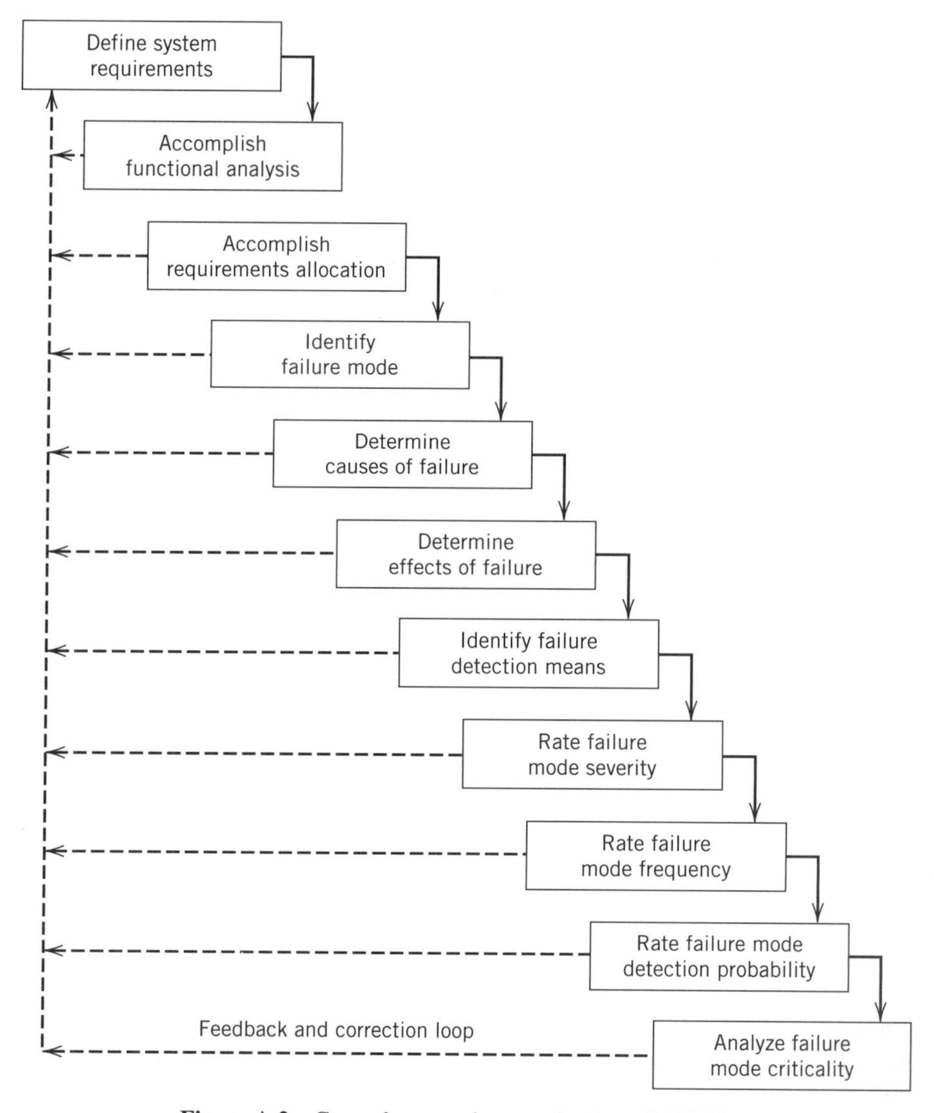

Figure A.2 General approach to conducting a FMECA.

1. Identifying the different failure modes; that is, the manner in which a system element fails to accomplish its function.
2. Determining the cause(s) of failure; that is, the cause(s) responsible for the occurrence of each failure. An Ishikawa *cause-and-effect,* or *fishbone,* dia-

Quality System, Saturn Corporation, 1990; and (3) *Potential Failure Mode and Effects Analysis (FMEA),* Reference Manual FMEA-1, developed by FMEA teams at Ford Motor Company, General Motors, Chrysler, Goodyear, Bosch, and Kelsey-Hayes, under the auspices of American Society of Quality Control (ASQC).

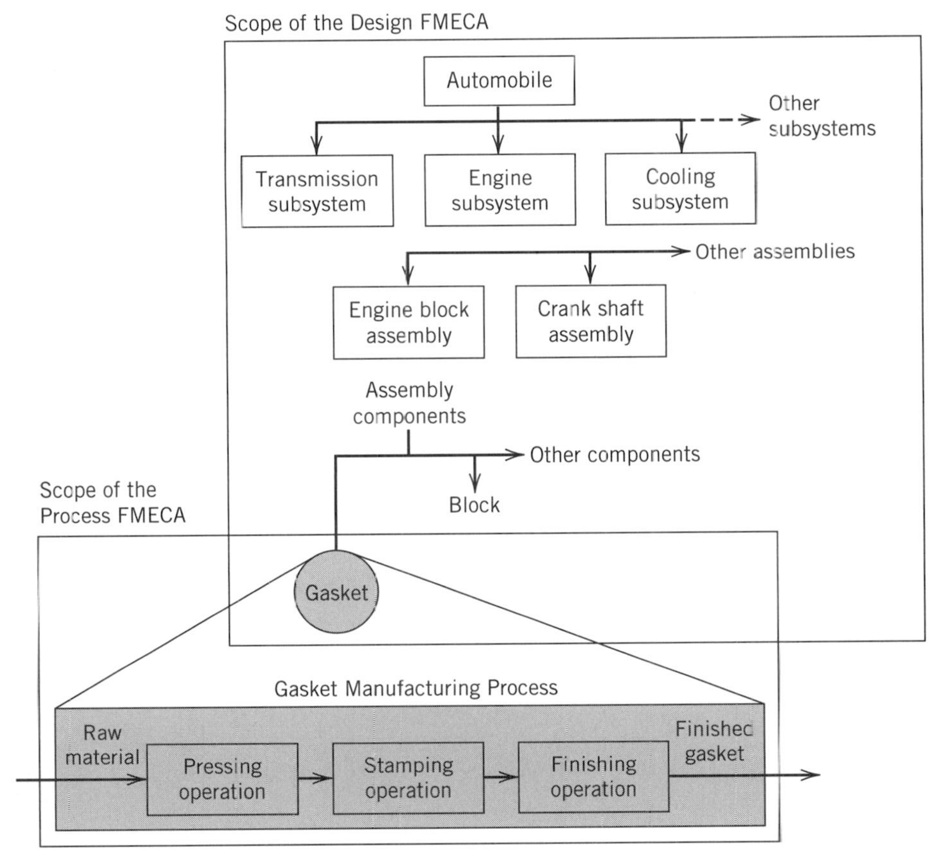

Figure A.3 Design and process FMECA focus and scope.

gram, as illustrated in Figure A.4, was utilized to help establish the relationships between failures and their possible causes.[3]

3. Determining the effects of failure; that is, the effects on subsequent functions/processes, on the next higher-level functional entity, and on the overall system.
4. Identifying failure detection means; that is, the current controls, design features, or verification procedures that will result in the detection of potential failure modes.
5. Determining the severity of a failure mode; that is, the seriousness of the effect or impact of a particular failure mode. The degree of severity was converted quantitatively on a scale of 1 to 10 with *minor* effects being 1, *low* effects being 2 to 3, *moderate* effects being 4 to 6, *high* effects being 7 to 8,

[3] K. Ishikawa, *Introduction to Quality Control,* Chapman and Hall, London, 1991.

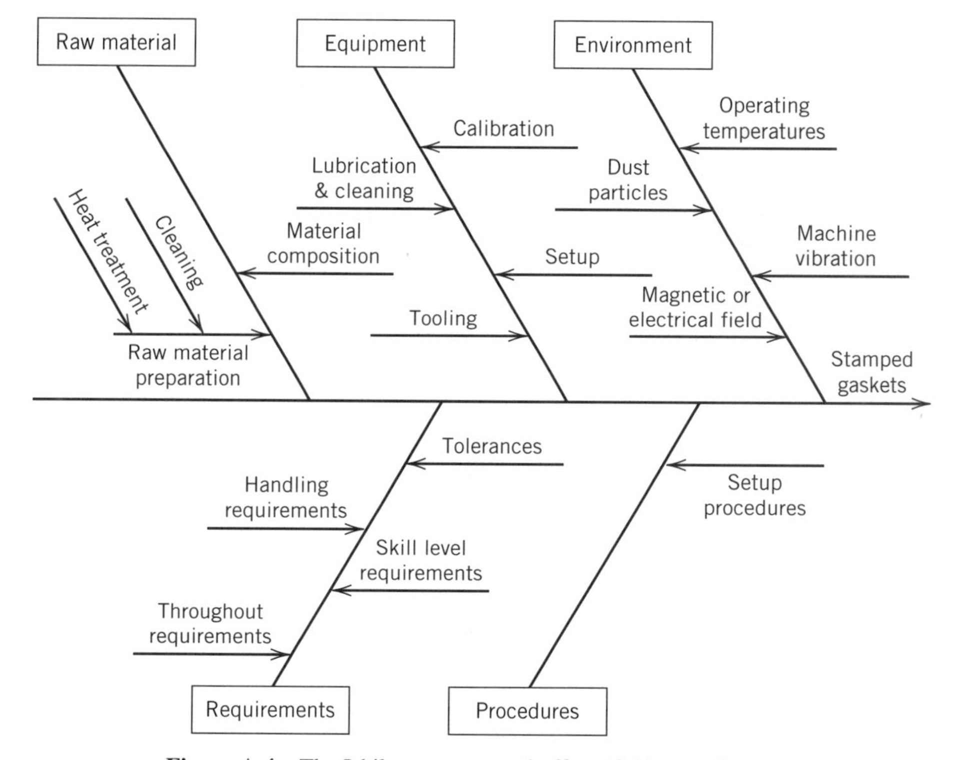

Figure A.4 The Ishikawa cause-and-effect (fishbone) diagram.

and *very high* effects being 9 to 10. The level of severity was related to issues pertaining to safety and the degree of customer dissatisfaction.

6. Determining the frequency of occurrence; that is, the frequency of occurrence of each individual failure mode or the probability of failure. A scale of 1 to 10 was applied with *remote* (failure is unlikely) being 1, *low* (relatively few failures) being 2 to 3, *moderate* (occasional failures) being 4 to 6, *high* (repeated failures) being 7 to 8, and *very high* (failure is almost inevitable) being 9 to 10. These rating factors were based on the number of failures per segment of operating time.
7. Determining the probability that a failure will be detected; that is, the probability that the design features/aids and/or verification procedures will detect potential failure modes in time to prevent a system-level failure. For a process application, this refers to the probability that a set of process controls currently in place will be in a position to detect and isolate a failure before it gets transferred to the subsequent processes or to the ultimate product output. This probability is once again rated on a scale of 1 to 10 with *very high* being 1 to 2, *high* being 3 to 4, *moderate* being 5 to 6, *low* being 7 to 8, *very low* being 9, and *absolute certainty of nondetection* being 10.

8. Analyzing failure mode criticality, that is, a function of severity (item 5), the frequency of occurrence of a failure mode (item 6), and the probability that it will be detected in time to preclude its impact at the system level (item 7). This resulted in the determination of the *risk priority number (RPN)* as a metric for evaluation. RPN can be expressed as:

$$\text{RPN} = (\text{severity rating})(\text{frequency rating})(\text{probability of detection rating}) \quad (\text{A.1})$$

   The RPN reflects failure-mode criticality. On inspection, one can see that a failure mode with a high frequency of occurrence, with significant impact on system performance, and that is difficult to detect is likely to have a very high RPN.
9. Identifying critical areas and recommending modifications for improvement; that is, the iterative process of identifying areas with high RPNs, evaluating the causes, and initiating recommendations for process/product improvement.

Figure A.5 shows a partial example of the format used for recording the results of the FMECA. The information was derived from the functional flow diagram and expanded to include the results from the steps presented in Figure A.2. Figure A.6 lists the resulting RPNs in order of priority (relative to requiring attention), and Figure A.7 presents the results in the form of a Pareto analysis.

A.1.3 The Analysis Results

After having completed the FMECA on the function identified in Figure A.3, Company ABC proceeded to evaluate each of its other 12 major functions/processes in a similar manner, utilizing a *team* approach. The activity was very beneficial overall, the individuals participating in the effort learned more about their own activities, and numerous changes were initiated for the purposes of improvement.

A.2 FAULT-TREE ANALYSIS (FTA)

A.2.1 Definition of the Problem

During the very early stages of the system design process, and in the absence of the information required to complete a FMECA (discussed in Section A.1), a fault-tree analysis (FTA) was conducted to gain insight into critical aspects of system design. The fault-tree analysis is a deductive approach involving the graphical enumeration and analysis of the different ways in which a particular system failure can occur, and the probability of its occurrence. A separate fault tree is developed for every critical failure mode, or undesired top-level event. Emphasis is on this top-level event and the first-tier causes associated with it. Each of these causes is next investigated for *its* causes, and so on. This top-down hierarchy, illustrated in Figure A.8, and the associated probabilities, is called a *fault tree.* Figure A.9 presents some of the symbology used in the development of such a structure.

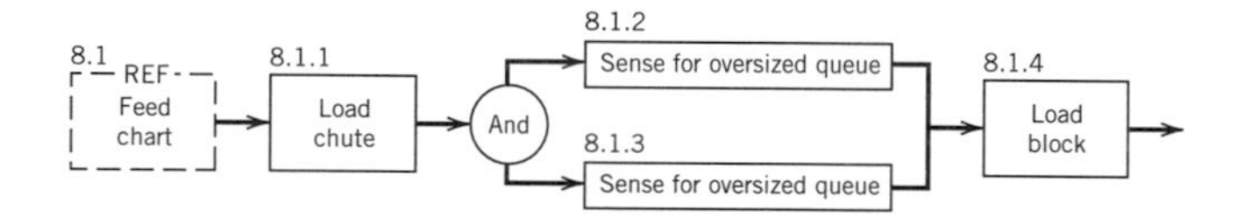

Process Failure Mode and Effects Analysis

Reference Number	Process Description	Potential Failure Mode	Potential Cause of Failure	Potential Effect(s) of Failure at Company ABC	Potential Effect(s) of Failure at Customer	Current Controls	Occurrence	Severity FM	Severity C	Detection	RPN	Recommended Action(s) and Status	Responsible Activity
8.1.2	Sense for oversized queue	A) Change in free spread	1) Sensor falls	a) Up process jams		a) Machine shops	1	1		1	1		
				b) Height variations		b) 1 Pc/5 min	1	7		3	21		
					c) Bearing loose when installed in engine	c) 2 Pcs/half hr	1		7	5	35		
			1) Sensor dirty	a) Up process jams		a) Machine shops	1	1		1	1		
				b) Height variations		b) 1 Pc/5 min	1	7		3	21		
					c) Bearing loose when installed in engine	c) 2 Pcs/half hr	1		7	5	35		
			1) Improper setup	a) Up process jams		a) Machine shops	1	1		1	1		
				b) Height variations		b) 1 Pc/5 min	1	7		3	21		
					c) Bearing loose when installed in engine	c) 2 Pcs/half hr	1		7	5	35		
8.1.3	Sense for undersized queue	A) Up mislocated	1) Sensor fails	a) Up process jams		a) 100% visual	2	1		1	2		
					b) Fillet ride	b) 5 Pcs/half hr	2		3	7	42		
			2) Sensor dirty	a) Up process jams		a) 100% visual	2	1		1	2		
					b) Fillet ride	b) 5 Pcs/half hr	2		3	7	42		
			3) Improper setup	a) Up process jams		a) 100% visual	2	1		1	2		
					b) Fillet ride	b) 5 Pcs/half hr	2		3	7	42		
		B) Facing/back damage	1) Sensor fails		a) Rejected at assembly	a) 100% visual	3		5	4	60		
			2) Sensor dirty		a) Rejected at assembly	a) 100% visual	3		5	4	60		
			3) Improper setup		a) Rejected at assembly	a) 100% visual	3		5	4	60		
						~~a) 5 Pcs/half hr~~	~~3~~	~~7~~		~~7~~	~~147~~		
8.1.4	Load block	A) Up mislocated	1) "Hold down" not set properly	a) Up smashed in broache		b) 100% visual	3	1		7	21		
				b) Up process jams	c) Fillet ride	c) 5 Pcs/half hr	3		3	7	63		
			2) Loose load block	a) Up smashed in broache		a) 5 Pcs/half hr	2	7		7	96		
						b) 100% visual	2	1		7	14		
				b) Up process jams	c) Fillet ride	c) 5 Pcs/half hr	2		3	7	42		
		B) Facing/back damage	1) Misaligned pusher		a) Rejected at assembly	a) 100% visual	4		5	4	80		

Figure A.5 Sample FMECA worksheet.

Causes	Risk Priority Numbers (RPNs)
Chip breaker angle ground incorrectly	273
Hold-down not set correctly	210
Undersize sensor fails	200
Undersize sensor dirty	200
Undersize sensor not positioned properly	200
Loose load block	161
Sharp die edge	120
Improper projection angle/resharpening of punch	108
Oversize sensor fails	105
Oversize sensor dirty	105
Oversize sensor not positioned properly	105
Improper sharpening of insert	93
Misaligned pusher	80
Worn tooling	72
Adapter reground to wrong dimension	60
Insert loose	60
Slivers in adapter	60
Insert off location	60
Worn/loose insert	60
Burrs from punch process caught	40
Ram stroke too long	36
Ram stroke too short	21
Broken/loose punch	12
Set screw fault	12
Insufficient stroke by ram/punch	12
Broken pressure spring	10
Total	2475

Figure A.6 Risk priority numbers (RPNs).

A.2.2 The Analysis Process

One of the outputs from the FTA is the probability of occurrence of the top-level event or failure. If the probability factor is unacceptable, the causal hierarchy developed provides engineers with insight into aspects of the system to which redesign efforts may be directed or compensatory provisions provided. The logic used in developing and analyzing a fault tree has its foundations in Boolean algebra.

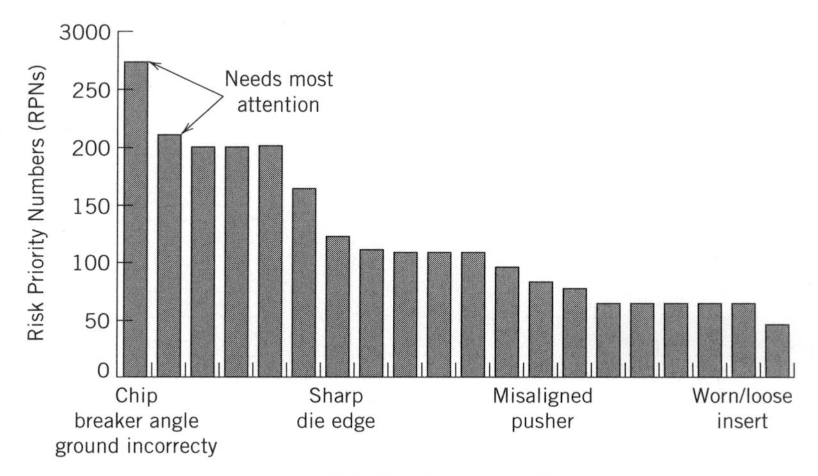

Figure A.7 Partial Pareto analysis.

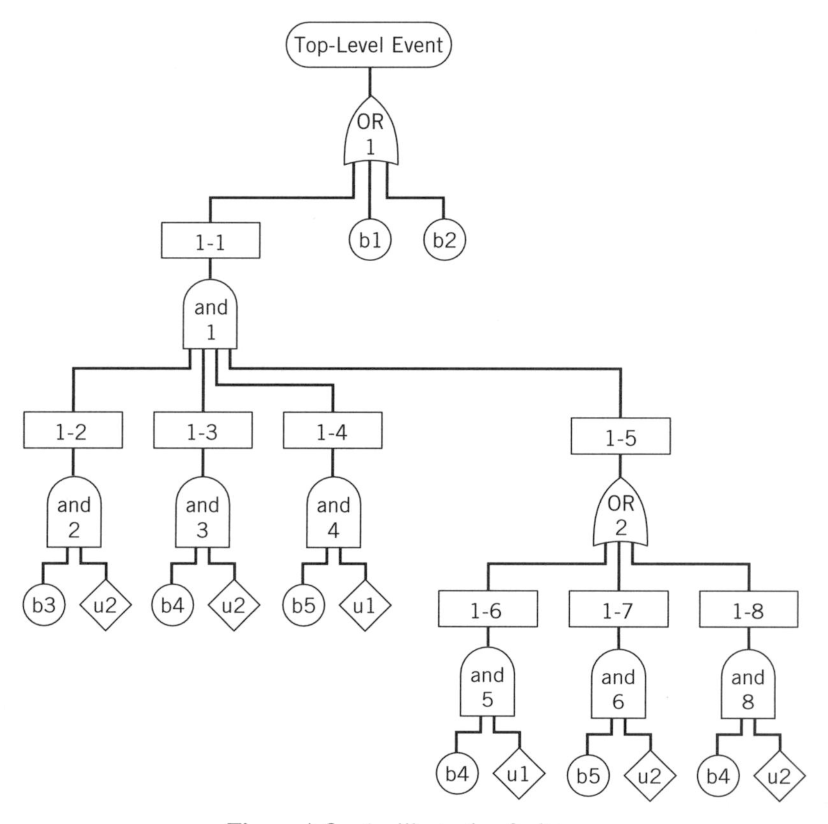

Figure A.8 An illustrative fault tree.

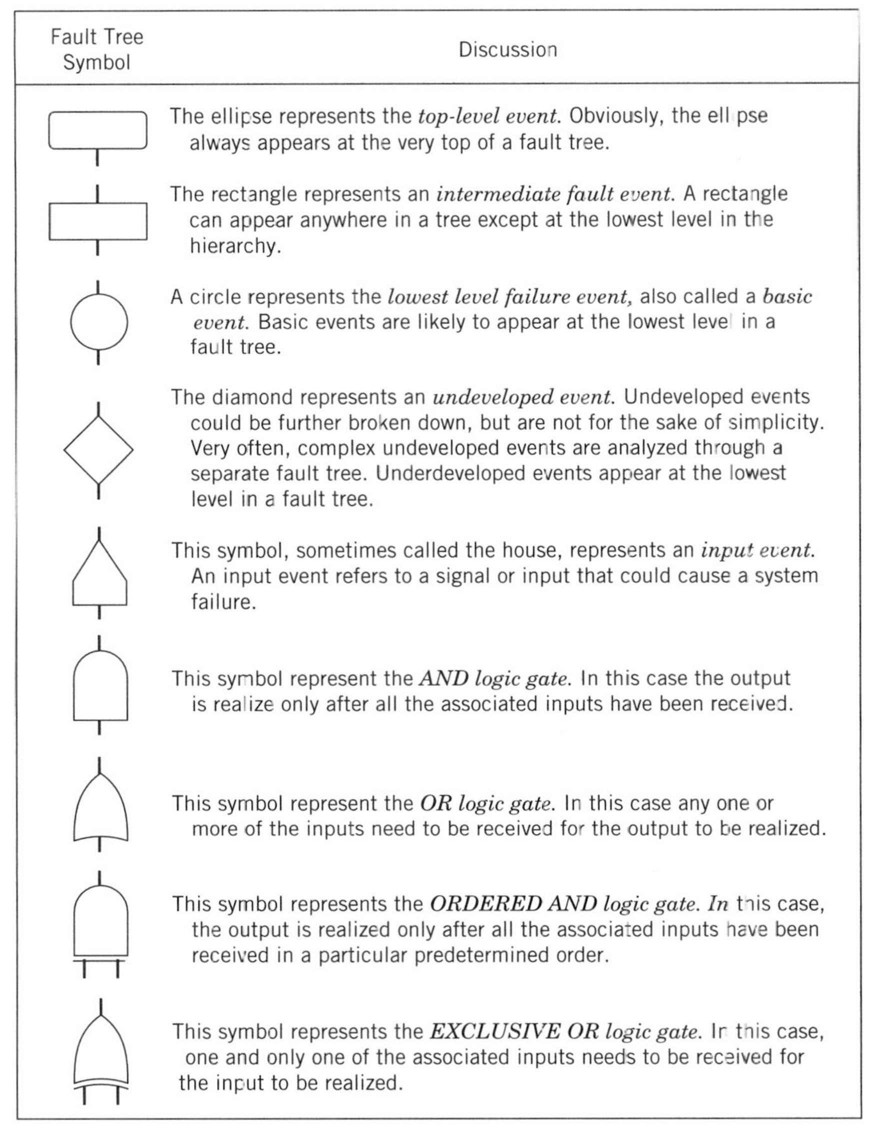

Fault Tree Symbol	Discussion
	The ellipse represents the *top-level event*. Obviously, the ellipse always appears at the very top of a fault tree.
	The rectangle represents an *intermediate fault event*. A rectangle can appear anywhere in a tree except at the lowest level in the hierarchy.
	A circle represents the *lowest level failure event,* also called a *basic event*. Basic events are likely to appear at the lowest level in a fault tree.
	The diamond represents an *undeveloped event*. Undeveloped events could be further broken down, but are not for the sake of simplicity. Very often, complex undeveloped events are analyzed through a separate fault tree. Underdeveloped events appear at the lowest level in a fault tree.
	This symbol, sometimes called the house, represents an *input event*. An input event refers to a signal or input that could cause a system failure.
	This symbol represent the *AND logic gate*. In this case the output is realize only after all the associated inputs have been received.
	This symbol represent the *OR logic gate*. In this case any one or more of the inputs need to be received for the output to be realized.
	This symbol represents the *ORDERED AND logic gate. In* this case, the output is realized only after all the associated inputs have been received in a particular predetermined order.
	This symbol represents the *EXCLUSIVE OR logic gate*. In this case, one and only one of the associated inputs needs to be received for the input to be realized.

Figure A.9 Fault-tree constructive symbology.

Axioms from Boolean algebra are used to collapse the initial version of the fault tree to an equivalent reduced tree with the objective of deriving *minimum cut sets*. Minimum cut sets are unique combinations of basic failure events that can cause the undesired top-level event to occur. These minimum cut sets are necessary to evaluate a fault tree from a qualitative and quantitative perspective. The basic steps in conducting a FTA are as follows:

1. Identify the top-level event. It is essential that the analyst be quite specific in defining this event. For example, it could be delineated as the "system catches fire," rather than the "system fails." Further, the top-level event should be clearly observable, and unambiguously definable and measurable. A generic and nonspecific definition is likely to result in a broad-based fault tree with too wide a scope and lacking in focus.
2. Develop the fault tree. Once the top-level event has been satisfactorily defined, the next step is to construct the initial causal hierarchy in the form of a fault tree. Once again, a technique such as Ishikawa's *cause-and-effect* diagram can prove to be beneficial (refer to Figure A.4). While developing the fault tree, all hidden failures must be considered and incorporated.

   For the sake of consistency and communication, a standard symbology to develop the fault tree is recommended. Figure A.9 depicts and defines the symbology to comprehensively represent the causal hierarchy and interconnects associated with a particular top-level event. In Figure A.8, the symbols OR1 and OR2 represent the two OR logic gates, and 1 through and 8 represent eight AND logic gates, 1-1 through 1-8 represent eight intermediate fault events, b1 through b5 represent five basic events, and u1 and u2 represent two undeveloped failure events. While constructing a fault tree, it is important to break every branch down to a reasonable and consistent level of detail.
3. Analyze the fault tree. The third step in conducting the FTA is to analyze the initial fault tree developed. A comprehensive analysis of a fault tree involves both a quantitative and a qualitative perspective. The important steps in completing the analysis of a fault tree are as follows:

(a) *Delineate the minimum cut sets.* As part of the analysis process, the minimum cut sets in the initial fault tree are first delineated. These are necessary to evaluate a fault tree from a qualitative and/or quantitative perspective. The objective of this step is to reduce the initial tree to a simpler equivalent reduced fault tree. The minimum cut sets can be derived using two different approaches. The first approach involves a graphical analysis of the initial tree, an enumeration of all the cut sets, and the subsequent delineation of the minimal cut sets. The second approach, on the other hand, involves translating the graphical fault tree into an equivalent Boolean expression. This Boolean expression is then reduced to a simpler equivalent expression by eliminating all the redundancies and so on. As an example, the fault tree depicted in Figure A.8 can be translated into a simpler and equivalent fault tree, through Boolean reduction, as depicted in Figure A.10.

(b) *Determine the reliability of the top-level event.* This is accomplished by first determining the probabilities of all relevant input events, and the subsequent consolidation of these probabilities in accordance with the underlying logic of the tree. The reliability of the top-level event is computed by taking the product of the reliabilities of the individual minimum cut sets.

(c) *Review analysis output.* If the derived top-level probability is unacceptable, necessary redesign or compensation efforts will need to be initiated. The development of the fault tree and subsequent delineation of minimum cut

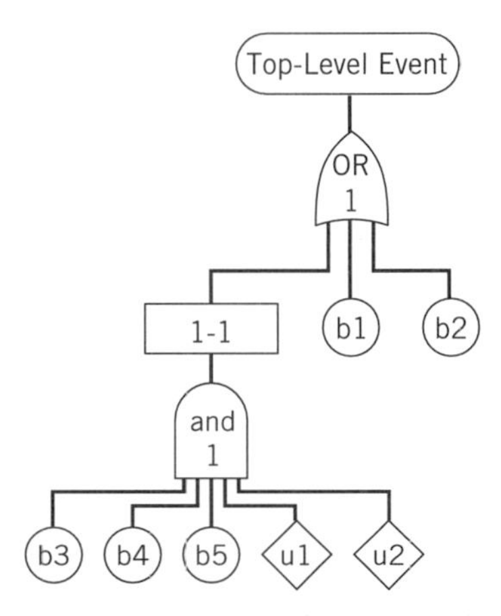

Figure A.10 A reduced equivalent fault tree (refer to Figure A.8).

sets provides engineers and analysts with the kind of foundation needed for making sound decisions.

A.2.3 The Analysis Results

The FTA can be effectively applied in the early phases of design to specific areas where potential problems are suspected. It is narrow in focus and easier to accomplish than the FMECA, requiring less input data to complete. For large and complex systems, which are highly software-intensive and where there are many interfaces, the use of the FTA is often preferred in lieu of the FMECA. The FTA is most beneficial if conducted, not in isolation, but as part of an overall system analysis process.[4]

A.3 RELIABILITY-CENTERED MAINTENANCE (RCM)

A.3.1 Definition of the Problem

Reliability-centered maintenance (RCM) is a systematic approach to develop a focused, effective, and cost-efficient preventive maintenance program and control plan for a system or product. This technique is best initiated during the early system design process and evolves as the system is developed, produced, and deployed. However, this technique also can be used to evaluate preventive maintenance programs for existing systems, with the objective of continuous product/process improvement.

The RCM technique was developed in the 1960s primarily through the efforts of

[4]Reliability Analysis Center (RAC), *Fault Tree Analysis Application Guide,* Rome Air Development Center, Rome, NY, is an excellent "how-to" source for the application of FTA depicting numerous case studies.

the commercial airline industry.[5] The approach is through a structured decision tree that leads the analyst through a "tailored" logic in order to delineate the most applicable preventive maintenance tasks (their nature and frequency). The overall process involved in implementing the RCM technique is illustrated in Figure A.11. Note that the functional analysis and the FMECA are necessary inputs to the RCM, and that there are trade-offs resulting in a balance between preventive maintenance and the accomplishment of corrective maintenance. Figure A.12 presents a simplified RCM decision logic, where system saftely is a prime consideration along with performance and cost.

A.3.2 The Analysis Process

The major steps in accomplishing an RCM analysis follow:

1. Identify the critical system functions and/or components. The first step in this analysis is to identify critical system functions and/or components; for example, airplane wings, car engine, printer head, video head, and so on. Criticality in terms of this analysis is a function of the failure frequency, the failure effect severity, and the probability of detection of the relevant failure modes. The concept of criticality is discussed in more detail in Section A.1. This step is facilitated through outputs from the system functional analysis (see Section 2.7) and the failure mode, effects, and criticality analysis (FMECA). This is also depicted in Figure A.11, Blocks 1.0 to 4.0.

2. Apply the RCM decision logic and PM program development approach. The critical system elements are next subjected to the tailored RCM decision logic. The objective here is to better understand the nature of failures associated with the critical system functions or components. In each case, and whenever feasible, this knowledge is translated into a set of preventive maintenance tasks, or a set of redesign requirements. A simplified illustrative RCM decision logic is depicted in Figure A.12. Numerous decision logics, with slight variations to the original MSG-3 logic and tailored to better address certain types of systems, have been developed and are currently being utilized.[6]

[5] A maintenance steering group (MSG) was formed in the 1960s that undertook the development of this technique. The result was a document entitled *747 Maintenance Steering Group Handbook: Maintenance Evaluation and Program Development (MSG-1)* published in 1968. This effort, focused toward a particular aircraft, was next generalized and published in 1970 as *Airline/Manufacturer Maintenance Program Planning Document-MSG2.* The MSG-2 approach was further developed and published in 1978 as *Reliability Centered Maintenance,* Report Number A066-579, prepared by United Airlines, and in 1980 as *Airline/Manufacturer Maintenance Program Planning Document-MSG3.* The MSG-3 report has been revised and is currently available as *Airline/Manufacturer Maintenance Program Development Document (MSG-3), 1993.* These reports are available from the Air Transport Association.

[6] RCM decision logics, with some variations, have also been proposed in (1) MIL-STD-2173(AS), "Reliability-Centered Maintenance Requirements for Naval Aircraft, Weapons Systems, and Support Equipment"; (2) AMC-P-750-2, *Guide to Reliability-Centered Maintenance;* and (3) John Moubray, *Reliability-Centered Maintenance,* Butterworth-Heinemann, London, 1991.

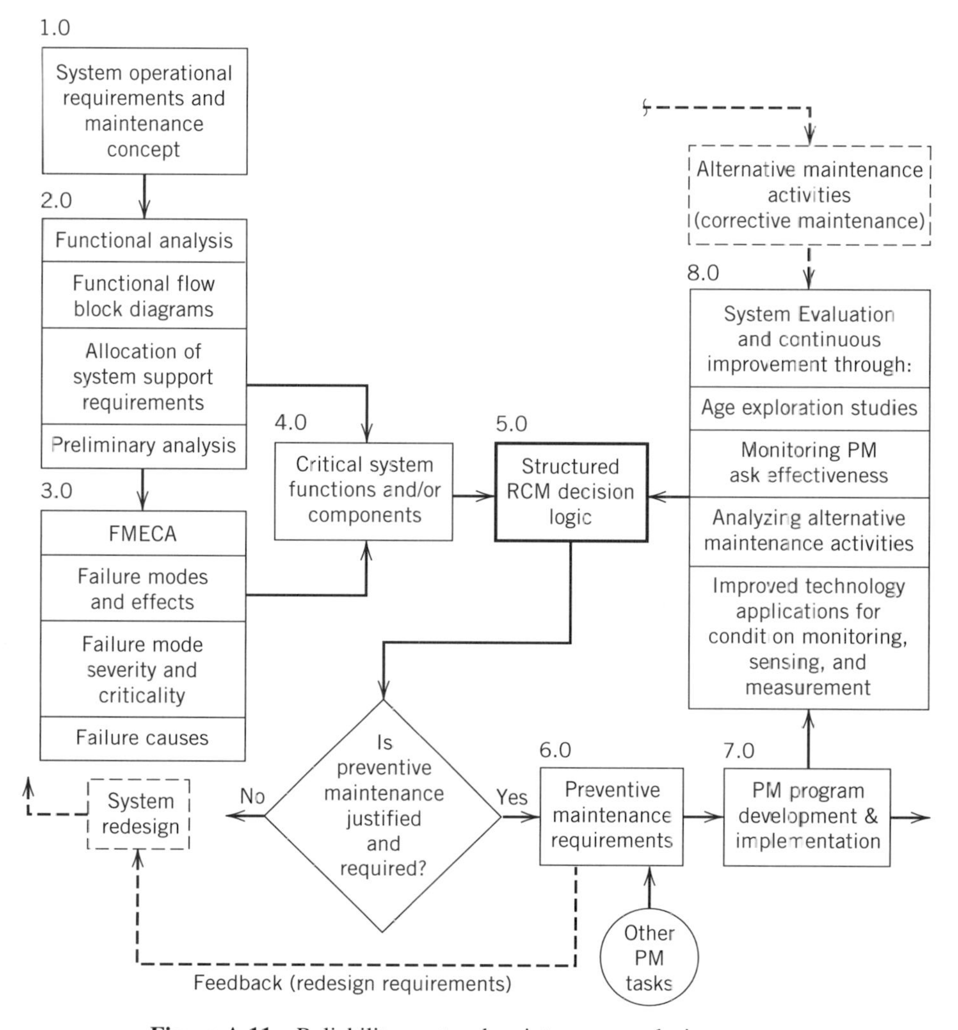

Figure A.11 Reliability-centered maintenance analysis process.

These slight variations notwithstanding (as illustrated in Figure A.12), the first concern is whether a *failure is evident or hidden.* A failure could become evident through the aid of certain color-coded, visual gauges and/or alarms. It may also become evident if it has a perceptible impact on system operation and performance. On the other hand, a failure may not be evident (i.e., hidden) in the absence of an appropriate alarm, and even more so if it does not have an immediate or direct impact on system performance. For example, a leaking engine basket is not likely to reflect an immediate and evident change in the automobile's operation, but it may in time and, after most of the engine oil has leaked, cause engine seizure. In the event that a failure is not immediately evident, it may be necessary to either institute

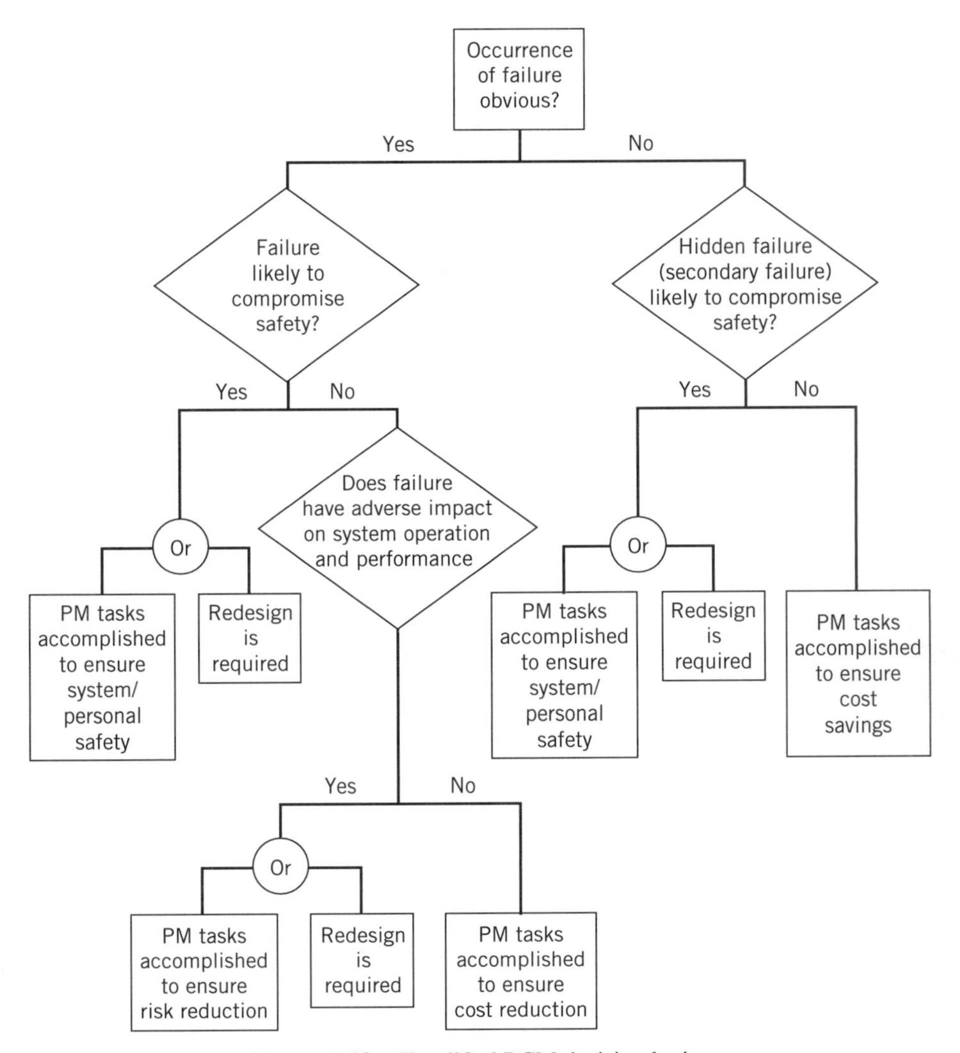

Figure A.12 Simplified RCM decision logic.

a specific fault-finding task as part of the overall PM program or design in an alarm that signals a failure (or pending failure).

The next concern is whether the failure is likely to compromise personal safety or system functionality. Queries exist in the decision logic to clarify this and other likely impacts of failures. This step in the overall process can be facilitated by the results of the FMECA (Section A.1). The objective is to better understand the basic nature of the failure being studied. Is the failure likely to compromise the system or personnel safety? Does it have an operational or economic impact? For example, a failure of an aircraft wing may be safety-related, whereas a certain failure in the case of an automobile engine may result in increased oil consumption without any operational degradation, and will therefore have an economic impact. In another case, a failed printer head may result in a complete loss of printing capability and is said to have an operational impact, and so on.

Once the failure has been identified as a certain type, it is then subjected to another set of questions. However, in order to answer this next set of questions adequately, the analyst must thoroughly understand the nature of the failure from a *physics-of-failure* perspective. For example, in the event of a crack in the airplane wing, how fast is this crack likely to propagate? How long before such a crack causes a functional failure?

These questions have an underlying objective of delineating a feasible set of compensatory provisions or preventive maintenance tasks. Is a lubrication or servicing task applicable and effective, and, if so, what is the most cost-effective and efficient frequency? Will a periodic check help preclude the failure, and at what frequency? Periodic inspections or checkouts are likely to be most applicable in situations where a failure is unlikely to occur immediately, but is likely to develop at a certain rate over a period of time. The frequency of inspections can vary from very infrequently to continuously, as in the case of condition monitoring. Some of the more specific queries are presented in Figure A.12. In each case, the analyst must not only respond with a "yes" or "no," but should also give specific reasons for each response. Why would lubrication either make, or not make, any difference? Why would periodic inspection be a *value-added* task? It could be that the component's wearout characteristics have a predictable trend, in which case inspections at predetermined intervals could preclude corrective maintenance. Would it be effective to discard and replace certain system elements in order to upgrade the overall inherent reliability? And, if so, at what intervals or after how many hours of system operation (e.g., changing the engine oil after 3000 miles of driving)? Further, in each case a trade-off study, in terms of the benefit/cost and overall impact on the system, needs to be accomplished between performing a task and not performing it.

In the event that a set of applicable and effective preventive maintenance requirements are delineated, they are input to the preventive maintenance program development process and subsequently implemented, as shown in Figure A.11, Blocks 5.0 to 7.0. If no feasible and cost-effective provisions or preventive maintenance tasks can be identified, a redesign effort may need to be initiated.

3. Accomplish PM program implementation and evaluation. Very often, the PM program initially delineated and implemented is likely to have failed to consider certain aspects of the system, delineated a very conservative set of PM tasks, or both. Continuous monitoring and evaluation of preventive maintenance tasks along with all other (corrective) maintenance actions is imperative in order to realize a cost-effective preventive maintenance program. This is depicted in Figure A.11, Block 8.0. Further, given the continuously improving technology applications in the field of condition monitoring, sensing, and measurement, PM tasks need to be reevaluated and modified whenever necessary.

Often, when the RCM technique is conducted in the early phases of the system design and development process, decisions are made in the absence of ample data. These decisions may need to be verified and modified, whenever justified, as part of the overall PM evaluation and continuous improvement program. Age exploration studies are often conducted to facilitate this process. Tests are conducted on samples of unique system elements or components with the objective of better understanding their reliability and wear-out characteristics under actual operating conditions. Such studies can aid the evaluation of applicable PM tasks, and help delineate any dominant failure modes associated with the component being monitored and/or any correlation between component age and reliability characteristics. If any significant correlation between age and reliability is noticed and verified, the associated PM tasks and their frequency may be modified and adapted for greater effectiveness. Also, redesign efforts may be initiated to account for some, if any, of the dominant component failure modes.

A.3.3 The Analysis Results

Quite often in the early design process, as system components are being selected, the issue of maintenance is ignored altogether. If addressed, however, the designer may tend to specify components requiring some preventive maintenance (usually recommended by the manufacturer). By doing so, the perception is that such PM recommendations are based on actual knowledge of the component in terms of its physical characteristics, expected modes of failure, and so on. It is also believed that the more preventive maintenance required, the better the reliability. In any event, there is often a tendency to overspecify the need for PM because of the reliability issue, particularly if the component *physics-of-failure* characteristics are not known and the designer assumes a conservative approach just in case!

Experience has indicated that although the accomplishment of some selective preventive maintenance is essential, the overspecification of PM activities can actually cause a degradation of system reliability and can be quite costly! The objective is to specified the correct amount of PM, to the depth required, and at the proper frequency; that is, *not too much or too little!* Further, as systems age, the required amount of PM may shift from one level to another. The application of RCM methods on a continuing basis is highly recommended, particularly when evaluating systems from a life-cycle cost perspective.

A.4 MAINTENANCE TASK ANALYSIS (MTA)

A.4.1 Definition of the Problem

Company DEF has been manufacturing Product 12345 for the past few years. The costs have been higher than anticipated and international competition has been increasing! As a result, company management has decided to conduct an evaluation of the overall production capability, identify "high-cost" contributors through the accomplishment of a life-cycle cost analysis, and identify possible problem areas where improvement can be realized. One area for possible improvement is the manufacturing test function where frequent failures have occurred during Product 12345 test. By reducing maintenance costs, it is likely that one can reduce the overall cost of the product and improve the company's competitive position in the marketplace. With the objective of identifying some "specifics," a detailed maintenance task analysis of the manufacturing test function is accomplished. Specific recommendations for improvement are being solicited.

A.4.2 The Analysis Process

In response, a detailed maintenance task analysis is accomplished using the format included in Appendix B of B. S. Blanchard, *Logistics Engineering and Management,* 4th Ed., Prentice Hall, Upper Saddle River, NJ, 1992. The format, as adapted for the purpose of this evaluation, includes the following general steps:

1. Review of historical information covering the performance of the manufacturing test capability indicated the frequent loss of power during the final testing of Product 12345. From this, a typical "symptom of failure" was identified, and a sample logic troubleshooting flow diagram was developed as shown in Figure A.13.
2. The applicable "go/no-go" functions, identified in Figure A.13, are converted to the task analysis format in Figures A.14 and A.15. The functions are analyzed on the basis of determining task requirements (task durations, parallel-series relationships, sequences), personnel quantity and skill-level requirements, spare/repair part requirements, test and support equipment requirements, special facility requirements, technical data requirements, and so on. The intent of Figure A.14 is to lay out the applicable maintenance tasks required, determine the anticipated frequency of occurrence, and identify the logistic support resources that are likely to be necessary for the performance of the required maintenance. This information, in turn, can be evaluated on the basis of cost.
3. Given the preliminary results of the analysis in terms of the layout of expected maintenance functions/tasks, the next step is evaluate the information presented in Figures A.14 and A.15, and suggest possible areas where improvement can be made.

A.4.3 The Analysis Results

Review of the information presented in Figures A.14 and A.15 suggests that the following areas be investigated further:

1. With the extensive resources required for the repair of Assembly A-7 (e.g., the variety of special test and support equipment, the necessity for a "cleanroom" facility for maintenance, the extensive amount of time required for the removal and replacement of CB-1A5, etc.), it may be feasible to identify Assembly A-7 as being nonrepairable! In other words, investigate the feasibility of whether the assemblies of Unit B should be classified as "repairable" or "discard at failure."
2. From Tasks 01 and 02, a "built-in test" capability exists at the organizational level for fault isolation to the Subsystem. However, fault isolation to the Unit re-

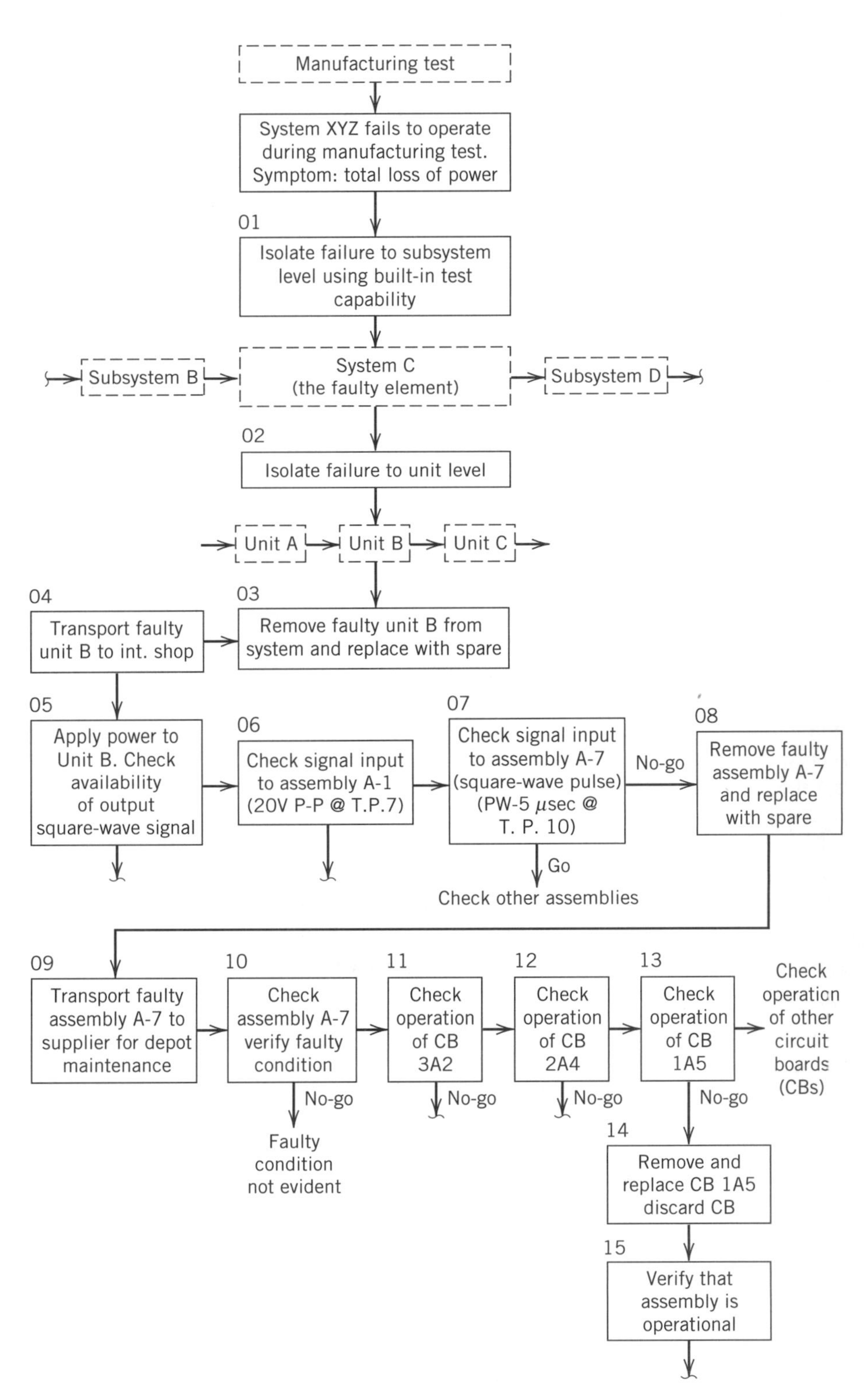

Figure A.13 Abbreviated logic troubleshooting flow diagram.

1. System: XYZ	2. Item name/part no.: Manufacturing Test/A4321		3. Next higher Assy.: Assembly and test		4. Description of requirement: During manufacturing and test of Product 12345 (Serial No. 654), System XYZ failed to operate. The symptom of failure was "loss of total power output." Requirement: Troubleshoot and repair system.
5. Req. No.: 01	6. Requirement: Diag./Repair	7. Req. Freq.: 0.00450	8. Maint. Level: Org/Inter.	9. Ma. Cont. No.: A12B100	

10. Task Number	11. Task Description	12. Elapsed time - minutes (2 4 6 8 10 12 14 16 18 20 22 24 26 28 30 32 34 36 38)	13. Total elap time	14. Task Freq	Personnel Man-min 15. B	16. I	17. S	18. Total
01	Isolate failure to subsystem level		5	0.00450	5	-	-	5
	(Subsystem C is faulty)							
02	Isolate failure to unit level	(1)	25		-	25	-	25
	(Unit B is faulty)	(2)					25	25
03	Remove Unit B from system and	(2nd cycle)	15		15	-	-	15
	replace with a spare Unit B							
04	Transport faulty unit to int. shop		30		30	-	-	30
05	Apply power to faulty unit. Check	(3rd cycle)	20		-	20	-	20
	for output squarewave signal							
06	Check signal input to Assembly A-1		15		-	15	-	15
	(20v P-P @ T.P.7)							
07	Check signal input to Assembly A-7	(4th cycle)	20		-	20	-	20
	(squarewave, PW-5μsec @T.P. 2)							
08	Remove faulty A-7 & replace		10		10	-	-	10
09	Transport faulty Assembly A-7 to	14 calendar days in transit						
	supplier for depot maintenance							
10	Check A-7 & verify faulty condition	(5th cycle)	25		-	-	25	25
11	Check operation of CB-3A2		15		-	-	15	15
12	Check operation of CB-2A4		10		-	-	10	10
13	Check operation of CB-1A5	(6th cycle)	20		-	-	20	20
14	Remove and replace faulty CB-1A5	(7th cycle)	40		-	40	-	40
	Discard faulty circuit board							
15	Verify that assembly is operational		15		-	-	15	15
	and return to inventory							
		Total	265	0.00450	60	120	110	290

Figure A.14 Maintenance task analysis (Part 1).

1. Item name/Part No.: Manufacturing Test/A4321		2. Req No.: 01		3. Requirement: Diagnostic Troubleshooting and repair		4. Req. Freq: 0.00450	5. Maint level: Organization, Intermediate, Depot	6. Ma. Cont. No.: A12B100
		Replacement Parts		Test & support/Handling equipment				
7. Task No.	8. Qty per Assy	9. Part nomenclature 11. Part number	10. Rep Freq	12. Qty	13. Item part nomenclature 15. Item part number	14. Use time (min)	16. Description of Facility Requirements	17. Special Technical Data Instructions
01	-	-	-	1	Bulit-in test equip. A123456	5	-	Organizational Maintenance
02				1	Special system tester 0-2310B	25	-	
03	1	Unit B B180265X	0.01866	1	Standard tool kit STK-100-B	15	-	
04	-	-	-	1	Standard cart (M-10)	30	-	Intermediate maintenance
05	-	-	-	1	Special system tester I-8891011-A	20	-	
06	-	-	-	1	Special system tester I-8891011-A	15	-	
07	-	-	-	1	Special system tester I-8891011-A	20	-	
08	1	Assembly A-7 MO-2378A	0.00995	1	Special extractor tool EX20003-4	10	- -	Refer to special removal instructions
09	-	-	-	1	Container, special handling T-300A	14 days	-	Normal trans. environment
10	-	-	-	1	Special system tester I-8891011-B	25	Clean room environment	Supplier (depot) maintenance
11	-	-	-	1	C.B. test set D-2252-A	15		
12	-	-	-	1	C.B. test set D-2252-A	10		
13	-	-	-	1	C.B. test set D-2252-A	20		
14	1	CB-1A5 GDA-221056C	0.00450	1	Special Extractor tool/EX45112-63 Standard tool kit STK-200	40		
15	-	-	-	1	Special system tester I-8891011-B	15		Return operating assy. to inventory

Figure A.15 Maintenance task analysis (Part 2).

quires a Special System Tester (0-2310B), and it takes 25 minutes of testing plus a highly skilled (supervisory skill) individual to accomplish the function. In essence, one should investigate the feasibility of extending the built-in test down to the unit level and eliminate the need for the special system tester and the high-skill level individual.

3. The physical removal and replacement of Unit B from the system take 15 minutes, which seems rather extensive. Although perhaps not a major item, it would be worthwhile investigating whether the removal/replacement time can be reduced (to less than 5 minutes, for example).

4. From Tasks 10 to 15, a special cleanroom facility is required for maintenance. Assuming that the various assemblies of Unit B are repaired (versus being classified as "discard at failure"), then it would be worthwhile to investigate changing the design of these assemblies such that a cleanroom environment is not required for maintenance. In other words, can the expensive maintenance facility requirement be eliminated?

5. There is an apparent requirement for a number of new "special" test equipment/tool items, that is, special system tester (0-2310B), special system tester (I-8891011-A), special system tester (I-8891011-B), CB test set (D-2252-A), special extractor tool (EX20003-4), and special extractor tool (EX45112-6). Usually, these *special* items are limited as to general application for other systems, and are expensive to acquire and maintain. Initially, one should investigate whether or not these items can be eliminated; if test equipment/tools are required, can *standard* items be utilized (in lieu of special items)? Also, if the various special testers are required, can they be integrated into a "single" requirement? In other words, can a single item be designed to replace the three special testers and the CB test set? Reducing the overall requirements for special test and support equipment is a major objective.

6. From Task 09, there is a special handling container for the transportation of Assembly A-7. This may impose a problem in terms of the availability of the container at the time and place of need. It would be preferable if normal packaging and handling methods could be utilized.

7. From Task 14, the removal and replacement of CB-1A5 takes 40 minutes and requires a highly skilled individual to accomplish the maintenance task. Assuming that Assembly A-7 is repairable, it would be appropriate to simplify the circuit board removal/replacement procedure by incorporating plug-in components, or at least simplify the task to allow one with a basic skill level to accomplish.

A.5 LEVEL-OF-REPAIR ANALYSIS (LORA)

A.5.1 Definition of the Problem

In the design of system components, one of the decision factors relates to the question: Should the component be designed to be "repairable" or should it be designed to be "discarded" in the event of failure? If it is designed to be repairable, at what level of maintenance should the repair be accomplished? Although these questions can be applied to any component of the system (e.g., equipment, unit, assembly, module, and element of software), this case study applies to the design of Assembly A-1. This assembly is one of 15 assemblies in Unit B of System XYZ. The objective is to evaluate design alternatives for the assembly on the basis of economic criteria.

A.5.2 The Analysis Process

The accomplishment of a level-of-repair analysis requires that the item being evaluated be presented in terms of a system operational requirement, a maintenance concept, and a program plan. In this instance, it is assumed that System XYZ is installed in an aircraft. When a maintenance action is required, there is a built-in test capability within the aircraft that allows one to isolate the fault to Unit A, Unit B, or Unit C. The applicable unit is removed, replaced with a spare, and the faulty item is transported to the intermediate-level maintenance shop for corrective maintenance. In the maintenance shop, fault isolation is accomplished within the unit to the assembly level. The faulty assembly is removed, replaced with a spare, and the unit is checked out and returned to the inventory as an operational spare. The basic question pertains to the disposition of the assembly?

In approaching this problem, the first step is to accomplish a level-of-repair analysis on Assembly A-1 as an individual entity. Subsequently, the results of this part of the analysis need to be viewed in the context of the whole; that is, the results of similar analyses involving Assemblies A-2, A-3, . . . , and A-15, and the applicable assemblies of Unit A and Unit C. There is usually a feedback effect among the individual assembly analysis,, the unit-level analysis, and the overall maintenance concept for the system as a whole.

In completing the level-of-repair analysis on Assembly A-1, the following information is provided:

1. System XYZ is installed in each of 60 aircraft, which are distributed equally at five operating sites over an 8-year time period. System utilization is on the average of 4 hours per day, and the total operating time for all systems is 452,600 hours.

2. As stated before, System XYZ includes three units: Unit A, Unit B, and Unit C. Unit B includes 15 assemblies, one of which is Assembly A-1. The estimated acquisition cost for Assembly A-1 (including design and development cost and production cost) is $1,700 each if the assembly is designed to be repairable, and $1,600 each if the assembly is to be designed to be discarded at failure. The design for repairability considers the incorporation of diagnostic provisions, accessibility, internal labeling, and so on, which is apt to cost more in terms of design and production costs.

3. The estimated failure rate (or corrective maintenance rate) of Assembly A-1 is 0.00045 failure per hour of system operation. When failures occur, repair is accomplished by a single technician who is assigned for the duration of the allocated active maintenance time. The estimated $\bar{M}ct$ is 3 hours. The loaded labor rate is $20 per labor hour for intermediate-level maintenance and $30 per labor hour for depot-level maintenance.

4. Supply support includes three categories of cost: the cost of spare assemblies in inventory, the cost of spare components to enable the repair of faulty assemblies, and the cost of inventory management and maintenance. Assume that 5 spare assemblies will be required in inventory when maintenance is accomplished at the intermediate level, and that 10 spare assemblies will be required when maintenance is accomplished at the depot level. For component spares, assume that the average cost of material consumed per maintenance action is $50. The estimated cost of inventory maintenance is assumed to be 20% of the inventory value (the summation of the costs for assembly and component spares).

5. When assembly repair is accomplished, special test and support equipment is required for fault diagnosis and assembly checkout. The cost per test station is $12,000, which includes acquisition cost and amortized maintenance cost. This cost is that part of the total cost that is attributed to the maintenance requirement for Assembly A-1, and there are five test stations required for intermediate-level maintenance.

6. Transportation and handling cost is considered as being negligible when maintenance is accomplished at the intermediate level. However, assembly maintenance accomplished at the depot level will involve an extensive amount of transportation. For depot maintenance, assume $150 per 100 pounds per one-way trip (independent of distance), and that the packaged assembly weighs 20 pounds.

7. The allocation for Assembly A-1 relative to maintenance facility cost is categorized in terms of an initial fixed cost, and a sustaining recurring cost proportional to facility utilization requirements. The initial fixed cost is $1,000 per installation, and the assumed usage cost allocation is $1.00 per direct maintenance labor hour at the intermediate level and $1.50 per direct labor hour at the depot level.

8. Technical data and maintenance software requirements constitute the maintenance instructions to be included in the technical manuals to support assembly repair activities, and the failure reporting and maintenance data covering each maintenance action in the field. Assume that the cost for preparing and distributing maintenance instructions (and supporting computer software) is $1,000, and that the cost for field maintenance data is $25 per maintenance action.

9. There will be some initial formal training costs associated with maintenance personnel when considering the assembly repair option. Assume 30 student-days of formal training for the intermediate level of maintenance (for the five sites in total) and 6 student-days for depot-level maintenance. The cost of training is $150 per student-day. The requirement for replenishment training as a result of attrition or turnover is considered as being negligible.

10. As a result of maintenance, there will be a requirement for disposal and/or the recycling of material. The assumed disposal cost is $20 per assembly and $2 per component.

The objective is to evaluate Assembly A-1 based on the information provided. Should Assembly A-1 be designed for (1) repair at the intermediate level of maintenance, (2) repair at the depot level of maintenance, or (3) discarded at failure?

A.5.3 The Analysis Results

Figure A.16 presents a worksheet with the results from the evaluation of Assembly A-1. Based on the information shown, it is recommended that the assembly be *repaired at the depot level of maintenance.*

Prior to making a final decision, however, one should review the data in Figure A.16 in terms of "high-cost" contributors and the sensitivities of various input factors. Some of the initial assumptions may have a great impact on the analysis results and, perhaps, should be challenged. Also, the analyst may wish to review the source of prediction data covering reliability, maintainability, and some of the input cost factors.

Given that the repair policy decision for Assembly A-1 is verified in terms its evaluation in an "isolated" sense (i.e., a decision has been made relative to the results of the individual analysis in Figure A.16), then it is essential that this decision be reviewed in context with other assemblies of System XYZ and with the maintenance concept. Figure A.17 reflects the results of individual level-of-repair analyses accomplished for each of the major assemblies in Unit B. The same approach used for Assembly A-1 is used for the evaluation of Assemblies A-2 through A-15.

From Figure A.17, there are several major choices: (1) adopt the individual repair policy for each assembly (i.e., a "mixed" overall policy), and (2) adopt a uniform overall policy for *all* assemblies based on the lowest total policy cost (i.e., repair at depot). Both options must be reviewed in terms of the feedback effects that occur, life-cycle cost implications, and associated risks.

Figure A.18 illustrates the basic process that has been discussed herein. There are many candidate items that can be evaluated in terms of repair versus discard decisions. Quite often, such decisions will be made based on "noneconomic" criteria. It may not be technically feasible to repair an item at the intermediate level. Safety criteria and/or the need for a specialized repair facility dictates that repair must be accomplished at the depot level. The proprietary aspects of a product dictate that an item must be repaired at the producer's facility (i.e., depot). The approach used in this example deals with those components where economic evaluation is feasible. From the figure, there are some decisions that may initially be "clear-cut," and there are other decisions where a more in-depth analysis is required.

A.6 DESIGN EVALUATION OF ALTERNATIVES

A.6.1 Definition of the Problem

Company DEF is responsible for the design and development of a major system, which, in turn, is comprised of a number of large subsystems. Subsystem XYZ is to be procured from an outside supplier, and there are three different configurations being evaluated for selection. Each of the configurations represents an existing design, with some redesign and additional development necessary to be compatible with the requirements for the new system. The evaluation criteria include different parameters such as performance, operability, effectiveness, design characteristics,

Evaluation Criteria	Repair at Intermediate Cost ($)	Repair at Depot Cost ($)	Discard at Failure Cost ($)	Description and Justification
1. Estimated acquisition cost for Assembly A-1 (to include design and development, production cost)	1,700/Assembly or 102,000 (47.8%)	1,700/Assembly or 102,000 (54.7%)	1,600/Assembly or 96,000 (19.5%)	Acquisition cost is based on 60 systems. Assembly design and production costs are less in the discard case (simplified configuration).
2. Maintenance labor cost	12,240 (5.7%)	18,360 (9.8%)	Not applicable	Based on 452,600 hours of operation and a maintenance rate of 0.00045, the estimated quantity of maintenance actions is 204. When repair is accomplished, one (1) technician is assigned on a full-time basis. The $\overline{\text{M}}$ct is 3 hours. The labor rate is $20/hour for intermediate and $30/hour for depot.
3. Supply support – spare assemblies	8,500 (4%)	17,000 (9.1%)	326,400 (66.4%)	For intermediate maintenance, 5 spare assemblies are required to compensate for turnaround time, the maintenance que, etc. 10 spares are required for depot maintenance. 100% spares are required for the discard case.
4. Supply support – spare components	10,200 (4.8%)	10,200 (5.5%)	Not applicable	Assume $50 per maintenance action.
5. Supply support – inventory maintenance	3,740 (1.8%)	5,440 (2.9%)	65,280 (13.3%)	Assume 20% of the inventory value (spare assemblies and spare components).
6. Special test and support equipment	60,000 (28.1%)	12,000 (6.4%)	Not applicable	Special test equipment is required in the repair case. The acquisition cost is $12,000 per installation. There are five (5) installations at intermediate and one(1) at depot.
7. Transportation and handling	Negligible	12,240 (6.6%)	Not applicable	Transportation costs at the intermediate level are negligible. For depot maintenance, assume 408 one-way trips at $150/100 pounds. One assembly weighs 20 pounds.
8. Maintenance training	4,500 (2.1%)	900 (0.5%)	Not applicable	Assume 10 students for 3 days at $150 per student day for intermediate, and 2 students for 3 days at $150 per student day for depot.
9. Maintenance facilities	5,612 (2.6%)	1,918 (1%)	Not applicable	Assume $1.00 per direct maintenance manhour for intermediate, and $1.50 per direct manhour for depot. Also, assume an initial fixed cost of $1,000 per installation.
10. Technical data	6,100 (2.9%)	6,100 (3.3%)	Not applicable	For repair case, assume $1,000 for the cost of preparation of maintenance instructions. Also, assume $25 per maintenance action for maintenance data.
11. Disposal	408)0.2%)	408 (0.2%)	4,080 (0.8%)	Assume $20 per assembly and $2 per component as the cost of disposal.
Total Estimated Cost	$213,300	$186,566	$491,760	

Figure A.16 Repair versus discard evaluation (Assembly A-1).

Assembly Number	Repair Policy			Decision
	Repair at Intermediate	Repair at Depot	Discard at Failure	
A-1	$213,300	$186,566	$491,760	Repair-Depot
A-2	130,800	82,622	75,440	Discard
A-3	215,611	210,420	382,452	Repair-Depot
A-4	141,633	162,912	238,601	Repair-Intermediate
A-5	132,319	98,122	121,112	Repair-Depot
A-6	112,189	96,938	89,226	Discard
A-7	125,611	142,206	157,982	Repair-Intermediate
A-8	99,812	131,413	145,662	Repair-Intermediate
A-9	128,460	79,007	66,080	Discard
A-10	167,400	141,788	314,560	Repair-Depot
A-11	185,850	142,372	136,740	Discard
A-12	135,611	122,453	111,502	Discard
A-13	105,667	113,775	133,492	Repair-Intermediate
A-14	111,523	89,411	99,223	Repair-Depot
A-15	142,119	120,813	115,723	Discard
Policy Cost	$2,147,905	$1,920,808	$2,679,555	Repair-Depot

Figure A.17 Summary of repair-level decisions.

schedule, and cost. Both qualitative and quantitative considerations are covered in the evaluation process.

A.6.2 The Analysis Process

The analyst commences with the development of a list of evaluation parameters, as depicted in Figure A.19. In this instance, there is no single parameter (or figure of merit) that is appropriate by itself, but there are 11 factors that must be considered on an integrated basis. Given the evaluation parameters, the next step is to determine the level of importance of each. Quantitative weighting factors from zero to 100 are assigned to each parameter in accordance with the degree of importance. The Delphi method, or some equivalent evaluation technique, may be used to establish the weighting factors. The sum of all weighting factors is 100.

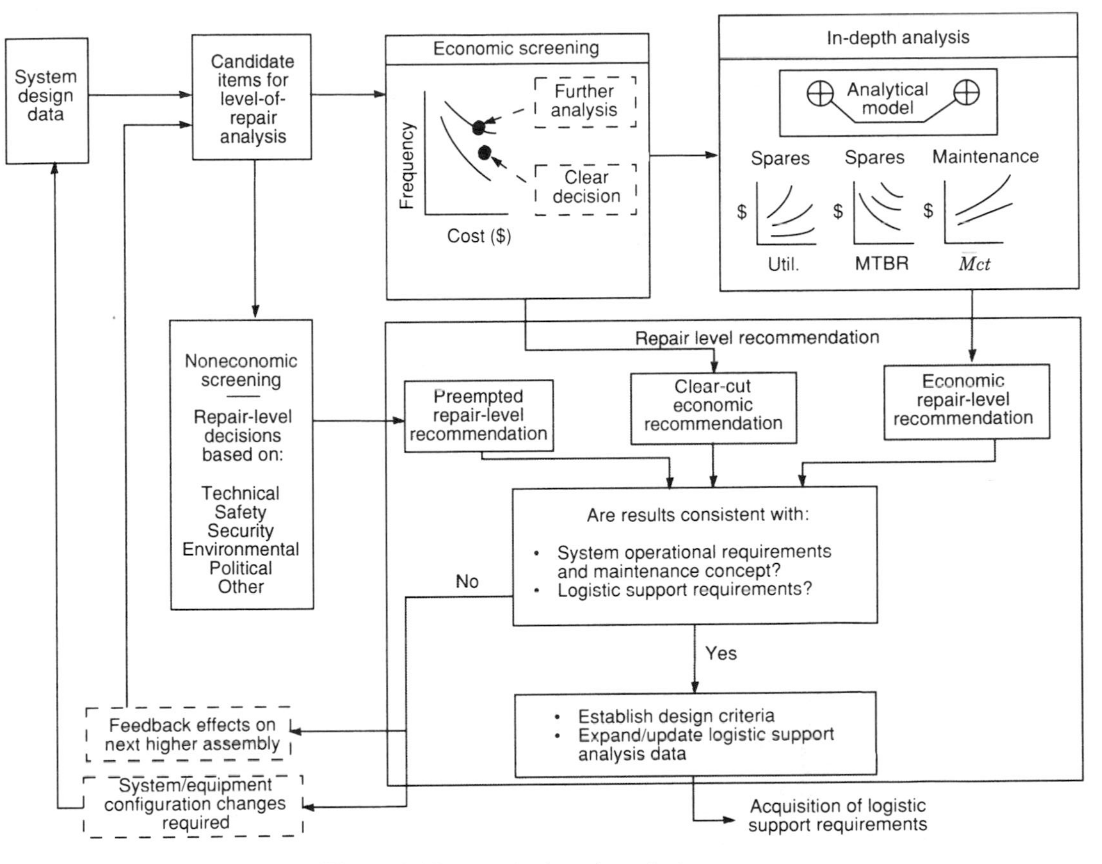

Figure A.18 Level-of-repair analysis process.

Item	Evaluation parameter	Weighting factor	Configuration A		Configuration B		Configuration C	
			Base rate	Score	Base rate	Score	Base rate	Score
1	Performance – input, output, accuracy, range, compatibility	14	6	84	9	126	3	42
2	Operability – simplicity and ease of operation	4	10	40	7	28	4	16
3	Effectiveness – Ao, MTBM, $\overline{M}ct$, $\overline{M}pt$, MDT, MLH/OH	12	5	60	8	96	7	84
4	Design characteristics – reliability, maintainability, human factors, supportability, producibility, interchange-ability	9	8	72	6	54	3	27
5	Design data – design drawings, specifications, logistics data, operating and maintenance procedures	2	6	12	8	16	5	10
6	Test aids – common and standard test equipment, calibration standards, maintenance and diagnostic computer programs	3	5	15	8	24	3	9
7	Facilities and utilities – space, weight, volume, environment, power, heat, water, air conditioning	5	7	35	8	40	4	20
8	Spare/repair parts – part type and quantity, standard parts, procurement time	6	9	54	7	42	5	30
9	Flexibility/growth potential – for reconfiguration, design change acceptability	3	4	12	8	24	6	18
10	Schedule – research and development, production	17	7	119	8	136	9	153
11	Cost – life cycle (R & D, investment, O & M)	25	10	250	9	225	5	125
Subtotal				753		811		534
Derating factor (development risk)				113 15%		81 10%		197 20%
Grand Total		100		640		730		427

Figure A.19 Evaluation summary (three alternatives).

For each of the 11 parameters identified in Figure A.19, the analyst may wish to develop a special checklist including criteria against which to evaluate the three proposed configurations. For instance, the parameter "PERFORMANCE" may be described in terms of degrees of desirability; that is, "highly desirable," "desirable," or "less desirable." Although each configuration must comply with a minimum set of requirements, one may be more desirable than the next when looking at the proposed performance characteristics. In other words, the analyst should break down each evaluation parameter into "levels of goodness"!

Each of the three proposed configurations of Subsystem XYZ is evaluated independently using the special checklist criteria. Base rating values from zero to 10 are applied according to the degree of compatibility with the desired goals. If a "highly desirable" evaluation is realized, a rating of 10 is assigned.

The base-rate values are multiplied by the weighting factors to obtain a score. The total score is then determined by adding the individual scores for each configuration. Because some redesign is required in each instance, a special derating factor is applied to cover the risk associated with the failure to meet a given requirement. The resultant values from the evaluation are summarized in Figure A.19.

A.6.3 The Analysis Results

From Figure A.19, Configuration B represents the preferred approach based on the highest total score of 730 points. This configuration is recommended on the basis of its inherent features relating to performance, operability, effectiveness, design characteristics, design data, and so on.

APPENDIX B
Life-Cycle Cost-Analysis Process

Many of our day-to-day decisions, as they pertain to the design and development of new systems and/or the reengineering of existing systems, are based on *technical* performance-related factors alone. *Economic* considerations, if addressed at all, have dealt primarily with initial procurement and acquisition costs only and not the "downstream" costs associated with system operation and maintenance support. Yet, these downstream costs, which often constitute a significant portion of the total life-cycle cost of a system, are highly influenced by the decisions made in the early phases of system development. In other words, the early decision-making process must consider the *total* spectrum of costs if economic benefits are to be gained in the long term. The consequences of the short-term approach often practiced in the past have been rather detrimental overall, as conveyed in Section 1.2 (Chapter 1). Total cost visibility, as illustrated in Figure B.1, is a *must* if the risks associated with the decision-making process are to be properly assessed.

Life-cycle costing includes the consideration of *all* future costs associated with research and development (i.e., design), construction, production, distribution, system operation, sustaining maintenance and support, system retirement, and material disposal and/or recycling. It involves the costs of all technical and management activities throughout the system life cycle; that is, customer activities, producer and/or contractor activities, supplier activities, and consumer or user activities. Although the influencing of these costs can be best realized during the early phases in the development of a *new* system, as conveyed in Figure B.2, benefits also can be gained through the evaluation and the identification of high-cost contributors for *existing* systems already in use. In other words, the applications and benefits that can be gained through the accomplishment of life-cycle cost analyses are numerous, as shown in Figure 3.40 (Chapter 3).

When performing a life-cycle cost analysis, there are a series of steps that one may follow. These steps are briefly described in Figure 3.38, and are conveyed in the context of the overall process in Figure 3.41. The purpose of this appendix is to provide some additional explanation covering each of the steps identified in Figure 3.38.

B.1 DEFINE SYSTEM REQUIREMENTS

The first step in the performance of a life-cycle cost analysis is to define the problem, identify the proposed technical solution, describe the operational requirements

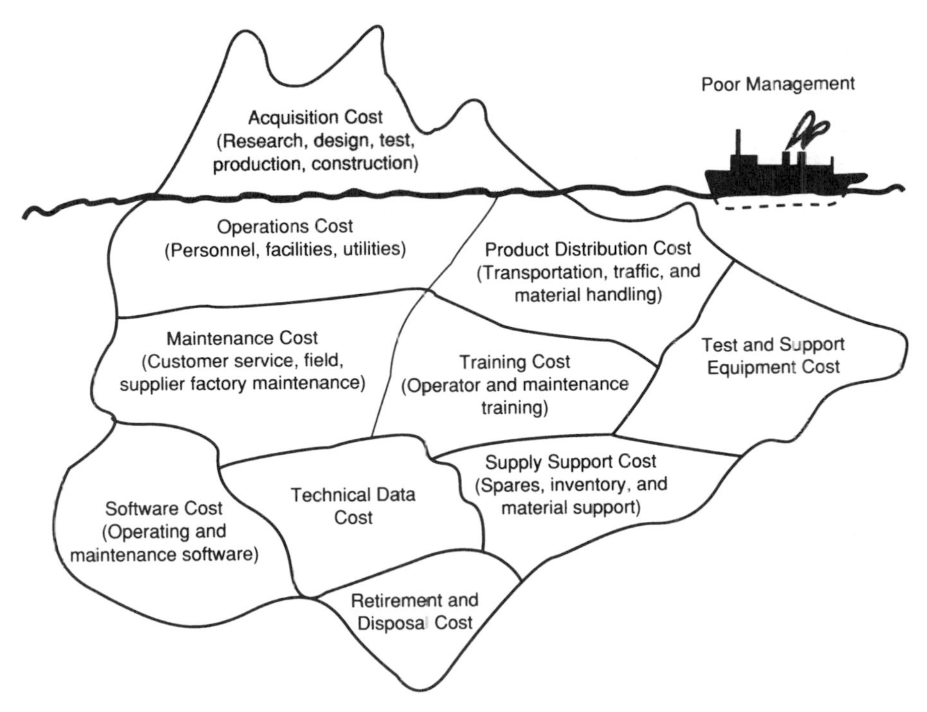

Figure B.1 Total cost visibility.

and the maintenance concept for the system, identify the critical technical performance measures (TPMs), and describe the system configuration in functional terms; that is, the process described in Sections 2.1 through 2.7 (Chapter 2). Depending on where one is in the system life cycle, the degree of definition may be rather cursory in nature or more in-depth. In any event, the basic system requirements must be defined in order to provide the necessary structure for the analysis, and the assumptions that are made at this point may have a significant impact on the results.

In Figure B.3, it is assumed that a ground vehicle in development requires the incorporation of a communications capability. Multiple quantities of the vehicle will be deployed to three different geographical locations, performing a variety of different mission scenarios (i.e., 20, 20, and 25 at each location, respectively). Although there are variations from one location to the next, it is assumed that each vehicle will be utilized on the average of 4 hours per day, 360 days per year. The equipment must enable communication with other vehicles at a range of at least 200 miles, overhead aircraft at an altitude of up to 10,000 feet, and with a centralized area communications facility. The system must have a reliability MTBF of 450 hours, a $\bar{M}ct$ of 30 minutes, a MLH/OH requirement of 0.2, and a unit life-cycle cost not to exceed \$20,000. The equipment will be functionally packaged in units (i.e., Units A, B, and C) and, in the event of failure, the problem will be isolated to the unit level, faulty units will be removed and replaced with a spare and sent back to the intermediate level of maintenance for corrective action, and so on.

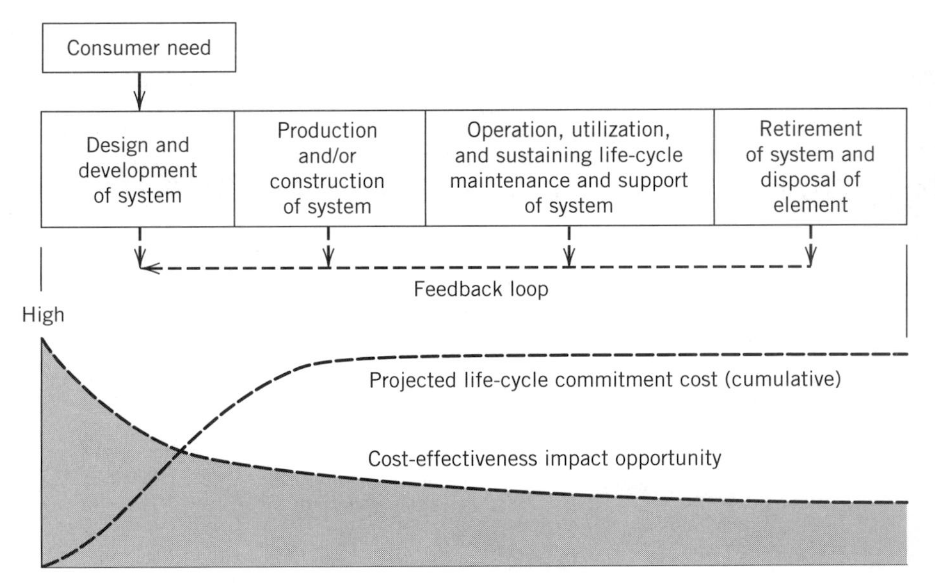

Figure B.2 Opportunity for impacting cost-effectiveness in the system life cycle.

From the figure, the system operational requirements and the maintenance concept have been defined to the depth that will allow for the accomplishment of a life-cycle cost analysis during the late conceptual design or early preliminary design phase. The next step is to describe the system, and the mission(s) that is to be performed, in *functional* terms by accomplishing a top-level functional analysis. From Figure 2.12 (Chapter 2), the communication system can be described in a similar manner, followed with an evaluation of each functional block to determine the resource requirements that will provide the basis for functional costing (see Figure 2.17).

B.2 DESCRIBE THE SYSTEM LIFE CYCLE AND IDENTIFY THE MAJOR ACTIVITIES IN EACH PHASE

Given the definition of system requirements and the identification of functions, it is appropriate to provide a timeline for these requirements in terms of the life cycle. In Figure B.3, the planned life cycle is 12 years. In other words, it is assumed that there is a need for the communication system and the functions that are to be performed for a 12-year period. Although this planning horizon may change (as requirements change), a baseline must be established. Thus, the 12-year period and the major activities identified in the figure will be assumed herein. The activity categories identified in the figure (i.e., research and development, investment/production, and operations and maintenance) form the basis for the development of a cost breakdown structure (CBS).

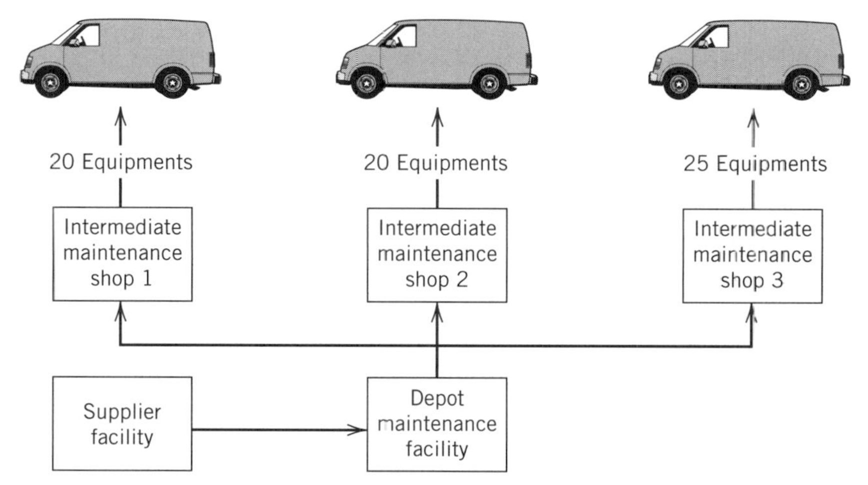

Deployment: Three geographical areas (flat and mountainous terrain)
Utilization: Four (4) hr/day throughout year (average)

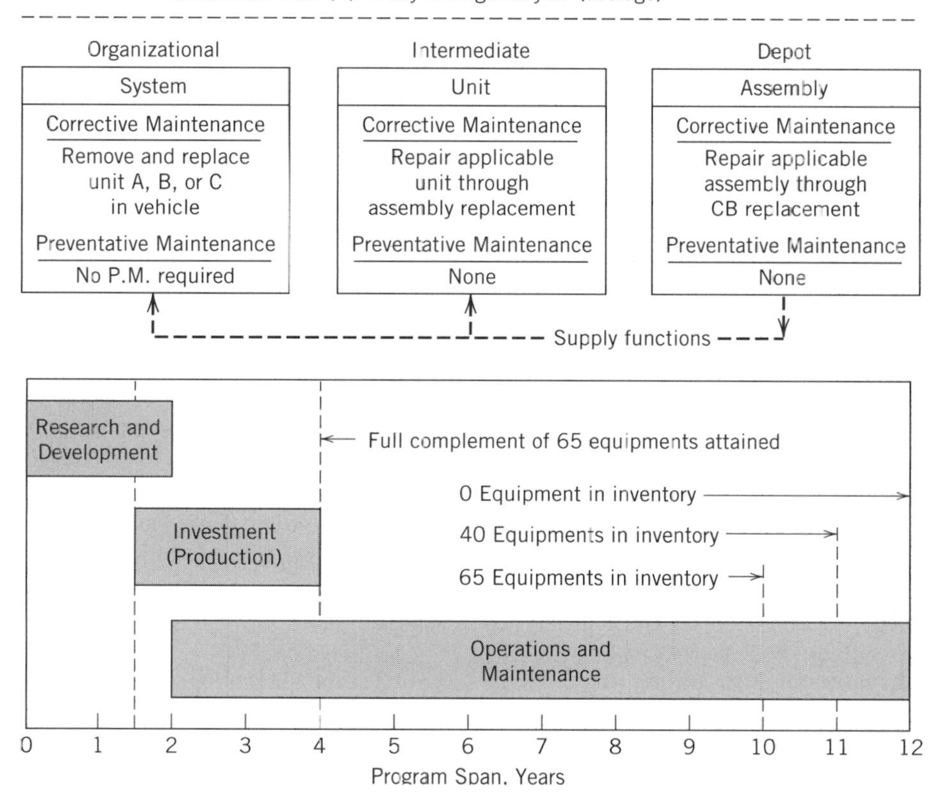

Figure B.3 Communication system requirements.

B.3 DEVELOP A COST BREAKDOWN STRUCTURE (CBS)

The functions described through the functional analysis can be broken down into subfunctions, categories of work, work packages, and ultimately the identification of physical elements. From a planning and management perspective, it is necessary to establish a top-down framework that will allow for the initial allocation and subsequent collecting, accumulating, organizing, and computing of costs. For a typical project, this may lead to the development of a work breakdown structure (WBS) prepared to show, in a hierarchical manner, all of the elements of work that are necessary to complete a given program. From Section 6.2.3 (Figure 6.11), a summary work breakdown structure (SWBS) may be developed initially, followed by one or more individual contract work breakdown structures (CWBS) designed to address specific elements of work that are covered through some form of a contractual arrangement. It is the SWBS that provides a good basis for the development of a cost breakdown structure (CBS) used in life-cycle cost analyses, primarily because its intent is to cover *all* future activities and associated costs; that is, research and development, construction/production, distribution, operation and maintenance support, and retirement activities.

The CBS is intended to show all future functions/activities, broken down to the depth necessary to provide the appropriate level of visibility, and "tailored" to the system configuration in question. Ultimately, the CBS will lead to the identification of a product and/or a process, with the objective of establishing a structure that can be initially used for the top-down allocation of costs during the conceptual design phase (refer to Section 2.8) and subsequently for the bottom-up collection of costs for the purposes of accomplishing a life-cycle cost analysis. Figure B.4 provides an illustration of a sample cost breakdown structure (CBS), and Figure B.5 provides an abbreviated example showing how each category of the CBS should be described in terms of what is included, how the costs are calculated, and the basis for accomplishing such! The CBS provides a vehicle for looking at costs from a *functional* perspective. As one proceeds with the life-cycle cost analysis, costs are estimated for each year in the planned life cycle and are summarized for each category in the CBS.[1]

B.4 ESTIMATE THE COSTS FOR EACH PHASE OF THE LIFE CYCLE

The next step is to estimate the costs, by category in the CBS, for each year in the system life cycle. Such estimates must consider the effects of inflation, learning

[1] The cost breakdown structure (CBS) should be tailored to the system in question. From Figure 3.39, another example is presented. If the system is very "software-intensive," then Category *Crs* should be broken down to show more detail. If the system is very "operator-intensive" (e.g., a ground radar tracking station requiring a large number of operating personnel), then Category *Cop* should be expanded. On the other hand, if Category *Cm* is too detailed for the purposes of a given analysis, then one can summarize the costs accordingly. The objective is to provide *visibility* relative to key functional activities.

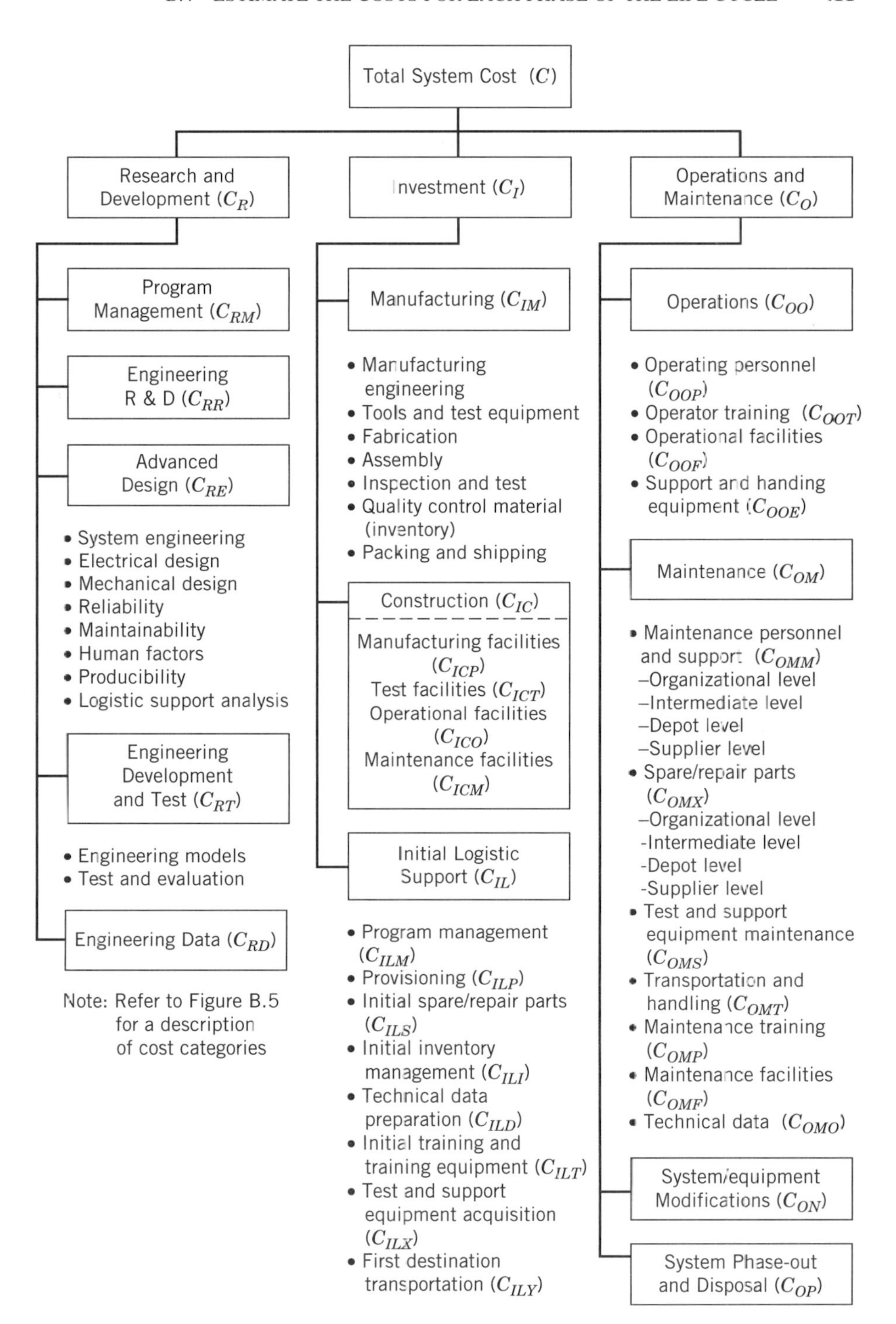

Figure B.4 Cost breakdown structure (CBS). (*Source:* B. S. Blanchard, *Logistics Engineering and Management,* 4th Ed., Prentice Hall, Englewood Cliffs, NJ, 1992.)

Cost Category (Figure B.4)	Method of Determination (Quantitive Expression)	Cost Category Description and Justification
Total system cost (C)	$C = C_R + C_I + C_O$ $C_R = R$ and D cost C_I = Investment cost C_O = Operations and maintenance cost	Includes all future costs associated with the acquisition, utilization, and subsequent disposal of system/equipment.
Research and development (C_R)	$C_R = C_{RM} + C_{RR} + C_{RE} + C_{RT} + C_{RD}$ C_{RM} = Program management cost C_{RR} = Advanced R&D cost C_{RE} = Engineering design cost C_{RT} = Equipment development/ test cost C_{RD} = Engineering data cost	Includes all costs associated with conceptual/feasibility studies, basic research, advanced research and development, engineering design, fabrication and test of engineering prototype models (hardware) and associated documentation. Also covers all related program management functions. These cost are basically nonrecurring.
Investment (C_I)	$C_I = C_{IM} + C_{IC} + C_{IL}$ C_{IM} = System/equipment manufacturing cost C_{IC} = System construction cost C_{IL} = Cost of initial logistic support	Includes all costs associated with the acquisition of systems/ equipment (once design and development has been completed). Specifically, this covers manufacturing (recurring and nonrecurring), manufactureing management, system construction, and initial logistic support.
Operations and maintenance (C_O)	$C_O = C_{OO} + C_{OM} + C_{ON} + C_{OP}$ C_{OO} = Cost of system/equipment life-cycle operations C_{OM} = Cost of system/equipment life-cycle maintenance C_{ON} = Cost of system/equipment life-cycle modifications C_{OP} = Cost of system/equipment phase-out and disposal	Includes all costs associated with the operation and maintenance support of the system throughout its life cycle subsequent to equipment delivery in the field. Specific categories cover the cost of system operation, maintenance, sustaining logic support, equipment modifications, and system/ equipment phase-out and disposal. Costs are generally determined for each year throughout the life cycle.

Figure B.5 Description of cost categories (partial).

Cost Category (Reference Figure B.4)	Method of Determination (Quantitive Expression)	Cost Category Description and Justification
Transportation and handling cost (C_{OMT})	$C_{OMT} = [(C_T)(Q_T) + (C_P)(Q_T)]$ C_T = Cost of transportation C_P = Cost of packing Q_T = Quantity of one-way shipments $C_T = [(W)(C_{TS})]$ W = Weight of item (lb) C_{TS} = Shipping cost ($/lb) C_{TS} will of course vary with the distance (in miles) of the one-way shipment. $C_P = [(W)(C_{TP})]$ C_{TP} = Packing cost ($/lbs) Packing cost and weight will vary depending on whether reuseable containers are employed.	Initial (first destination) transportation and handling costs are covered in C_{ILY}. This category includes all sustaining transportation and handling (or packing and shipping) between organizational, intermediate, depot, and supplier facilities in support of maintenance operations. This includes the return of faulty material items to a higher echelon; the transportation of items to a higher echelon for preventive maintenance (overhaul, calibration); and the shipment of spare/repair parts, personnel, data, etc., from the supplier to forward echelons.
Maintenance training cost (C_{OMP})	$C_{OMP} = [(Q_{SM})(T_T)(C_{TOM})]$ Q_{SM} = Quantity of maintenance students C_{TOM} = Cost of maintenance training ($/student-week) T_T = Duration of training program (weeks)	Initial maintenance training cost is included in C_{ILT}. This category covers the *formal* training of personnel assigned to maintain the prime equipment, test and support equipment, and training equipment. Such training is accomplished on a periodic basis throughout the system life-cycle to cover personnel replacements due to attrition. Total costs include instructor time; supervision; student pay and allowances while in school; training facilities (allocation of portion of facility required specifically for formal training); training aids and data; and student transportation as applicable.
Operational facilities cost (C_{OOF})	$C_{OOF} = [(C_{PPE} + C_U)(\%\ \text{Allocation}) \times (N_{OS})]$ C_{PPE} = Cost of operational facility support ($/site) C_U = Cost of utilities ($/site) N_{OS} = Number of operational sites *Alternative Approach* $C_{OOF} = [(C_{PPF})(N_{OS})(S_O)]$ C_{PPF} = Cost of operational facility space ($/square foot/site). Utility cost allocation is included. S_O = Facility space requirements (square feet)	Initial acquisition cost for operational facilities is included in C_{ICO}. This category covers the annual recurring costs associated with the occupancy and maintenance (repair, paint, etc.) of operational facilities throughout the system life-cycle. Utility costs are also included. Facility and utility costs are proportionately allocated to each system.

Figure B.5 *(Continued)*

curves when repetitive processes or activities occur, and any other factors that are likely to cause changes in cost, either upward or downward. Cost estimates may be derived from a combination of accounting records, cost projections, supplier proposals, and predictions in one form or another.

In Figure B.2, the early stages in the system life cycle is the preferred time to commence with the estimation of costs, because it is at this point when the greatest impact on total system life-cycle cost can be realized. However, the availability of good historical cost data at this time is almost nonexistent in most organizations,

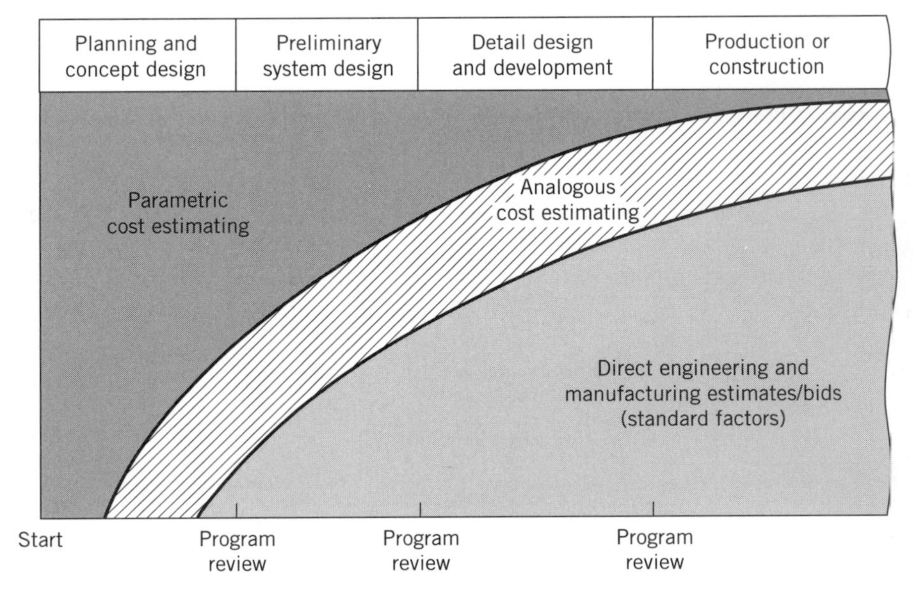

Figure B.6 Cost estimation by program phase.

particularly the type of data that pertain to the downstream activities of operations and support for similar systems in the past. Thus, one must depend heavily on the use of various cost-estimating methods in order to accomplish the end objectives.

From Figure B.6, as the system configuration becomes better defined in a developmental effort, the use of direct engineering and manufacturing standard factors based on past experience can be applied as is the case for any "cost-to-complete" projection on a typical project today (e.g., cost per labor hour). On the other hand, in the earlier stages of the life cycle when the system configuration has not been well-defined, the analyst must rely on the use of a combination of analogous and/or parametric methods developed from experience on similar systems in the past. The objective is to collect data on a "known entity," identify the major functions that have been accomplished and the costs associated with these functions, relate the costs in terms of some functional or physical parameter of the system, and then use this relationship in attempting to estimate the costs for a new system. As a goal, one should identify the applicable technical performance measures (TPMs) for the system in question and estimate the cost per a given level of performance (e.g., cost per unit of product output, cost per mile of range, cost per unit of weight, cost per volume of capacity used, cost per unit of acceleration, cost per functional output, etc.). Costs can be related to the appropriate blocks in the functional description of the system. Figures B.7 and B.8 provide some simple illustrations of considerations in cost estimating. However, care must be exercised to ensure that the historical information used in the development of cost-estimating relationships (CERs) is relevant to the system configuration being evaluated today. CERs based on the mission

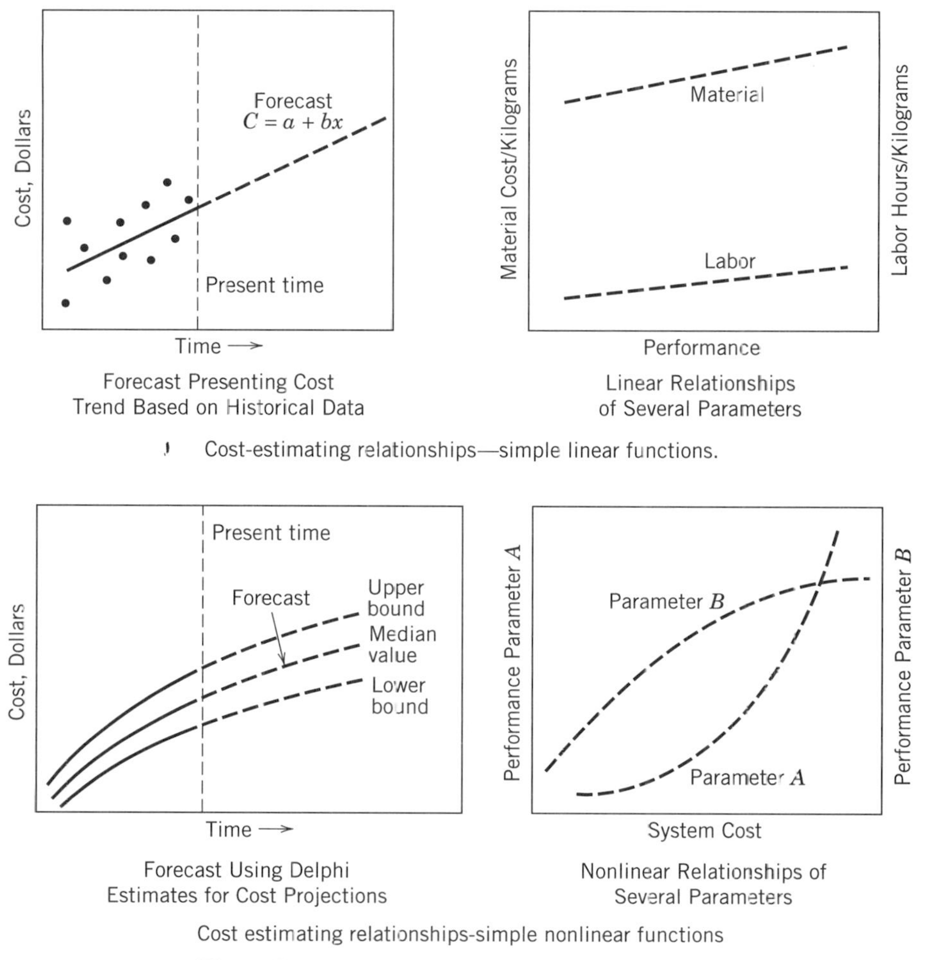

Figure B.7 Cost-estimating relationships (CERs).

and performance characteristics of one system may not be appropriate for another system configuration, even if the configuration is similar in a physical sense. Thus, costs must be related from a *functional* perspective.

To be effective in total cost management (and in the accomplishment of cost-effectiveness analyses) requires full-cost visibility allowing for the traceability of all costs back to the activities, processes, and/or products that generate these costs. In the traditional accounting structures employed in most organizations, a large percentage of the total cost cannot be traced back to the "causes"! For example, "overhead" or "indirect" costs, which often constitute greater than 50% of the total, include a lot of management costs, supporting organization costs, and other costs that are difficult to trace and assign to specific objects (refer to the overhead costs in Figure 6.26). With these costs being allocated across the board, it is impossible to

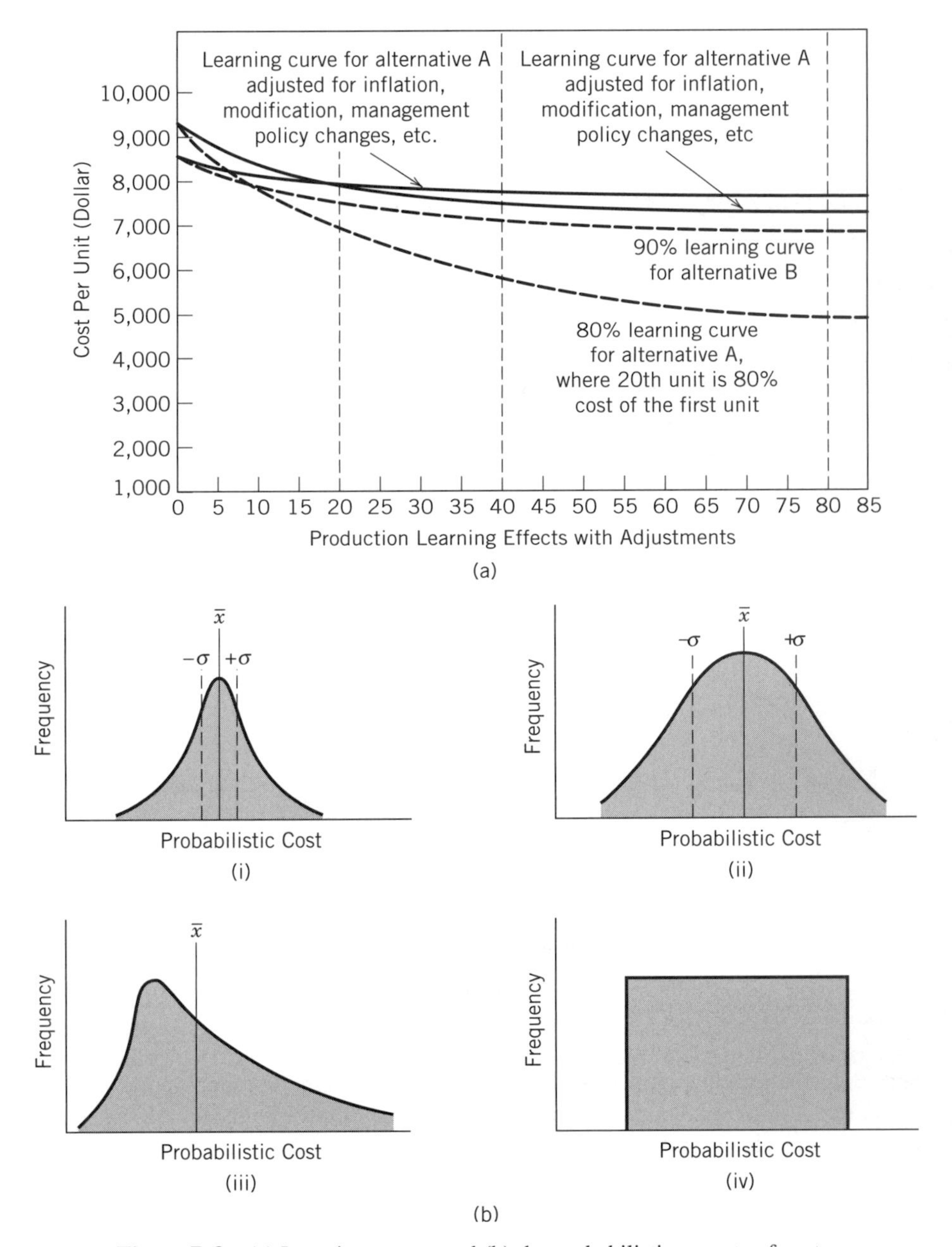

Figure B.8 (a) Learning curves and (b) the probabilistic aspects of costs.

identify the actual "causes" and to pinpoint the *true* high-cost contributors. As a result, the concept of *activity-based costing* (ABC) has been introduced.[2]

Activity-based costing is a methodology directed toward the detailing and assignment of costs to the items that cause them to occur. The objective is to enable the "traceability" of *all* applicable costs to the process or product that generates these costs. The ABC approach allows for the initial allocation and later assessment of costs by function, and was developed to deal with the shortcomings of the traditional management accounting structure where large overhead factors are assigned to all elements of the enterprise across the board without concern for whether they directly apply or not. More specifically, the principles of ABC follow:

1. Cost are directly traceable to the applicable cost-generating process, product, and/or a related object. Cause-and effect relationships are established between a cost factor and a specific process or activity.
2. There is no distinction between direct and indirect (or overhead) costs. While 80 to 90% of all costs are traceable, those nontraceable costs are not allocated across the board, but are allocated directly to the organizational unit(s) involved in the project.
3. Costs can be easy allocated on a *functional* basis; that is, the functions identified in Figures 2.13 and 2.16 (Chapter 2) It is relatively easy to develop cost-estimating relationships in terms of the cost of activities per some activity measure (i.e., the cost per unit output).
4. The emphasis in ABC is on "resource consumption" (versus "spending"). Processes and products consume activities, and activities consume resources. With resource consumption being the objective, the ABC approach facilitates the evaluation of day-to-day decisions in terms of their impact on resource consumption downstream.
5. The ABC approach fosters the establishment of cause-and-effect relationships and, as such, enables the identification of the "high-cost contributors." Areas of risk can be identified with some specific activity and the decisions that are being made within.
6. The ABC approach tends to eliminate some of the cost doubling (or double counting) that occurs when attempting to differentiate on what should be included as a "direct" cost or as an "indirect" cost. By not having the necessary visibility, there is the potential of including the same costs in both categories.

Implementation of the ABC approach, or something of an equivalent nature, is essential if one is to do a good job of total cost management. Costs are tied to objects and viewed over the long term, and the use of such facilitates the life-cycle cost-analysis process. An objective for the future is to convince the accounting

[2]J. R. Canada, W. G. Sullivan, and J. A. White, *Capital Investment Analysis for Engineering and Management,* 2nd Ed., Prentice Hall, Upper Saddle River, NJ, 1996; and P. T. Kidd, *Agile Manufacturing: Forging New Frontiers,* Addison-Wesley, Reading, MA, 1994.

organizations in various companies/agencies to supplement their current end-of-year financial reporting structure to include the objectives of ABC.

B.5 SELECT A COMPUTER-BASED MODEL TO FACILITATE THE ANALYSIS PROCESS

In the selection of a computer-based model, one must ensure that the tool selected does what is expected, is sensitive to the problem at hand, and allows for the visibility needed in addressing the system as an entity as well as any of its major components on an individual-by-individual basis. The model must enable the comparison of *many* different alternatives and aid in selecting the best among them rapidly and efficiently. The model must be *comprehensive* allowing for the integration of many different parameters; *flexible* in structure, enabling the analyst to look at the system as a whole or any part of the system; *reliable* in terms of repeatability of results; and be *user-friendly.* So often, one selects a computer model based on the advertised brochure material alone, purchases the necessary equipment and software, uses the model to manipulate data, and believes in the output results without having any idea as to how the model was put together, the internal analytical relationships established, whether it is sensitive to the variation of input parameters in terms of output results, and so on. The results of a recent survey indicates that there are over 350 computer-based tools available in the commercial marketplace and intended for use in accomplishing different levels of analysis. Each was developed on a relatively "independent" or "isolated" basis in terms of selected platform, the language used, input data needs, and interface requirements. In general, the models do not "talk to each other," are not user-friendly, and are too complex for use in early system design and development.

When using a model, it is essential that the analyst become thoroughly familiar with the tool, know how it was put together, and understand what it can do. For the purposes of accomplishing a life-cycle cost analysis, it may be appropriate to select a group of models, combined as illustrated in Figure B.9, and integrated in such a manner that will enable the analyst to look at not only the cost for the system overall, but at some of the key functional areas representing potential high-cost contributors. The model(s) must be structured around the cost breakdown structure (CBS) and in such a way that will allow the analyst to look at the costs associated with each of the major functions. Further, it must be *adaptable* for use during the early stages of conceptual design as well as in the detail design and development phase.

B.6 DEVELOP A "BASELINE" COST PROFILE

Through the application of various estimating methods, the costs for each CBS category and for each year in the system life cycle are projected in the form of a cost profile. The worksheet format presented in Figure B.10 can serve as a vehicle for recording costs, and the profile shown in Figure B.11 can represent the anticipated cost stream.

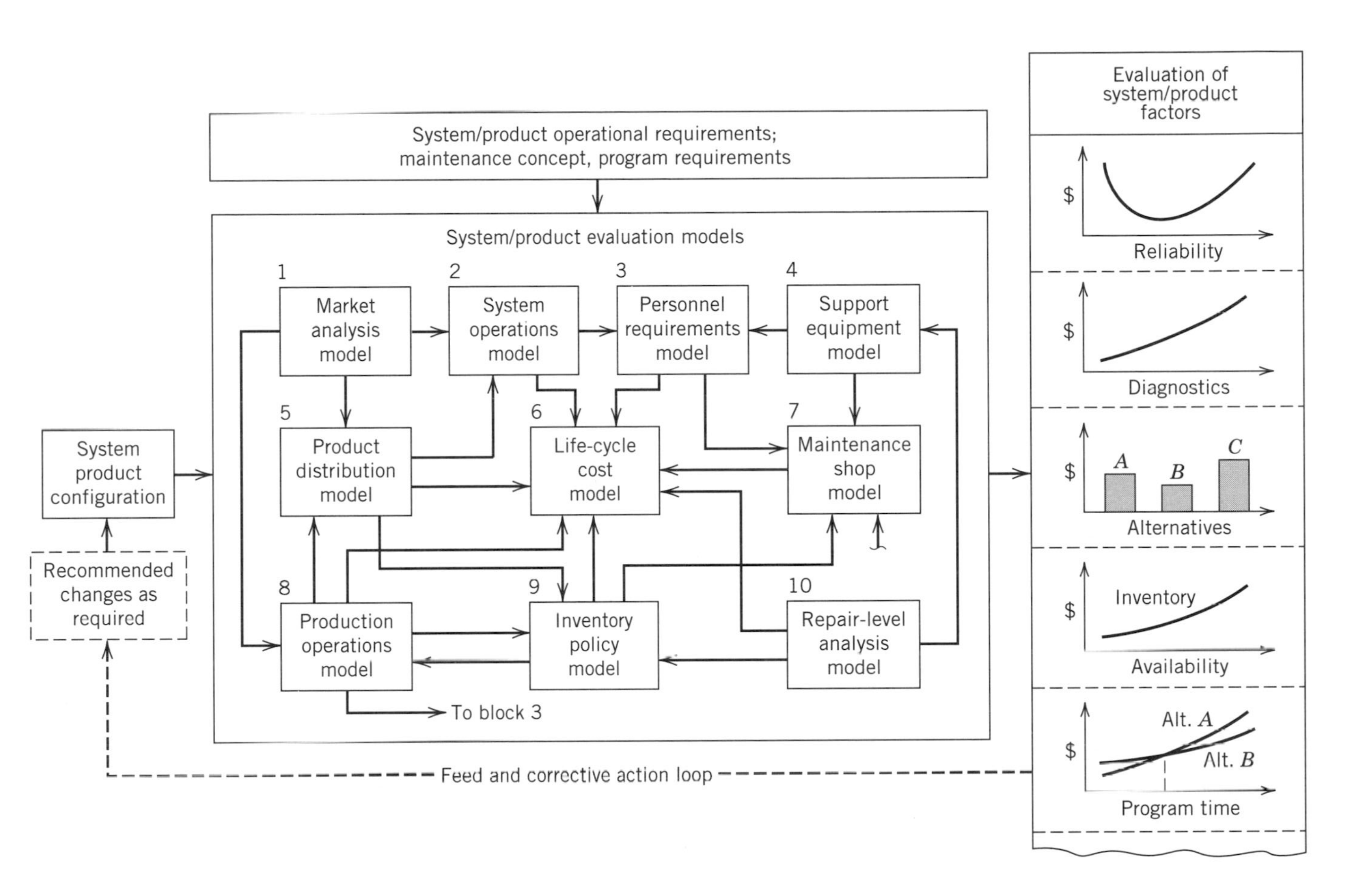

Figure B.9 Example models in life-cycle costing.

Program Activity	Cost Category Designation	Cost by Program Year ($)												Total cost ($)	Percent Contr. (%)
		1	2	3	4	5	6	7	8	9	10	11	12		
Alternative A															
1. Reserach and development cost	C_R														
a. Program management	C_{RM}														
b. Engineering design	C_{RE}														
c. Electrical design	C_{RED}														
d. Engineering data	C_{RD}														
2.															
3.															
Others															
Total Actual Cost	C														
Total P.V. Cost (10%)	$C_{(10)}$														
Alternative B															
1. Research and development cost	C_R														
a. Program management	C_{RM}														
b. Engineering design	C_{RE}														
Etc.															

Figure B.10 Cost collection worksheet.

In developing profiles, it may be feasible to start out with one presented in terms of *constant* dollars first (i.e., the costs for each year in the future presented in terms of today's dollars), and then develop a second profile by adding the appropriate inflationary factors for each year to reflect a *budgetary* stream. When comparing alternative profiles, the appropriate economic analysis methods must be applied in converting the various alternative cost streams to the *present value* or to the point in time when the decision is to be made in selecting a preferred approach. It is necessary to evaluate alternative profiles on the basis of some form of *equivalence.*[3]

B.7 DEVELOP A COST SUMMARY AND IDENTIFY THE HIGH-COST CONTRIBUTORS

In order to gain some insight pertaining to the costs for each major category in the CBS and to readily identify the *high-cost contributors,* it may be appropriate to

[3] The treatment of cost streams considering the "time value of money" is presented in most texts dealing with engineering economy. Two references are (1) G. J. Thuesen and W. J. Fabrycky, *Engineering Economy,* 8th Ed., Prentice Hall, Upper Saddle River, NJ, 1993; and (2) W. J. Fabrycky and B. S. Blanchard, *Life-Cycle Cost and Economic Analysis,* Prentice Hall, Upper Saddle River, NJ, 1991. Refer to Appendix F for additional references.

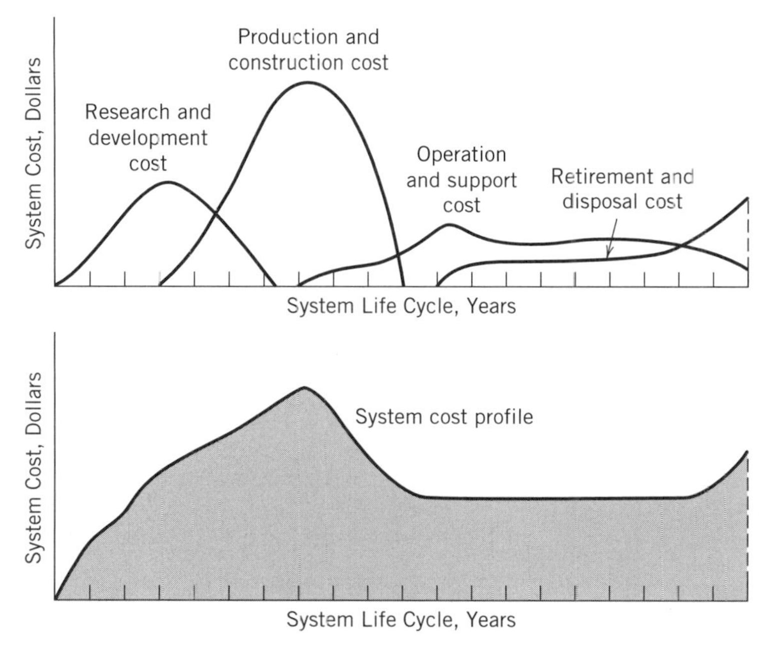

Figure B.11 Development of a cost profile.

view the results presented in a tabular form. In Figure B.12, the costs for each category are identified along with the percent contribution of each. Note that, in this example, the high-cost areas include the initial costs associated with "facilities" and "capital equipment," and the operating and maintenance costs related to the "inspection and test" function being accomplished within the production process. For the purposes of product and/or process improvement, the "inspection and test" area should be investigated further. Through the planned life cycle, 17% of the total cost is attributed to the operation and support of this functional area of activity, and the analyst should proceed with determining some of the reasons for this high cost.

B.8 DETERMINE THE CAUSE-AND-EFFECT RELATIONSHIPS PERTAINING TO HIGH-COST AREAS

Given the presentation of costs (and the percent contribution) as shown in Figure B.12, the next step is to determine the likely "causes" for these costs. The analyst will need to "revisit" the CBS, the assumptions made leading to the determination of the costs, and the cost-estimating relationships utililized in the process. Additionally, in order to ensure the proper "traceability," it is hoped that an *activity-based costing (ABC)* approach was used, or something of an equivalent nature. The application of an Ishikawa cause-and-effect diagram, as illustrated in Figure A.4 (Appendix A), may be used to assist in pinpointing the actual "causes!" The problem may

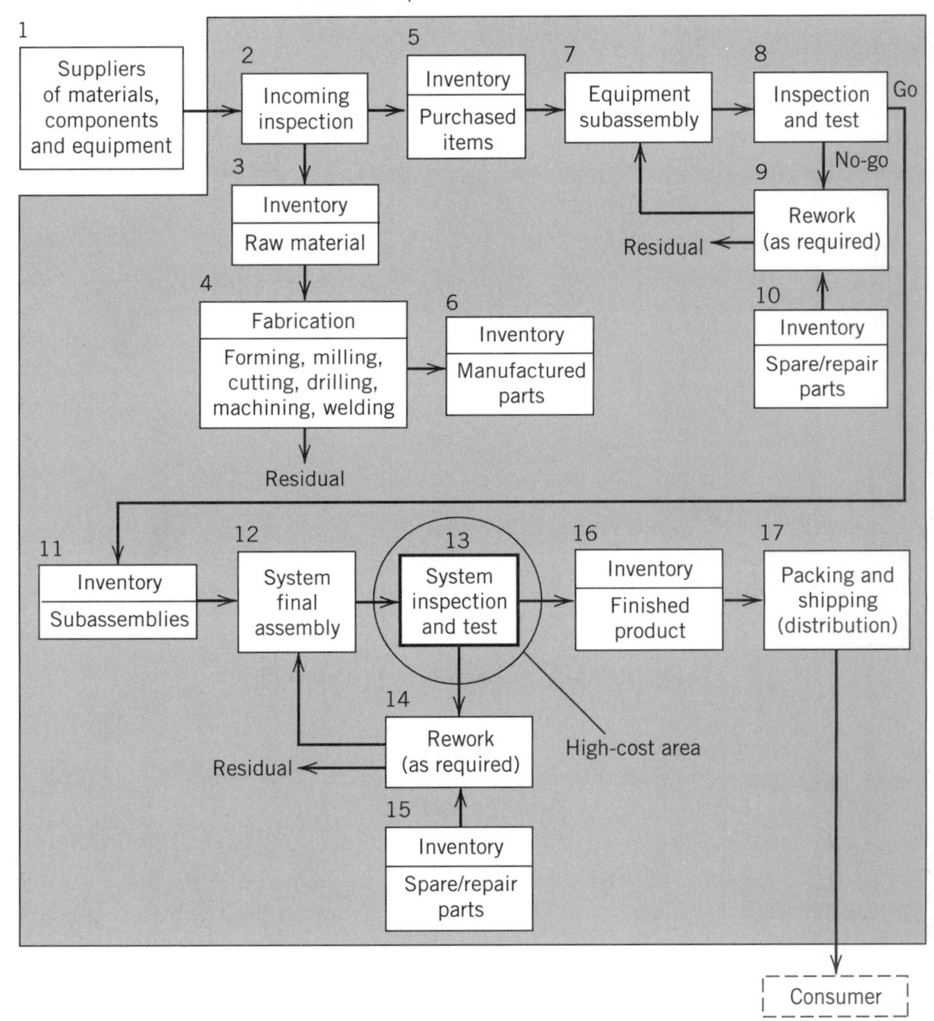

Cost Category	Cost × 1,000 ($)	% of Total
1. Architecture and design	2,248	7
2. Architecture and design	12,524	39
(a) Facilities	6,744	21
(b) Capital equipment	5,780	18
3. Future operation and maintenance	17,342	54
(a) Incoming inspection	963	3
(b) Fabrication	3,854	12
(c) Subassembly	1,927	6
(d) Final assembly	3,533	11
(e) inspection and test	5,459	17
(f) Packing and shipping	1,606	5
Grand Total	$32,114	100%

Figure B.12 Life-cycle cost breakdown summary.

relate to an unreliable product requiring a lot of maintenance, an inadequate procedure or poor process, a supplier problem, and so on.

B.9 CONDUCT A SENSITIVITY ANALYSIS

In order to properly assess the results of the life-cycle cost analysis, the validity of the data presented in Figure B.12, and the associated risks, the analyst needs to conduct a *sensitivity analysis.* One may challenge the accuracy of the input data (i.e., the factors used and the assumptions made in the beginning) and determine their impact on the analysis results. This may be accomplished by identifying the critical factors at the input stage (i.e., those parameters that are suspected as having a large impact on the results), introduce variations over a designated range at the input stage, and determine the differences in output. For example, if the initially predicted reliability MTBF value is "suspect," it may be appropriate to apply variations at the input stage and determine the changes in cost at the output. The object is to identify those areas where a small variation at the input stage will cause a large delta cost at the output. This, in turn, leads to the identification of potential high-risk areas, a necessary input to the risk management program described in Section 6.6 (Chapter 6).

B.10 CONDUCT A PARETO ANALYSIS TO IDENTIFY MAJOR PROBLEM AREAS

With the objective of implementing a program for *continuous process improvement,* the analyst may wish to rank the problem areas on the basis of relative "importance," the higher-ranked problems requiring immediate attention. This may be facilitated through the conductance of a Pareto analysis and the construction of a diagram, as shown in Figure B.13.

B.11 IDENTIFY AND EVALUATE FEASIBLE ALTERNATIVES

By referring to the requirements for the communication system described in Section B.1, two different potential suppliers were considered through a feasibility analysis; that is, Configuration A and Configuration B. Figure B.14 presents a *budgetary* profile for each of three configurations, with Configuration C being eliminated for noncompliance. For the purposes of comparison on an equivalent basis, the two remaining profiles have been converted to reflect *present value* costs. Figure B.15 presents a breakdown summary of these present value costs by major CBS category, and identifies the relative percent contribution of each category in terms of the total. A 10% interest rate was used in determining present value costs.

Although a review of Figure B.15 might lead one to immediately select Configuration A as being preferable, prior to making such a decision the analyst needs to project the two cost streams in terms of the life cycle and determine the point in time

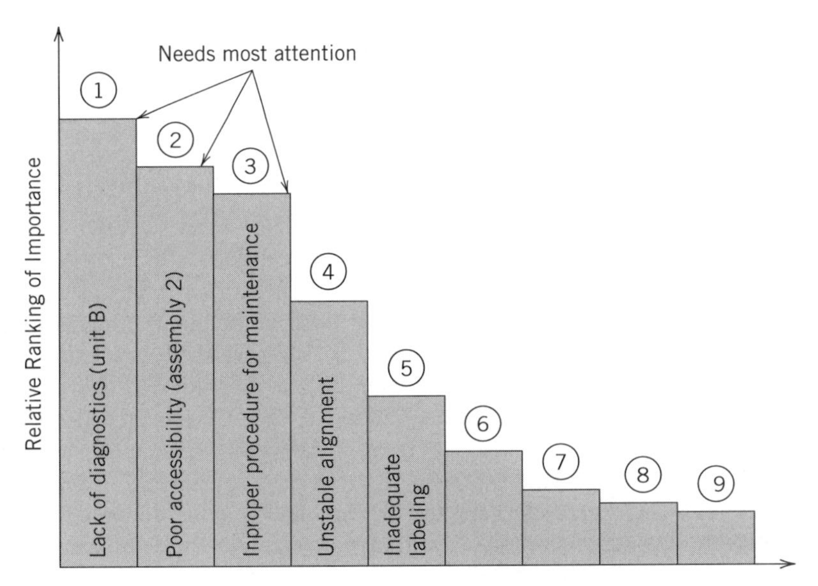

Figure B.13 Pareto ranking of major problem areas.

when Configuration A assumes the position of *preference.* Figure B.16 shows the results of a break-even analysis, and it appears that A is preferable after approximately 6.5 years into the future. The question arises as to whether this break-even point is reasonable when considering the type of system and its mission, the technologies being utilized, the length of the planned life cycle, and the possibilities of obsolescence. For systems where the requirements are changing constantly and obsolescence could become a problem 2 to 3 years hence, the selection of Configuration B may be preferable. On the other hand, for larger systems with longer life cycles (e.g., 10 to 15 years and greater), the selection of Configuration A may represent the best choice.

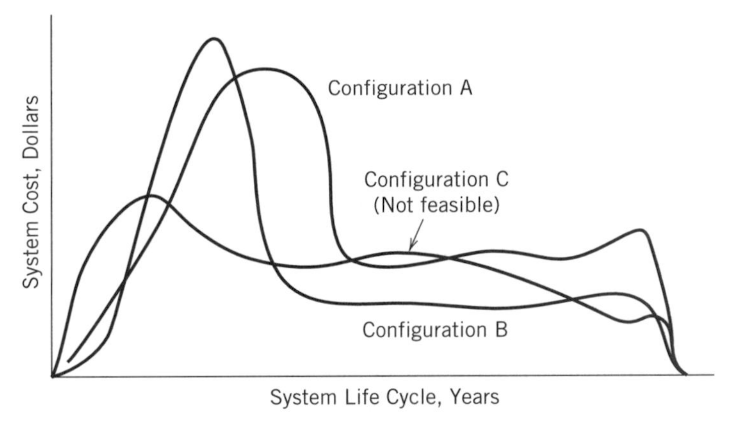

Figure B.14 Alternative cost profiles.

Cost Category	Configuration A		Configuration B	
	Present Cost	% of Total	Present Cost	% of Total
1. Research and development	$70,219	7.8	$53,246	4.2
(a) Management	9,374	1.1	9,252	0.8
(b) Engineering	45.552	5.0	28,731	2.3
(c) Test and evaluation	12,176	1.4	12,153	0.9
(d) Technical data	3,117	0.3	3,110	0.2
2. Production (investment)	407,114	45.3	330,885	26.1
(a) Construction	45,553	5.1	43,227	3.4
(b) Manufacturing	362,261	40.2	287,658	22.7
3. Operations and maintenance	422,217	46.7	883,629	69.4
(a) Operations	37,811	4.2	39,301	3.1
(b) Maintenance	382,106	42.5	841,108	66.3
-maintenance personnel	210,659	23.4	407,219	32.2
-spares/repair parts	103,520	11.5	228,926	18.1
-Test equipment	47,713	5.3	131,747	10.4
-Transportation	14,404	1.6	51,838	4.1
-Maintenance training	1,808	0.2	2,125	0.1
-Facilities	900	0.1	1,021	Neg.
-Field data	3,102	0.4	18,232	1.4
4. Phaseout and disposal	2,300	0.2	3,220	0.3
Grand Total	$900,250	100%	$1,267,760	100%

Figure B.15 Life-cycle cost breakdown (evaluation of two alternative configurations).

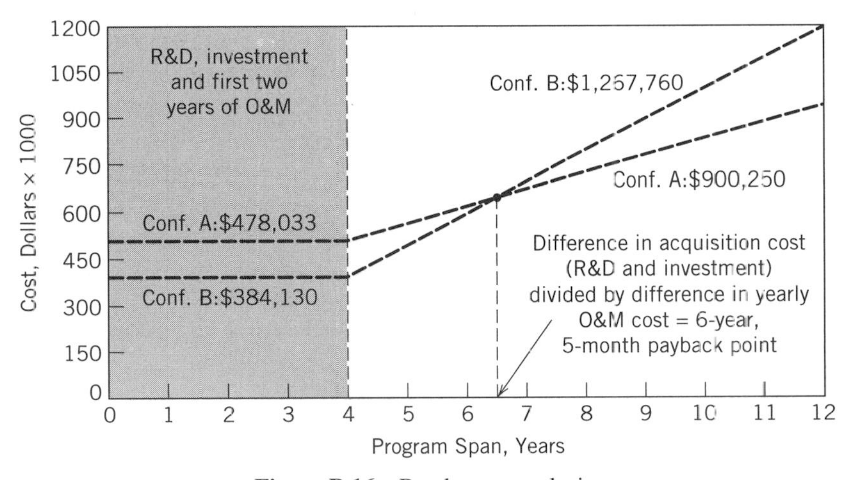

Figure B.16 Break-even analysis.

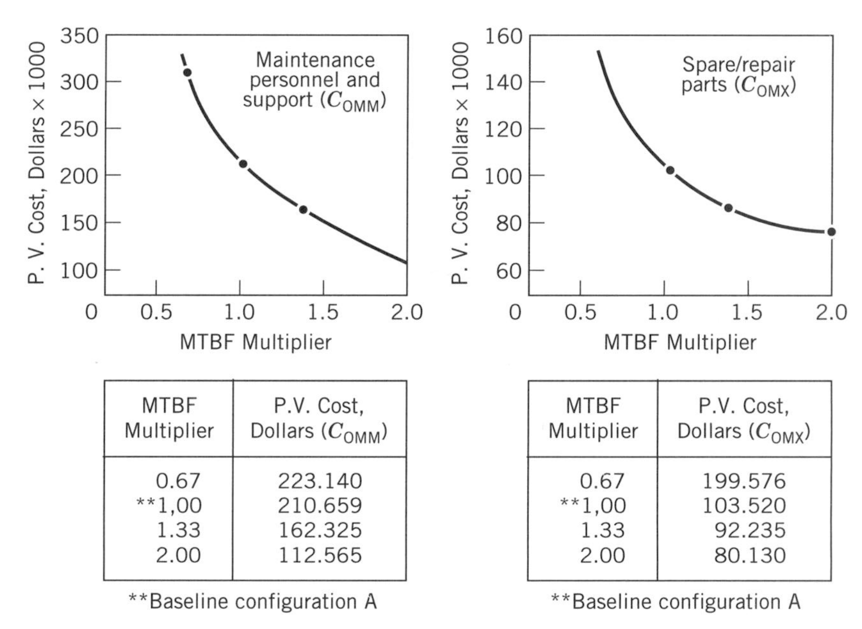

MTBF Multiplier	P.V. Cost, Dollars (C_{OMM})
0.67	223.140
**1,00	210.659
1.33	162.325
2.00	112.565

**Baseline configuration A

MTBF Multiplier	P.V. Cost, Dollars (C_{OMX})
0.67	199.576
**1,00	103.520
1.33	92.235
2.00	80.130

**Baseline configuration A

Figure B.17 Sensitivity analysis.

In this case, it is assumed that Configuration A is preferable. However, when the cost profile for this alternative is converted back to a *budgetary* projection, it is realized that a further reduction of cost is necessary. This, in turn, leads the analyst to Figure B.15 and the identification of potential *high-cost* contributors. Given that a large percentage of the total cost of a system is often in the area of maintenance

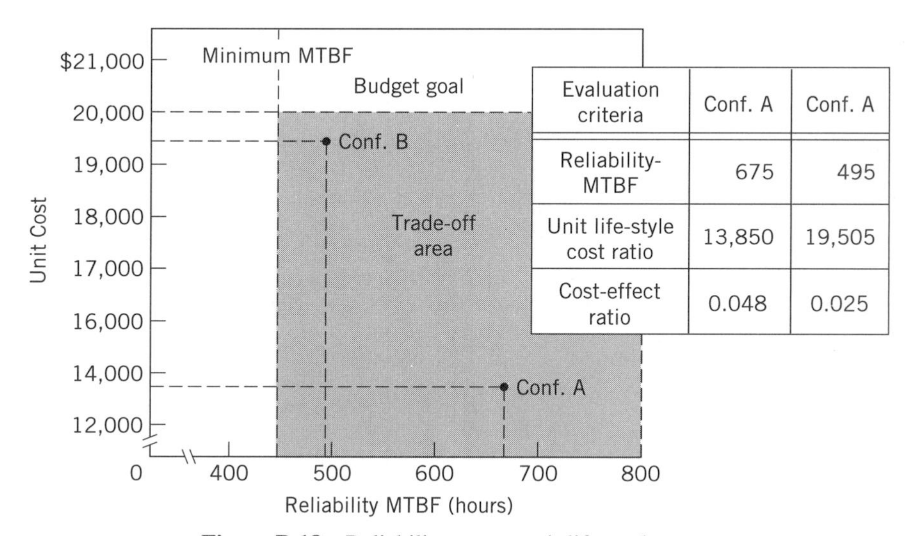

Evaluation criteria	Conf. A	Conf. A
Reliability-MTBF	675	495
Unit life-style cost ratio	13,850	19,505
Cost-effect ratio	0.048	0.025

Figure B.18 Reliability versus unit life-cycle cost.

and support, one might investigate the categories of "maintenance personnel" and "spares/repair parts," representing 23.4% and 11.5% of the total cost, respectively. The next step is to identify the applicable cause-and-effect relationships and to determine the actual "causes" for such high costs. This may be accomplished by being able to trace the costs back to a specific function, process, product design characteristic, or a combination thereof. The analyst also needs to refer back to the CBS and review how the costs were initially derived and the assumptions that were made at the input stage. In any event, the problem may be traced back to a specific function where the resource consumption is high, a particular component of the system with a low reliability and requiring frequent maintenance, a specific system operating function that requires a lot of highly skilled personnel, or something of an equivalent nature. Various design tools can be effectively utilized to aid in making visible these "causes" and to help identify areas where improvement can be made; for example, the failure mode, effects, and criticality analysis, the detailed task analysis, and so on.

As a final step, the analyst needs to conduct a sensitivity analysis in order to properly assess the risks associated with the selection of Configuration A. Figure B.17 illustrates this approach as it applies to the "maintenance personnel" and "spares/repair parts" categories addressed earlier. The objective is to identify those areas where a small variation at the input stage will cause a large delta cost at the output. This, in turn, leads to the identification of potential high-risk areas, a necessary input to the risk management program described in Section 6.6.

B.12 SELECT A PREFERRED DESIGN APPROACH

By having addressed the cost issue, it is necessary to view the results in the context of the overall cost-effectiveness balance illustrated in Figure 1.23 (Chapter 1). Although the emphasis here has been on cost, the ultimate decision-making process must consider both sides of the spectrum; that is, *cost* and *effectiveness*. As an example, the two alternative communication system configurations discussed earlier must meet the reliability and cost goals described in Section B.1. In Figure B.18, the shaded area represents the allowable design trade-off "space," and the alternatives must be viewed not only in terms of cost, but reliability as well. From Section 3.3.9, the ultimate decision may be based on an overall cost-effectivenesss ratio, or some equivalent metric.

APPENDIX C
Design Review Checklist

Periodically throughout the system design and development process, it is appropriate to conduct an informal review to (1) determine whether the necessary tasks have been completed on a given program to ensure optimum results overall, and (2) assess whether the appropriate characteristics have been considered and incorporated into the system/product design configuration. Questions, presented in the form of a checklist, have been developed to reflect certain features. System-level requirements and detailed design considerations are included. Not all questions are applicable in all reviews; however, the answer to those questions that are applicable should be "yes" in order to reflect desirable results. For some questions, it may be necessary to pursue a more in-depth study of the subject through review of selected references prior to arriving at a decision. These questions directly support the abbreviated checklist presented in Figure 5.4 (Chapter 5).

C.1 GENERAL REQUIREMENTS

1.0 Feasibility Analysis

1.1 Has the "need" for the system/product been defined and justified?

1.2 Has the overall technical design approach for the system been justified through a feasibility analysis?

1.3 In the conductance of the feasibility analysis, have *all* appropriate technology applications been considered and prioritized in the development of a technical design approach for the system?

1.4 Have *existing* technologies been selected where feasible, as compared to the selection of new state-of-the-art technologies?

1.5 In the evaluation of alternative technology applications, have life-cycle cost considerations been employed? Have peculiar support requirements been identified?

1.6 Will the technologies selected be available in time to meet program schedule requirements?

1.7 Are there multiple sources of supply for each of the technologies selected for application (i.e., more than one supplier)?

1.8 Have research-and-development activities been defined for areas where deficiencies exist?

1.9 Have areas of risk and uncertainty been identified?

2.0 Operational Requirements

2.1 Has the mission for the system been defined (i.e., primary and secondary missions with applicable scenarios or profiles)?

2.2 Have system performance requirements been defined?

2.3 Have the technical performance measures (TPMs) for the system/program been identified, described; and prioritized? Are they measurable?

2.4 Has the system/product life cycle, and the major activities within, been adequately defined (i.e., design and development, production and/or construction, distribution, operational use, sustaining support, retirement and disposal)?

2.5 Has the planned operational deployment, or the geographical distribution of system components, been defined (i.e., customer requirements, quantity of items per "user" location, distribution schedule)?

2.6 Have system utilization requirements been defined? This may include projected hours of system operations, or quantity of operational cycles in a given time period. A "dynamic" operational scenario is desired.

2.7 Has the projected operational environment been adequately described in terms of temperature cycles and extremes, humidity, vibration and shock, storage, transportation and handling?

3.0 Maintenance Concept

3.1 Have the anticipated levels of maintenance been identified and defined?

3.2 Have the basic maintenance functions been identified for each level?

3.3 Have the organizational responsibilities for system maintenance and support been assigned (i.e., "user" support, contractor support, supplier support, third-party maintenance)?

3.4 Have level-of-repair policies been established (i.e., repair versus discard)? Have the criteria for level-of-repair decisions been adequately defined?

3.5 Have the requirements for "standardization" been established (as it applies to the overall system support capability)?

3.6 Have criteria been established for personnel quantities and/or skills at each level of maintenance?

3.7 Have criteria been established for test and support equipment at each level of maintenance? Built-in versus external test equipment? Diagnostic requirements?

3.8 Have software requirements for system and/or component testing been defined? Test language requirements?

3.9 Have criteria been established for maintenance facilities?

3.10 Have criteria been established for packaging, transportation, and handling?

3.11 Have the appropriate effectiveness factors been established for design for the overall support capability (i.e., spare part demand rates, inventory locations and levels, test equipment availability and utilization, maintenance shop que and process time, facility utilization, turnaround time, and so on)?

3.12 Have the maintenance and support environments been defined in terms of temperature cycle and extremes, humidity, vibration and shock, transportation and handling, and storage?

4.0 Effectiveness Factors

4.1 Have the appropriate system effectiveness and cost effectiveness figures of merit (FOMs) been defined for the system/product (i.e., availability, dependability, capability, readiness, life-cycle cost, design to cost)?

4.2 Have applicable quantitative factors for reliability, maintainability, human factors, and supportability been specified (i.e., MTBM, MTBR, MTBF, λ, fpt, MDT, $\bar{M}$, ADT, LDT, $\bar{M}$ct, $\bar{M}$pt, MLH/OH, Cost/OH, Cost/MA, TAT)?

4.3 Are the effectiveness factors that have been specified directly traceable to either the system operational requirements (mission scenario) or the maintenance concept?

4.4 Can each of the effectiveness factors identified for the system be measured? Have test and evaluation provisions been incorporated in the Test and Evaluation Master Plan (TEMP) for the purposes of verification?

4.5 In the event that two or more effectiveness measures are applicable, are the measures properly weighted to indicate degree of significance or relative level of importance?

4.6 Are all significant effectiveness FOMs properly integrated into the TPM evaluation and reporting capability?

5.0 Functional Analysis and Allocation

5.1 Has the system/product been adequately defined in *functional* terms utilizing the functional block diagram approach?

5.2 Have all major system operational functions and maintenance functions been defined?

5.3 Does the functional analysis evolve directly from the system operational requirements and the maintenance concept? Are the functions directly traceable to top system-level requirements (i.e., the mission scenario)?

5.4 Do the maintenance functions evolve directly from the operational functions?

5.5 Is the functional analysis presented in enough detail to allow for the proper development of the reliability block diagram, FTA, FMECA, maintainability prediction, maintenance analysis, detailed operator task analysis, operational sequence diagrams, safety hazard analysis, and the logistic support analysis (LSA)?

5.6 Is the functional analysis presented in enough detail for development of the system specification?

5.7 Have the appropriate system-level requirements been allocated to the depth necessary for adequate design definition (i.e., subsystem level and below)? This may include the allocation of reliability requirements, maintainability requirements, supportability factors, and cost parameters.

5.8 In the allocation of factors from the system to the subsystem, unit, and below, are the parameters traceable from one level to the next? Are the parameters meaningful in terms of being good measures for the level of the system specified?

6.0 System Specification

6.1 Has a program/project specification tree been developed (showing governing specifications presented in a hierarchal manner)?

6.2 Has a system specification been prepared (i.e., Type "A" Specification)?

6.3 Have the appropriate development, procurement, process, and material specifications been prepared (i.e., Types "B," "C," "D," and "E")?

6.4 Does the system specification include operating requirements, the maintenance concept, a functional definition of the system, and effectiveness requirements (i.e., reliability, maintainability, human factors, safety, supportability, economic, and quality factors)?

6.5 Are the various specifications that are applicable compatible with each other? Have conflicting specification requirements been eliminated? If not, has precedence been established?

7.0 Supplier Requirements[1]

7.1 Have the criteria and procedures for the initial identification, evaluation, and selection of component suppliers been established?

7.2 Have all component suppliers deen identified? Has an on-site evaluation been conducted for each potential supplier?

7.3 Have supplier specifications been prepared and properly applied through the appropriate contractual arrangements?

7.4 Have individual supplier program plans been prepared and implemented?

7.5 Are supplier design data and documentation compatible with the requirements for the overall program?

7.6 Have the appropriate configuration management procedures been imposed on supplier design and development activities?

[1] Also refer to Appendix D.

7.7 Have the appropriate quality control procedures been established for the ongoing monitoring and control of supplier activities? Has a supplier rating system been implemented?

8.0 System Engineering Management Plan (SEMP)

8.1 Has the System Engineering Management Plan (SEMP) been developed?

8.2 Does the SEMP address the overall system life cycle and its phases/activities?

8.3 Does the plan adequately describe the system engineering process?

8.4 Does the plan convey the proper integration of the different engineering specialties involved in the system/product design process?

8.5 Does the SEMP properly integrate other plans such as the Reliability Program Plan, Maintainability Program Plan, Human-Factors Program Plan, Safety Engineering Plan, ILS Plan, Logistic Support Analysis Plan, Configuration Management Plan, Test and Evaluation Master Plan (TEMP), and so on?

8.6 Have major system trade-off studies been adequately documented, and are they appropriately referenced in the SEMP?

8.7 Does the SEMP adequately support the system specification?

8.8 Are program tasks, organizational structure and responsibilities, WBS, schedules, cost projections, and program monitoring and control functions included?

8.9 Have a personnel development plan and an organizational training plan been included?

8.10 Have supplier program requirements been covered?

8.11 Have formal design reviews been covered? Has a formal system evaluation and corrective-action procedure been described and implemented?

Additional questions pertaining to the SEMP are included in Section 6.2 (Chapter 6).

C.2 DESIGN FEATURES

1.0 Accessibility

1.1 Are key system components directly accessible for the performance of both operator and maintenance tasks?

1.2 Is access easily attained?

1.3 Are access requirements compatible with the frequency of maintenance (or criticality of need)? Accessibility for items requiring frequent maintenance should be greater than for items requiring infrequent maintenance.

1.4 Are access doors provided where appropriate? Are hinged doors utilized? Can access doors that are hinged be supported in the open position?

1.5 Are access openings adequate in size and optimally located for the access required?

1.6 Are access door fasteners minimized?

1.7 Are access door fasteners of the quick-release variety?

1.8 Can access be attained without the use of tools?

1.9 If tools are required to gain access, are the number of tools held to a minimum? Are the tools of the standard variety?

1.10 Are access provisions between modules and components adequate?

1.11 Are access doors and openings labeled in terms of items that are accessible from within?

2.0 Adjustments and Alignments

2.1 Have adjustment/alignment requirements been minimized, if not eliminated?

2.2 Are adjustment requirements and frequencies known where applicable?

2.3 Are adjustment points accessible?

2.4 Are adjustment-point locations compatible with the maintenance level at which the adjustment is made?

2.5 Have adjustment/alignment interaction effects been eliminated?

2.6 Are factory adjustments specified?

2.7 Are adjustment points adequately labeled?

2.8 Can adjustments/alignments be made without the requirement for special tools?

3.0 Cables and Connectors

3.1 Are cables fabricated in removable sections?

3.2 Are cables routed to avoid sharp bends?

3.3 Are cables routed to avoid pinching?

3.4 Is cable labeling adequate?

3.5 Is cable clamping adequate?

3.6 Are the connectors used of the quick-disconnect variety?

3.7 Are connectors that are mounted on surfaces far enough apart so that they can be firmly grasped for connecting and disconnecting?

3.8 Are connectors and receptables labeled?

3.9 Are connectors and receptables keyed?

3.10 Are connectors standardized?

3.11 Do the connectors incorporate provisions for moisture prevention?

4.0 Calibration

4.1 Have calibration requirements been minimized, if not eliminated?

4.2 Are calibration requirements known where applicable?

4.3 Are calibration frequencies and tolerances known?

4.4 Have the facilities for calibration been identified?

4.5 Are the necessary standards available for calibration?

4.6 Have calibration procedures been prepared?

4.7 Is traceability to the National Institute of Standards and Technology (NIST) possible?

4.8 Are calibration requirements compatible with the maintenance concept and the logistic support analysis (LSA)?

5.0 Data Requirements

5.1 Has the design been properly defined through good data and documentation (i.e., layouts, drawings, functional diagrams, and materials and parts lists)?

5.2 Are system components adequately covered through good up-to-date design data?

5.3 Have the results of significant design trade-off studies been properly recorded through good documentation?

5.4 Have all data requirements been defined for each applicable program?

5.5 Have all supplier data requirements been defined and properly integrated into the overall data requirements for the program?

5.6 Have the procedures for data collection, distribution, and processing been developed and described?

5.7 Is standardization employed, where appropriate, in data formatting, processing, and reporting?

5.8 Are the data properly controlled in accordance with approved configuration management procedures?

6.0 Disposability

6.1 Has the equipment been designed for disposability (e.g., selection of materials, packaging)?

6.2 Have procedures been prepared to cover system/equipment/component disposal?

6.3 Can the components or materials used in system/equipment design be recycled for use in other products?

6.4 If component/material recycling is not feasible, can decomposition be accomplished?

6.5 Can recycling and/or decomposition be accomplishing using existing logistic support resources?

6.6 Are recycling and/or decomposition methods and results consistent with environmental, ecological, safety, political, and social requirements?

6.7 Is the method(s) used for recycling and/or decomposition economically feasible?

7.0 Ecological Requirements

7.1 Has an environmental impact study been completed (to determine if the system will have an adverse impact on the environment)?

7.2 Are the required standards associated with air quality, water quality, noise level(s), solid-waste processing, and so on, being maintained in spite of the introduction, operation, and sustaining support of the system/product?

7.3 Have potentially degrading ecological effects been identified? Is corrective action being taken to eliminate problems in this area?

7.4 Have the appropriate handling and transportation methods/procedures been described for the processing of solid waste?

8.0 Economic Feasibility

8.1 Has the system/product been justified in terms of total *life-cycle* revenues and costs?

8.2 Are all cost elements considered?

8.3 Are all cost categories adequately defined?

8.4 Are cost estimates relevant?

8.5 Are variable and fixed costs separately identifiable?

8.6 Are escalation factors specified and employed where applicable?

8.7 Are learning curves specified and employed where applicable?

8.8 Is the project economically feasible, considering all possible alternatives?

9.0 Environmental Requirements

9.1 Has system/product design considered all possible phases of activity from an environmental standpoint, for example, environmental requirement during system and handling operation/utilization, transportation, storage, and maintenance?

9.2 Has system/product design considered the following: temperature, humidity, vibration, shock, pressure, wind, salt spray, sand, and dust? Have the ranges and extreme conditions been specified and properly addressed in design? Have the proper environmental profiles been addressed?

9.3 Is the design compatible with air and water quality standards?

9.4 Have provisions been made to specify and control noise, illumination, temperature, and humidity in areas where personnel are required to perform operating and maintenance tasks?

10.0 Facility Requirements

10.1 Have facility requirements (space, volume, capital equipment, utilities, etc.) necessary for system operation been defined?

10.2 Have facility requirements (space, volume, capital equipment, utilities, etc.) necessary for system maintenance at each level been defined?

10.3 Have operational and maintenance facility requirements been minimized to the greatest extent possible?

10.4 Have environmental system requirements (e.g., temperature, humidity, and dust control) associated with operational and maintenance facilities been identified?

10.5 Have storage or shelf-space requirements for spare/repair parts been defined?

10.6 Have storage environments been defined?

10.7 Are the designated facility and storage requirements compatible with the logistic support analysis and human-factors data?

11.0 Fasteners

11.1 Are quick-release fasteners used on doors and access panels?

11.2 Are the total number of fasteners minimized?

11.3 Are the number of different type of fasteners held to a minimum? This relates to standardization.

11.4 Have fasteners been selected based on the requirement for standard tools rather than special tools?

12.0 Handling

12.1 For heavy items, are hoist lugs (lifting eyes) or base-fitting provisions for forklift-truck application incorporated? Hoist lugs should be provided on all items weighing more than 150 pounds.

12.2 Are hoist and base-lifting points identified relative to lifting capacity?

12.3 Are weight labels provided?

12.4 Are packages, units, components, or other items weighing over 10 pounds provided with handles? Are the proper-sized handles used and are they located in the right position? Are the handles optimally located from the weight-distribution standpoint? Handles should be located over the center of gravity.

12.5 Are packages, units, or other items weighing more than 40 pounds provided with two handles (for a two-man carrying capability)?

12.6 Can normal packing materials be used for shipping? If not, are special containers, cases, or covers provided to protect component-vulnerable areas from damage during handling?

13.0 Human Factors

13.1 Has a system analysis been accomplished to verify optimum human–machine interfaces? Are automated and manual functions adequately identified?

13.2 Are the identified automated/manual functions consistent with the results of the overall system-level functional analysis?

13.3 Have operational sequence diagrams (OSDs) been prepared where appropriate?

13.4 Has a detailed *operator* task analysis been accomplished to verify task sequence, task complexities, personnel skills, and so on?

13.5 Has a detailed *maintenance* task analysis been accomplished to verify maintenance task sequences, task complexities, personnel skills, and so on?

13.6 Is the detailed maintenance task analysis compatible with reliability data, maintainability data, and logistic support analysis (LSA) data?

13.7 Are the detailed operator and maintenance task analyses compatible with system/product operating and maintenance procedures (e.g., task sequences, depth of explanatory material based on task complexity)?

13.8 For human-interface functions, is the system/product design optimum when considering anthropometric factors, human sensory factors, psychological factors, and physiological factors? For manual tasks, does the design reflect "ease of operation" by low-skilled personnel? Is the design such that potential human error rates are minimized?

13.9 Has a detailed training plan for operator and maintenance personnel been prepared? Have training facility, equipment, material, software, and data requirements been identified?

13.10 Is the human-factors effort compatible with safety engineering requirements?

13.11 Has an approach been established for personnel test and evaluation?

14.0 Interchangeability

14.1 Are equipment, modules, and/or components that perform similar operations electrically, functionally, and physically interchangeable?

14.2 Can replacements of like items be made without adjustments and/or alignments?

15.0 Maintainability

15.1 Is the system/product maintainable in terms of troubleshooting and diagnostic provisions, accessibility, ease of replacement, handling capabilities, accuracy of test and verification, and economics in the performance of maintenance (corrective and preventive)? Actually, many of the other items in this checklist may be appropriately included under maintainability depending on the organization involved in the design.

15.2 Have maintainability requirements for the system/equipment been adequately defined? Are they compatible with system performance, reliability, supportability, and effectiveness factors?

15.3 Have maintainability requirements been allocated to the appropriate level (e.g., MTBM, MDT, MMH/OH, $\bar{M}ct$, $\bar{M}pt$, $/MA to the unit assembly, subassembly, and/or other appropriate component of the system)?

15.4 Have anticipated system/product corrective and preventive maintenance requirements been identified through a detailed maintenance engineering analysis? Have the proper trade-off studies been conducted to attain the proper balance between corrective and preventive maintenance? Too much preventive maintenance can be costly and can impact significantly correct maintenance requirements. Are the results compatible with logistic support analysis (LSA) data?

15.5 Has a level-of-repair analysis (LORA) been completed? Are the results consistent with the maintenance concept and the logistic support analysis (LSA)?

15.6 Have maintainability predictions been accomplished to assess the design in terms of the specified requirements? Do the predictions indicate compliance with the requirements?

15.7 Have maintainability demonstrations been conducted? Do the results indicate compliance with the requirements? Are these requirements included in the Test and Evaluation Master Plan (TEMP)?

16.0 Mobility

16.1 Can the equipment/component be easily transported? Moved from one location to another?

16.2 Can the system component be moved utilizing common and standard support/handling equipment? The use of special handling equipment should be avoided.

17.0 Operability

17.1 Is the system designed for ease of operation?

17.2 Can the system be operated effectively by individuals with basic skills and with a minimum of special training?

17.3 Can system operation be accomplished with a minimum of error?

18.0 Packaging and Mounting

18.1 Is the packaging design attractive from the standpoint of consumer appeal (e.g., color, shape, size)?

18.2 Is functional packaging incorporated to the maximum extent possible? Interaction effects between packages should be minimized. It should be possible to limit maintenance to the removal of one module (the one containing the failed part) when a failure occurs and not require the removal of two, three, or four modules in order to resolve the problem.

18.3 Is the packaging design compatible with level-of-repair analysis decisions? Repairable items are designed to include maintenance provisions such as test points, accessibility, and plug-in components. Items classified as "discard at failure" should be encapsulated and relatively low in cost. Maintenance provisions within the disposable module are not required.

18.4 Are disposable modules incorporated to the maximum extent practical? It is highly desirable to reduce overall support through a no-maintenance-design concept as long as the items being discarded are relatively high in reliability and low in cost.

18.5 Are plug-in modules and components utilized to the maximum extent possible (unless the use of plug-in components significantly degrades the equipment reliability)?

18.6 Are accesses between modules adequate to allow for hand grasping?

18.7 Are modules and components mounted such that the removal of any single item for maintenance will not require the removal of other items? Component stacking should be avoided where possible.

18.8 In areas where module stacking is necessary because of limited space, are the modules mounted in such a way that access priority has been assigned in accordance with the predicted removal and replacement frequency? Items that require frequent maintenance should be more accessible.

18.9 Are modules and components, not of a plug-in variety, mounted with four fasteners or less? Modules should be securely mounted, but the number of fasteners should be held to a minimum.

18.10 Are shock-mounting provisions incorporated where shock and vibration requirements are excessive?

18.11 Are provisions incorporated to preclude installation of the wrong module?

18.12 Are plug-in modules and components removable without the use of tools? If tools are required, they should be of the standard variety.

18.13 Are guides (slides or pins) provided to facilitate module installation?

18.14 Are modules and components labeled?

18.15 Are module and component labels located on top or immediately adjacent to the item and in plain sight?

18.16 Are the labels permanently affixed and unlikely to come off during a maintenance action or as a result of environment? Is the information on the label adequate? Disposable modules should be so labeled. In equipment racks, are the heavier items mounted at the bottom of the rack? Unit weight should decrease with the increase installation height.

18.17 Are operator panels optimally positioned? For personnel in the standing position, panels should be located between 40 and 70 inches above the floor. Critical or precise controls should be between 48 and 64 inches above the floor. For personnel in the sitting position, panels should be located 30 inches above the floor.

18.18 Are drawers in equipment racks mounted on roll-out slides?

19.0 Panel Displays and Controls

19.1 Are controls standardized?

19.2 Are controls sequentially positioned?

19.3 Is control spacing adequate?

19.4 Is control labeling adequate?

19.5 Have the proper control/display relationships been incorporated (based on good human factors criteria)?

19.6 Are the proper type of panel switches used?

19.7 Is the control panel lighting adequate?

19.8 Are the controls placed according to frequency and/or criticality of use?

20.0 Personnel and Training

20.1 Have operational and maintenance personnel requirements (quantity and skill levels) been defined?

20.2 Are operational and maintenance personnel requirements minimized to the greatest extent possible?

20.3 Are operational and maintenance personnel requirements compatible with logistic support analysis and with human-factors data? Personnel quantities and skill levels should "track" both sources.

20.4 Are the planned personnel skill levels at each location compatible with the complexity of the operational and maintenance tasks specified?

20.5 Has maximum consideration been given to the use of existing personnel skills for new equipment?

20.6 Have personnel attrition rates been established?

20.7 Have personnel effectiveness factors been determined (actual time that work is accomplished per the total time allowed for work accomplishment)?

20.8 Have operational and maintenance training requirements been specified? This includes consideration of both initial training and replenishment training throughout the life cycle.

20.9 Have specific training programs been planned? The type of training, frequency of training, duration of training, and student entry requirements should be identified.

20.10 Are the planned training programs compatible with the personnel skill-level requirements specified for the performance of operational and maintenance tasks?

20.11 Have training equipment requirements been defined? Acquired?

20.12 Have maintenance provisions for training equipment been planned?

20.13 Have training data requirements been defined?

20.14 Are the planned operating and maintenance procedures (designated for support of the system throughout its life cycle) utilized to the maximum extent possible in the training program(s)?

21.0 Producibility

21.1 Does the design lend itself to economic production? Can simplified fabrication and assembly techniques be employed?

21.2 Has the design stabilized (minimum change)? If not, are changes properly controlled through good configuration management methods?

21.3 Is the design such that rework requirements are minimized? Are spoilage factors held to a minimum?

21.4 Has the design been verified through prototype testing, environmental qualification, reliability qualification, maintainability demonstration, and the like?

21.5 Is the design such that many models of the same item can be produced with identical results? Are fabrication steps, manufacturing processes, and assembly methods adequately controlled through good quality assurance procedures?

21.6 Has adequate consideration been given to the application of just-in-time (JIT), Taguchi, material requirements planning (MRP), and related methods in the production process?

21.7 Are production drawings, CAD/CAM/CALS data, material lists, and so on, adequate for production needs?

21.8 Can currently available facilities, standard tools, and existing personnel be used for fabrication, assembly, manufacturing and test operations?

21.9 Is the design such that automated manufacturing processes (e.g., CAM, numerical control techniques) can be applied for high-volume repetitive functions?

21.10 Is the design definition such that two or more suppliers can produce the system/product from a given set of data with identical results?

22.0 Reconfigurability

22.1 Is the design configuration such that it can be readily upgraded for improved capability?

22.2 Have preplanned product improvements been considered in the initial design of the system?

22.3 Can modifications for performance enhancement be incorporated at minimum cost?

23.0 Reliability

23.1 Is the design simple? Have the number of component parts been kept to a minimum?

23.2 Are standard high-reliability parts being utilized?

23.3 Are item failure rates known? Has the mean life been determined?

23.4 Have parts been selected to meet reliability requirements?

23.5 Have parts with excessive failure rates been identified (unreliable parts)?

23.6 Have adequate derating factors been established and adhered to where appropriate?

23.7 Have the shelf life and wearout characteristics of parts been determined?

23.8 Have all critical-useful-life items been eliminated from the design? If not, have they been identified with inspection/replacement requirements specified? Has a critical-useful-life analysis been accomplished?

23.9 Have critical parts that require special procurement methods, testing, and handling provisions been identified?

23.10 Has the need for the selection of "matching" parts been eliminated?

23.11 Have fail-safe provisions been incorporated where possible (protection against secondary failures resulting from primary failures)?

23.12 Has the use of "adjustable" components been minimized?

23.13 Have safety factors and safety margins been used in the application of parts?

23.14 Have component failure modes and effects been identified? Has a FMEA, a FMECA, and/or a fault-tree analysis (FTA) been accomplished?

23.15 Has a stress–strength analysis been accomplished?

23.16 Have cooling provisions been incorporated in design "hot spot" areas? Is cooling directed toward the most critical items?

23.17 Has redundancy been incorporated in the design where needed to meet specified reliability requirements?

23.18 Are the best available methods for reducing the adverse effects of operational and maintenance environments on critical components being incorporated?

23.19 Have the risks associated with critical-item failures been identified and accepted? Is corrective action in design being taken?

23.20 Have reliability requirements for spares and repair parts been considered?

23.21 Have reliability predictions been accomplished? Have reliability testing requirements been defined? Test requirements in design? Test requirements in production/construction? Have they been covered in the Test and Evaluation Master Plan (TEMP)?

23.22 Has a reliability failure analysis and corrective action capability been installed?

24.0 Safety

24.1 Has an integrated safety engineering plan been prepared and implemented?

24.2 Has a hazard analysis been accomplished to identify potential hazardous conditions? Is the hazard analysis compatible with the reliability FMECA/FMEA (where applicable)?

24.3 Have system/product hazards from heat, cold, thermal change, barometric change, humidity change, shock, vibration, light, mold, bacteria, corrosion, rodents, fungi, odors, chemicals, oils, greases, handling and transportation, and so on, been eliminated?

24.4 Have fail-safe provisions been incorporated in the design?

24.5 Have protruding devices been eliminated or are they suitably protected?

24.6 Have provisions been incorporated for protection against high voltages? Are all external metal parts adequately grounded?

24.7 Are sharp metal edges, access openings, and corners protected with rubber, fillets, fiber, or plastic coating?

24.8 Are electrical circuit interlocks employed?

24.9 Are standoffs or handles provided to protect system components from damage during the performance of shop maintenance?

24.10 Are tools that are used near high-voltage areas adequately insulated at the handle or at other parts of the tool that the maintenance man is likely to touch?

24.11 Are the environments such that personnel safety is ensured? Are noise levels with a safe range? Is illumination adequate? Is the air clean? Are the temperatures at a proper level? Are OSHA requirements being maintained?

24.12 Has the proper protective clothing been identified for areas where the environment could be detrimental to human safety? Radiation, intense cold or heat, gas, loud noise, and so on, are examples.

24.13 Are safety equipment requirements identified in areas where ordinance devices (and the like) are activated?

25.0 Selection of Parts/Materials

25.1 Have appropriate standards been consulted for the selection of components and materials?

25.2 Have all component parts and materials selected for the design been adequately evaluated prior to their procurement and application? Evaluation should consider performance parameters, reliability, maintainability, supportability, human factors, quality, and cost.

25.3 Have supplier sources for component-part and material procurement been established?

25.4 Are the established supplier sources reliable in terms of quality level, ability to deliver on time, and willingness to accept component-warranty provisions? There is an ongoing concern for control specifications, process variations, stresses, tolerances, item interchangeability, and so on.

25.5 Have alternative supplier sources been identified in the event that the prime source fails to deliver?

26.0 Servicing and Lubrication

26.1 Have servicing requirements been held to a minimum?

26.2 When servicing is indicated, are the specific requirements identified?

26.3 Are procurement sources for servicing materials known?

26.4 Are servicing points accessible?

26.5 Have personnel and equipment requirements for servicing been identified? This includes handling equipment, vehicles, carts, and so on.

26.6 Does the design including servicing indicators?

27.0 Societal Requirements

27.1 Does the system/product satisfy societal needs?

27.2 Have the societal effects from introducing the system/product into the inventory been evaluated (to determine the impact of system operation and support on community life)?

27.3 Have all adverse societal effects caused by the introduction, operation, and support of the system/product been minimized, if not eliminated?

28.0 Software

28.1 Have all system software requirements for operating and maintenance functions been identified? Have these requirements been developed through the system-level functional analysis (i.e., is there traceability indicated)?

28.2 Is the software complete in terms of scope and depth of coverage?

28.3 Is the software compatible relative to the equipment with which it interfaces? Is operating software compatible with maintenance software? With other elements of the system?

28.4 Are the language requirements for operating software and maintenance software compatible?

28.5 Is all software adequately covered through good documentation (i.e., logic functional flows and coded programs)?

28.6 Has the software been adequately tested and verified for accuracy (performance), reliability, and maintainability?

29.0 Standardization

29.1 Are standard "off-the-shelf" components and parts incorporated in the design to the maximum extent possible (except for items not compatible with effectiveness factors)? Maximum standardization is desirable.

29.2 Are the same items and/or parts used in similar applications?

29.3 Are identifying equipment labels and nomenclature assignments standardized to the maximum extent possible?

29.4 Are equipment-control-panel positions and layouts (from panel to panel) the same or similar when a number of panels are incorporated and provide comparable functions?

30.0 Storage

30.1 Can the equipment be stored for extended periods of time without excessive degradation (beyond specification limits)?

30.2 Have scheduled maintenance requirements for stored equipment been defined?

30.3 Have scheduled maintenance requirements for stored equipment been eliminated or minimized?

30.4 Have the required maintenance resources necessary to service stored equipment been identified?

30.5 Have storage environments been defined?

31.0 Supportability

31.1 Have spare/repair part requirements been minimized to the greatest extent possible? Are the number of different part types used throughout the design minimized?

31.2 Are the types and quantity of spare/repair parts compatible with the system maintenance concept, the logistic support analysis (LSA), and level-of-repair analysis data?

31.3 Are the types and quantity of spare/repair parts designated for a given location appropriate for the estimated demand at that location? Too many or too few spares can be costly.

31.4 Have the distribution channels and inventory points for spare/repair parts been established?

31.5 Are spare/repair part provisioning factors (e.g., replacement frequencies) directly traceable to reliability and maintainability predictions?

31.6 Are the specified logistics pipeline times compatible with effective supply support? Long pipeline times place a tremendous burden on logistic support.

31.7 Have spare/repair parts been identified and provisioned for pre-operational support activities (e.g., interim producer or supplier support, test programs)?

31.8 Have test and acceptance procedures been developed for spare/repair parts? Spare/repair parts should be processed, produced, and accepted on a similar basis with their equivalent components in the prime equipment.

31.9 Have the consequences (risks) of stock-out been defined in terms of effect on mission requirements and cost?

31.10 Has an inventory safety stock level been defined?

31.11 Has a provisioning or procurement cycle been defined (procurement or order frequency)? Have EOQ factors been determined?

31.12 Has a supply-availability requirement been established (the probability of having a spare available when required)?

31.13 Have the test and support equipment requirements been defined for each level of maintenance?

31.14 Have standard test and support equipment items been selected? Newly designed equipment should not be necessary unless standard equipment is unavailable.

31.15 Are the selected test and support equipment items compatible with the prime equipment? Does the test equipment do the job?

31.16 Are the test and support equipment requirements compatible with maintenance concept, logistic support analysis (LSA), and level-of-repair analysis data?

31.17 Have test and support equipment requirements (both in terms of variety and quantity) been minimized to the greatest extent possible?

31.18 Are the reliability and maintainability features in the test and support equipment compatible with those equivalent features in the prime equipment? It is not practical to select an item of support equipment that is not as reliable as the item it supports.

31.19 Have logistic support requirements for the selected test and support equipment been defined? This includes maintenance tasks, calibration equipment, spare/repair parts, personnel and training, data, and facilities.

31.20 Is the test and support equipment selection process based on cost-effectiveness considerations (i.e., life-cycle cost)?

31.21 Have test and maintenance software requirements been adequately defined?

31.22 Have operational and maintenance personnel requirements (quantity and skill levels) been defined?

31.23 Are operational and maintenance personnel requirements minimized to the greatest extent possible?

31.24 Are operational and maintenance personnel requirements compatible with logistics support analysis (LSA) and with human-factors data? Personnel quantities and skill levels should track both sources.

31.25 Are the planned personnel skill levels at each location compatible with the complexity of the operational and maintenance tasks specified?

31.26 Has maximum consideration been given to the use of existing personnel skills for the new system?

31.27 Have personnel attrition rates been established?

31.28 Have personnel effectiveness factors been determined (actual time that work is accomplished per the total time allowed for work accomplishment)?

31.29 Have operational and maintenance training requirements been specified? This includes consideration of both initial training and replenishment training throughout the life cycle.

31.30 Have specific training programs been planned? The type of training, frequency of training, duration of training, and student entry requirements should be identified.

31.31 Are the planned training programs compatible with the personnel skill-level requirements specified for the performance of operational and maintenance tasks?

31.32 Have training equipment requirements been defined? Procured?

31.33 Have maintenance provisions for training equipment been planned?

31.34 Have training data requirements been defined?

31.35 Are the planned operating and maintenance procedures (designated for support of the system throughout its life cycle) utilized to the maximum extent possible in the training program(s)?

32.0 Support Equipment Requirements

Refer to Questions 31.15 through 31.21, Category 31—Supportability, for coverage of test and support equipment.

33.0 Survivability

33.1 Has an integrated survivability engineering plan been prepared and implemented?

33.2 Have survivability measures been established, and are they appropriately integrated with other system TPMs?

33.3 Has a test and evaluation approach been defined for the verification of system survivability? Is this covered in the Test and Evaluation Master Plan (TEMP)?

34.0 Testability

34.1 Have self-test provisions been incorporated where appropriate?

34.2 Is reliability degradation due to the incorporation of built-in test minimized? The BIT capability should not significantly impact the reliability of the overall system.

34.3 Is the extent or depth of self-testing compatible with the level-of-repair analysis?

34.4 Are self-test provisions automatic?

34.5 Have direct fault indicators been provided (a fault light, an audio signal, or a means of determining that a malfunction positively exists)? Are continuous condition monitoring provisions incorporated where appropriate?

34.6 Are test points provided to enable check-out and fault isolation beyond the level of self-test? Test points for fault isolation within an assembly should not be incorporated if the assembly is to be discarded at failure. Test point provisions must be compatible with the level of repair analysis.

34.7 Are test points accessible? Accessibility should be compatible with the extent of maintenance performed. Test points on the operator's front panel are not required for a depot maintenance action.

34.8 Are test points functionally and conveniently grouped to allow for sequential testing (following a signal flow), testing of similar functions, or frequency of use when access is limited?

34.9 Are test points provided for a direct test of all replaceable items?

34.10 Are test points adequately labeled? Each test point should be identified with a unique number, and the proper signal or expected measured output should be specified on a label located adjacent to the test point.

34.11 Are test points adequately illuminated to allow the technician to see the test point number and labeled signal value?

34.12 Can every component malfunction (degradation beyond specification tolerance limits) which could possibly occur be detected through a no-go indication at the system level? Are false alarm rates minimized? This is a measure of test thoroughness.

34.13 Will the prescribed maintenance software provide adequate diagnostic information?

35.0 Transportability

35.1 Have transportation and handling requirements been defined?

35.2 Have transportation requirements been considered in the equipment design? This includes consideration of temperature ranges, vibration and shock, humidity, and so on. Has the possibility of equipment degradation been minimized if transported by air, ground vehicle, ship, or rail?

35.3 Can the equipment be easily disassembled, packed, transported from one location to another, reassembled, and operated with a minimum of performance and reliability degradation?

35.4 Have container requirements been defined?

35.5 Have the requirements for ground handling equipment been defined?

35.6 Was the selection of handling equipment based on cost-effectiveness considerations?

36.0 Quality

36.1 Has a total quality management (TQM) plan been prepared and implemented? Does it include coverage of customer, contractor (producer), and supplier activities and interfaces?

36.2 Are formal quality (i.e., TQM) training programs being conducted within the customer/contractor/supplier organization?

36.3 Are statistical process control (SPC) methods/techniques being implemented where appropriate?

36.4 Are quality control requirements specified and imposed on all suppliers?

The preceding questions are representative of what may be considered in conducting a program/design review. They should not be considered as being all-inclusive by any means. In fact, it is appropriate for the reviewer to develop a list covering applicable issues "tailored" to the program in question.

APPENDIX D
Supplier Evaluation Checklist

This checklist, to be applied in the evaluation of suppliers, is "tailored" and is a supplemental version of the in-depth design review checklist presented in Appendix C. Not all of the questions are applicable in all situations; however, the answer to those questions that are applicable should be "yes" in order to reflect the desired results.

D.1 General Criteria

D.1.1 Has a technical performance specification been prepared covering the product being acquired? Is this specification "supportive" of and "traceable" from the System Specification?

D.1.2 Is the product a commercial off-the-shelf (COTS) item requiring no adaptation, modification, and/or rework for installation?

D.1.3 Has the COTS item been assessed in terms of effectiveness and life-cycle cost?

D.1.4 If the product is a COTS item and requires some modification for installation,

1. Has the degree of modification been clearly defined and minimized to the extent possible?
2. Has the impact of the modification been assessed in terms of effectiveness and life-cycle cost? Has the life-cycle cost been minimized to the extent possible?
3. Can the modification be accomplished easily and with a minimum of interaction effects?
4. Have common and standard parts, reusable software, recycleable material, and so on, been incorporated in the modification/interface package or kit?

D.1.5 Have alternative sources of supply for the same product been identified?

D.1.6 If new design is required, has it been justified to the extent that the COTS, modified COTS, and comparable options are not feasible?

D.2 Product Design Characteristics

D.2.1 Technical Performance Parameters

1. Does the product fully comply with the functional performance specification (i.e., development, product, process, and/or material specification as applicable)?
2. Has the applicable mission scenario (or operational/utilization profile) been defined for the product? Refer to Appendix C, Item C.1.2.
3. Are the product's design characteristics responsive to the prioritized technical performance measures (TPMs)? Does the design reflect the most important features?
4. Were the design characteristics derived through the use of a QFD (or equivalent) approach?
5. Are the performance requirements easily traceable from those specified for the system level?
6. Are the performance requirements measurable? Can they be verified or validated?

D.2.2 Technology Applications

1. Does the design utilize state-of-the-art and commercially available technologies?
2. Do the technologies utilized have a life cycle that is at least equivalent to the product life cycle?
3. Have "short-life" technologies been eliminated? If not, have such applications been minimized?
4. Has an "open-architecture" approach been utilized in the design such that new technologies can be inserted without causing a redesign of other elements of the product?
5. Have alternative sources for each of the technologies being utilized been identified?
6. Have the technologies being utilized reached a point of maturity/stability relative to their applications?

D.2.3 Physical Characteristics

1. Is the product both functionally and physically interchangeable?
2. Can the product be physically removed and replaced with a like item without requiring any subsequent adjustments or alignments? If not, have such interaction effects been minimized?
3. Does the product design comply with the physical requirements in the technical specification (i.e., size, shape, and weight)?

D.2.4 Effectiveness Factors

1. Have the appropriate effectiveness factors been defined and included in the technical specification (i.e., TPMs applicable to the product being acquired)? Refer to Appendix C, Item C.1.4.
2. Can the effectiveness requirements be traced back to comparable requirements specified at the system level?

3. Has the supplier provided a measure of reliability for the product (e.g., *R*, failure rate, and/or MTBF)? Is this figure of merit based on actual field experience?
4. Have the applicable reliability requirements been considered in the product design? Refer to Appendix C, Item C.2.23.
5. Has the supplier provided a measure of maintainability for the product (e.g., MTBM, MLH/OH, $\bar{M}ct$, $\bar{M}pt$, MDT, and/or equivalent)? Is this figure or merit based on actual field experience?
6. Have the applicable maintainability requirements been considered in the product design? Refer to Appendix C, Items C.2.1, C.2.2, C.2.3, C.2.4, C.2.11, C.2.12, C.2.15, C.2.16, C.2.18, C.2.26, C.2.34, and C.2.35.
7. Have the applicable human-factors requirements been considered in the product design? Refer to Appendix C, Items C.2.13, C.2.17, and C.2.19.
8. Have the applicable safety requirements been considered in the product design? Refer to Appendix C, Item C.2.24.
9. Have the applicable supportability/serviceability requirements been considered in the product design? Refer to Appendix C, Items C.1.3 and C.1.31.
10. Have the applicable quality requirements been considered in the product design? Refer to Appendix C, Item C.2.36.

D.2.5 Producibility Factors

1. Has the product been designed for producibility? Refer to Appendix C, Item C.2.21.
2. Is the design data/documentation such that any other supplier with comparable facilities/equipment, capabilities, and experience can manufacture the product in accordance with the specification?

D.2.6 Disposability Factors

1. Has the product been designed for disposability? Refer to Appendix C, Item C.2.6.
2. Has the supplier developed the appropriate planning documentation and procedures covering the disposal and/or recycling of the product?

D.2.7 Environmental Factors

1. Has the product been designed with ecological and environmental requirements in mind? Refer to Appendix C, Items C.2.7 and C.2.9
2. Has the supplier prepared an environmental impact statement for the introduction of the product?

D.2.8 Economic Factors

1. Has the product been designed with economic considerations in mind. Refer to Appendix C, Item C.2.8.

2. Has the supplier conducted a life-cycle cost analysis for the product? Are the results realistic? Refer to Appendix B.

D.3 Product Maintenance and Support Infrastructure

D.3.1 Maintenance and Support Requirements

1. Does the supplier have an established maintenance and support infrastructure in place?
2. Has the supplier defined the maintenance concept/plan for the product. Refer to Appendix C, Item C.1.3.
3. Have the appropriate supportability "metrics" been established for the product and included in the maintenance concept/plan (i.e., response time, turnaround time, maintenance process time, test equipment reliability and maintainability factors, facility utilization, spare parts demand rates and inventory levels, transportation rates and times, etc.)?
4. Does the maintenance concept/plan facilitate or allow for the required degree of *responsiveness* on the part of the supplier?
5. Have the preventive maintenance requirements been established for the product (if any)? Have these requirements been justified through a reliability-centered maintenance (RCM) approach?
6. Have the product maintenance and support resource requirements been defined (i.e., spares, repair parts, and associated inventories; personnel quantities, skill levels, and training; test and support equipment; facilities; packaging, transportation and handling; technical data; and computer resources)? Have these requirements been adequately justified through a maintenance engineering analysis (MEA), a logistics support analysis (LSA), or equivalent? Refer to Appendix C, Item C.2.3.1.

D.3.2 Data/Documentation

1. Does the supplier have a computerized maintenance management data capability in place? Is this capability being effectively utilized for the purposes of *continuous product/process improvement?* Does it provide the visibility relative to how well the product is performing in the field?
2. Does the supplier have in place a reliability data collection, analysis, feedback, and corrective-action process? Are product failures properly recorded and are they traceable to the "cause?"
3. Is the supplier monitoring and measuring the effectiveness of its preventive maintenance program? Where applicable, have the preventive maintenance requirements been revised to reflect a more cost-effective approach?

D.3.3 Warranty/Guarantee Provisions

1. Have product warranties/guarantees been established?
2. Have the established warranty provisions been adequately defined through some form of a contractual mechanism?
3. Are the warranty provisions consistent with the defined maintenance concept?

D.3.4 Customer Service

1. Does the supplier have an established customer service capability in place?
2. Will the supplier provide assistance in the installation and checkout of the product at the producer's site and/or the user's site (if required)?
3. Will the supplier provide on-site field service support if required?
4. Does the supplier provide operator and maintenance training at the producer's site and/or the user's site when necessary? Is this training available "on call?" Will it be available throughout the product life cycle?
5. In support of training activities, will the supplier provide the necessary data, training manuals, software, aids, equipment, simulators, and so on? Will the supplier provide updates/revisions to the training material as applicable?
6. Does the supplier have a program for measuring training effectiveness?

D.3.5 Economic Factors

1. Is the product support infrastructure cost-effective?
2. Have the requirements been based on life-cycle cost objectives?

D.4 Supplier Qualifications

D.4.1 Planning/Procedures

1. Does the supplier have a standard policies and procedures manual/guide?
2. Are the appropriate management procedures properly documented and followed on a day-to-day basis?
3. Are the procedures/processes periodically reviewed, evaluated, and revised as necessary for the purposes of *continuous process improvement?*
4. Has the supplier identified the activities and tasks that are essential in the successful accomplishment of system engineering requirements?

D.4.2 Organizational Factors

1. Has the supplier's organization been adequately defined in terms of activities, responsibilities, interface requirements, and so on?

2. Does the organizational structure support the overall program objectives for the system? Is it compatible with the producer's organizational structure?
3. Has the supplier identified the organizational element responsible for the accomplishment of system engineering tasks (as applicable)?

D.4.3 Available Personnel and Resources

1. Does the supplier have the available personnel and associated resources to assign to the task(s) being contracted? Will these personnel/resources be available for the duration of the program?
2. Do the personnel assigned have the proper background, experience, and training to do the job effectively?

D.4.4 Design Approach

1. Has the supplier implemented the system engineering process in the design of its products?
2. Has an effective design database been established, and is it compatible with the system-level database established by the producer (prime contractor)?
3. Does the supplier have in place a configuration management program, along with a disciplined change-control process? Has a configuration "baseline" approach been implemented in the development and growth of the product?
4. Has the supplier's design process been enhanced through the use of such tools as CAD, simulation, rapid prototyping, and so on?

D.4.5 Manufacturing Capability

1. Does the supplier have a well-defined manufacturing process in place?
2. Does the process incorporate the latest technologies and computer-aided methods (i.e., robotics, the use of CAD or CIM technology, etc.)?
3. Is the process flexible and does it support an "agile" and/or "lean" manufacturing approach?
4. Does the supplier utilize materials requirements planning (MRP), capacity planning (CP), shop floor control (SFC), just-in-time (JIT), master production scheduling (MPS), statistical process control (SPC), and other such methods in the manufacturing process?
5. Has the supplier implemented a formal quality program in accordance with ISO-9000 (or equivalent)? Is the supplier ISO-9000-certified? Does the supplier have a formal procedure in place for correcting deficiencies?
6. Has the supplier implemented a total productive maintenance (TPM) program within its manufacturing plant? Has a TPM measure of ef-

fectiveness been established (i.e., OEE, or overall equipment effectiveness)?

D.4.6 Test and Evaluation Approach

1. Has the supplier developed an integrated test and evaluation plan for the product?
2. Have the requirements for testing been derived in a logical manner, and are they compatible with the identified technical performance measures (TPMs) for the system, and as allocated for the product?
3. Does the supplier have the proper facilities and resources to support all product testing requirements (i.e., people, facilities, equipment, data)?
4. Does the supplier have in place a data collection, analysis, and reporting capability covering all testing activities?
5. Does the supplier have a plan for "retesting" if required?

D.4.7 Management Controls

1. Has the supplier incorporated the necessary controls for monitoring, reporting, providing feedback, and intiating corrective action with regard to technical performance measurement, cost measurement, and scheduling?
2. Has the supplier implemented a configuration management capability?
3. Has the supplier implemented an integrated data management capability?
4. Has the supplier developed a risk management plan?

D.4.8 Experience Factors

1. Has the supplier had experience in designing, testing, manufacturing, handling, delivering, and supporting this product before?
2. Has the supplier utilized experiences from other projects to help respond to the requirements for this program; that is, the transfer of "lessons learned"?

D.4.9 Past Performance

1. Has the supplier successfully completed similar projects in the past?
2. Has the supplier been responsive to all of the requirements for past projects?
3. Has the supplier been successful in delivering products in a timely manner and within cost?
4. Has the supplier delivered reliable and high-quality products?
5. Has the supplier been responsive in initiating any corrective action that has been required to correct deficiencies?
6. Has the supplier stood behind all product warranties/guarantees?

7. Does the supplier's organization reflect stability, growth, and high quality?
8. Is the supplier's business posture good?
9. Does the supplier enjoy an excellent reputation?

D.4.10 Maturity

1. Has the supplier established a process for benchmarking?
2. Has the supplier implemented an organizational assessment program (i.e., SE-CMM or equivalent)?

D.4.11 Economic Factors

1. Has the supplier implemented a life-cycle cost-analysis approach for all of its functions, products, processes, and so on?
2. Has the supplier implemented an "activity-based costing (ABC)" approach with the objective of acquiring full visibility relative to the high-cost contributors, cause-and-effect relationships, and leading to the implementation of improvements for cost-reduction purposes?

APPENDIX E
Glossary of Selected Terms

Throughout this text, there are numerous terms and definitions that are important to the understanding of *System Engineering* and its objectives. As many of these are fairly well explained in the context of text material (or through a recommended reference), it is not intended to be repetitive here. However, in the interest of providing a final summary of the material, a few terms considered as being "key" to the concepts described have been chosen. Through a quick review of these, the reader can acquire a feeling for those considered as being important relative to acquiring a thorough understanding of the subject matter.

1. *Acquisition process:* the process of bringing a system into being. This includes the phases of conceptual design and advance planning, preliminary system design, detail design and development, and production and/or construction. Given a defined need, the process includes those steps leading from the requirements definition stage to the delivery of the system for consumer use. Refer to Chapter 1.

2. *Allocation:* the top-down distribution, or apportionment, of system-level requirements to the subsystem, equipment, software, unit, or below, to the depth necessary for providing criteria as an input to design. This process tends to promote a top-down "systems approach" in helping to establish specific design requirements for all levels of the system hierarchy as appropriate. Refer to Section 2.8.

3. *Computer-aided logistics support* (CALS): the application of computerized technology and available computer software to the entire spectrum of logistics. This includes the use of computer methods/tools in the design for system supportability (integrated into CASE/CAE/CAD activities), in the development of logistics support analysis data in determining logistics resource requirements, in the provisioning and acquisition of the identified elements of support (e.g., spare/repair parts, test and support equipment), and in the assessment of the system support capability in the user's environment. CALS also includes the development of technical manuals and the processing of design data automatically and using a digital data format. Refer to Section 4.5.

4. *Computer-aided design* (CAD): the process of utilizing computer capabilities and available software to support detailed engineering design activities. CAD tends to deal primarily with three-dimensional graphics, circuit board layouts, the accomplishment of various categories of analyses, and the like. Refer to Section 4.3.

5. *Computer-aided engineering* (CAE): the process of utilizing computer capabilities and available software to support engineering design activities. CAE tends to deal primarily with engineering analyses and lower-level design activity, similar to the functions of CAD.

6. *Computer-aided manufacturing* (CAM): the process of utilizing computer capabilities, available software, numerical control equipment, robotics, and related resources to manufacture and produce products through automated means. CAM tends to deal with production process planning, materials handling, manufacturing, inventory control, and production management. Refer to Section 4.4.

7. *Computer-aided systems engineering* (CASE): the process of integrating system engineering concepts and constructs, computer capabilities, the use of analytical methods, and available software in such a manner as to complete system engineering functions in an effective and efficient manner. CASE represents a broader level of capability than either CAD or CAE, and is best described in H. Eisner's text, *Computer-Aided Systems Engineering* (refer to Appendix F, Item C.3).

8. *Computer-integrated manufacturing* (CIM): the process of utilizing computer capabilities and available software to manufacture products via automated means. CIM is used in a manner similar to CAM, except that CIM tends to emphasize the use of microcomputers and a common database.

9. *Concurrent engineering:* "a systematic approach to the integrated, concurrent design of products and their related processes, including manufacture and support. This approach is intended to cause the developers, from the outset, to consider all elements of the product life cycle from conception through disposal, including quality, cost, schedule, and user requirements" (Institute for Defense Analysis Report, R-338). Refer to Section 1.41.

10. *Configuration baselines:* designated points in the system design and development process where the system configuration is defined in detail. Common points include the (1) *Functional Baseline,* where the system configuration is described through the definition of operational requirements and the maintenance concept, the System Type "A" Specification, and feasibility study reports; (2) *Allocated Baseline,* where the system configuration is defined through a combination of Development, Process, Product, and Material Specifications (Types "B," "C," "D," and "E"), selected design trade-off study reports, and system/subsystem design data; and (3) *Product Baseline,* where the system configuration is defined through a combination of Process, Product, and Material Specifications (Types "C," "D," and "E"), trade-off study reports, detailed design data (drawings, parts lists), supplier data, and so on. The relative program timing for each of these "Baselines" is illustrated in Figure 1.12 (Chapter 1).

11. *Configuration control:* deals with the categorization and control of proposed design changes, that is, Class 1 Changes that affect form, fit, and/or function, and Class 2 Changes that are relatively minor in nature. Given a designated configuration baseline, all changes applied to that baseline must be closely evaluated and controlled. Refer to Section 5.4.

12. *Configuration control board* (CCB): a board consisting of expertise representing different design disciplines, responsible for the reviewing and approving of changes to a given configuration baseline. Refer to Section 5.4.

13. *Configuration management* (CM): a management process used to identify the functional and physical characteristics of an item in the early phases of its life cycle, control changes to those characteristics, and record and report change processing and implementation status. CM involves four functions: (1) configuration identification, (2) configuration control, (3) configuration status accounting, and (4) configuration audits. CM is the concept of "baseline" management. Refer to Section 1.4.4.

14. *Contract structure:* the type of contract negotiated between the customer and the contractor, and/or between the contractor and the supplier. Major categories of contracts include firm-fixed-price (FFP), fixed-price-with-escalation, fixed-price-incentive, cost-plus-fixed-fee (CPFF), cost-plus-incentive-fee (CPIF), cost sharing, time and materials, and letter agreement. The nature and type of contract negotiated are major considerations in the implementation of system engineering requirements. Refer to Section 8.3.

15. *Cost breakdown structure* (CBS): a breakdown of cost in functional terms. All future costs associated with activities throughout all phases of the system life cycle must be included, and costs must be broken out to the depth required to provide the necessary visibility relative to different elements of the system and/or different program activities. An example is presented in Appendix B and in Figure 3.39 (Chapter 3).

16. *Cost effectiveness* (CE): the measure of a system in terms of its technical characteristics and life-cycle cost. Technical characteristics may include a combination of performance, capacity, range, weight and size, reliability, maintainability, supportability, producibility, quality, and related parameters. Life-cycle cost may include all future costs associated with research, design and development, production and/or construction, distribution, system utilization, sustaining maintenance and support, retirement, disposal, and the recycling of materials as necessary. These technical characteristics may be combined in some manner to provide a measure of "system effectiveness." Cost effectiveness can be expressed as the ratio of system effectiveness to life-cycle cost (LCC). Refer to Section 3.3.9.

17. *Design review:* a formal review of the system configuration as it is defined at a specific point in time. Formal design reviews include the Conceptual Design Review, one or more System (Preliminary) Design Reviews, one or more Equipment/Software Design Reviews, and a Critical Design Review. Formal design reviews include the "checks and balances" necessary in the implementation of system engineering functions. Refer to Figures 1.21 and 5.1.

18. *Design-to-cost* (DTC): a quantitative "design to" figure of merit specified as a system requirement during conceptual design and included in the System Type "A" Specification. The DTC figure of merit should be specified in terms of life-cycle cost. This can, in turn, be broken down into "Design to Unit Acquisition Cost" and "Design to Unit Operation and Support Cost" (or something equivalent).

The basic categories of cost should be defined in terms of what is (or is not) included.

19. *Effectiveness:* a measure of the system in terms of its technical characteristics. This measure, or figure of merit, which will vary depending on the type of system and its mission, may be derived from a combination of performance factors, weight and size, capacity, reliability, maintainability, supportability, quality, and so on. Refer to Section 3.3.9.

20. *Feasibility analysis:* the early investigation, study, and determination of possible technical design approaches in response to a defined need for a new system configuration. This includes the evaluation and comparison of new technologies, as well as the accomplishment of applied research in areas where additional knowledge is desired. Refer to Section 2.3.

21. *Functional analysis:* the process of translating system-level requirements into detailed design criteria leading to the development of system components. Given the definition of system operational requirements and the maintenance concept, the next step is to define the system in *functional terms,* identifying the WHATs in terms of specific requirements. This can be accomplished through a series of functional block diagrams. These functions (to include both operational and maintenance functions) are broken down into subfunctions, trade-off studies are accomplished to determine the HOWs associated with the accomplishment of the subfunctions, and resources are identified in terms of human requirements, equipment requirements, software, data, facilities, and so on. Again, this is a top-down process stemming from system-level requirements and leading to the identification of specific design requirements. Refer to Section 2.7.

22. *Human factors:* the characteristics of system design that relate to the human element of the system. Considerations in design must include anthropometric factors, human sensory factors, physiological factors, and psychological factors. Refer to Section 3.3.3.

23. *Integrated logistics support* (ILS): a management function that provides the initial planning, funding, and controls that help to ensure that the ultimate consumer (or user) will receive a system that will not only meet performance requirements, but one that can be expeditiously and economically supported throughout its programmed life cycle. The basic program elements include the initial planning for logistics support, the design for supportability, the analysis and acquisition of the various elements of support, and the assessment of the system's support capability in the field. Refer to Section 3.3.8.

24. *Life cycle:* the planned life cycle of the system from the initial identification of a need through the retirement and phaseout of that system. It usually includes the phases of conceptual design and advance planning, preliminary system design, detail design and development, production and/or construction, system operation (utilization), sustaining maintenance and support, and retirement. Although the life cycle (and its phases) may change because of budgetary limitations, the introduction of obsolescence, and so on, it is still essential that a life-cycle approach be assumed. Refer to Section 1.2.3.

25. *Life-cycle cost* (LCC): the composite of all costs associated with the activities planned and/or accomplished throughout the system life cycle. This includes the costs of research and development, design, production/construction, operation use, maintenance and support, and system retirement. Refer to Sections 1.4.6 and 3.3.9, and Appendix B.

26. *Management information system* (MIS): a data collection, processing, and handling capability that supports management in the implementation of program requirements. A prime objective for system engineering is the establishment of a good communications network that will allow for a rapid assessment of program status, and for the initiation of corrective action in an expeditious manner.

27. *Maintainability:* the characteristics of design that deal with the ease, accuracy, safety, and economy in the performance of maintenance actions. The measures most commonly associated with maintainability are maintenance elapsed times ($\bar{M}ct$, MTTR, $\bar{M}pt$, $\bar{M}$, MDT), maintenance frequency factors (MTBR, MTBM), maintenance personnel labor-hour factors (MLH/OH, MLH/MA, MLH/Month, and maintenance cost ($/MA). When addressing maintenance elapsed time only, maintainability can be defined as the probability that a system can be retained in or restored to a satisfactory operating condition, when maintenance is performed by personnel with specified skills, using approved procedures and resources, at each designated level of maintenance. Refer to Section 3.3.2.

28. *Performance:* the characteristics of system design that relate to such measures as input-output requirements, throughput, capacity, size and weight, range, accuracy, power output, data transmitted per designated increment of time, and so on. Throughout this text, this term has been used to cover all of the technical characteristics of a system, with the possible exception of reliability, maintainability, supportability, human factors, quality, and so on.

29. *Producibility:* the characteristics of system design that relate to the ease, accuracy, and economy associated with the follow-on manufacture of system elements in multiple quantities as required. The objective is to design products that can be easily produced in multiple quantities, using conventional manufacturing processes. Refer to Section 3.3.6.

30. *Reliability:* the characteristics of design that relate to the ability of a system to perform for a designated period of time. More specifically, it can be defined as the probability that a system or product will perform in a satisfactory manner for a given period of time when used under specified operating conditions. The measures of reliability include MTBF, MTBM, MTTF, R, and λ. Refer to Section 3.3.1.

31. *Risk management:* an organized method for identifying and measuring risk, and for selecting and developing options for the handling of risk. Risk management must be addressed in the System Engineering Management Plan (SEMP), and it includes the functions of risk assessment, risk analysis, and risk abatement. Refer to Section 6.6.

32. *Supportability:* the characteristics of system design that deal with the ability of a system to be supported in an effective and economic manner. These characteris-

tics pertain not only to the prime elements of the system, but to the design of test and support equipment, the supply support capability, training equipment, facilities, maintenance software, and so on. Many of the principles of reliability, maintainability, and human factors are included. Refer to Section 3.3.8.

33. *System:* a set of components working together with the common objective of performing a function in response to a designated need. A system constitutes a complex set of resources integrated so as to fulfill a defined mission scenario. Such resources may take the form of human beings, materials, equipment, software, facilities, data, and so on. The system must have purpose, it must be functional, be able to respond to some identified need, and it should be able to achieve its overall objective in a cost-effective manner. Refer to Section 1.2.1.

34. *System analysis:* an ongoing iterative analytical process, included as part of the system engineering process, involving the evaluation of design approaches, the accomplishment of trade-off studies, and so on. System analysis is accomplished through the appropriate use of various operations research methods to assist in problem resolution (simulation, queuing theory, linear and dynamic programming, networking, etc.). Refer to Section 1.2.7.

35. *System engineering:* the effective application of scientific and engineering efforts to transform an operational need into a defined system configuration through the top-down iterative process of requirements definition, functional analysis, allocation, synthesis, design optimization, test, and evaluation. It involves the design engineering process of bringing a system into being, with emphasis on an *integrated, top-down, life-cycle* approach. Refer to Section 1.2.4.

36. *System engineering management:* the management activities necessary for the implementation of system engineering requirements. This includes the initial planning, organization for system engineering, and the ongoing program evaluation and control activities to ensure that system engineering objectives are met. Refer to Section 1.3 and Chapters 6, 7, and 8.

37. *System Engineering Management Plan* (SEMP): the principle management-oriented document covering the implementation of system engineering program requirements. This plan is developed during the conceptual design phase of a program, includes the results of some advanced planning, and leads into the requirements for the subsequent phases of system acquisition. Refer to Sections 1.3 and 6.2.

38. *System integration:* involves the technical integration of system elements as the design and development effort progresses, along with the integration of the various design and supporting disciplines into the overall design effort from a managerial perspective. Later, during detail design and development, the integration process often involves a bottom-up approach relative to the combining of the various system components into subassemblies, subassemblies into assemblies, assemblies into units, and so on, until a totally integrated system is functioning in accordance with the initially specified requirements. Refer to Section 8.5.

39. *System specification:* the top-level technical document that defines the basic requirements for *system* design and development. Refer to Section 3.1 and Figures 3.2 and 3.3 (Chapter 3).

40. *System synthesis:* the combining and structuring of components in such a way as to represent a feasible system configuration. This may be accomplished on a number of occasions throughout the system design and development process, and the particular configuration structured may not reflect the final design approach. In essence, one needs to define a configuration in such a way that it can be evaluated. Refer to Section 2.9.

41. *Technical performance measures* (TPMs): those measures of the system, or of program activities, that are considered as being critical for the successful accomplishment of system engineering objectives. Specific quantitative parameters, reflecting the basic design characteristics of the system, are specified initially in the System Specification (Type "A"). Then, as design and development progresses, these factors need to be integrated into the periodically scheduled program/design review process for comparison against the initially specified requirements. Finally, an evaluation of the system, in terms of compliance with these requirements, is accomplished through the final system evaluation and test activity. The objective is to identify the critical factors that are *performance*-related. Refer to Section 2.6 and Figures 2.10 and 5.2.

42. *Test and Evaluation Master Plan* (TEMP): a key planning document, developed during the conceptual design and advance planning phase, covering the proposed integration and testing requirements for the system as an entity. A *total integrated test approach* is essential. Refer to Figure 1.21 and Section 2.11.2.

43. *Total quality management* (TQM): a total integrated management approach that addresses system/product quality during all phases of the system life cycle and at each level in the overall system hierarchy. It provides a "before-the-fact" orientation to quality, and it focuses on system design and development activities, as well as production, manufacturing, assembly, construction, and related functions. It includes activities associated with the "design for producibility," quality engineering, quality control, statistical process control (SPC), quality assurance, and supplier evaluation and control. Refer to Section 1.4.3.

APPENDIX F
Selected Bibliography

A selected bibliography has been compiled to include some references that are specific to systems engineering and a number of additional references covering related areas. By becoming familiar with this literature, the reader should gain additional insight into the systems engineering approach for the design and development of new systems and/or the reengineering of existing systems. Specific areas covered by these references include the following:

A. Systems, Systems Theory, Systems Engineering, and Systems Analysis
B. Concurrent/Simultaneous Engineering
C. Software/Computer-Aided Systems
D. Reliability Engineering
E. Maintainability and Maintenance
F. Human Factors and Safety
G. Logistics, Logistics Engineering, and Logistics Management
H. Quality, Quality Engineering, and Quality Assurance
I. Operations Analysis and Operations Research
J. Engineering Economy, Life-Cycle Costing, and Cost Estimating
K. Management and Supporting Areas

A. SYSTEMS, SYSTEMS THEORY, SYSTEMS ENGINEERING, AND SYSTEMS ANALYSIS

1. Ackoff, R. L., S. K. Gupta, and J. S. Minas, *Scientific Method: Optimizing Applied Research Decisions,* John Wiley, New York, 1962.
2. Beam, W. R., *Command, Control and Communications Systems Engineering,* McGraw-Hill, New York, 1989.
3. Beam, W. R., *Systems Engineering: Architecture and Design,* McGraw-Hill, New York, 1990.
4. Belcher, R., and E. Aslaksen, *Systems Engineering,* Prentice Hall, Sydney, Australia, 1992.
5. Blanchard, B. S., *System Engineering Management,* John Wiley, New York, 1991.
6. Blanchard, B. S., and W. J. Fabrycky, *Systems Engineering and Analysis,* 2nd Ed., Prentice Hall, Upper Saddle River, NJ, 1990.
7. Blanchard, B. S., W. J. Fabrycky, and D. Verma (Eds.), *Application of the System Engineering Process to Define Requirements for Computer-Based Design Tools,* Technical Monograph, International Society of Logistics (SOLE), Hyattsville, MD, 1994.
8. Boardman, J., *Systems Engineering: An Introduction,* Prentice-Hall International, London, 1990.
9. Defense Systems Management College, *Systems Engineering Management Guide,* DSMC, Fort Belvoir, VA, various years.
10. Department of Defense, *Mandatory Procedures for Major Defense Acquisition Programs (MDAPs) and Major Automated Information System (MAIS) Acquisition Programs,* DODR 5000.2-R, DOD, Washington, DC, 1996.
11. Drew, D. R., and C. H. Hsieh, *A Systems View of Development: Methodology of Systems Engineering and Management,* Cheng Yang Publishing, Taipei, ROC, 1984.
12. Electronic Industries Association, *Systems Engineering,* EIA/IS-632, EIA, Arlington, VA, 1994.
13. Forrester, J. W., *Principles of Systems,* The MIT Press, Cambridge, MA, 1968.
14. Gheorghe, A., *Applied Systems Engineering,* John Wiley, New York, 1982.
15. Goode, H. H., and R. E. Machol, *System Engineering: An Introduction to the Design of Large-Scale Systems,* McGraw-Hill, New York, 1957.
16. Grady, J. O., *System Engineering Planning and Enterprise Identity,* CRC Press, Boca Raton, FL, 1995.

17. Grady, J. O., *Systems Requirements Analysis,* McGraw-Hill, New York, 1993.
18. Hall, A. D., *A Methodology for Systems Engineering,* D. Van Nostrand, Princeton, 1962.
19. Institute of Electrical and Electronics Engineers, *Standard for the Application and Management of the System Engineering Process,* IEEE P1220, IEEE, Piscataway, NJ, 1994.
20. *Journal of the International Council on Systems Engineering,* INCOSE, Seattle, WA, various years.
21. *Proceedings of the Annual Conference,* INCOSE, Seattle, WA, various years.
22. Lacy, J. A., *Systems Engineering Management: Achieving Total Quality,* McGraw-Hill, New York, 1992.
23. Pugh, S., *Total Design: Integrated Methods for Successful Product Engineering,* Addison-Wesley, Reading, MA, 1991.
24. Rechtin, E., *Systems Architecting: Creating and Building Complex Systems,* Prentice Hall, Upper Saddle River, NJ, 1991.
25. Rechtin, E., and M. Maier, *The Art of Systems Architecting,* CRC Press, Boca Raton, FL, 1996.
26. Reilly, Norman B., *Successful Systems Engineering for Engineers and Managers,* Van Nostrand Reinhold, New York, 1993.
27. Rouse, W. B., *Systems Engineering Models of Human-Machine Interaction,* Elsevier/North Holland, New York, 1980.
28. Sage, A. P., *Decision Support Systems Engineering* John Wiley, New York, 1991.
29. Sage, A. P., *Economic System Analysis: Microeconomics for Systems Engineering, Engineering Management, and Project Selection,* Elsevier, New York, 1983.
30. Sage, A. P., *Methodology for Large Scale Systems,* McGraw-Hill, New York, 1977.
31. Sage, A. P., *Systems Engineering,* John Wiley, New York, 1992.
32. Sage, A. P., *Systems Management for Information Technology and Software Engineering,* John Wiley, New York, 1995.
33. Sandquist, G. M., *Introduction to System Science,* Prentice Hall, Upper Saddle River, NJ, 1985.
34. Shishko, R., *NASA Systems Engineering Handbook,* NASA, Washington, DC, 1995.
35. Singh, M. G. (Ed.), *Systems and Control Encyclopedia: Theory, Technology, Applications,* Pergamon, Elmsford, NY, 1987.
36. "A Systems Engineering Capability Maturity Model (SE-CMM)," Version 1.1., SECMM-95-01, SEI, Carnegie Mellon University, Pittsburgh, 1995.
37. Software Productivity Consortium, "Systems Engineering Maturity and Benchmarking," Version 01.00.06, SPC-95075-CMC, SPC, Herndon, VA, 1996.
38. Thomé, Bernhard (Ed.), *Systems Engineering: Principles and Practice of Computer-Based Systems Engineering,* John Wiley, New York, 1993.
39. Truxal, J. G., *Introductory System Engineering,* McGraw-Hill, New York, 1972.
40. Von Bertalanffy, L., *General Systems Theory,* George Braziller, New York, 1968.
41. Weinberg, G. M., *An Introduction to General Systems Thinking,* John Wiley, New York, 1975.
42. Weinberg, G. M., *Rethinking Systems Analysis and Design,* Dorset House, New York, 1988.
43. Wymore, A. W., *A Mathematical Theory of Systems Engineering,* John Wiley, New York, 1967.

44. Wymore, A. W., *Model-Based Systems Engineering,* CRC Press, Boca Raton, FL, 1993.
45. Wymore, A. W., *Systems Engineering Methodology for Interdisciplinary Teams,* John Wiley, New York, 1976.

B. CONCURRENT/SIMULTANEOUS ENGINEERING

1. Carter, D. E., and B. S. Baker, *Concurrent Engineering: The Product Development Environment for the 1990's,* Addison-Wesley, New York, NY, 1992.
2. Gu, P., and A. Kusiak (Eds.), *Concurrent Engineering: Methodology and Applications,* Elsevier, Amsterdam, 1993.
3. Kusiak, A., *Concurrent Engineering: Automation, Tools, and Techniques,* John Wiley, New York, 1992.
4. Miller, L. C. G., *Concurrent Engineering Design: Integrating the Best Practices for Process Improvement,* Society of Manufacturing Engineers, Dearborn, MI, 1993.
5. Nevins, J. A., and D. E. Whitney (Eds.), *Concurrent Design of Products and Processes,* McGraw-Hill, New York, 1989.
6. Parsaei, H. R., and W. G. Sullivan (Eds.), *Concurrent Engineering: Contemporary Issues and Modern Design Tools,* Chapman and Hall, London, 1993.
7. Prasad, B., *Concurrent Engineering Fundamentals: Integrated Product and Process Organization,* Prentice Hall, Upper Saddle River, NJ, 1996.
8. Shina, S. G., *Concurrent Engineering and Design for Manufacture of Electronics Products and Processes,* Van Nostrand Reinhold, New York, 1991.
9. Shina, S. G. (Ed.), *Successful Implementation of Concurrent Engineering Products and Processes,* Van Nostrand Reinhold, New York, 1994.
10. Syan, C. S., and U. Menon, *Concurrent Engineering: Concepts, Implementation, and Practice,* Chapman and Hall, London, 1994.
11. Turino. J., *Managing Concurrent Engineering: Burying Time to Market,* Van Nostrand Reinhold, New York, 1992.
12. Winner, R. I., J. P. Pennell, H. E. Bertrand, and M. M. G. Slusarczuk, *The Role of Concurrent Engineering in Weapons Systems Acquisition,* Report R-338, Institute for Defense Analysis, Arlington, VA, 1988.

C. SOFTWARE/COMPUTER-AIDED SYSTEMS

1. Boehm, B. W., *Software Engineering Economics,* Prentice Hall, Upper Saddle River, NJ, 1981.
2. Defense Systems Management College, *Mission Critical Computer Resources Management Guide,* DSMC, Fort Belvoir, VA, latest edition.
3. Department of Defense, "Software Development and Documentation," MIL-STD-498, DOD, Washington, DC, 1994.
4. Eisner, H., *Computer-Aided Systems Engineering,* Prentice Hall, Upper Saddle River, NJ, 1988.
5. Fairley, R., *Software Engineering Concepts,* McGraw-Hill, New York, 1985.

6. Hoover, S. V., and R. F. Perry, *Simulation: A Problem Solving Approach,* Addison-Wesley, Reading, MA, 1990.
7. Humphrey, W. S., *A Discipline for Software Engineering,* Addison-Wesley, Reading, MA, 1995.
8. Krouse, J. K., *What Every Engineer Should Know About Computer-Aided Design and Computer-Aided Manufacturing,* Marcel Dekker, New York, 1982.
9. Musa, J. D., A. Lannino, and K. Okumoto, *Software Reliability: Measurement, Prediction, Application,* McGraw-Hill, New York, 1987.
10. Pidd, M., *Computer Simulation in Management Science,* John Wiley, New York, 1992.
11. Pressman, R. S., *Software Engineering: A Practitioner's Approach,* 3rd Ed., McGraw-Hill, New York, 1992.
12. Sage, A. P., and J. D. Palmer, *Software Systems Engineering,* John Wiley, New York, 1990.
13. Shannon, R. E., *Systems Simulation: The Art and the Science,* Prentice Hall, Upper Saddle River, NJ, 1975.
14. Shere, K. D., *Software Engineering and Management,* Prentice Hall, Upper Saddle River, NJ, 1988.
15. Shooman, M., *Software Engineering Design, Reliability and Management,* McGraw-Hill, New York, 1983.
16. Vick, C. R., and C. V. Ramamcorthy, *Handbook of Software Engineering,* Van Nostrand Reinhold, New York, 1984.
17. Zeid, I., *CAD/CAM Theory and Practice,* McGraw-Hill, New York, 1991.

D. RELIABILITY ENGINEERING

1. *Annual Reliability and Maintainability Symposium, Proceedings,* Evans Associates, Durham, NC, various years.
2. Barlow, R. E., *Mathematical Theory of Reliability,* John Wiley, New York, 1965.
3. Chrysler, Ford, and General Motors Corp., *Potential Failure Mode and Effects Analysis,* Instruction Manual, Revised, Detroit, 1993.
4. Department of Defense, "Environmental Stress Screening Process for Electronic Equipment," MIL-STD-2164, DOD, Washington, DC.
5. Department of Defense, "Failure Reporting, Analysis, and Corrective Action System (FRACAS)," MIL-STD-2155, DOD, Washington, DC.
6. Department of Defense, "Procedures for Performing a Failure Mode, Effects and Criticality Analysis," MIL-STD-1629A, DOD, Washington, DC.
7. Department of Defense, *Reliability Growth Management,* MIL-HDBK-189, DOD, Washington, DC.
8. Department of Defense, *Reliability Predictions of Electronic Equipment,* MIL-HDBK-217F, DOD, Washington, DC.
9. Department of Defense, "Reliability Program for Systems and Equipment Development and Production," MIL-STD-785B, DOD, Washington, DC.
10. Department of Defense, "Reliability Testing for Engineering Development, Qualification, and Production," MIL-STD-781D, Washington, DC.

11. Doty, L. A., *Reliability Engineering for Electronics Design,* 2nd Ed., Industrial Press, New York, 1989.
12. Fuqua, N. B., *Reliability Engineering for Electronic Design,* Marcel Dekker, New York, 1987.
13. Gnedenko, B., and I. Ushakov, *Probabilistic Reliability Engineering,* John Wiley, New York, 1995.
14. Grosh, D. L., *A Primer of Reliability Theory,* John Wiley, New York, 1989.
15. *Failure Mode and Effects Analysis,* Instruction Manual, Saturn Corporation, Saturn Quality System, 1990.
16. Ireson, W. G., and C. F. Coombs (Eds.), *Handbook of Reliability Engineering and Management,* McGraw-Hill, New York, 1988.
17. Kapur, K. C., and L. R. Lamberson, *Reliability in Engineering Design,* John Wiley, New York, 1977.
18. Kececioglu, D., *Reliability Engineering Handbook,* 2 vols., Prentice Hall, Upper Saddle River, NJ, 1991.
19. Klion, J., *Practical Electronic Reliability Engineering,* Von Nostrand Reinhold, New York, 1992.
20. Knezevic, J., *Reliability, Maintainability, and Supportability: A Probabilistic Approach,* McGraw-Hill, New York, 1993.
21. Lewis, E. E., *Introduction to Reliability Engineering,* John Wiley, New York, 1987.
22. Lloyd, D. K., and M. Lipow, *Reliability: Management, Methods, and Mathematics,* 2nd Ed., Defense and Space Systems Group, TRW Systems and Energy, Redondo Beach, CA, 1984.
23. Modarres, M., *What Every Engineer Should Know About Reliability and Risk Analysis,* Marcel Dekker, New York, 1992.
24. Musa, J. D., A. Lannino, and K. Okumoto, *Software Reliability: Measurement, Prediction, Application,* McGraw-Hill, New York, 1987.
25. O'Connor, P. D. T., *Practical Reliability Engineering,* 3rd Ed., John Wiley, New York, 1991.
26. Pham, H., *Software Reliability and Testing,* IEEE Computer Society, Los Alamitos, CA, 1995.
27. Raheja, D. G., *Assurance Technologies: Principles and Practices,* McGraw-Hill, New York, 1991.
28. Rao, S. S., *Reliability-Based Design,* McGraw-Hill, New York, 1992.
29. Reliability Analysis Center, *Failure Mode, Effects and Criticality Analysis (FMECA),* RAC, Rome, NY, 1992.
30. Reliability Analysis Center, *Reliability Design Handbook,* RDH-376, RAC, Rome, NY, 1976.
31. Rome Air Development Center, *RADC, Reliability Engineer's Toolkit,* RADC/RBE, Rome, NY, 1988.
32. Shooman, M. L., *Probabilistic Reliability: An Engineering Approach,* 2nd Ed., Krieger, 1990.
33. Siewiorek, D. P., and R. B. Swarz, *The Theory and Practice of Reliable System Design,* Digital Press, Bedford, MA, 1982.

E. MAINTAINABILITY AND MAINTENANCE

1. Anderson, R. T., and L. Neri, *Reliability-Centered Maintenance,* Elsevier, London, 1990.
2. Blanchard, B. S., D. Verma, and E. Peterson, *Maintainability: A Key to Effective Serviceability and Maintenance Management,* John Wiley, New York, 1995.
3. Bray, D. E., and D. McBride (Eds.), *Nondestructive Testing Techniques,* John Wiley, New York, 1992.
4. Department of Defense,"Definitions of Effectiveness Terms for Reliability, Maintainability, Human Factors, and Safety," MIL-STD-72IC, DOD, Washington, DC.
5. Department of Defense, "Level of Repair Analysis," MIL-STD-1390D (Navy), DOD, Washington, DC.
6. Department of Defense, *Maintainability Prediction,* MIL-HDBK-472, DOD, Washington, DC.
7. Department of Defense, "Maintainability Program for Systems and Equipment," MIL-STD-470B, DOD, Washington, DC.
8. Department of Defense, "Maintainability Verification, Demonstration, Evaluation," MIL-STD-417A, DOD, Washington, DC.
9. Department of Defense,"Testability Program for Electronic Systems and Equipments," MIL-STD-2165, DOD, Washington, DC.
10. Faulkenberry, L. M. (Ed.), *Systems Troubleshooting Handbook,* John Wiley, New York, 1986.
11. Gotoh, F., *Equipment Planning for TPM,* Productivity Press, Portland, OR, 1991.
12. Higgins, L. R., *Maintenance Engineering Handbook,* 5th Ed., McGraw-Hill, New York, 1994.
13. Jardine, A. K. S., *Maintenance, Replacement, and Reliability,* John Wiley, New York, 1973.
14. Kelly, A., and M. J. Harris, *Management of Industrial Maintenance,* Newness-Butterworths, London, 1978.
15. Knezevic, J., *Reliability, Maintainability, and Supportability: A Probabilistic Approach,* McGraw-Hill, New York, 1993.
16. Mann, L. *Maintenance Management,* D. C. Heath, Lexington, MA, 1976.
17. Mobley, R. K., *An Introduction to Predictive Maintenance,* Van Nostrand Reinhold, New York, 1990.
18. Moss, M. A., *Designing for Minimal Maintenance Expense: The Practical Application of Reliability and Maintainability,* Marcel Dekker, New York, 1985.
19. Moubray, J., *Reliability-Centered Maintenance,* Butterworth-Heinemann, Boston, 1992.
20. Nakajima, S., *Total Productive Maintenance (TPM),* Productivity Press, Portland, OR, 1988.
21. Nakajima, S. (Ed.), *TPM Development Program: Implementing Total Productive Maintenance,* Productivity Press, Portland, OR, 1989.
22. Niebel, B. W., *Engineering Maintenance Management,* 2nd Ed., Marcel Dekker, New York, 1994.
23. Nowlan, F. S., and H. F. Heap, *Reliability-Centered Maintenance,* MDA 903-75-C-0349, United Airlines, San Francisco, 1978.

24. Patton, J. D., *Maintainability and Maintenance Management,* 2nd Ed., Instrument Society of America, Research Triangle Park, NC, 1994.
25. Reiche, H., *Maintenance Minimization for Competitive Advantage,* Gordon and Breach, Langhorne, PA, 1994.
26. Smith, A. M., *Reliability-Centered Maintenance,* McGraw-Hill, New York, 1993.
27. Suzuki, T. (Ed.), *TPM in Process Industries,* Productivity Press, Portland, OR, 1994.
28. Wireman, T., *World Class Maintenance Management,* Industrial Press, New York, 1990.

F. HUMAN FACTORS AND SAFETY

1. Cushman, W. H., and D. J. Rosenberg, *Human Factors in Product Design,* Elsevier, New York, 1991.
2. DeGreene, K. B. (Ed.), *Systems Psychology,* McGraw-Hill, New York, 1970.
3. Department of Defense, "Human Engineering Design Criteria for Military Systems, Equipment, and Facilities," MIL-STD-1472D, DOD, Washington, DC.
4. Department of Defense, "Human Engineering Requirements for Military Systems, Equipment, and Facilities," MIL-H-46855B, U.S. Army Missile R&D Command (DRDMI-ESD), Redstone Arsenal, AL.
5. Department of Defense, "Human Factors Engineering Performance Requirements for Systems," MIL-STD-1800, DOD, Washington, DC.
6. Department of Defense, *Human Factors Engineering Design for Army Material,* MIL-HDBK-759A, DOD, Washington, DC.
7. Department of Defense, "System Safety Program Requirements," MIL-STD-882B, DOD, Washington, DC.
8. Eastman Kodak Company, *Ergonomic Design for People At Work,* Van Nostrand Reinhold, New York, 1983.
9. Eberts, R. E., *User Interface Design,* Prentice Hall, Upper Saddle River, NJ, 1994.
10. Hammer, W., *Occupational Safety Management and Engineering,* 4th Ed., Prentice Hall, Upper Saddle River, NJ, 1988.
11. Kroemer, K., H. Kroemer, and K. Kroemer-Elbert, *Ergonomics—How to Design for Ease and Efficiency,* Prentice Hall, Upper Saddle River, NJ, 1994.
12. Meister, D., *Behavioral Analysis and Measurement Methods,* John Wiley, New York, 1985.
13. Roland, H. E., and B. Moriarty, *System Safety Engineering and Management,* 2nd Ed., John Wiley, New York, 1990.
14. Salvendy, G. (Ed.), *Handbook of Human Factors,* John Wiley, New York, 1987.
15. Sanders, M. S., and E. J. McCormick, *Human Factors in Engineering and Design,* 7th Ed., McGraw-Hill, New York, 1992.
16. Van Cott, H. P., and R. G. Kinkade (Eds.), *Human Engineering Guide to Equipment Design,* U.S. Government Printing Office, Washington, DC, 1972.
17. Weimer, J., *Handbook of Ergonomic and Human Factors Tables,* Prentice Hall, Upper Saddle River, NJ, 1993.
18. Woodson, W. E., B. Tillman, and P. Tillman, *Human Factors Design,* 2nd Ed., McGraw-Hill, New York, 1992.

G. LOGISTICS, LOGISTICS ENGINEERING, AND LOGISTICS MANAGEMENT

1. American Defense Preparedness Association, *Commercial-Off-the-Shelf (COTS) Supportability Study,* Technical Report, ADPA, Arlington, VA, 1994.
2. Air Force, *Air Force Journal of Logistics,* U.S. Government Printing Office, Washington, DC.
3. Air Force, *Compendium of Authenticated System and Logistics Terms, Definitions, and Acronyms,* AU-AFIT-LS-3-81, U.S. Air Force Institute of Technology, Wright-Patterson AFB, Dayton, OH.
4. Allen, M. K., and O. K. Helferich, *Putting Expert Systems to Work in Logistics,* Council of Logistics Management, Oak Brook, IL, 1990.
5. Ballou, R. H., *Business Logistics Management,* 3rd Ed., Prentice Hall, Upper Saddle River, NJ, 1992.
6. Banks, J., and W. J. Fabrycky, *Procurement and Inventory Systems Analysis,* Prentice Hall, Upper Saddle River, NJ, 1987.
7. Barnes, T. A., *Logistics Support Training: Design and Development,* McGraw-Hill, New York, 1992.
8. Blanchard, B. S., *Logistics Engineering and Management,* 4th Ed., Prentice Hall, Upper Saddle River, NJ, 1992.
9. Bowersox, D., D. Closs, and O. Helferich, *Logistical Management,* Macmillan, New York, 1986.
10. Council of Logistics Management, *Journal of Business Logistics,* Oak Brook, IL.
11. Council of Logistics Management, *Logistics Software,* Annual Ed., Anderson Consulting, New York, 1993.
12. Coyle, J. J., E. J. Bardi, and J. L. Cavinato, *Transportation,* 4th Ed., West Publishing, St. Paul, MN, 1993.
13. Coyle, J. J., E. J. Bardi, and C. J. Langley, *The Management of Business Logistics,* 5th Ed., West Publishing, St. Paul, MN, 1992.
14. Defense Systems Management College, *Integrated Logistics Support Guide,* DSMC, Fort Belvoir, VA.
15. Department of Defense, *Application of Reliability-Centered Maintenance to Naval Aircraft, Weapon Systems, and Support Equipment,* MIL-HDBK-226, DOD, Washington, DC.
16. Department of Defense, "Automated Interchange of Technical Information," MIL-STD-1840A, DOD, Washington, DC.
17. Department of Defense, *Computer-Aided Acquisition Logistic Support (CALS) Implementation Guide,* MIL-HDBK-59A, DOD, Washington, DC.
18. Department of Defense, "Department of Defense Requirements for a Logistic Support Analysis Record," MIL-STD-1388-2B, DOD, Washington, DC.
19. Department of Defense, "Logistic Support Analysis," MIL-STD-1388-1A, DOD, Washington, DC.
20. Glaskowsky, N. A., D. R. Hudson, and R. M. Ivie, *Business Logistics,* 3rd Ed., Harcourt Brace Jovanovich, Orlando, FL, 1992.
21. Green, L. L., *Logistics Engineering,* John Wiley, New York, 1991.

22. Hutchinson, N. E., *An Integrated Approach to Logistics Management,* Prentice Hall, Upper Saddle River, NJ, 1987.
23. John, J. C., and D. F. Wood, *Contemporary Logistics,* 4th Ed., Macmillan, New York, 1990.
24. Jones, J. V., *Integrated Logistics Support Handbook,* Tab, Blue Ridge Summit, PA, 1987.
25. Jones, J. V., *Logistic Support Analysis Handbook,* Tab, Blue Ridge Summit, PA, 1989.
26. Langford, J. W., *Logistics Principles and Practices,* McGraw-Hill, New York, 1995.
27. Magee, J. F., W. C. Copacino, and D. B. Rosenfield, *Modern Logistics Management,* John Wiley, New York, 1985.
28. O'Neil and Associates, *Integrated Logistics Support,* OAI, Dayton, OH.
29. Orsburn, Douglas K., *Introduction to Spares Management,* Academy Printing & Publishing, Paramount, CA, 1985.
30. Patton, Jr., Joseph D., *Logistics Technology and Management—The New Approach,* Solomon Press, New York, 1986.
31. Peppers, J. G., *History of United States Military Logistics—1935 to 1985,* Fairborn, OH, 1988.
32. Society of Logistics Engineers, *Annals,* SOLE, Hyattsville, MD.
33. Society of Logistics Engineers, *Annual Symposium, Proceedings,* SOLE, Hyattsville, MD.
34. Society of Logistics Engineers, *Logistics Spectrum,* SOLE, Hyattsville, MD.
35. Stock, James R., and Douglas M. Lambert, *Strategic Logistics Management,* 3rd Ed., Richard D. Irwin, Homewood, IL, 1992.
36. Tersine, R. J., *Principles of Inventory and Materials Management,* 4th Ed., Prentice Hall, Upper Saddle River, NJ, 1994.

H. QUALITY, QUALITY ENGINEERING, AND QUALITY ASSURANCE

1. Akao, Y. (Ed.), *Quality Function Deployment: Integrating Customer Requirements Into Product Design,* Productivity Press, Portland, OR, 1990.
2. American Productivity and Quality Center, *The Benchmarking Management Guide,* Productivity Press, Portland, OR, 1993.
3. Andrews, D. C., and S. K. Stalick, *Business Reengineering: The Survival Guide,* Prentice Hall, Upper Saddle River, NJ, 1994.
4. Asaka, T., and K. Ozeki, *Handbook of Quality Tools,* Productivity Press, Portland, OR, 1990.
5. Besterfield, D. H., *Quality Control,* 4th Ed., Prentice Hall, Upper Saddle River, NJ, 1993.
6. Besterfield, D. H., *Total Quality Management,* Prentice Hall, Upper Saddle River, NJ, 1994.
7. Cohen, L., *Quality Function Deployment: How to Make QFD Work for You,* Addison-Wesley, Reading, MA, 1995.
8. Crosby, P. B., *Quality is Free,* New American Library, New York, 1980.
9. Deming, W. E., *Out of the Crisis,* The MIT Press, Cambridge, MA, 1986.
10. Department of Defense, "Quality Program Requirements," MIL-Q-9858A, Washington, DC.

11. Duncan, A. J., *Quality Control and Industrial Statistics,* 5th Ed., Richard D. Irwin, Homewood, IL, 1986.
12. Feigenbaum, A. V., *Total Quality Control,* 3rd Ed., McGraw-Hill, New York, 1991.
13. Garvin, D. A., *Managing Quality: The Strategic and Competitive Edge,* The Free Press, New York, 1988.
14. Grant, E. L., and R. S. Leavenworth, *Statistical Quality Control,* 6th Ed., McGraw-Hill Book, New York, 1988.
15. Hansen, G. A., *Automating Business Process Reengineering: Breaking the TQM Barrier,* Prentice Hall, Upper Saddle River, NJ, 1994.
16. Hauser, J. R., and D. Clausing, "The House of Quality," *Harvard Business Review* (May–June 1988).
17. Ishikawa, K., *Introduction to Quality Control,* Chapman and Hall, London, 1991.
18. Juran, J. M., and F. M. Gryna (Eds.), *Quality Planning and Analysis,* 3th Ed., McGraw-Hill, New York, 1993.
19. Juran, J. M., and F. M. Gryna (Eds.), *Quality Control Handbook,* 4th Ed., McGraw-Hill, New York, 1988.
20. Lochner, R., and J. Matar, *Design for Quality,* ASQC, Milwaukee, WI, 1990.
21. Pall, G. A., *Quality Process Management,* Prentice Hall, Upper Saddle River, NJ, 1987.
22. Rome Air Development Center, *A Guide for Implementing Total Quality Management,* RAC SOAR-7, Rome, NY, 1990.
23. Ross, P. J., *Taguchi Techniques for Quality Engineering,* 2nd Ed., McGraw-Hill, New York, 1995.
24. Rothery, B., *ISO 9000,* 2nd Ed., Gower Press, Brookfield, VT, 1993.
25. Saylor, J. H., *TQM Field Manual,* McGraw-Hill, New York, 1992.
26. Scherkenback, W. W., *Deming's Road to Continuous Improvement,* SPC Press, 1991.
27. Scherkenback, W. W., *The Deming Route to Quality and Productivity: Road Maps and Roadblocks,* CEE Press, Washington, DC, 1986.
28. Taguchi, G., E. A. Elsayed, and T. C. Hsiang, *Quality Engineering in Production Systems,* McGraw-Hill, New York, 1989.
29. Wilton, P. S., *The Quality System Development Handbook with ISO 9002,* Prentice Hall, Upper Saddle River, NJ, 1994.
30. Winston, W. L., *Operations Research: Applications and Alogorithms,* 3rd Ed., PWS-Kent Publishing, Boston, MA, 1994.

I. OPERATIONS ANALYSIS AND OPERATIONS RESEARCH

1. Buffa, E. S., and R. K. Sarin, *Modern Production and Operations Management,* 8th Ed., John Wiley, New York, 1987.
2. Churchman, C. W., R, L. Ackoff, and E. L. Arnoff, *Introduction to Operations Research,* John Wiley, New York, 1957.
3. Fabrycky, W. J., P. M. Ghare, and P. E. Torgerson, *Applied Operations Research and Management Science,* Prentice Hall, Upper Saddle River, NJ, 1984.
4. Flagel, C. D., W. H. Huggins, and R. H. Roy, (Eds.), *Operations Research and Systems Engineering,* Johns Hopkins Press, Baltimore, MD, 1960.

5. Hillier, F. S., and G. J. Lieberman, *Introduction to Operations Research,* 6th Ed., McGraw-Hill, New York, 1995.
6. Nahmias, S., *Production and Operations Analysis,* 2nd Ed., Richard D. Irwin, Boston, 1993.
7. Stark, H., and J. W. Woods, *Probability, Random Processes and Estimation Theory for Engineering,* 2nd Ed., Prentice Hall, Upper Saddle River, NJ, 1984.
8. Taha, H. A., *Operations Research: An Introduction,* 5th Ed., Macmillan, New York, 1992.
9. Vollmann, T. E., W. L. Berry, and D. C. Whybark, *Manufacturing Planning and Control Systems,* 2nd Ed., Richard Irwin, Homewood, IL, 1988.

J. ENGINEERING ECONOMY, LIFE-CYCLE COSTING, AND COST ESTIMATING

1. Berliner, C., and J. Brimson, *Cost Management for Today's Advanced Manufacturing—The CAM-I Conceptual Design,* Harvard Business School Press, Boston, 1988.
2. Brown, R. J., and R. R. Yanuck, *Introduction to Life Cycle Costing,* AEE Energy Books, Atlanta, GA, 1985.
3. Burk, K. B., and D. W. Webster, *Activity Based Costing and Performance,* American Management Systems, Fairfax, VA, 1994.
4. Canada, J. R., and W. G. Sullivan, *Economic and Multiattribute Evaluation of Advanced Manufacturing Systems,* Prentice Hall, Upper Saddle River, NJ, 1989.
5. Canada, J. R., W. G. Sullivan, and J. A. White, *Capital Investment Analysis for Engineering and Management,* 2nd Ed., Prentice Hall, Upper Saddle River, NJ, 1996.
6. Department of Defense, "Design to Cost," MIL-STD-337, DOD, Washington, DC.
7. Department of Defense, *Design to Cost,* DOD-HDBK-766, DOD, Washington, DC.
8. Department of Defense, "Level of Repair," MIL-STD-1390C, DOD, Washington, DC.
9. Department of Defense, *Life Cycle Cost in Navy Acquisitions,* MIL-HDBK-259, DOD, Washington, DC.
10. Departments of the Army/Navy/Air Force, *Joint-Design-to-Cost Guide, Life Cycle Cost as a Design Parameter,* DARCOM P7000-6 (Army), NAVMAT P5242 (Navy), AFLCP/AFSCP 800-19 (Air Force), Washington DC.
11. Dhillon, B. S., *Life Cycle Costing: Techniques, Models and Applications,* Gordon and Breach, New York, 1989.
12. Fabrycky, W. J., and B. S. Blanchard, *Life-Cycle Cost and Economic Analysis,* Prentice Hall, Upper Saddle River, NJ, 1991.
13. Fabrycky, W. J., and G. J. Thuesen, *Economic Decision Analysis,* 2nd Ed. Prentice Hall, Upper Saddle River, NJ, 1980.
14. Fowler, T. C., *Value Analysis in Design,* Van Nostrand Reinhold, New York, 1990.
15. Grant, E. L., W. G. Ireson, and R. S. Levenworth, *Principles of Engineering Economy,* 8th Ed., Ronald Press, New York, 1990.
16. Gulledge, T. R., and L. A. Litteral, *Cost Analysis Applications of Economics and Operations Research,* Springer-Verlag, New York, 1989.
17. Gulledge, T. R., W. P. Hutzler, and J. S. Lovelace (Eds.), *Cost Estimating and Analysis: Balancing Technology and Declining Budgets,* Springer-Verlag, New York, 1992.

18. Jelen, F. C., and J. H. Black, *Cost and Optimization Engineering,* McGraw-Hill, New York, 1983.
19. McNichols, G. R. (Ed.), *Cost Analysis,* Operations Research Society of America, Baltimore, MD, 1984.
20. Michaels, J. V., and W. P. Wood, *Design to Cost,* John Wiley, New York, 1989.
21. Office of the Management of the Budget, *Cost Comparison Handbook,* OMB Circular A-76, OMB, Washington, DC.
22. O'Guin, M. C., *The Complete Guide to Activity-Based Costing,* Prentice Hall, Upper Saddle River, NJ, 1991.
23. Ostwald, P. F., *Engineering Cost Estimating,* 3rd Ed., Prentice Hall, Upper Saddle River, NJ, 1992.
24. Stewart, R. D., *Cost Estimating,* 2nd Ed., John Wiley, New York, 1991.
25. Stewart, R. D., and A. L. Stewart, *Cost Estimating with Microcomputers,* McGraw-Hill, New York, 1980.
26. Stewart, R. D., and R. M. Wyskida, *Cost Estimator's Reference Manual,* 2nd Ed., John Wiley, New York, 1995.
27. Thusesen, G. J., and W. J. Fabrycky, *Engineering Economy,* 8th Ed., Prentice Hall, Upper Saddle River, NJ, 1993.
28. Witt, P. R., *Cost Competitive Products: Managing Product Concept to Marketplace Reality,* Reston Publishing, Reston, VA, 1986.

K. MANAGEMENT AND SUPPORTING AREAS

1. Cleland, D. I., and W. R. King, *Project Management Handbook,* 2nd Ed., Van Nostrand Reinhold, New York, 1989.
2. Defense Systems Management College, *Manufacturing Management: Guide for Program Managers,* DSMC, Fort Belvoir, VA.
3. Defense Systems Management College, *Test and Evaluation Management Guide,* DSMC, Fort Belvoir, VA.
4. Department of Defense, "Technical Reviews and Audits for Systems, Equipments, and Computer Software," MIL-STD-1521B, DOD, Washington, DC.
5. Dieter, G. E., *Engineering Design: A Materials and Processing Approach,* 2nd Ed., McGraw-Hill, New York, 1991.
6. Griffin, R. W., and G. Moorhead, *Organizational Behavior,* Houghton Mifflin, Boston, 1986.
7. Hajek, V. J., *Management of Engineering Projects,* New York, 1984.
8. Johnson, R. A., F. E. Kast, and J. E. Rosenzweig, *The Theory and Management of Systems,* 3rd Ed., McGraw-Hill, New York, 1973.
9. Kerzner, H., *Project Management: A Systems Approach to Planning, Scheduling, and Controlling,* 5th Ed., Van Nostrand Reinhold, New York, 1995.
10. Kidd, P. T., *Agile Manufacturing: Forging New Frontiers,* Addison-Wesley, Reading, MA, 1994.
11. Koontz, H., C. O'Donnell, and H. Weihrich, *Essentials of Management,* 5th Ed., McGraw-Hill, New York, 1990.

12. Roman, D. D., *Managing Projects: A Systems Approach,* Elsevier, New York, 1986.
13. Shtub, A., J. F. Bard, and S. Globerson, *Project Management: Engineering, Technology, and Implementation,* Prentice Hall, Upper Saddle River, NJ, 1994.
14. Stewart, R. D., and A. L. Stewart, *Proposal Preparation,* 2nd Ed., John Wiley, New York, 1992.
15. Thamhain, H. J., *Engineering Management,* John Wiley, New York, 1992.
16. Ullmann, J. E., D. A. Christman, B. Holtje (Eds.), *Handbook of Engineering Management,* John Wiley, New York, 1990.
18. Weinberg, G. M., *Quality Software Management, Vol. 1: Systems Thinking,* Dorset House, New York, 1992.

INDEX